Opportunities and Challenges in Climate-Friendly Clean Water and Energy Technologies

J. Sadhik Basha
Department of Process Engineering, National University of Science & Technology, Oman

Jimoh Kayode Adewole
Department of Process Engineering, National University of Science & Technology, Oman

Anteneh Mesfin Yeneneh
Department of Process Engineering, National University of Science & Technology, Oman

Luqman Kolawole Abidoye
Department of Process Engineering, National University of Science & Technology, Oman

A volume in the Advances in Environmental Engineering and Green Technologies (AEEGT) Book Series

IGI Global
PUBLISHER of TIMELY KNOWLEDGE

Published in the United States of America by
IGI Global
Engineering Science Reference (an imprint of IGI Global)
701 E. Chocolate Avenue
Hershey PA, USA 17033
Tel: 717-533-8845
Fax: 717-533-8661
E-mail: cust@igi-global.com
Web site: http://www.igi-global.com

Library of Congress Cataloging-in-Publication Data

Names: Basha, J. Sadhik, 1977- editor.
Title: Opportunities and challenges in climate-friendly clean water and energy technologies / edited by J. Sadhik Basha, Jimoh Kayode Adewole, Anteneh Mesfin Yeneneh, Luqman Abidoye.
Description: Hershey, PA : Engineering Science Reference, [2023] | Includes bibliographical references and index. | Summary: "The goal of this book is to present the most recent technological advancements that are relevant to environmentally friendly methods of producing clean water and a sustainable supply of energy. In order to protect against climate changes, it will also highlight the most recent empirical research findings in the field of cutting-edge industrial and transportation emission reduction measures. This book will also be useful for professionals who wish to improve their knowledge on energy sustainability inline with the current methods to combat environmental changes cum practical and affordable methods for producing drinkable water for future generations on a worldwide scale. Thus, this book will serve as a wake-up call for all researchers and scientists to harness recent discoveries and developments in climate-friendly water and energy technology for future generations"-- Provided by publisher.
Identifiers: LCCN 2022049399 (print) | LCCN 2022049400 (ebook) | ISBN 9781668473030 (h/c) | ISBN 9781668473047 (s/c) | ISBN 9781668473054 (eISBN)
Subjects: LCSH: Climate change mitigation. | Water--Purification. | Clean energy.
Classification: LCC TD171.75 .O66 2023 (print) | LCC TD171.75 (ebook) | DDC 363.739/40286--dc23/eng/20221212
LC record available at https://lccn.loc.gov/2022049399
LC ebook record available at https://lccn.loc.gov/2022049400

This book is published in the IGI Global book series Advances in Environmental Engineering and Green Technologies (AEEGT) (ISSN: 2326-9162; eISSN: 2326-9170)

British Cataloguing in Publication Data
A Cataloguing in Publication record for this book is available from the British Library.

Advances in Environmental Engineering and Green Technologies (AEEGT) Book Series

Sang-Bing Tsai
Zhongshan Institute, University of Electronic Science and Technology of China, China & Wuyi University, China
Ming-Lang Tseng
Lunghwa University of Science and Technology, Taiwan
Yuchi Wang
University of Electronic Science and Technology of China Zhongshan Institute, China

ISSN:2326-9162
EISSN:2326-9170

Mission

Growing awareness and an increased focus on environmental issues such as climate change, energy use, and loss of non-renewable resources have brought about a greater need for research that provides potential solutions to these problems. Research in environmental science and engineering continues to play a vital role in uncovering new opportunities for a "green" future.

The **Advances in Environmental Engineering and Green Technologies (AEEGT)** book series is a mouthpiece for research in all aspects of environmental science, earth science, and green initiatives. This series supports the ongoing research in this field through publishing books that discuss topics within environmental engineering or that deal with the interdisciplinary field of green technologies.

Coverage

- Industrial Waste Management and Minimization
- Green Technology
- Waste Management
- Pollution Management
- Air Quality
- Alternative Power Sources
- Electric Vehicles
- Policies Involving Green Technologies and Environmental Engineering
- Radioactive Waste Treatment
- Contaminated Site Remediation

IGI Global is currently accepting manuscripts for publication within this series. To submit a proposal for a volume in this series, please contact our Acquisition Editors at Acquisitions@igi-global.com or visit: http://www.igi-global.com/publish/.

The Advances in Environmental Engineering and Green Technologies (AEEGT) Book Series (ISSN 2326-9162) is published by IGI Global, 701 E. Chocolate Avenue, Hershey, PA 17033-1240, USA, www.igi-global.com. This series is composed of titles available for purchase individually; each title is edited to be contextually exclusive from any other title within the series. For pricing and ordering information please visit http://www.igi-global.com/book-series/advances-environmental-engineering-green-technologies/73679. Postmaster: Send all address changes to above address.

Titles in this Series

For a list of additional titles in this series, please visit: www.igi-global.com/book-series

Transcending Humanitarian Engineering Strategies for Sustainable Futures
Yiannis Koumpouros (University of West Attica, Greece) Angelos Georgoulas (University of West Attica, Greece) and Georgia Kremmyda (University of Warwick, UK)
Engineering Science Reference • © 2023 • 338pp • H/C (ISBN: 9781668456194) • US $270.00

Food Sustainability, Environmental Awareness, and Adaptation and Mitigation Strategies for Developing Countries
Ahmad Ni'matullah Al-Baarri (Universitas Diponegoro, Indonesia) and Diana Nur Afifah (Universitas Diponegoro, Indonesia)
Engineering Science Reference • © 2023 • 293pp • H/C (ISBN: 9781668456293) • US $110.00

Perspectives on Ecological Degradation and Technological Progress
Veli Yilanci (Faculty of Political Sciences, Canakkale Onsekiz Mart University, Turkey)
Engineering Science Reference • © 2023 • 300pp • H/C (ISBN: 9781668467275) • US $225.00

Climate Change and the Economic Importance and Damages of Insects
Ahmed Karmaoui (Moulay Ismail University of Meknes, Morocco & Moroccan Center for Culture and Science, Morocco)
Engineering Science Reference • © 2023 • 279pp • H/C (ISBN: 9781668448243) • US $215.00

Implications of Nanoecotoxicology on Environmental Sustainability
Rafiq Lone (Central University of Kashmir, India) and Javid Ahmad Malik (Guru Ghasidas Vishwavidyalaya, India)
Engineering Science Reference • © 2023 • 360pp • H/C (ISBN: 9781668455333) • US $250.00

Human Agro-Energy Optimization for Business and Industry
Pandian Vasant (MERLIN, Faculty of Electrical and Electronics Engineering, Ton Duc Thang University, Vietnam) Roman Rodríguez-Aguilar (Panamerican University, Mexico) Igor Litvinchev (Universidad Autónoma de Nuevo León, Mexico) and Jose Antonio Marmolejo-Saucedo (Panamerican University, Mexico)
Engineering Science Reference • © 2023 • 322pp • H/C (ISBN: 9781668441183) • US $240.00

Applying Drone Technologies and Robotics for Agricultural Sustainability
Pethuru Raj (Reliance Jio Platforms Ltd., Bangalore, India) Kavita Saini (Galgotias University, India) and Vinicius Pacheco (Office of the Prosecutor General, Brazil)
Engineering Science Reference • © 2023 • 318pp • H/C (ISBN: 9781668464137) • US $250.00

IGI Global
PUBLISHER of TIMELY KNOWLEDGE

701 East Chocolate Avenue, Hershey, PA 17033, USA
Tel: 717-533-8845 x100 • Fax: 717-533-8661
E-Mail: cust@igi-global.com • www.igi-global.com

This book is dedicated to our beloved families and friends.

Table of Contents

Detailed Table of Contents

Water is very important for the survival of living organisms, which include both plants and animals. Sources of water available include rivers, oceans, seas, springs, groundwater, and rainwater. Increase in population, expanded urbanization, and establishment of industries put pressure on sources, quality, and quantity of available water; hence, demand for water increases. As water utilization increases, generation of wastewater increases. The wastewater treatments are described under the followings sub-heading, namely sources and types of wastewater, properties of wastewater, methods of treatment, and challenges of wastewater treatment methods. Information is sourced from the internet through Google search engine. The objective of this chapter is to compile information on emerging contaminants in wastewater, methods that can be adopted in the process of wastewater treatment that will minimize energy use and ensure safety of environment. Green methods of wastewater treatment and disposal are suggested.

Water scarcity has emerged at the forefront of the world's most critical environmental concerns. In the future decades, the rapid population development in emerging regions will continue to raise the need for clean water from residential, agricultural, industrial, and energy perspectives. Clean drinking water is the most significant factor that influences the quality of life for humans. Nanomaterials are ideal for wastewater treatment due to their potential properties, which include small size, large surface area, high porosity, high catalytic activity, tailorable physical and chemical properties, ease of separation, and reproducibility. These favorable properties of nanomaterials make them an attractive candidate for wastewater treatment. This chapter comprehensively overviews a broad spectrum of nanomaterials in wastewater treatment, including different carbon allotropes (graphene, CNTs, MWCNTs, etc.) and metal oxide nanoparticles. Finally, the prospects of these new generation materials in water and wastewater treatment are also discussed.

Challenges of insufficient freshwater for domestic and industrial uses in different parts of the world necessitate the development of desalination technology. However, brine wastes and CO2 emitted from the desalination process are of detrimental effects to the environments and living beings, owing to high-level of alkalinity and presence of toxic levels of various chemicals. This study acknowledged efforts being made by scientists to integrate carbon mineralization of brine wastes with desalination system to generate various carbonates: nesquehonite, ragonite, calcite, dolomite, etc. However, the carbonation process behaves differently under the effects of impurities in the CO2 stream. Studying and analyzing the process using PHREQC simulation software, the carbonation of brine using pure CO2 shows high drop in pH of the solution, thereby increasing the level of the acidity. The presence of impurity in the CO2 further worsens the acidity problem in the presence of N2 and O2.

The environmental problems such as global warming and contamination have prompted the usage of renewable energy sources (RES) as well as storage units in the modern power systems. Among several RES, the WECS is considered as a highly efficient system due to its numerous benefits such as ease of availability as well as pollution free. In this chapter, a switched-reluctance generator-based wind energy conversion system is proposed that reduces switching burdens in the grid and improves the average output power with reduced error. The system utilizes a DC-DC converter that generates improved efficiency with reduced switching stress thereby solving the problem of reverse voltage polarity. Further, an improved grey wolf optimization (GWO)-based PI controller is employed to reduce the error of mean power. The proposed approach is simulated as well as validated generating an enhanced efficiency of 98% and reduced THD of 2.2%. The obtained results clearly indicate a dynamic response with reduced harmonic content and low power factor that assures sustainable energy.

Electric vehicles (EVs) are becoming a popular alternative to gas-powered cars. These cars need "full" batteries to run. Solar-powered chargers are an exciting alternative to grid-based EV charging. These chargers give electric vehicles pollution-free electricity, which benefits the environment. Solar PV is the most popular renewable energy source. This chapter establishes a solar EV charging station, which charges EVs. Bi-directional batteries store photovoltaic (PV) energy for use during power outages. PV overproduction is transferred to the grid for later use. Cascaded interval type 2 fuzzy logic controller (CIT2FLC) boosts voltage using KY converter to track maximum photovoltaic power. To accomplish grid synchronization, a DC voltage is delivered to a grid-connected 1 phase VSI and optimized using a PI controller. During peak hours, EV gets power from the grid via 1 phase VSI, with the KY converter in buck mode. The suggested work ensures uninterrupted charging. The complete structure was tested using MATLAB Simulink and yielded 94.7% efficiency and 3.9% THD.

Coffee is regarded as highly consumed beverage throughout the world and has established a key spot in the world economy as an important commodity for trading. In general, it is produced by brewing its roasted and ground beans, which release aromatic coffee as well as produce an equivalent amount of spent coffee grounds (SCG) as waste. In recent times, they are valorized into SCG briquettes owing to their high calorific value, which gained wide attention for supplying energy renewably, especially to the rising energy demand; and also have been identified as an effective measure to reduce their pollution. With this in mind, this chapter focuses on reviewing the availability and chemistry involved in these SCG wastes, pre-treatments, and preparations required for their briquetting, compacting techniques followed, and fuel characteristics of their briquettes, from various available works of literature. Eventually, all the reported data were in accordance with their permissible standards and suggested these SCG wastes as a highly renewable solid biofuel.

Cashew nut shells (CNS) are the primary waste produced during the processing of cashew nuts and need constant attention to handle or valorize these wastes effectively. As a result, these CNS wastes are processed into solid briquettes citing their significant calorific content, thus making them a promising renewable biofuel for combustion-based applications. In most cases, these wastes are pre-treated either through de-oiling or carbonizing prior to compaction, thus removing the harmful hydrocarbons present in them in form of CNS liquids. Presently, this chapter focuses on summarising various data related to these CNS wastes and their briquettes in terms of their availability, chemical characteristics, pre-treatment, processing technique, and fuel and combustion properties as reported in various works of literature. Moreover, all the reported results and data included in this present study were in accordance with the international testing standards and ranged in between their permissible range.

Preface

The world is facing numerous challenges in the realm of environmental sustainability. From climate change to resource depletion, the impact of human activities is becoming increasingly evident. Climate change and environmental degradation have become major concerns in today's world. The need for clean water and sustainable energy has become more important than ever. The book "Opportunities and Challenges in Climate-Friendly Clean Water and Energy Technologies" explores the latest advancements in technologies that address these challenges. The book aims to provide an in-depth analysis of the emerging technologies in the field of clean water and energy that can contribute towards mitigating climate change and preserving the environment. It offers insights into the opportunities and challenges associated with the adoption of these technologies, their economic feasibility, and their potential impact on society. The book is intended mainly for researchers, scientists, academics, industry professionals and students working in the areas of clean water and energy. It offers a comprehensive overview of the current state of the air technologies and provides a roadmap for future research and development. The authors of this book are experts in the fields of water and energy technologies, and they have come together to present a collection of insightful and informative chapters. They have collaborated to bring together their diverse perspectives on the challenges and opportunities that arise from the adoption of climate-friendly technologies.

This book delves into various innovative and sustainable approaches to address the challenges associated with wastewater treatment. The chapters cover a range of topics, from nanomaterials-based sustainable wastewater treatment strategies and eco-friendly integration of seawater desalination with carbon mineralization of brine wastes to a high-performance grid integrated wind energy conversion system (WECS) for sustainable energy development. Furthermore, the book explores the use of GIS-based maritime spatial planning and exergy efficiency in offshore wind turbine site selection.

In addition, the book addresses topic of advanced fuels for engine emission control, the use of alternative fuels and emission control methods for internal combustion engines, and the production of biogas from biowaste and its potential as a future alternative energy for power generation. It also presents a novel Type 2 intelligent controller for grid-tied solar electric vehicle charging stations and critical reviews of spent coffee ground briquettes and bio-briquettes from processed cashew nutshell wastes.

The chapter on "Nanomaterials-Based Sustainable Wastewater Treatment Strategies for a Sustainable Planet" discusses the use of nanomaterials in wastewater treatment, which offers several benefits, including high efficiency, low cost, and reduced environmental impact. The chapter also highlights the significant progress that has been made in wastewater treatment technologies and the potential for further innovation in this field. The chapter explores potential solutions to these challenges, including the use of energy-efficient treatment technologies, the adoption of advanced treatment processes. The authors explore the various challenges associated with traditional wastewater treatment methods and the potential of nanomaterials in addressing these challenges. It also discusses the different types of nanomaterials that have been studied for wastewater treatment, including carbon nanotubes, nanofibers, and nanoparticles. The authors highlight the unique properties of nanomaterials, such as their high surface area, catalytic activity, and chemical reactivity, which make them ideal for enhancing the efficiency and effectiveness of wastewater treatment processes. The chapter on "Eco-Friendly Integration of Seawater Desalination With Carbon Mineralization of Brine Wastes" explores the integration of seawater desalination and carbon mineralization of brine wastes, which can help to reduce fossil fuel-based energy consumption and the environmental impact of desalination processes. The chapter on "A High-Performance Grid-Integrated Wind Energy Conversion System (WECS) for Sustainable Energy Development" discusses the design and implementation of a high-performance grid-integrated wind energy conversion system, which can contribute to sustainable energy development. The chapter on "Offshore Wind Turbine Site Selection in the Northern Aegean Sea by Utilizing GIS-Based Maritime Spatial Planning and Exergy Efficiency" explores the use of GIS-based maritime spatial planning and exergy efficiency analysis for offshore wind turbine site selection in the Northern Aegean Sea. The chapter on "Factors Affecting Farm Income and Climate Adaptation Strategies in Farming Sector of Gujarat State of India" examines the factors affecting farm income and climate adaptation strategies in the farming sector of Gujarat State of India. The chapters on "Advanced Fuels for Engine Emission Control" and "The Use of Alternative Fuels and Emission Control Methods for Compression Ignition Engines" discuss the use of advanced and alternative fuels and emission control methods for engines, which can help reduce emissions and improve air quality. The chapter on "Production of Biogas From Biowaste and its Potential as a Future Alternative Energy for Power Generation" explores the potential of biogas produced from biowaste as a future alternative energy source for power generation. The chapter on "Type 2 Intelligent Controller for Grid Tied Solar Electric Vehicle Charging Station: Novel Controller for PV Fed Grid Tied EV Charger" discusses the design and implementation of a type 2 intelligent controller for grid-tied solar electric vehicle charging stations. The chapters on "Spent Coffee Ground Briquettes: A Critical Review" and "Bio-Briquettes From Processed Cashew Nut Shell Waste: A Review" examine the use of spent coffee grounds and processed cashew nut shell wastes for the production of bio-briquettes, which can be used as a sustainable energy source.

Each chapter provides a critical review of a specific topic, offering insights into the current state of knowledge and identifying areas for future research. Overall, this book aims to provide a comprehensive overview of the challenges associated with wastewater treatment and sustainable energy and the innovative approaches being developed to address these challenges. It is hoped that the insights provided in this book will help to inform future research in this critical area of water, environmental and energy sustainability. It is hoped that this book will inspire further research and development in the field of clean water and energy technologies and contribute to a more sustainable future for us.

J. Sadhik Basha
Department of Process Engineering, National University of Science & Technology, Oman

Jimoh Kayode Adewole
Department of Process Engineering, National University of Science & Technology, Oman

Anteneh Mesfin Yeneneh
Department of Process Engineering, National University of Science & Technology, Oman

Luqman Kolawole Abidoye
Department of Process Engineering, National University of Science & Technology, Oman

Acknowledgment

The Process Engineering Department, College of International Maritime, National University of Science & Technology has three research groups: Advanced Fuels Technology and Systems; Bioprocess and Catalysis; and Process Design and Computational Engineering. This book is the latest product of collaboration between these research groups. While this book bear our names as editors, it would not have been possible without the support and contributions of our colleagues, students, and collaborators worldwide. Our sincere gratitude to all of them for making this book project a reality.

Our special thanks go to reviewers who (despite their busy schedules) spent time reviewing the book chapters. We sincerely appreciate all the authors for their dedication, timely responses, and willingness to accept editorial comments and suggestions.

We would also like to acknowledge the efforts of the IGI team, especially Jocelynn Hessler and Elizabeth Barrantes, who were always there to provide the necessary guidance and supports in taking this book project to its final stage. Finally, our heartfelt gratitude to the management and staff of the National University of Science & Technology for their support, encouragement, and well wishes throughout the book project.

Editorial Advisory Board

Chapter 1
Wastewater Treatment and Challenges

Hassan Ayedun
Olusegun Agagu University of Science and Technology, Okitipupa, Nigeria

Abdulrasaq Olalekan Oyedeji
Federal Polytechnic, Ilaro, Nigeria

Peter Odunayo Ayadi
Olusegun Agagu University of Science and Technology, Okitipupa, Nigeria

ABSTRACT

Water is very important for the survival of living organisms, which include both plants and animals. Sources of water available include rivers, oceans, seas, springs, groundwater, and rainwater. Increase in population, expanded urbanization, and establishment of industries put pressure on sources, quality, and quantity of available water; hence, demand for water increases. As water utilization increases, generation of wastewater increases. The wastewater treatments are described under the followings subheading, namely sources and types of wastewater, properties of wastewater, methods of treatment, and challenges of wastewater treatment methods. Information is sourced from the internet through Google search engine. The objective of this chapter is to compile information on emerging contaminants in wastewater, methods that can be adopted in the process of wastewater treatment that will minimize energy use and ensure safety of environment. Green methods of wastewater treatment and disposal are suggested.

INTRODUCTION

Good quality water is needed for cooking, drinking, bathing and industries. Polluted water is not suitable for human and animal consumption as it can lead to adverse health effect. Wastewater composition depends on human activities around the sources and the system that generated it. The presence of several emerging contaminants, such as plastics, pharmaceuticals, pathogenic organisms, personal care products and endocrine disruptor compounds in wastewaters is a source of concern (Gogate and Pandit,

DOI: 10.4018/978-1-6684-7303-0.ch001

2004). The wastewater industry is in a state of transition due to the recent wastewater effluent standards and complexity of the emerging contaminants in water bodies (Moji and Bashir, 2022; Soares, 2020).

These compounds have been identified but the severity of adverse effect associated with some of them is yet to be fully unraveled. The final destination of these compounds is a source of concern as it ends-up in food chain which may initiates adverse health condition and aggravates it in predisposed individuals. Plants, animals and human beings are mostly affected with devastating consequences.

Several factors affect wastewater treatment namely, heterogeneity of influents composition (Wang et al., 2017). Different water pollutant requires different technology for treatment. Nanosilver and micro-plastic in wastewater portend danger to human health, hence, cannot be treated in the same way as other pollutants. Such pollutants lack a well-defined quality standard that can be tolerated in water environment (Heiss and Kuster, 2015). The temperature at a particular climatic region may affect the activity of microorganisms required for degradation in cases of microbial treatment of wastewater. Other factors are space and energy demand. The increasing level of pollutants from cities and industries compel the need for treatment for appropriate technologies for wastewater treatment (Agyemang *et al.*, 2013).

Sewage sludge that are produced from wastewater treatment contains pathogenic organisms, heavy metals and partially degraded organic matter need to be treated before disposal (European Commission, 2018). The appropriate steps to take is to dewatered, incinerate, disposed professionally as land fill or used for soil amendment. However, nutrients such as phosphorus can be recovered from sludge ash to concentrate in large quantities, elements of interest useful for agricultural purposes. Organics and nutrients in wastewater from agricultural activities has been removed using membrane and biological methods (Gutu *et al.*, 2021) while different types of membrane techniques were found to be useful for the treatment of poultry slaughterhouse wastewater (Fatima *et al*, 2021).

The increase in uncontrolled peri – urban development, unapproved extension, creation of slums has been making people live in housing structures without facilities like sewerage, electricity, water or paved roads (Harremoes, 1997). Social amenities such as water supply, communication network, access road, transport, energy, health care services and schools are often inadequately provided. In such a situation, people will rely on stream, wells and boreholes for their water sources. There will be no channel for erosion, the wastewater generated will flow freely into the available streams, contaminate the soil and groundwater, exposing people to different types of health related diseases. Mojiri *et al.* (2020), reported that conventional wastewater treatments have failed to remove emerging contaminants from water bodies. Therefore, researchers have tried to propose new systems with maximum performance in removing emerging pollutants.

SOURCES AND TYPES OF WASTEWATER

There are two major sources of wastewater namely domestic and industrial sources. The domestic wastewaters are those from kitchen, lavatories, septic tanks, run off from roofs when rainfalls and human wastes. The quality and quantity of wastewater generated determines the extent of damage caused to plants, and animals in the receiving water bodies. This may affect the usage of such a water body as recreational facilities. It may increase microbial load which may cause diseases when consumed.

Different industries have different wastewater type such as chemical, food and beverages, mining, oil exploration, textiles and slaughterhouses, agriculture, pesticides and nutrients. Crude oil processing wastewater may contain poly-aromatic hydrocarbon (PAHS), and waste gas. The presence of chemicals in water bodies will affect productivity of fish and even caused the extinction of vulnerable species. Similarly, introduction of nutrients from food, detergent, cosmetics industries, and agricultural field may cause alga boom. Industrial wastewater varied in composition and quantity depends on needed water in daily operations (Table 1). The treatment and disposal methods depends on the volume and the type of specific chemicals used (Gumel *et al.*, 2015). The food industries wastewater may consist of brewing spent, diaries, cyanide, milk, carbohydrates and alcohol.

Table 1. Different types of effluent and their sources

S/N	Effluent	Industrial Source
1.	Suspended solid, organic matter and nutrients	Pulp and paper industry
2.	Reactive dyes, Cadmium, chromium and molybdenum	Textile manufacturing industry
3.	Blood, and gut contents	Slaughtering industry
4.	Acids, alkali, salt and organic matter	Food industry
5.	Plasticizers and aerated lagoons	Plastic industry
6.	Selenium and ammoniacal nitrogen	Diary processing industry
7.	Nitrate as N, selenium and zinc	Breweries and distilleries industry
8.	Salt, surfactants, cationic materials, hydrogen peroxide, urea, biocides	Colour industry
9.	Pb, As, Cr Hg	Smelting, electroplating industries

Source: Gumel *et al.* (2015); modified

Some food industries engaged in soaking of grains, washing, filtering, pressing and molding which generates viscous liquid with repulsive smell. The resulting wastewater contains proteins, suspended solids which in most cases generates toxic compounds. The colour becomes brown or black with smell of rotten egg. The air around the vicinity becomes polluted and can results in respiratory diseases. If the wastewater is discharged into water body, it will make the water unsuitable for drinking.

The generation of wastewater in mining industries occur as a result of use of water for extraction of soil minerals. The process of quarrying, milling, crushing, screening, washing, and flotation also generates wastewater in different ways.

Water contaminants are classified as physical, chemical, radiological, and biological (Faust and Aly, 2018). Physical contaminants are suspended, and organic matter found in the water of lakes, rivers, and streams. They originated from soil and erosion. Chemical contaminants are salts, metals, nitrogen, bleach, toxins produced by bacteria, and animal, drugs residues and pesticides (Lamastra et al., 2016).

Biological contaminants such as microorganisms, viruses, bacteria, protozoans, and parasites are present in water and wastes water (Ashbolt, 2015). Ionizing radiation such as uranium, plutonium, and cesium constituted the radiological contaminants in water and wastewater. The adverse effect is great which may be may cause genetic disorder if the background dosage is exceeded (NRC, 2016).

The wastewater from textile industries are turbid and contains mainly reactive dyes which are mostly dark in colour. This prevent plant photosynthesis and causes species depletion in water ecosystem (Joseph, 2007).

PROPERTIES OF WASTEWATER

The physical and chemical properties of wastewaters is very important because of its effect on receiving water bodies and soil. It actually affects the survival of flora, fauna and micro-organisms in the soil. Similarly, wastewater discharged into waterbodies may affect the fish and other aquatic organisms. The wastes water released from human habitations is responsible for diverse sources and contaminants load. The physical and chemical properties of untreated wastewater from industries will be discussed under the sub-headings Biochemical Oxygen Demand (BOD), Chemical Oxygen Demand (COD), phosphates (PO_4-P), ammonia (NH_3-N), Turbidity, microbes, colour and pH. The limits of different parameters of wastewater as found in literature for South Africa, Nigeria and Ghana are presented in Table 2 (DWARF, 2010; Hamad and Oludele, 2022; Agymang *et al.*, 2013). The developed world has zero tolerance for majority of the pollutants in water while African Countries still tolerate most of them.

Biochemical Oxygen Demand (BOD): BOD is the quantity of dissolved oxygen (DO)) required by micro-oganisms for the biological degradation of organic matter. The oxygen consumed is related to the amount of biodegradable organics. When organic substances are broken down in water, oxygen is consumed. In most cases, microorganisms are added in the dilution water (seeded water) so as to have enough microorganisms for biodegradation of organic waste. In this case, the oxygen demand of seeded water is subtracted from the demand of mixed sample of waste and dilution water.

Chemical Oxygen Demand (COD): COD is a measure of the oxygen equivalent of the organic matter content of a sample that is susceptible to oxidation by a strong chemical oxidant. It is an indirect method to measure the amount of organic compounds in water. It is carried out on the sewage to determine the extent of readily oxidizable organic matter. It is expressed in milligrams per liter (mg/L).

A sample is refluxed in strongly acidic solution with a known excess of potassium dichromate ($K_2Cr_2O_7$) for 2-3 h. After digestion, the remaining unreduced $K_2Cr_2O_7$ is titrated with ferrous ammonium sulphate to determine the amount of $K_2Cr_2O_7$ consumed. The oxidizable matter is calculated in terms of oxygen equivalent. This procedure is applicable to COD values between 40 and 400 mg/L (Russell, 2006).

Phosphates (PO_4-P): Phosphorus are found in wastewater bodies through agricultural practices, industries, mining operations and sewage disposal. High concentration causes enrichment of water natural bodies with nutrients (eutrophication). It results in excessive plant growth and depletion of oxygen which may cause ecological disequilibrium (Xu, *et al.*, 2010; Loganathan *et al.*, 2014).

Ammonia (NH_3-N): The source of nitrogen compounds found in water could be domestic (septic), agricultural based fertilizer, decaying plants and animals, and industries. Bacteria and algae transformed atmospheric nitrogen to organic nitrogen (Peechattukudy and Dhoble, 2017). Wastewater contains nitrogen compounds as ammonia (NH_3), ammonium (NH_4), nitrate (NO_3), organic nitrogen and nitrites (NO_2)

Turbidity: Suspended particles of various sizes and colloidal particles in water bodies impact turbidity to water. Both of them may be of mineral or organic origin, but, their presence in water scatter or absorb electromagnetic radiation in infrared and visible range (Mucha and Kułakowski, 2016). Suspended organic matter is usually found in wastewater treatment effluent which is measured in Formazin Nephelometric Unit (FNU) while ordinary water turbidity is measured in Nephelometric Turbidity Unit (NTU) (Oregon Water 2014).

Microbes: Microorganisms exist in aggregates form such as biofilms, flocs and granules in most wastewater treatment plants. This occur as a results of diverse constituents of biologically degraded products. Extracellular polymeric mixture of high-molecular-weight were revealed in pure cultures of micro-organisms found in wastewater sludge using electron microscopes (Sheng et al, 2010). The extracellular polymeric substances are formed after the bacteria have consumed all the organic components of the sludge. The extracellular polymeric mixture consists of polysaccharides, proteins, nucleic acids, phospholipase, humic, uronic, and none polymeric substances

Colour: Wastewater from dye, textiles and food industries are usually imparted by colour of different types. The colour needs to be removed for wastewater to be acceptable for use in irrigation and toilet flushing.

pH: pH is very important in assessing the hydrogen ion concentration of wastewater. It has been established that pH affects the presence of trace elements, heavy metals and micronutrients in wastewater (Agoro *et al.*, 2018). Acidic or alkaline wastewater discharge affect aquatic plants, animals and microbes in surface water thereby disrupting the ecosystem.

Table 2. Comparison of limit of parameters in wastewater of South Africa, Nigeria, and Ghana

Parameter	DWARF, South Africa, 2010 General Limit	Federal Ministry of Environment Limit, Nigeria (Hamad and Oludele, 2022)	EPA Ghana (2000) (Agyemang *et al.*2013)
DO mg/L	7.5	2-7.5	< 1
pH	5.5 -9.5	6 – 9	6 – 9
EC µS/ Cm^2	75	15	750
TDS mg/L	450	3,500	< 1,000
Temp °C	35	40	30
SO_4^{2-}mg/L	200	100	250
PO_4^-mg/L	10	5	NA
NO_3^- mg/L	15	10	NA
NH_4^{2-} mg/L	3.0	1	1
COD mg/L	75	40	< 250
BOD mg/L	3	4	< 50

Key: NA- Not Available

GENERAL METHODS OF WASTEWATER TREATMENT

Two major water treatment adopted in Africa is the decentralized and centralized treatment. The decentralized treatment is the one in which individuals attempt to treat and dispose the wastewater. Treatment is based on individual capacity, affordability and desire to treat the water before re-use. Rural, semi urban and unplanned settlement create emergency wastewater channels when it tends to obstruct the roads, affect buildings and produced repulsive odour. In most cases, until it forms stagnant pools of ponds. In such environments, many people showed less interest in the treatment of wastewater except few educated individuals (Idris-Nda *et al.*, 2013). The major objective of their actions is to drive away the wastewater from the vicinity.

However, the centralized system is the one in which a water board is in charge of treatment. The raw water is collected, coagulated, filtered and disinfected before circulation for re-use. Three major level of treatment are preliminary, primary, secondary and tertiary treatments while the sub-units of wastewater treatment are physical, chemical and biological methods.

Preliminary Treatment: This is a physical sub-unit of wastewater treatment. It consists of grits for screening to remove coarse and suspended solids. The plastic materials, nylon bags and debris of wood are removed at this stage (Miruka, 2016).

Primary Treatment: This level is used to remove some settled organic and inorganic matter such as organic phosphorus, organic nitrogen and some heavy metals.

Secondary Treatment: This is used to remove residual organic and inorganic from primary treatment effluents. The dissolved organic and inorganic matter cannot be removed through primary treatment, hence, it is necessary to carry out secondary treatments (Jhansi and Mishra, 2013). The treatment can be achieved by using trickling filters, activated sludge or anaerobic treatments.

In Kenya, wastewater stabilization ponds are used for wastewater treatment but overloading occur due increase in population, high cost and poor management strategy (Omosa et al., 2012). The BOD, TSS, TDS and COD are reduced substantially from wastewater.

Wetland system of wastewater treatment has been found to be efficient in the removal of pollutants for its end product water to be useful for landscaping. It consumed less energy, easy to maintain, and very cost effective (Makopondo *et al.*, 2020).

Tertiary Treatment: This is an advanced method of wastewater treatment which remove significant amounts of nitrogen, phosphorus, bacteria and viruses. Chlorine, UV radiation or Ozone may be introduced to ensure complete removal of microorganisms. Tertiary treatment is carried out when the secondary treatment cannot remove all the contaminants and in most cases water meant for domestic reuse (Oji *et al.*, 2018).

Advanced Oxidation Process (AOP) is used for treating wastewater containing coloring matters, pesticides, and inorganic substances. It consists of two stages namely the creation of strong oxidants (Hydroxyl Radicals) and reaction of the oxidants produced with organic contaminants in wastewater (Sathasivam *et al.*, 2019). The most common combination are hydrogen peroxide with UV light process, hydroxyl radicals, titanium dioxide with UV light process, and hydrogen peroxide with iron catalyst (Fenton's Oxidation).

Figure 1 illustrates laboratory treatment of wastewater where the inlet represents parameters in wastewater before treatment and outlet represents treated wastewater. The pH changed from 6.1 to 7.2, Temperature changed from 27.3^{0}C to 27.4^{0}C, Electrical conductivity reduced from 312 μScm^{-1} to 231.3 μScm^{-1}, Alkalinity reduced from 100 mgL^{-1} to 10 mgL^{-1}, hardness reduced from 92.1 mgL^{-1} to 23.7 mgL^{-1}, Chloride reduced from 260 mgL^{-1} to 107.2 mgL^{-1}, Total Dissolved Solid reduced from 117 mgL^{-1} to 162.7 mgL^{-1}, Total suspended solid reduced from 250 mgL^{-1} to 26.8 mgL^{-1}, Dissolved Oxygen reduced from 4.5 mgL^{-1} to 6.4 mgL^{-1}, Biochemical Oxygen Demand reduced from 186 mgL^{-1} to 10.6 mgL^{-1}, Chemical Oxygen Demand reduced from 248 to 20.4 mgL^{-1}, Phosphates reduced from 11 mgL^{-1} to 3.4 mgL^{-1}, Sulfates reduced from 24 to 9, Total Coliform Count reduced from 1200 CFU /100 mL to 5 CFU/100 mL, Total Plate Count reduced from 3000 CFU /100 mL to 6 CFU /100 mL

Figure 1. Illustrates results of wastewater treatment in the laboratory
Key: *Temp-Temperature, EC Electrical conductivity, Cl- Chloride, TDS- Total Dissolve Solid, TSS- Total Suspended Solid, DO- Dissolved Oxygen, BOD- Biochemical Oxygen Demand, COD- Chemical Oxygen Demand, PO_4-Phosphate, SO_4- Sulphate, TCC-Total Coliform Count, TPC-Total Plate Count*
Source: Ayedun et al. (2022)

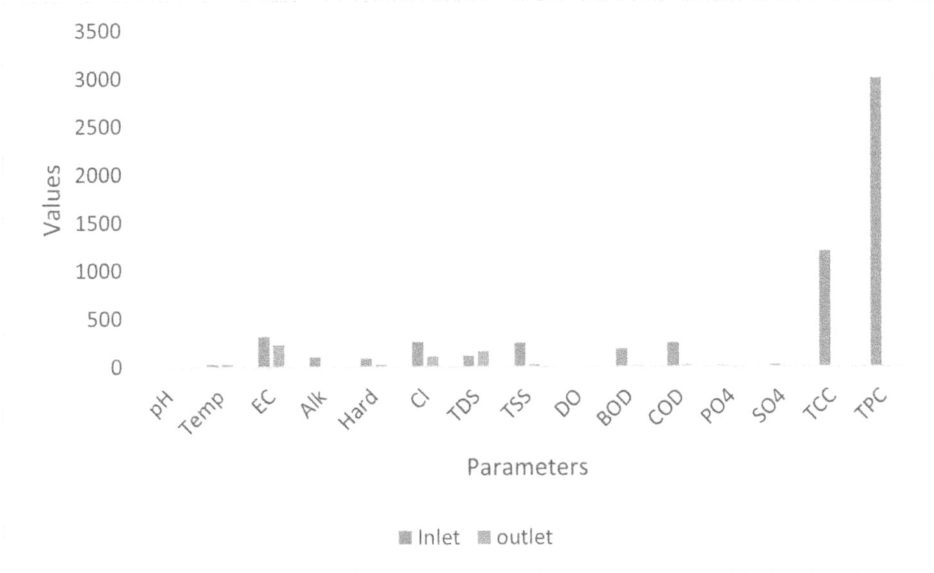

INDUSTRIAL WASTEWATER TREATMENT

The industrial wastewater treatment required special techniques based on the composition of the wastes, for example, wastes water from textile industries cannot be treated in the same way as wastewater from food industries. The food industries wastewater commonly enriched with high BOD, COD, oil, grease, potassium, phosphorus, and nitrogen. Slaughterhouse wastewater were treated by dissolve air floatation method (Physical), sedimentation, coagulation - flocculation and electrocoagulation process (membrane) methods and were found to be effective (Musa and Idrus, 2021).

Most pharmaceuticals are not degradable during conventional wastewater treatment, therefore, there is need for advanced technologies to remove them. The quality of surface water can be affected by the presence of pharmaceutical wastes (Zaman & Akter, 2015). Most of them have severe toxic and mutagenic effect on humans, therefore treatment before disposal is compulsory. Cavitation method has been used in recent past (Zupanc, et al., 2022). It involves the growth and collapse of small bubbles clusters at the same time to liberate high amounts of energy. Cavitation is an advanced oxidation method based on formation of hydroxyl radicals that can oxidize recalcitrant organic compound (Braeutigam *et al*, 2012).

Effluents from plastic Industries consists of rubber and latex which makes them have high BOD, suspended solids, variable pH and characteristics colour. Treatment method include activated carbon adsorption, chlorination and aeration. (Ghafoor *et al.*, 1994).

The colour related wastes water do not affect humans alone but also the environment through the hydrological cycle. It seeps into ground water through the soil and pollute the water resources meant for drinking (Glazer, 1997). Some wastes water contains physiological poisons such as cyanides, resins and phenol whose presence in drinking water can deteriorate human health and affects the physiological function of plants.

The discovery of new kinds of contaminants in wastes water as a results of advancement in technology is one of the challenges to environmental complexity and sustainability in the world. Green method such as use of nanoparticles embedded membrane for wastewater treatment which will lead to sustainable water re-use will be discussed.

Agro-industrial and industrial wastewaters were simultaneously treated with respect to Cr(VI) and winery effluents (Tatoulis *et al.*, 2018). A pilot-scale biological trickling filters were used to reduce Cr(VI) and remove dissolved chemical oxygen demand (COD) from the winery effluents (Zouboulis and Katsoyiannis, 2019). The study showed that carbon obtained from agro-industrial wastewater can be used to reduce Cr(VI) and at the same time reduce operational cost.

In Nigeria, it is a common practice that sewage water was used for the irrigation of farmlands during dry season, especially in northern part of the Country. Few studies reported investigation into quality analysis of such water before used for cultivation of vegetables. Sarkinnoma *et al.* (2013), investigated water from drainage channels of Bauchi metropolis and reported high alkaline pH, conductivity and nitrates-nitrogen. The effluents of sewage water of Wupa, Abuja, Nigeria was assessed for compliance with standard set by the Federal Ministry of Environment, it was found that the microbial count and the physicochemical parameters are within the limit of acceptance (Ahmad and Oludele, 2022).

In Tanzania, Wawa (2020), reported that Dodoma Urban Water supply and Sewerage Authority was faced with problem of city population surge accompanied with lack of funds for rehabilitation of water infrastructure. This has led to the build-up squatter settlements, inadequate waste disposal, poor sanitation and spread of diseases.

INNOVATIVE METHODS OF WASTEWATER TREATMENT

The emergence of nanoscience provides alternative to conventional water treatment methods, due to its efficiency and versatility of size dependent action. The concentration of toxic substances in wastewater are reduced drastically by nanotechnology thereby making water treatment easier to meet regulatory requirements (Savage and Diallo, 2005). Novel technologies have been used to desalinate water by targeting specific particles based on size, incorporated into a membrane or other structural media (Zhang et al. 2016). Carbon based nanomaterials such as graphene, graphene oxide (GO), and nanotubes (NTs), have been used for advanced water treatments and purification (Smith and Rodrigues 2015). The novel application of nano-compounds is possible due to distinctive properties such as large surface area, effectiveness in microbial disinfection, possible chemical and physical modifications, efficiency in the removal of both organic and inorganic contaminants (Smith and Rodrigues 2015).

Nanomaterials have great benefits over other materials because of large surface area, good reactivity, and a better performance (Durgalakshmi *et al.* 2019). The size effect of nano - materials makes them effective in wastewater treatment, water remediation, antibacterial coatings, and biosensor applications

The removal of Phosphate from water using a hybrid nanocomposite of aluminum has been carried out. The results obtained showed a good selectivity for phosphate which are regenerated with the addition of HCl dilute solution. Calcium-rich dolomite was used as a filter to remove phosphate from wastewater, 30% of the phosphate were removed in two hours (Table 2) (Faris et al., 2021).

Table 3. Few examples of innovative wastewater treatment

Method	Items Removed	Author
HPN nanocomposite	0.80 mg/g P	Oliveira *et al.* (2012)
Calcium-rich dolomite filter	30% P, 2hrs	Faris *et al.* (2021)
Nano-ferrites adsorption	Pb, Hg, Cd, Cr, Cd	Bharti *et al.* (2020); Rahimzadeh *et al.* (2017)
Cavitation (advanced oxidation method)	Naproxen 86%, Carbamazepine 72% and Diclofenac 77% Clofibric acid, Ketoprofen, Ibuprofen,	Zupanc *et al.* (2022)
Hybrid trickling filters	Agro-industrial wastes, Cr (VI)	Tatoulis *et al.* (2018)

CHALLENGES OF WASTEWATER TREATMENT METHODS

No availability of appropriate infrastructure and inadequate infrastructure makes wastewater treatment a difficult task. The dissolved air floatation, sedimentation, coagulation–flocculation and electrocoagulation process suggested for slaughterhouse wastewater is very expensive and required a large space. Similarly, the anaerobic, activated sludge and trickling filters required much time to start-up. The problem of infrastructure and expertise in the area of environmental engineering can be solved by funding. If appropriate fund is available, the infrastructure will be procured and manpower to oversee them will be employed. It is a matter of priority for the government, owners of private organization and individuals to have a rethink about wastewater treatment and reuse for sustainable development.

The blockage of drainage system caused by lack of sanitation and sewerage system causes flooding in the Cities. Absence of sewerage and environmental engineer to guide and direct the part of wastewater is one of the challenges facing wastewater treatment. The flood plain topographic nature of an area may cause mixing of wastewater with septic tank content thereby making wastewater very risky for human consumption (Wawa, 2020).

Individual attitude to wastewater reuse is also a factor that can affect investment on wastewater treatment especially in the developing nations of the world. Peoples disposition to reuse of treated wastewater has been narrowed down to agricultural and flushing of toilets (Oji *et al.*2018). The acceptability of treated wastewater by residents of Federal Capital City of Nigeria was investigated (Iheukwumere et al., 2021). It was revealed that there is positive perception for using treated wastewater to flush toilet, lawn wetting, construction work, dust control, watering of parks, firefighting, and watering of golf courses, however, for commercial purpose, it is used for car wash, block production, paper mills and paint production.

In Australia, people perceived desalinated water as environmental unfriendly, and treated wastewater as constituting human health risk if used (Robinson *et al*, 2005). After years of serious drought and acute water shortages the disposition of people changed to acceptance of desalinated water and recycled water. People's attitudinal change to wastewater utilization can be achieved through awareness campaign and reassurance on the quality of the wastewater under consideration. The guidelines for wastewater, use, disposal and treatment should be well defined by the stakeholders.

Most of the challenges of African Countries is lack of application of renewable energy such as solar, wind and thermal to wastewater treatment. There is abundance of resources to embrace renewable energy but the technology used presently is mainly for provision of electricity for domestic purposes but is yet to be adopted for large scale.

The policies on wastewater treatment are developed without input from the stakeholders especially the masses. There is need for public participation to achieve public support (Wang et al., 2014). Health implications, training and public awareness is necessary to ensure acceptance for the use of wastewater treated. The global environment can be protected if individual nation embraced optimization of wastewater management practices which reduces health hazard and diseases which may emanates from consuming or using wastewater by humans, animals and plant s (Faisal *et al.* 2015).

CONCLUSION

The holistic method of wastewater treatment and management will be needed to overcome the challenges and ensure sustainability. More awareness is needed for wastewater treatment for domestic use. People should be assured of the quality of treated wastewater and must be properly educated on possible uses of wastewater. Appropriate technologies must be put in place for industrial wastewater treatment. Policy frame work must be put in place on disposal of sludge, treatment and use of wastewater. A cleaner technology and greener economy policies should be adopted and applied to wastewater treatment. The emerging contaminants identification, removal, and disposal as well as simple, cost effective techniques should be investigated and adopted.

REFERENCES

Adeleye, A. S., Conway, J. R., Garner, K., Huang, Y., Su, Y., & Keller, A. A. (2016). Engineered nanomaterials for water treatment and remediation: Costs, benefits, and applicability. *Chemical Engineering Journal, 286*, 640–662. doi:10.1016/j.cej.2015.10.105

Agoro, M. A., Okoh, O. O., Adefisoye, M. A., & Okoh, A. I. (2018). Physicochemical Properties of Wastewater in Three Typical South African Sewage Works. *Journal of Environmental Studies (Northborough, Mass.), 27*, 1–9.

Agyemang, E. O., Awuah, E., Darkwah, L., Arthur, R., & Osei, G. (2013). Water quality assessment of a wastewater treatment plant in a Ghanaian Beverage Industry. *International Journal of Water Resources & Environmental Engineering, 5*(5), 272–279.

Ahmad, H., & Oludele, J. A. (2022). Assessment of the level of conformity of Wupa sewage treatment plant to environmental standards. *Sokoto Journal of the Social Sciences, 12*, 47–57.

Ashbolt, N. J. (2015). Microbial contamination of drinking water and human health from community water systems. *Current Environmental Health Reports, 2*(1), 95–106. doi:10.100740572-014-0037-5 PMID:25821716

Ayedun, H., Daudu, O. O., & Eje, R. E. (2022). Laboratory treatment of industrial wastewater. Research Project of Olusegun Agagu University of Science and Technology, Ondo State, Nigeria.

Bharti, M. K., Gupta, S., Chalia, S., Garg, I., Thakur, P., & Thakur, A. (2020). Potential of Magnetic Nano-ferrites in Removal of Heavy Metals from Contaminated Water: Mini Review. *Journal of Superconductivity and Novel Magnetism*, *33*(12), 3651–3665. doi:10.100710948-020-05657-1

Braeutigam, P., Franke, M., Schneider, R. J., Lehmann, A., Stolle, A., & Ondruschka, B. (2012, May). Degradation of carbamazepine in environmentally relevant concentrations in water by Hydrodynamic-Acoustic-Cavitation (HAC). *Water Research*, *46*(7), 2469–2477. doi:10.1016/j.watres.2012.02.013 PMID:22365175

Durgalakshmi, D., Rajendran, S., & Naushad, M. (2019). Current Role of Nanomaterials in Environmental Remediation. In *Advanced Nanostructured Materials for Environmental Remediation. Environmental Chemistry for a Sustainable World* (pp. 25–30). doi:10.1007/978-3-030-04477-0_1

DWAF-WRC. (2010). *Department of Water Affairs and Forestry and Water Research Commission. South African water quality management series*. Procedures to Assess Effluent Discharge Impacts.

European Commission. (2018). *Sewage Sludge*. Available at: https://ec.europa.eu/environment/waste/sludge/index.htm

Faisal, M., Mulana, F., Gani, A., & Daimon, H. (2015). Physical and chemical properties of wastewater discharged from Tofu industries in Banda Aceh city, Indonesia. *Research Journal of Pharmaceutical, Biological and Chemical Sciences*, *6*, 1053–1058.

Faris, M. H., Mohammed, M. I., Alwash, R. S., & Kot, P. (2021). Phosphate removal from water using low-cost dolomite filters. IOP Conf. Series. *Materials Science and Engineering*, *1058*, 012017.

Faust, S. D., & Aly, O. M. (2018). *Chemistry of water treatment*. CRC.

Ghafoor, A., Rauf, A., Arif, M., & Muzaffar, W. (1994). Chemical composition of effluents from different industries of the Faisalabad city. *Pakistan Journal of Agricultural Sciences*, *31*, 367–369.

Glazer, A. N. (1997). *Microbial biotechnology*. W.H. Freeman and Company.

Gogate, P. R., & Pandit, A. B. (2004). A review of imperative technologies for wastewater treatment I: Oxidation technologies at ambient conditions. *Advances in Environmental Research*, *8*(3-4), 501–551. doi:10.1016/S1093-0191(03)00032-7

Gumel, S. M., Usman, M. T., & Ado, A. (2015). Colouration Industry Wastewater Treatments in Nigeria–Hazard and Treatment: A Review. *International Journal of Chemical and Biomolecular Science*, *1*, 27–33. http://www.aiscience.org/journal/ijcbs

Gutu, L., Basitere, M., Harding, T., Ikumi, D., Njoya, M., & Gaszynski, C. (2021). Multi-Integrated Systems for Treatment of Abattoir Wastewater: A Review. *Water (Basel)*, *13*(18), 2462. doi:10.3390/w13182462

Harremöes, P. (1997). Integrated water and waste management. *Water Science and Technology*, *35*(9), 11–12. doi:10.2166/wst.1997.0327

Heiss, A., & Kuster, T. (2015). A regulatory perspective on prioritization of emerging pollutants in the context of the Water Framework Directive. *Environmental Toxicology and Chemistry, 34*(10), 2181–2187. doi:10.1002/etc.3047 PMID:26414540

Idris-Nda, A., Aliyu, H. K., & Dalil, M. (2013). The challenges of domestic wastewater management in Nigeria: A case study of Minna, central Nigeria. *International Journal of Development and Sustainability, 2*, 1169–1182. www.isdsnet.com/ijds

Iheukwumere, S. O., Phil-Eze, P. O., & Nkwocha, K. F. (2021). Water resource management in Nigeria: Acceptability of treated municipal wastewater reuse in Federal Capital City, Abuja. *British Journal of Environmental Science, 9*, 26–39.

IOP Publishing doi:10.1088/1757-899X/1058/1/01201

Jhansi, S. C., & Mishra, S. K. (2013). Wastewater treatment and reuse: Sustainability options. *Journal of Sustainable Development, 10*, 1–15.

Joseph, E. I. (2007). Wastewater treatment in the textile industry. *Dyeing Printing Finishing, 10*, 60-66.

Kilingo, F., Bernard, Z., & Hong–bin, C. (2021). The Analysis of Wastewater Treatment System Efficiencies in Kenya: A Review. *International Journal of Scientific and Research Publications, 11*(5), 204–214. doi:10.29322/IJSRP.11.05.2021.p11322

Lamastra, L., Balderacchi, M., & Trevisan, M. (2016). Inclusion of emerging organic contaminants in groundwater monitoring plans. *MethodsX, 3*, 459–476. doi:10.1016/j.mex.2016.05.008 PMID:27366676

Loganathan, P., Kandasamy, V. J., & Bolan, N. S. (2014). Removal and Recovery of Phosphate from Water Using Sorption. *Critical Reviews in Environmental Science and Technology, 44*(8), 847–907. do i:10.1080/10643389.2012.741311

Makopondo, R. O., Rotich, L. K., & Kamau, C. G. (2020). Potential Use and Challenges of Constructed Wetlands for Wastewater Treatment and Conservation in Game Lodges and Resorts in Kenya. *TheScientificWorldJournal, 6*, 4–15. doi:10.1155/2020/9184192 PMID:33122972

Mojiri, A., & Bashir, M. J. K. (2022). Wastewater Treatment: Current and Future Techniques. *Water (Basel), 14*(3), 448. Advance online publication. doi:10.3390/w14030448

Mojiri, A., Zhou, J., Vakili, M., & Van Le, H. (2020). Removal performance and optimisation of pharmaceutical micropollutants from synthetic domestic wastewater by hybrid treatment. *Journal of Contaminant Hydrology, 235*, 103736. doi:10.1016/j.jconhyd.2020.103736 PMID:33142212

Mucha, Z., & Kułakowski, P. (2016). Turbidity measurements as a tool of monitoring and control of the SBR effluent at the small wastewater treatment plant – preliminary study. *Archives of Environmental Protection, 42*(3), 33–36. doi:10.1515/aep-2016-0030

Musa, M. A., & Idrus, S. (2021). Physical and Biological Treatment Technologies of Slaughterhouse Wastewater: A Review. *Sustainability, 13*(9), 46–56. doi:10.3390u13094656

NRC. (2016). *Backgrounder on tritium, radiation protection limits, and drinking water standards*. https://www.nrc.gov/reading-rm/doc-collections/factsheet

Oji, I. S., Chukwuma, N. P., & Friday, N. K. (2018). Domestic Wastewater Treatment and Reuse in Awka Urban, Anambra State, Nigeria. *International Journal of Geography and Environmental Management, 4*, 16–24. www.iiardpub.org

Oliveira, M., Machado, A. V., & Nogueira, R. (2012). Phosphorus Removal from Eutrophic Waters with an Aluminium Hybrid Nanocomposite. *Phosphorus Removal from Eutrophic Waters with an Aluminium Hybrid Nanocomposite Water Air Soil Pollut, 2012*(223), 4831–4840. doi:10.100711270-012-1239-9

Omosa, I. B., Wang, H., Cheng, S., & Li, F. (2012). *Sustainable tertiary wastewater treatment is required for water resources pollution control in Africa.* Oregon Water Science Center. http://or.water.usgs.gov/grapher/fnu. html

Peechattukudy, L. A., & Dhoble, R. M. (2017). Removal of Nitrate from Water by Adsorption – A Review. *International Journal of Science Technology & Engineering, 3*, 322–325.

Rahimzadeh, M. R., Rahimzadeh, M. R., Kazemi, S., & Moghadamnia, A. A. (2017). Cadmium toxicity and treatment: An update. *Journal of Internal Medicine, 8*, 135–145. PMID:28932363

Rakesh Singh Asiwal, R. S., Sar, S. K., Singh, S., & Sahu, M. (2016). Wastewater Treatment by Effluent Treatment Plants. *SSRG International Journal of Civil Engineering, 3*, 29-35. www.internationaljournalssrg.org

Ravina, M., Galletta, S., Dagbetin, A., Kamaleldin, O. A. H., Mng'ombe, M., Mnyenyembe, L., Shanko, A., & Zanetti, M. (2021). Urban Wastewater Treatment in African Countries: Evidence from the Hydroaid Initiative. *Sustainability, 13*(22), 8–28. doi:10.3390u132212828

Robinson, K. G., Robinson, C. H., & Hawkins, S. A. (2005). Assessment of public perception regarding wastewater reuse. *Water Science and Technology: Water Supply, 5*(1), 59–65. doi:10.2166/ws.2005.0008

Sarkinnoma, A., Yarkasuwa, C. I., & Modu, K. A. (2013). Analysis of Physicochemical Parameters of Sewage Water used for Irrigation in Bauchi Metropolis – Nigeria. *Journal of Environment and Earth Science, 10*, 37–41.

Sathasivam, M., Shanmugapriya, S., Yogeshwaran, V., & Priya, A. K. (2019). Industrial Wastewater Treatment Using Advanced Oxidation Process – A Review. *International Journal of Engineering and Advanced Technology, 8*, 485–488.

Savage, N., & Diallo, M. S. (2005). Nanomaterials and water purification: Opportunities and challenges. *Journal of Nanoparticle Research, 7*(4-5), 331–342. doi:10.100711051-005-7523-5

Sheng, G. P., Yu, H.-Q., & Li, X.-Y. (2010). Extracellular polymeric substances (EPS) of microbial aggregates in biological wastewater treatment systems: A review. *Biotechnology Advances, 28*(6), 882–8946. doi:10.1016/j.biotechadv.2010.08.001 PMID:20705128

Smith, S. C., & Rodrigues, D. F. (2015). Carbon-based nanomaterials for removal of chemical and biological contaminants from water: A review of mechanisms and applications. *Carbon, 91*, 122–143. doi:10.1016/j.carbon.2015.04.043

Soares, A. (2020). Wastewater treatment in 2050: Challenges ahead and future vision in a European context. *Environ. Sci. Ecotechnol., 2*, 100030. doi:10.1016/j.ese.2020.100030 PMID:36160927

Tatoulis, T. I., Michailides, M. K., Tekerlekopoulou, A. G., Akratos, C. S., Pavlou, S., & Vayenas, D. V. (2018). Simultaneous Treatment of Agro-Industrial and Industrial Wastewaters: Case Studies of Cr(VI)/Second Cheese Whey and Cr(VI)/Winery Effluents. *Water (Basel)*, *10*(4), 382–393. doi:10.3390/w10040382

Wang, H., Wang, T., & Zhang, B. (2017). Constructed wetlands for wastewater treatment in cold climate. A review. *Journal of Environmental Sciences (China)*, *57*, 112–127. doi:10.1016/j.jes.2016.12.019 PMID:28647250

Wang, H., Wang, T., Zhang, B., Li, F., Toure, B., Omosa, I. S., Chiramba, T., Abdel-Monem, M., & Pradhan, M. (2014). Water and Wastewater Treatment in Africa – Current Practices and Challenges Clean – *Soil, Air. Water (Basel)*, *42*, 1029–1035. ww.clean-journal.com

Wawa, A. I. (2020). Challenges facing Wastewater Management in Fast Growing Cities in Tanzania: A Case of Dodoma City Council. *Huria Journal*, *27*, 168–185.

Xu, Y., Dai, Y., Zhou, J., Xu, Z. P., Qian, G., & Lu, G. Q. M. (2010). Removal efficiency of arsenate and phosphate aqueous solution using layered double hydroxide materials: Intercalation vs. precipitation. *Journal of Materials Chemistry*, *20*(22), 4684–4691. doi:10.1039/b926239c

Zaman, M. F., & Akter, S. M. (2015). Pharmaceutical Wastewater Treatment and the Efficiency of ETP in Context of Bangladesh. *Proceedings of the Waste Safe*, 1-4.

Zhang, Y., Wu, B., Xu, H., Liu, H., Wang, M., He, Y., & Pan, B. (2016). Nanomaterials-enabled water and wastewater treatment. *NanoImpact*, *3*, 22–39. doi:10.1016/j.impact.2016.09.004

Zouboulis, A. I., & Katsoyiannis, I. A. (2019). Recent Advances in Water and Wastewater Treatment with Emphasis in Membrane Treatment Operations. *Water (Basel)*, *11*(1), 45–51. doi:10.3390/w11010045

Zupanc, M., Kosjek, T., Kompare, B., Blažeka, Z., Ješe, U., Dular, M., Širok, B., & Heath, E. (2022). *Hydrodynamic cavitation: A technique for augmentation of removal of persistent pharmaceuticals?* Assessed on 03/09/2022 https://www.researchgate.net/publication/267554856

Chapter 2
Nanomaterials-Based Sustainable Wastewater Treatment Strategies for a Sustainable Planet

Kamal Prajapat
Medi-Caps University, Indore, India

Mahesh Dhonde
Medi-Caps University, Indore, India

Kirti Sahu
Shri Neelkantheshwar Government Post Graduate College, Khandwa, India

V. V. S. Murty
Government Holkar Science College, Indore, India

Parasharam M. Shirage
Indian Institute of Technology, Indore, India

ABSTRACT

Water scarcity has emerged at the forefront of the world's most critical environmental concerns. In the future decades, the rapid population development in emerging regions will continue to raise the need for clean water from residential, agricultural, industrial, and energy perspectives. Clean drinking water is the most significant factor that influences the quality of life for humans. Nanomaterials are ideal for wastewater treatment due to their potential properties, which include small size, large surface area, high porosity, high catalytic activity, tailorable physical and chemical properties, ease of separation, and reproducibility. These favorable properties of nanomaterials make them an attractive candidate for wastewater treatment. This chapter comprehensively overviews a broad spectrum of nanomaterials in wastewater treatment, including different carbon allotropes (graphene, CNTs, MWCNTs, etc.) and metal oxide nanoparticles. Finally, the prospects of these new generation materials in water and wastewater treatment are also discussed.

DOI: 10.4018/978-1-6684-7303-0.ch002

INTRODUCTION

For the survival of all forms of life on Earth, water is an absolute necessity (Morrison et al., 2020). Two major issues that make water management challenging are the growing need for water and the pollution of water supplies (Elbeltagi et al., 2021). Water scarcity is the most pressing environmental issue of our day. The need for clean water is expected to increase in the next decades for various reasons, including residential, agricultural, industrial, and energy uses. The reason for this is the ever-increasing population of developing nations (Savage & Diallo, 2005). Only Earth qualifies as a habitable planet with its 70% water content (Pradeep & Anshup, 2009). But only 2.5% of Earth's water is potable, with the rest being salty (Oki & Kanae, 2006). The amount of freshwater that is accessible as surface water (rivers, lakes, etc.) is extremely low. That's why we can't do without surface water, and it's essential to our survival (Bhat et al., 2017). By 2025, half of the global population is expected to live in areas with inadequate access to clean water (WHO, 2014).

Prior to 2015, only 20% of the world's wastewater was treated. According to the UN, 70% of industrial wastewater is dumped untreated in underdeveloped countries. Wastewater treatment and safe water supply in both industrialized and developing nations are falling behind stricter rules and rising demand for clean water (Narayan, 2010). A material's smallest dimension must be less than 100 nm to be considered a nanomaterial. These materials are typically distinguished from their more substantial equivalents by distinct physical or chemical characteristics. Due to their larger surface area, nanoparticles contain more active sites per unit mass than any other particle size. Furthermore, the surface free energy of nanoparticles is higher, resulting in better surface reactivity (Yadav et al., 2017). The problem is exacerbated by industrialization and the disposal of electronic waste. Water bodies' pollution includes lead, mercury, arsenic, cobalt, nickel, and other toxic elements (Khan & Malik, 2019). Additionally, new pollutants can harm the ecological system and human health, even at low concentrations. Low quantities of emerging pollutants may also harm the environment and people's health.

Most developing contaminants in wastewater bodies are chemicals that are not biodegradable, accumulate in food webs, and are harmful to humans, animals, plants, and ecosystems (Kumar et al., 2015; Larramendy et al., 2015). One of the world's major problems is the unregulated release of wastewater into waterways such as streams and rivers. As a result, the water quality in these bodies of water is severely compromised (Naidoo & Olaniran, 2013). The largest cities in India generate 38,354 MLD of manure and wastewater daily, yet the country only has 11,786 MLD worth of sewage treatment capacity. Also, 13,468 MLD of industrial wastewater is produced in India, but only 8080.8 MLD of that volume can be handled (Kaur et al., 2012.). Pakistan (2301 MLD), Nepal (133 MLD), Sri Lanka (163 MLD), and Bangladesh (150 MLD) are some of the other Asian nations that produce a lot of wastewater every year (Batool & Shahzad, 2021). Methods such as coagulation-flocculation, electrochemical treatment, adsorption, membrane filtration, etc., have been used for decades to treat contaminated water (Batool & Shahzad, 2021; Shah et al., 2020). Modern techniques for treating contaminated water include ultrafiltration, adsorption, chemical precipitation, biological oxidation, coagulation, etc. Heavy metals, some of which are harmful, are frequently found in industrial effluent (F. Liu et al., 2020). As a result, efficient and cost-effective energy-saving technology is needed for wastewater treatment.

Nanotechnology has made it possible to treat wastewater very efficiently and can be used for many different things. This novel treatment process may make it feasible to use alternative water sources to meet rising demand profitably. Treating wastewater with nanotechnology is a hot issue that brings academics from several disciplines. Nanomaterials based on multidimensional covalent organic frameworks (COF) are currently the subject of research to determine their viability as prospective materials for wastewater treatment (X. Liu et al., 2021). Additionally, the material that is being developed, graphitic carbon nitride, has various applications, one of which is the withdrawal of toxic metals from the water phase (Hu et al., 2021). In addition to being affordable and environmentally benign, the material possesses a high degree of chemical and thermal resilience. Furthermore, graphene, a nanomaterial made of carbon, has the potential to reduce pollution. However, there are concerns about its production cost, effectiveness, and environmental impact. On the other hand, Graphene oxide (GO)-based environmental advances have the same properties as pure graphene but are much cheaper to produce. Two basal planes for the adsorption of impurities are present in the initial layer of graphene nanomaterials. In contrast, the innermost walls of carbon nanotubes (CNTs) are impermeable to adsorbents (Fenati et al., 2021). Three different forms of CNTs may be distinguished: composite carbon nanotubes (CNC), multi-walled carbon nanotubes (MWCNT), and single-walled carbon nanotubes (SWCNT) (Mudasir & Naqash, 2021; Thostenson et al., 2001). Because of their one-dimensional structure, CNTs offer exceptional mechanical and thermal characteristics (Den et al., 2006; Huang et al., 2021). Dimensional parameters substantially affect the determination of the CNTs' characteristics. CNT adsorption occurs by a similar process as graphene. Adsorption depends on CNTs' functional dimensions, even in some cases when their porous nature is crucial (Madannejad et al., 2019).

Numerous materials have been developed for water purification, including nanomaterials, carbon-based materials, polymers, pillared clays, zeolites, highly porous oxides, and metal-organic frameworks (MOFs) (Kadam et al., 2019). It has been established that carbon-based materials work best as adsorbents to remove both organic and inorganic contaminants (Kalaitzidou et al., 2020; J. Wang et al., 2020; S. Wang et al., 2012). Carbon-based nanomaterials have large specific surface areas, high porosities, adjustable pore diameters, excellent electrical conductivities, and good chemical stability in both alkaline and acidic environments (Hao et al., 2021; Lam & Luong, 2014). Carbon-based MOFs include carbon skeletons, non-metal-doped carbons, metal-free porous carbons, and composites (Singh et al., 2022). The removal efficiency, affordability, and environmental consequences of using graphene to deal with pollution are all factors that must be considered. Graphene oxide (GO) is an inexpensive alternative to pure graphene that may be used in environmental technology. CNT can be replaced with graphene in the treatment of sewage. Graphene-based adsorbents are superior to CNTs in several ways. Two adsorption planes are present in first-layer graphene nanomaterials. CNTs' inner walls are impenetrable to adsorbents (Gopakumar et al., 2019).

Furthermore, nano-scale metals and oxides are also used to remove contaminants from wastewater. Zero-valent iron (ZVI), ferric oxides, aluminum oxides, manganese oxides, titanium oxides, magnesium oxides, and cerium oxides are common nano-scale materials for water treatment (Botes & Cloete, 2010). Recent developments include nanosized magnetic adsorbents. Nano-scale metals and metal oxides show strong removal ability towards arsenic, cadmium, chromium, and uranium and outperform organics and phosphate in capacity and selectivity (Kidd et al., 2021; Prasad et al., 2014; Tsarev et al., 2016).

In addition, several studies have examined the effects of incorporating nanoparticles into inorganic or polymer membranes to achieve multi-functional or synergistic effects (Prasad et al., 2017). Their applicability extends to many other areas, including medicine (through drug-delivery systems), agriculture (via improved soil quality), engineering (via wastewater treatment), and biology (by DNA sequencing). The use of nanocomposites for the removal of toxins from polluted water supplies has received a lot of attention in recent years. Nanocomposites are valuable instruments for cleaning up contaminated environments because of their enormous surface areas and other properties. Researchers are starting to work on nanocomposite materials to get the most out of the nanoparticles and the hosts. As wastewater treatment involves the testing and exploring of a wide variety of materials with different chemical and physical properties, including different carbonaceous materials (Graphene, CNTs, MWCNTs, etc.), zero-valent metal nanoparticles, metal oxide nanoparticles (NPs) and nanocomposites this chapter attempts to provide an overview of the materials utilized in wastewater treatment.

VARIOUS NANOMATERIALS FOR WASTEWATER TREATMENT

Zero-Valent Metal Nanoparticles

Zero-valent metal nanoparticles generally signify metallic iron elements lacking charged Fe atoms. Usually, it is used as a large-scale reducing agent or to change oxidized compounds, which can sometimes be dangerous (Prasad et al., 2017). Zero-valent metal nanoparticles are utilized to clean up polluted soils, sediments, water bodies, and wastewater because of their unique optical, mechanical, magnetic, electrical, and catalytic capabilities (Nowack & Bucheli, 2007; Sharma et al., 2015; Yan et al., 2012). Nanoparticles made of Zn, Ag, or Fe are examples of zero-valent metals (Fu et al., 2014).

Zinc Nanoparticles (Zn NPs)

Zn NPs are becoming more common in sunscreens, packaging, paints, plastics, nutritional supplements, and cosmetics. This implies they are gradually leaching into ecosystems (Ghosh et al., 2016). Zn NPs are a possible replacement for Fe nanoparticles in wastewater treatment because Zn is a more powerful reducing agent relative to Fe. Moreover, it has a strong negative reduction potential and has widespread use (Puay et al., 2015). Compared to wastewater treated using zero-valent iron (ZVI) nanoparticles, the rate of pollutant breakdown in wastewater treated with zinc particles is significantly greater. Most research on how Zn NPs can be used to clean up wastewater has been based on dehalogenation (Bokare et al., 2013; Zheng et al., 2011). Several pieces of research evaluated the powerful function of Zinc NPs in traditional activated sludge procedures for sewage treatment (Anjum et al., 2019). Furthermore, it has been demonstrated that Zn NPs efficiently degrade octachlorodibenzo-p-dioxin into its substituted chlorinated congeners at room temperature (Rudakiya & Pawar, 2017). In various investigations, Zn NPs have been shown to reduce halogenated and chlorinated organic molecules. It has been demonstrated that Zn NPs efficiently remediate a small subset of contaminants; however, this is unusual. This means there has been a lack of sophisticated full-scale research of Zn NPs in contaminated field settings.

Iron Nanoparticles (Fe NPs)

Since ZVI nanoparticles have been demonstrated an excellent removal efficiency for hexachlorocyclohexane (lindane), carbon tetrachloride, vinyl chloride, etc., their use in creating remedial solutions for wastewater treatment has shown encouraging results. Most research on applying Fe NPs in sewage water has been predicated on a reductive dehalogenation process since these particles serve as efficient and cheap electron donors (Fu et al., 2014). When exposed to water molecules or protons under anaerobic circumstances, they oxidize, producing Fe(II) and H_2-two powerful reducing agents for many wastewater pollutants. By transferring a couple of electrons to an O_2 molecule, ZVI NPs create hydrogen peroxide,

Figure 1. Cationic removal mechanism of ZVI in aqueous media
Source: Tarekegn et al. (2021)

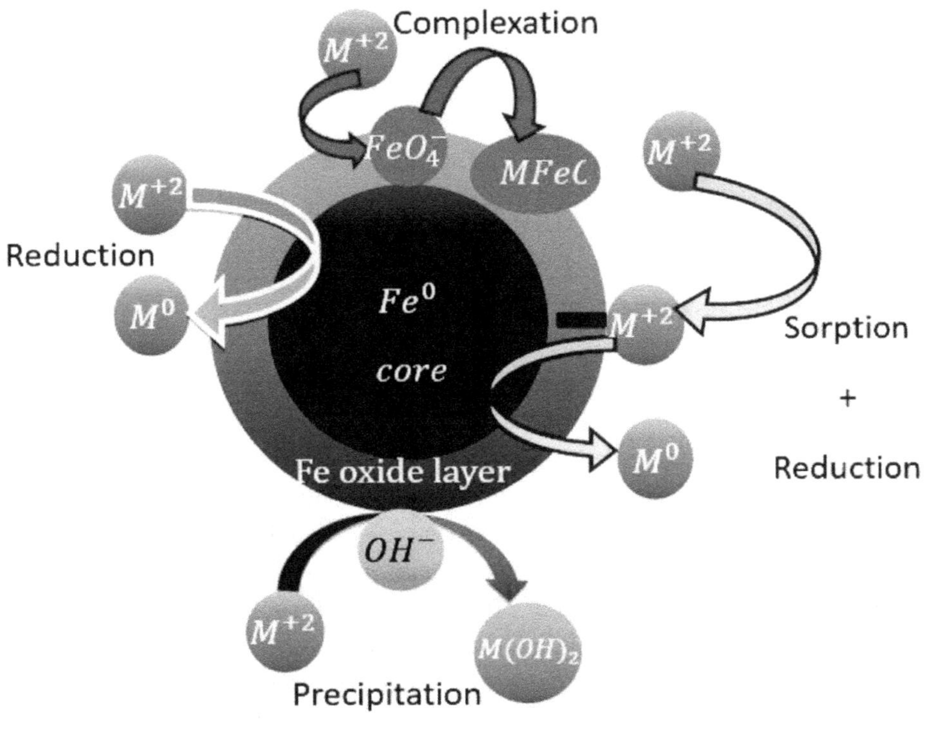

which is then converted to water, oxidizing and degrading a wide variety of organic pollutants (Yan et al., 2012). Hydroxyl radicals, produced by the interaction of hydrogen peroxide with Fe(II), are highly oxidizing to a wide range of organic molecules. Nanomaterials are utilized all over the world to clean up contaminated environments thanks to the oxidation, precipitation, reduction, and adsorption capabilities that they possess. In addition ZVI has shown promising applications in the removal of heavy metal ions such as Cd^{2+}, Cu^{2+}, and Pb^{2+}(Tarekegn et al., 2021). Figure1 illustrates the mechanism of heavy ion removal of ZVI from a aqueous medium.

Despite their usefulness, ZVI nanoparticles have a few drawbacks that need to be considered. These include their agglomeration, oxidation, and segregation difficulties in the pollutant degrading system. To boost the effectiveness of ZVI nanoparticles, scientists have turned to various novel techniques, such as surface treatment, emulsification, doping with crucial metallic ions, and hybridization with supports.

Silver Nanoparticles (Ag NPs)

The effectiveness of silver nanoparticles (Ag NPs) against a wide variety of pathogens, including bacteria, viruses, and fungi, has led to their widespread usage for water disinfection (Aziz et al., 2016). They bind to bacterial cell membranes, creating free radicals to increase permeability (Aziz et al., 2014). They destroy cell membranes, causing apoptosis. Ag NPs degrade DNA's phosphorus and sulfur (Aziz et al., 2015, 2019). Dissolving Ag NPs can release Ag+ ions, which can bind to the thiol group of important enzymes and render them inactive, interfering with their ability to carry out their regular biological tasks

(Aziz et al., 2015; Ju-Nam & Lead, 2016). However, their ability to build up in liquid medium limits their usefulness over time. Polyethersulfone (PES) MF membranes have also been modified to include Ag NPs produced using several chemical reductions.

Over the past two decades, there has been a great deal of interest in immobilizing Ag NPs on membranes and ceramic materials due to their potential utility in biofouling mitigation and sterilizing in sewerage and wastewater treatment. Nanoparticles make up a large percentage of the nanomaterials mentioned above. Increased usage of nanoparticles in wastewater treatment raises inherent challenges, such as aggregation, separation problems, and the risk of damage to humans and the environment (Crini & Lichtfouse, 2019).

Carbon Allotropes-Based Nanomaterials

Several carbonaceous materials have proven helpful as adsorbents for treating organic and inorganic contaminants (Nasrollahzadeh et al., 2021a). Carbonaceous nanomaterials like graphene, graphene oxide (GO), reduced graphene (rGO), fullerenes, graphitic carbon nitrides, and carbon nanotubes (CNT) have large surface areas, high porosities, changeable pore sizes, high electrical conductivities, and high chemical stability in both acidic and alkaline environments (Yaqoob et al., 2020). MOFs made from carbons include carbon skeletons, materials with non-metal doping in the carbon, metal-free nanoporous carbons, and different composites (Irshad et al., 2023). Organic ligands serve as the backbone of these three-dimensional structures, whereas metal clusters are used as the coupling elements. Due to their high specific surface area, copious pore structure, configurable morphologies, and flexibility, carbon-based MOFs have attracted a great deal of scientific attention for their potential use in the fields of gas sensing, energy storage, adsorption, and catalysis (Burakov et al., 2018; Gallareta-Olivares et al., 2023; J. Li et al., 2018). Figure 2, demonstrates various types of carbon allotropes, including 0D, 1D, 2D and 3D nanostructures.

Graphene-Based Nanomaterials (GNMs)

Graphene is a single nanoscale-sized graphite sheet with significant promise for environmental protection. More and more researchers are using GNMs to manage the aquatic environment. The production cost, efficacy, and ecological effects of graphene for mitigating pollution are factors to consider. Because of their cheap manufacturing costs, environmental innovations based on graphene oxide (GO) are similar to pure graphene. Graphene is an excellent nanomaterial and an alternative to CNT to treat wastewater. While CNTs are commonly used as adsorbents, graphene-based materials have many benefits. Many procedures, such as mechanical peeling, chemical oxidation-reduction, and carbon nanotube conversion, are used to create graphene. In our present aquatic environment, surfactant pollution is a severe public health issue. Due to their functional surface properties, surfactants can easily penetrate waterbodies, potentially leading to excessive nutrient enrichment of waterbodies. Both graphene oxide (GO) and its reduced counterpart, i.e., reduced graphene (rGO), were utilized for the removal of nonionic surfactant (TX-100). TX-100 adsorption efficiency was measured, and the findings showed that GO and rGO would perform the best among the materials tested. GO suppresses anaerobic denitrification while encouraging the formation of anaerobic ammonia (Brazdis et al., 2023). Two base planes in the first layer of graphene nanomaterials allow impurities to stick to them. In contrast, the innermost walls of CNTs are impermeable to adsorbents (Zhu et al., 2010).

Figure 2. Various nanoforms of carbon allotropes with examples for 0D, 1D, 2D, and 3D carbon nanostructures
Source: Gaur et al. (2021)

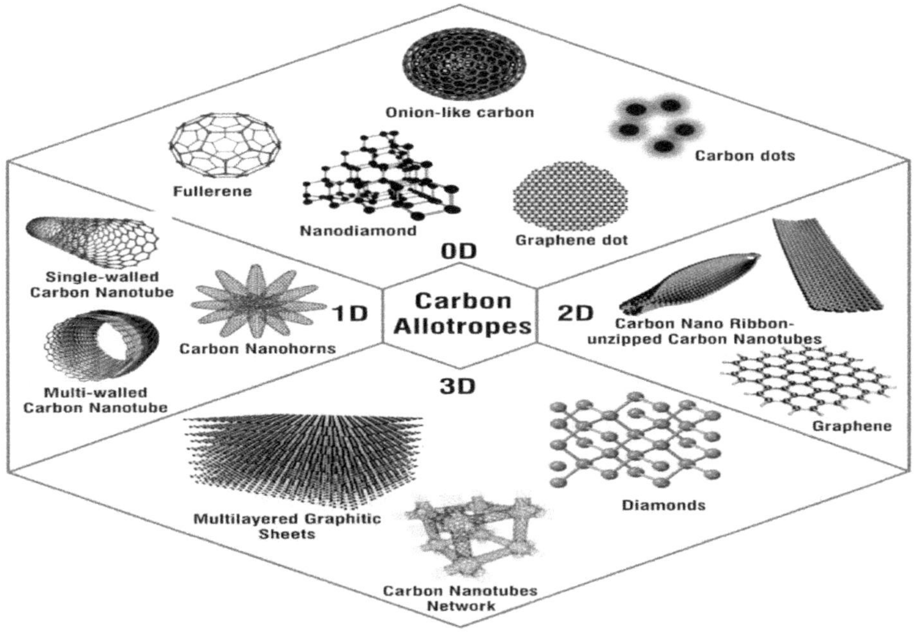

Additionally, GO and rGO can be synthesized by the chemical exfoliation of graphite without the need for specialized equipment or metallic catalysts. As a result, no catalyst remnants are present in the graphene, thus not requiring further processing. Researchers in chemistry and environmental water quality management are devoting more time and energy to studying the potential benefits of graphene-based nanomaterials. In addition to its huge surface area, graphene also possesses many π-bonds and functional groups with enough oxygen (Man et al., 2019). Five distinct interactions might occur during graphene-based nanomaterials' adsorption: π-π bonds, hydrophobic effects, electrostatic, hydrogen, and covalent bonds (Lu et al., 2018; Zhu et al., 2010). An essential factor in adsorption is the availability of surface area, and another is the presence of functional groups that can bind contaminants. However, functional groups like the hydroxyl-OH and carbonyl-COOH groups are used to chemically link organic molecules to CNTs and Graphene (Sitko et al., 2013). Physiochemically, the hydrophilic character and adsorption power of some contaminants are enhanced by graphene and its oxides. The stability of graphene's physical and chemical properties is aided by the oxidation process (Y. Yang et al., 2013). The PVP-rGO has demonstrated excellent absorption capabilities relative to other similar compounds. PVP-rGO was developed with the express purpose of removing copper from the environment. PVP-rGO was shown to have a high Cu target contaminant adsorption value because it increased the surface area. The interaction between the "Cu" cation and the functional group "π " bond led to a greater adsorption rate and surface area (Tiwari et al., 2013; Zhang et al., 2014).

Graphitic Carbon Nitride (g-C_3N_4)

Using thermal polymerase containing a pre-bundled C-N core is the most common method for producing g-C_3N_4. Melamine, urea, thiourea, and cyanamide are just a few of the many precursors that have been investigated(X. Yang et al., 2020; Y. Yang et al., 2013). To create g-C_3N_4, a thermal polymerization process typically converts precursors to melamine. The elimination of ammonia then follows condensation to make g-C_3N_4 at temperatures between 500 and 530 °C. Nevertheless, g-C_3N_4 is extremely volatile and dissociates into nitrogen and cyano fragments above 700 °C and just above 600 °C, respectively (Qamar et al., 2023). Favorable surface morphology, high surface area, porous structure, g-C_3N_4 C/N ratio, and photoluminescence are just a few of the features that can be altered by using different materials and adapting to the experiment's needs.

Nanoporous Carbon (NPCs)

NPCs can be prepared by burning organic precursor molecules, such as coal, wood, polymers, or fruit peels, and stimulating them chemically, physically, or both. Different nanoporous materials may be distinguished from one another based on their pore size, ranging from microporous to macroporous. The shortcomings, limited conductivity, and higher temperature graphitization make these technologies less practical (Y. Yang et al., 2013). Soft template synthesis depends on organic molecules coming together on their own to make nanostructures. Another widely used technique for creating mesoporous carbon compounds was established by utilizing a stiff template. Well-structured NPCs with ordered mesoporous assemblies may be produced via the hard template method. The following are the steps involved in the procedure: A controlled microstructure solid gel is utilized to make a template, which is then incorporated into a precursor, cross-linked, carbonized, and finally dissolved (Yin et al., 2020). However, when extracting from a hard template synthesis, NPC mesoporous nanostructures are often damaged, rendering the synthesis method ineffective. These constraints might be overcome by template synthesis. It is possible because there are thermodynamic constraints on how the molecules interact. Usually, amphiphilic templates are employed in this method(Thines et al., 2017). The precursor must have extensive interaction with the pore-producing component for the soft template-driven synthesis of NPCs. It must be thermally stable yet degradable or easily retrievable at experimental conditions (Ai & Jiang, 2012). Moreover, after carbonizing or removing the pore-forming component, a strongly cross-linked polymer must be generated to sustain the nanostructure.

Carbon Nanofibers (CNFs)

Carbon filaments with a nano-scale diameter are called carbon nanofibers. They all have fascinating applications and are distinguished by high surface area to volume ratios, large aspect ratios, and exceptional mechanical stability. Electrospinning, CVD, and tessellation are the three most typical synthetic approaches for synthesizing carbon nanofibers. The most popular and economical method for creating premium carbon nanofibers is electrospinning. In this method, the polymer gel is stretched and pumped into a syringe pump at a high voltage. This makes fine carbon filaments on an electrode receiver (Chen et al., 2014). The polymer gel or solution is pumped at a high voltage and steady pace using a spinneret. A few nanometer-wide nanofibers are formed when the solvent evaporates and the stretched material solidifies. These nanofibers are collected on the electrode collector. Electrospun carbon nanofibers may be fabricated from a wide variety of materials, including polyacrylonitrile, phenolic resins, cellulose polyvinylpyrrolidone, polyvinyl alcohol, and polybenzimidazole (Guo et al., 2009). Figure 3 depicts the magnified TEM images of carbonized polyacrylonitrile (PAN) nanofibers produced by electrospinning.

Figure 3. TEM images of carbonized polyacrylonitrile (PAN) nanofibers
Source: Guo et al. (2009) with permission

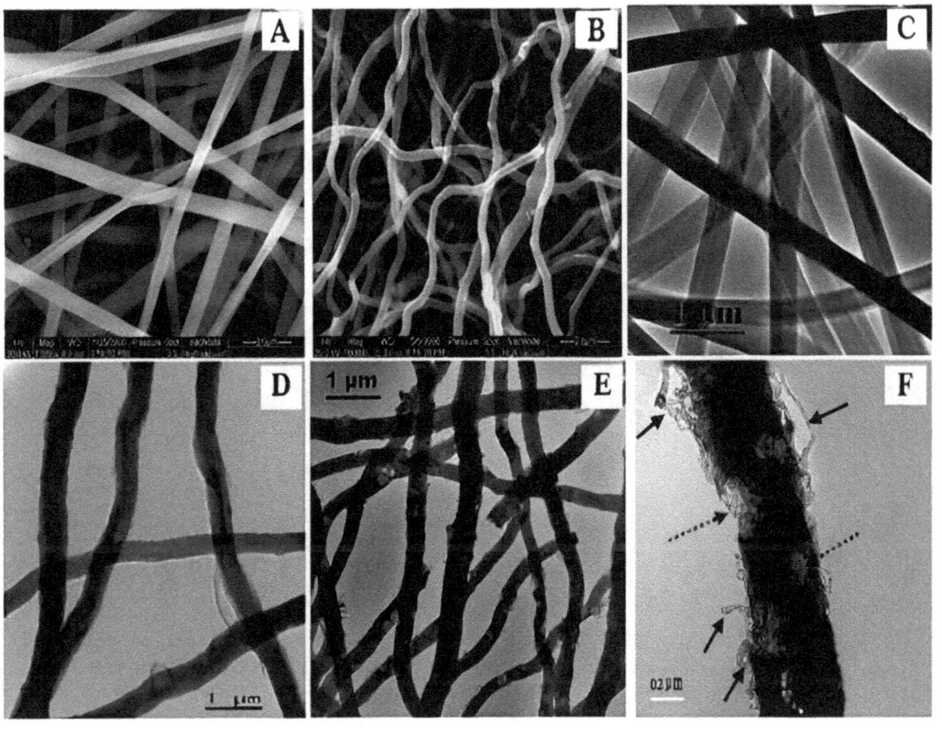

Metal and Metal Oxides-Based Nanoparticles

Metal oxide nanoparticles (MNPs) offer great promise as an affordable, sustainable, and ecologically sound wastewater treatment method. In recent decades, research into using nano-scale metals and metal oxides for environmental protection has gained much attention. The majority of nano-scales are made up of ZVI, titanium oxides, ferric oxides, manganese oxides, magnesium oxides, ceria oxides, and aluminum oxides (Mohamed, 2011.; J. Wang et al., 2011). The development of magnetic adsorbents on the nano-scale is another recent development. Several reports have shown that nano-scale metals and metal oxides are better at removing arsenic, cadmium, chromium, and uranium than other well-known contaminants, such as organics and phosphate, because they have a higher capacity and are more selective (Q. Li et al., 2008).

These NPs are easily recyclable due to their high specific surface area, low intraparticle diffusion distance, ability to be compressed without sacrificing surface area, increased adsorption sites, and ease of usage. Moreover, several of them have super-paramagnetic characteristics that make them even more effective absorbers than activated carbon (Nasrollahzadeh et al., 2021b). The major mechanism for the adsorption is the complexation between dissolved metals and oxygen molecules in metal oxide groups. Furthermore, the effective adsorption of heavy metals outside of MNPs starts with rate-limiting intraparticle diffusion along the walls (Xue et al., 2017). With excellent adsorption capability and fast kinetics, these nanoparticles result from their vast active sites, higher surface area, and shorter intra-particle diffusion distance (Ambashta & Sillanpää, 2010). Water pollution and its remediation strategies are depicted in Figure 4.

Figure 4. Water contamination source and wastewater treatment methods

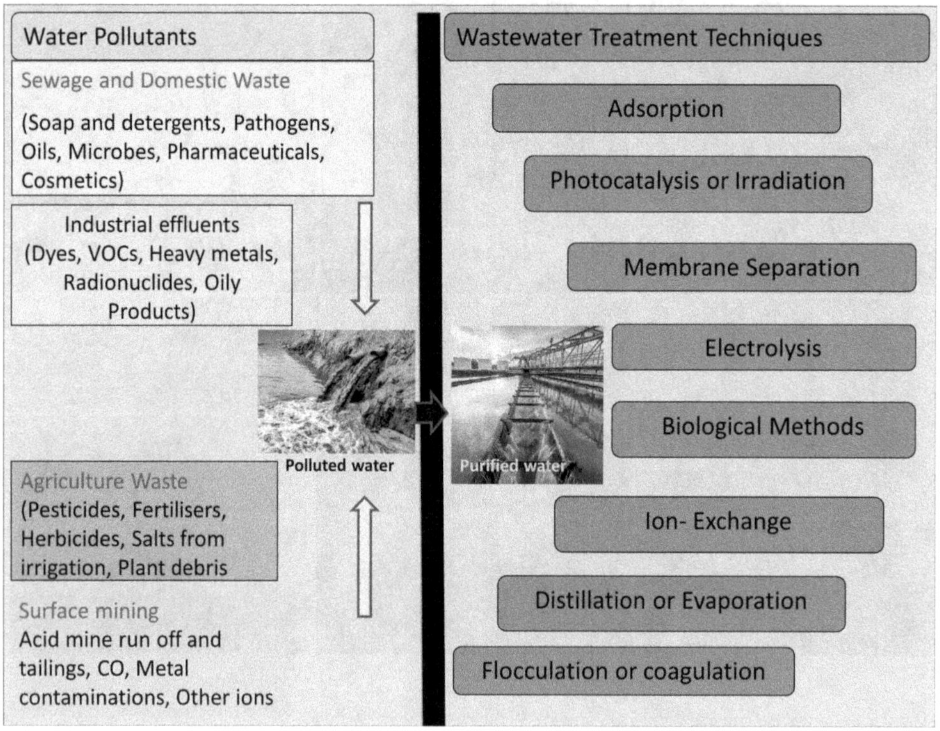

Because of its increased specific surface area, nanomagnetite's adsorption capability rises by a factor of up to 100 as particle size decreases (Pradeep & Anshup, 2009). This phenomenon has been attributed to nanoparticle alteration on the magnetite surface, resulting in new adsorption sites forming. Metal oxides of Fe, Ti, Zn, Al, and Cu are excellent and cheap ways to eliminate radionuclides and heavy metals (Corsi et al., 2018). In addition, zero-valent metals are thought to be the ideal adsorbent for removal when it comes to removing heavy metals from contaminated water during treatment owing to their high reactivity, capacity for degradation, and magnetic characteristics. Further, an increase in oxidation and subsequent clustering of zero-valent metals reduces their catalytic activity. Fe°, Sn°, Ti°, Ni°, and Al° Mg°, Pd°, and Zn° are all zero-valent metals (Dave & Chopda, 2014). Zero-valent iron, surprisingly the fourth most common metal in the crust of the Earth, is most frequently employed because of its strong reactivity and adsorption abilities. Treating wastewater with a heterogeneous catalyst such as zero-valent iron effectively removes both organic and inorganic contaminants from the wastewater.

Titanium Dioxide Nanoparticles (TiO_2 NPs)

TiO_2 nanoparticles are the most widely utilized photocatalyst nanoparticles, especially in wastewater treatment owing to their superior chemical stability, affordability, excellent photocatalytic activity, and minimal toxicity (Ilankoon, 2014). Several researchers have looked into and used TiO_2 as a catalyst in ozonation, which totally and firmly mineralizes organic contaminants (Guerra et al., 2018). The lowest amount of TiO_2 required to eradicate microorganisms in polluted water varies from 0.1 to 1 g per liter and depends entirely on the wavelength intensity and particle size (Qu et al., 2013). TiO_2's inhibitory actions may be traced back to the production of reactive oxygen species (ROS), which are created during ultraviolet (UV)- Reductive and oxidative irradiation (Das et al., 2017). When ROS are generated, they can cause harm to microbes by oxidizing their proteins and DNA, rupturing their plasmic membranes to release harmful ions, disrupting electron transport, and interfering with their ability to breathe. In addition, studies have demonstrated that a coating of TiO_2 on plastic containers will rapidly kill fecal coliforms at a concentration of 3,000 cfu per 100 mL (Panahi et al., 2018). A schematic illustration of the TiO_2 photocatalytic process is demonstrated in Figure 5.

Figure 5. Photocatalytic Activity of TiO_2
Source: Moon et al. (2014)

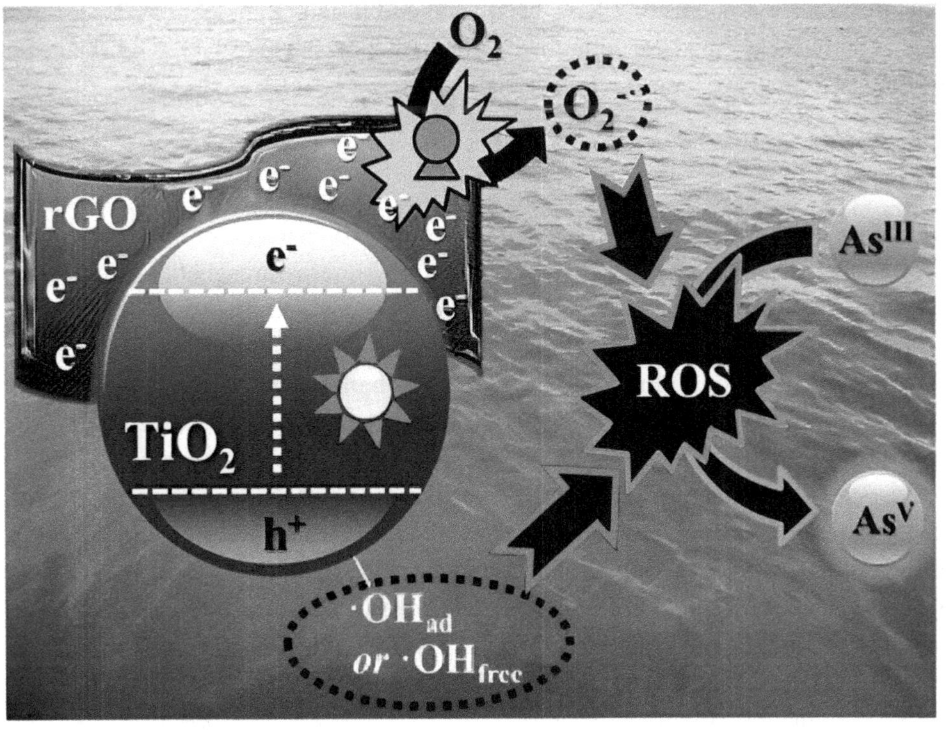

Moreover, the adsorption capacity may be impacted by the higher porosity and crystallinity of TiO_2 NPs. Ni(II), Pb(II), Cd (II), Cu (II), and Zn(II) are better adsorbed by nanosized TiO_2 than at smaller particle sizes (M. Li et al., 2014). Because it has more surface reactive atoms that are not stable, a smaller NP has more dense surface active sites (Ibhadon & Fitzpatrick, 2013). The higher porosity and crystallinity of TiO_2 NPs may also impact adsorption efficiency. A commercial TiO_2 Hombikat UV100 outperforms Degussa P25 in terms of As(V) and As(III) adsorption capability.

Zinc Oxide Nanoparticles (ZnO NPs)

In the field of photocatalytic degradation, ZnO NPs have recently gained much attention due to their great visible light transparency and efficient UV light absorption. High oxidation capacity, outstanding photocatalytic activity, and a sizable bandgap are only a few of their desirable characteristics, making them promising candidates for wastewater treatment. (Rambabu et al., 2021). They can be used in various contexts, such as in formulations of paints, packaging, cosmetic items, etc., and are immediately discharged into the environment (Rana et al., 2018). Additionally, it has been shown that ZnO NPs significantly reduce the amount of biological nitrogen removed from wastewater (M. M. Wang et al., 2017). They show promising antibacterial properties against a diverse set of harmful bacteria, while the mechanisms behind these effects are still up for debate. One leading theory suggests that photocatalytic production of reactive oxygen species is responsible (Beyth et al., 2015).

Furthermore, ZnO NPs impede bacterial development by causing membrane instability and cellular envelope penetration. (Bhuyan et al., 2015). One typical method to enhance the performance of ZnO NPs is the addition of metal dopants. Anionic dopants, cationic dopants, rare-earth dopants, and co-dopants are just some of the metal dopants that have been tried and tested (Rana et al., 2018). Several reports have proven that combining ZnO NPs with other semiconductors, such as GO, rGO, TiO_2, SnO_2, etc., is a workable strategy to improve their photodegradation properties (Das et al., 2017).

Iron Oxide Nanoparticles

Nanoparticles made of iron oxide have lately risen in favor due to their ease of utilization and abundance. Since iron oxide nanoparticles have a high surface volume, superparamagnetism, and nano-scale size, they have been the subject of substantial research into their use and manufacture for novel purposes and properties (A. Ali et al., 2016). The relative simplicity of their synthesis, customization, or coating, as well as the possibility of controlling or altering matter on the atomic scale, may make for unrivaled flexibility. Also, iron oxide nanoparticles have a lot of potentials when combined with biotechnology because they don't react with chemicals and are safe for living things (Dinali et al., 2017).

Hematite (a-Fe_2O_3), a nonmagnetic nanosorbent, and maghemite, a magnetic nanosorbent (c-Fe_2O_3). The nano-scale nature of sorbent materials makes extraction and recovery from polluted water challenging. However, magnetite (Fe_3O_4) and magnetic maghemite (C-Fe_2O_3) can be recovered in the presence of an external magnetic field, and both minerals have proven effective as sorbents for removing heavy metals from water (Voisin et al., 2017). For a flexible ligand shell to successfully encapsulate Fe_3O_4 nanoparticles, the nanoparticles must contain numerous functional groups. The identification of a polymer coating to increase the nanostructures' dispersal stability was also a key finding (Gutierrez et al., 2017). Polymer molecules could form bonding with metal ions, turning the polymer chain into a "carrier" of the metal ions. It has been suggested that hematite (α-Fe_2O_3) could be integrated into various fields, such as inexpensive sensors, efficient catalysts, and environmental applications (Saharan et al., 2014). Also, studies have demonstrated that nano hematite can effectively absorb ionized metals from polluted drinking water.

Furthermore, creating 3D-Fe_2O_3 microstructures in the shape of flowers has opened the door to a new water purification method. Due to its porous nature, α-Fe_2O_3 may effectively agglomerate pollutants and provide access through its active sites. As synthesized Cr(VI) and As(V) α-Fe_2O_3 had the highest adsorption properties, surpassing many other nanomaterials. Both laboratory and field investigations have demonstrated that iron oxide nanoparticles may effectively remove various contaminants. (Tang & Wang, 2020). Adsorption and photocatalysis are today's common uses for iron oxide nanoparticles in the wastewater treatment industry (Dinali et al., 2017). Iron oxide nanoparticles must be chemically altered or modified to enhance their adsorption capability and usefulness (Gutierrez et al., 2017).

Nanocomposites

Nanocomposites can potentially lessen the environmental impact of nanotechnology by minimizing the discharge of nanoparticles and increasing the compatibility of nanotechnology with current systems. The material comprises multiphase solids such as colloids, porous media, copolymers, and gels. To some extent, the overall performance of nanocomposites may depend on the hosts that are used to create them. Free nanoparticles are more dispersed, stable, and recyclable than nanocomposites. Therefore, composite nanomaterials could be capable of linking the nano-scale and mesoscale. Until now, it was thought that water nanotechnology could only be used on a big scale by using nanocomposites (Ahmed et al., 2022).

Figure 6. Schematic representation of nanocomposite-enabled wastewater treatment technique
Source: Emam et al. (2022)

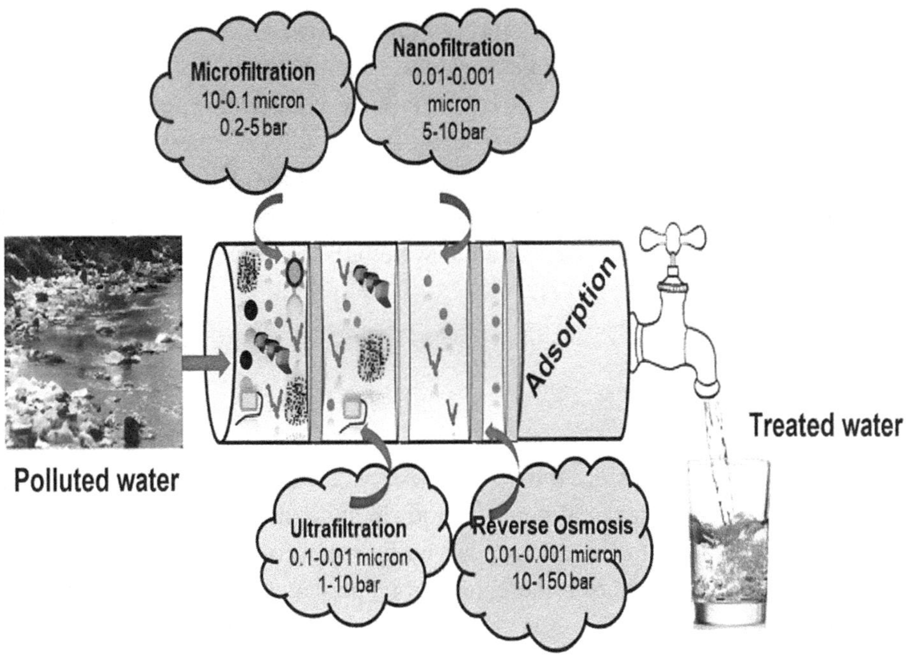

The aforementioned nanomaterials are mostly composed of nanoparticles. There are a variety of challenges that must be overcome as wastewater treatment increasingly incorporates nanoparticles. These include agglomeration, segregation challenges, and the risk of harm to humans and the environment. Researchers are developing nanocomposite materials to maximize the benefits of both the loaded nanoparticles and the hosts (Solis et al., 2020). In addition, nanocomposites can potentially lessen the environmental impact of nanotechnology by reducing the number of nanoparticles emitted into the atmosphere. The material is a solid multiphase of gels, copolymers, porous media, and colloids. The performance of nanocomposites may be affected by selecting suitable hosts. Free nanoparticles offer more excellent distribution, stability, and reusability than nanocomposites. So, nanocomposite materials might bridge the gap between the nano- and meso-levels (X. Yang et al., 2020). Numerous research projects have been carried out with the goal of adding nanoparticles to inorganic or polymeric membranes to provide multifunction or synergism. TiO_2 and bimetallic nanoparticles, as well as antibacterial nanoparticles like CNTs and nano-Ag and MNPs, are examples of photocatalytic nanomaterials (e.g., zeolite Al_2O_3 and TiO_2) (I. Ali et al., 2023). They are used in many fields, such as biological sciences, wastewater treatment, improving the quality of soil, and drug delivery systems (Solis et al., 2020). Adsorption of various compounds onto nanocomposites from polluted wastewater, including dyes, heavy metal ions, and pesticides, has recently garnered much attention (Thoniyot et al., 2015). Nanocomposites are good tools for removing pollutants because of their large surface areas and other qualities (Bogue, 2011). When certain photoactive or antibacterial nanomaterials are added to nanocomposite membranes, they make the membranes more reactive. This makes fouling less likely to happen (Rasalingam et al., 2014). In addition, polymeric adsorbents and membranes also have shown great promise in recent years for wastewater treatment. Polymer nanocomposites involve the incorporation of nanofillers and/or nanoparticles into polymeric matrices as a dendrimer, cellulose, or resins to enhance the resulting material's mechanical, thermal, physical, and chemical properties (Emam et al., 2022). Figure 6, illustrates the polymeric nanocomposites-based membranes for treating wastewater.

Similar to how nanocomposite membranes containing antibacterial, photocatalytic, and catalytic nanoparticles may be created (Amin et al., 2014), it may be conceivable to fabricate nanocomposite membranes with a variety of functional nanoparticles. Photocatalytic nanoparticles included in nanocomposites deliver UV light to the system. A submerged ultrafiltration (UF) membrane or microfiltration (MF) reactor setup is recommended for simple access to UV-based photocatalytic light. Furthermore, the potential application of various optical fibers in underwater UF membranes or MF reactors is intriguing. (Guerra et al., 2018; Rodrigues et al., 2017).

WASTEWATER CHALLENGES

Wastewater treatment is challenging since there is a lack of necessary infrastructure or no available infrastructure. Slaughterhouse wastewater treatment methods include dissolved air flotation, sedimentation, coagulation-flocculation, and electrocoagulation are both prohibitively expensive and space-intensive. Similar amounts of time were needed to initiate anaerobic, activated sludge operation and trickling filters. Finance solves the environmental engineering field's infrastructural and talent gaps. If sufficient funds are available, the necessary facilities will be built, and the essential personnel will be hired to run them. Rethinking wastewater treatment and reuse is critical for long-term sustainability and should be a top concern for governments, business owners, and common citizens. In addition, inadequate waste management and sewage infrastructure obstruct drainage systems, leading to urban floods. Wastewater treatment has difficulties due, in part, to the lack of a sewerage and environmental engineer to oversee and coordinate the process. Wastewater is highly hazardous for human consumption if exposed to septic tank content due to the floodplain topography of a location (Wawa, 2021).

Moreover, a person's perspective on reusing wastewater can influence the amount of money spent on wastewater treatment, which is especially important in developing countries. Most people are only willing to reuse treated wastewater for irrigation and toilet flushing (Oji et al., 2018). There is a lack of participation from key stakeholders, notably the general public, in formulating wastewater treatment policies. Public support requires people's input (H. Wang et al., 2014). Wastewater treatment has health concerns; thus, education and public awareness are crucial for its widespread adoption. Each country can do its part to maintain the planet's ecosystem by adopting wastewater management systems that maximize efficiency and minimize the risks to human, animal, and plant health posed by wastewater.

CONCLUSION AND FUTURE PROSPECTS

Treatment and management of wastewater raise issues as well as services to manage and maintain the water balance. To date, several projects have been carried out to enhance human and animal health and safety. Nanomaterials with special physical and chemical characteristics offer much potential for cleaning pollutants. The concept of developing nanomaterials has been put up to prioritize opportunities for their use. They featured several advantageous qualities, such as fast throughput, broad-spectrum photochemicals, targeted selectivity, rapid kinetics, effectiveness, and potent antimicrobial action. Metal oxide nanoparticles can be used to build powerful photocatalysts that show improved performance under visible light. Nano-sorbents with the property of metal oxides are effective at removing organic and heavy metal pollution. When dealing with nanomaterials, engineers face several obstacles. Nanomaterials, notably van der Waals interactions, are to blame for the phenomenon of clustering. Whether or not magnetics are incorporated, recycling nanomaterials from treated water utilizing nanoparticles is a significant challenge. Nanoparticles' unique properties are ideal for a quick and efficient water treatment. Their physical and chemical qualities have contributed to their success, but they have only recently found a niche use case in wastewater treatment. In this article, we have looked at several nanomaterials technologies that have been suggested or are on the cusp of active development. Research on metal oxide-containing nanomaterials for in vivo and in vitro applications is still in its early stages, and many approaches are still in the pilot or experimental stages. Iron oxides come in various chemical and physical forms, each having significant potential for use in cutting-edge environmental applications. Further study is needed to investigate concerns with nanotechnology. Only a few nanomaterials are now available to the public. Future research must focus on developing and optimizing the usage of low-cost nanomaterials for use in water and wastewater treatment.

In conclusion, tailored nanoparticles are widely used nowadays and offer a noninvasive method for treating wastewater. However, before being employed in different applications, the destiny of nanomaterials in the environment must be examined due to the ambiguity surrounding their possible impacts on human and environmental health. More and more studies are showing that releasing nanoparticles into the environment can have serious consequences, and this trend is only anticipated to accelerate as the number of industries that utilize nanotechnology expands. Nanomaterials have the potential to revolutionize wastewater treatment, but present restrictions prevent their widespread application. Future studies addressing safety concerns, system affordability, and scalability should help remove these obstacles.

REFERENCES

Ahmed, S. F., Mofijur, M., Ahmed, B., Mehnaz, T., Mehejabin, F., Maliat, D., Hoang, A. T., & Shafiullah, G. M. (2022). Nanomaterials as a sustainable choice for treating wastewater. *Environmental Research, 214*, 113807. doi:10.1016/j.envres.2022.113807 PMID:35798266

Ai, L., & Jiang, J. (2012). Removal of methylene blue from aqueous solution with self-assembled cylindrical graphene–carbon nanotube hybrid. *Chemical Engineering Journal, 192*, 156–163. doi:10.1016/j.cej.2012.03.056

Ali, A., Zafar, H., Zia, M., ul Haq, I., Phull, A. R., Ali, J. S., & Hussain, A. (2016). Synthesis, characterization, applications, and challenges of iron oxide nanoparticles. *Nanotechnology, Science and Applications, 9*, 49–67. doi:10.2147/NSA.S99986 PMID:27578966

Ali, I., Alyona, S., Tatiana, K., Anastasiya, G., Albishri, H. M., & Alshitari, W. H. (2023). Facile adsorption-electroflotation method for the removal of heavy metal ions from water using carbon nanomaterials. *Environmental Science and Pollution Research International, 30*(13), 1–12. doi:10.100711356-022-24509-0 PMID:36593318

Ambashta, R. D., & Sillanpää, M. (2010). Water purification using magnetic assistance: A review. *Journal of Hazardous Materials, 180*(1–3), 38–49. doi:10.1016/j.jhazmat.2010.04.105 PMID:20488616

Amin, M. T., Alazba, A. A., & Manzoor, U. (2014). A review of removal of pollutants from water/wastewater using different types of nanomaterials. *Advances in Materials Science and Engineering, 2014*, 1–24. Advance online publication. doi:10.1155/2014/825910

Anjum, M., Miandad, R., Waqas, M., Gehany, F., & Barakat, M. A. (2019). Remediation of wastewater using various nano-materials. *Arabian Journal of Chemistry, 12*(8), 4897–4919. doi:10.1016/j.arabjc.2016.10.004

Aziz, N., Faraz, M., Pandey, R., Shakir, M., Fatma, T., Varma, A., Barman, I., & Prasad, R. (2015). Facile Algae-Derived Route to Biogenic Silver Nanoparticles: Synthesis, Antibacterial, and Photocatalytic Properties. *Langmuir, 31*(42), 11605–11612. doi:10.1021/acs.langmuir.5b03081 PMID:26447769

Aziz, N., Faraz, M., Sherwani, M. A., Fatma, T., & Prasad, R. (2019). Illuminating the anticancerous efficacy of a new fungal chassis for silver nanoparticle synthesis. *Frontiers in Chemistry, 7*(FEB), 65. doi:10.3389/fchem.2019.00065 PMID:30800654

Aziz, N., Fatma, T., Varma, A., & Prasad, R. (2014). Biogenic Synthesis of Silver Nanoparticles Using Scenedesmus abundans and Evaluation of Their Antibacterial Activity. *Journal of Nanoparticles, 2014*, 1–6. doi:10.1155/2014/689419

Aziz, N., Pandey, R., Barman, I., & Prasad, R. (2016). Leveraging the attributes of mucor hiemalis-derived silver nanoparticles for a synergistic broad-spectrum antimicrobial platform. *Frontiers in Microbiology, 7*(DEC), 1984. doi:10.3389/fmicb.2016.01984 PMID:28018316

Batool, M., & Shahzad, L. (2021). An analytical study on municipal wastewater to energy generation, current trends, and future prospects in South Asian developing countries (an update on Pakistan scenario). *Environmental Science and Pollution Research International*, *28*(25), 32075–32094. doi:10.100711356-021-14029-8 PMID:33914246

Beyth, N., Houri-Haddad, Y., Domb, A., Khan, W., & Hazan, R. (2015). Alternative antimicrobial approach: Nano-antimicrobial materials. *Evidence-Based Complementary and Alternative Medicine*, *2015*, 1–16. Advance online publication. doi:10.1155/2015/246012 PMID:25861355

Bhat, S. A., Pandit, B. A., Khan, J. N., Kumar, R., & Jan, R. (2017). Water Requirements and Irrigation Scheduling of Maize Crop using CROPWAT Model. *International Journal of Current Microbiology and Applied Sciences*, *6*(11), 1662–1670. doi:10.20546/ijcmas.2017.611.199

Bhuyan, T., Mishra, K., Khanuja, M., Prasad, R., & Varma, A. (2015). Biosynthesis of zinc oxide nanoparticles from Azadirachta indica for antibacterial and photocatalytic applications. *Materials Science in Semiconductor Processing*, *32*, 55–61. doi:10.1016/j.mssp.2014.12.053

Bogue, R. (2011). Nanocomposites: A review of technology and applications. *Assembly Automation*, *31*(2), 106–112. doi:10.1108/01445151111117683

Bokare, V., Jung, J., Chang, Y.-Y., & Chang, Y.-S. (2013). Reductive dechlorination of octachlorodibenzo-p-dioxin by nanosized zero-valent zinc: Modeling of rate kinetics and congener profile. *Journal of Hazardous Materials*, *250–251*, 397–402. doi:10.1016/j.jhazmat.2013.02.020 PMID:23500419

Botes, M., & Cloete, T. E. (2010). *The potential of nanofibers and nanobiocides in water purification.* doi:10.3109/10408410903397332

Brazdis, R.-I., Fierascu, R. C., Avramescu, S.-M., & Fierascu, I. (2023). Nanomaterials for the removal of organic pollutants from agrofood wastewaters. *Advanced Technologies in Wastewater Treatment*, 355–381. doi:10.1016/B978-0-323-88510-2.00008-7

Burakov, A. E., Burakova, I. v., Galunin, E. v., & Kucherova, A. E. (2018). New Carbon Nanomaterials for Water Purification from Heavy Metals. In *Handbook of Ecomaterials* (pp. 1–20). Springer International Publishing. doi:10.1007/978-3-319-48281-1_166-1

Chen, C., Li, D., & Shao, X. (2014). High-performance nanocomposite films: Reinforced with chitosan nanofiber extracted from prawn shells. *Journal of Materials Science*, *49*(3), 1215–1221. doi:10.100710853-013-7803-x

Corsi, I., Winther-Nielsen, M., Sethi, R., Punta, C., della Torre, C., Libralato, G., Lofrano, G., Sabatini, L., Aiello, M., Fiordi, L., Cinuzzi, F., Caneschi, A., Pellegrini, D., & Buttino, I. (2018). Ecofriendly nanotechnologies and nanomaterials for environmental applications: Key issue and consensus recommendations for sustainable and ecosafe nanoremediation. *Ecotoxicology and Environmental Safety*, *154*, 237–244. doi:10.1016/j.ecoenv.2018.02.037 PMID:29476973

Crini, G., & Lichtfouse, E. (2019). Advantages and disadvantages of techniques used for wastewater treatment. *Environmental Chemistry Letters*, *17*(1), 145–155. doi:10.100710311-018-0785-9

Das, R., Vecitis, C. D., Schulze, A., Cao, B., Ismail, A. F., Lu, X., Chen, J., & Ramakrishna, S. (2017). Recent advances in nanomaterials for water protection and monitoring. *Chemical Society Reviews*, *46*(22), 6946–7020. doi:10.1039/C6CS00921B PMID:28959815

Dave, P. N., & Chopda, L. (2014). Application of iron oxide nanomaterials for the removal of heavy metals. *Journal of Nanotechnology*, *2014*, 1–14. Advance online publication. doi:10.1155/2014/398569

Den, W., Liu, H.-C., Chan, S.-F., Kin, K. T., & Huang, C. (2006). Adsorption of phthalate esters with multiwalled carbon nanotubes and its application. *Journal of Environmental Engineering and Management, 16*(4). https://www.researchgate.net/publication/255583426

Dinali, R., Ebrahiminezhad, A., Manley-Harris, M., Ghasemi, Y., & Berenjian, A. (2017). *Iron oxide nanoparticles in modern microbiology and biotechnology*. doi:10.1080/1040841X.2016.1267708

Elbeltagi, A., Azad, N., Arshad, A., Mohammed, S., Mokhtar, A., Pande, C., Etedali, H. R., Bhat, S. A., Islam, A. R. M. T., & Deng, J. (2021). Applications of Gaussian process regression for predicting blue water footprint: Case study in Ad Daqahliyah, Egypt. *Agricultural Water Management*, *255*, 107052. doi:10.1016/j.agwat.2021.107052

Emam, H. E., Badawi, A. K., Adeola, A. O., & Nomngongo, P. N. (2022). Advanced Polymeric Nanocomposites for Water Treatment Applications: A Holistic Perspective. *Polymers, 14*(12), 2462. doi:10.3390/polym14122462

Fenati, R. A., Sherrell, P. C., Khodakov, D. A., Shapter, J. G., & Ellis, A. (2021). Spatially isolated redox processes enabled by ambipolar charge transport in multi-walled carbon nanotube mats. *Materials Horizons*, *8*(4), 1304–1313. doi:10.1039/D0MH01967D PMID:34821923

Fu, F., Dionysiou, D. D., & Liu, H. (2014). The use of zero-valent iron for groundwater remediation and wastewater treatment: A review. *Journal of Hazardous Materials*, *267*, 194–205. doi:10.1016/j.jhazmat.2013.12.062 PMID:24457611

Gallareta-Olivares, G., Rivas-Sanchez, A., Cruz-Cruz, A., Hussain, S. M., González-González, R. B., Cárdenas-Alcaide, M. F., Iqbal, H. M. N., & Parra-Saldívar, R. (2023). Metal-doped carbon dots as robust nanomaterials for the monitoring and degradation of water pollutants. *Chemosphere*, *312*, 137190. doi:10.1016/j.chemosphere.2022.137190 PMID:36368530

Gaur, M., Misra, C., Yadav, A. B., Swaroop, S., Maolmhuaidh, F., Bechelany, M., & Barhoum, A. (2021). Biomedical Applications of Carbon Nanomaterials: Fullerenes, Quantum Dots, Nanotubes, Nanofibers, and Graphene. *Materials, 14*(20), 5978. doi:10.3390/ma14205978

Ghosh, A., Nayak, A. K., & Pal, A. (2016). Nano-Particle-Mediated Wastewater Treatment: a Review. *Current Pollution Reports, 3*(1), 17–30. doi:10.1007/s40726-016-0045-1

Gopakumar, D. A., Abdul, A. K., Pai, A. R., Thomas, S., Pasquini, D., & Ben, L. (2019). Nanomaterials—State of Art, New Challenges, and Opportunities. *Nanoscale Materials in Water Purification*, 1–24. doi:10.1016/B978-0-12-813926-4.00001-X

Guerra, F. D., Attia, M. F., Whitehead, D. C., & Alexis, F. (2018). Nanotechnology for Environmental Remediation: Materials and Applications. *Molecules, 23*(7), 1760. doi:10.3390/molecules23071760

Guo, Q., Zhou, X., Li, X., Chen, S., Seema, A., Greiner, A., & Hou, H. (2009). Supercapacitors based on hybrid carbon nanofibers containing multiwalled carbon nanotubes. *Journal of Materials Chemistry*, *19*(18), 2810–2816. doi:10.1039/b820170f

Gutierrez, A. M., Dziubla, T. D., & Hilt, J. Z. (2017). Recent advances on iron oxide magnetic nanoparticles as sorbents of organic pollutants in water and wastewater treatment. *Reviews on Environmental Health*, *32*(1–2), 111–117. doi:10.1515/reveh-2016-0063 PMID:28231068

Hao, M., Qiu, M., Yang, H., Hu, B., & Wang, X. (2021). Recent advances on preparation and environmental applications of MOF-derived carbons in catalysis. *The Science of the Total Environment*, *760*, 143333. doi:10.1016/j.scitotenv.2020.143333 PMID:33190884

Hu, B., Wang, H., Liu, R., & Qiu, M. (2021). Highly efficient U(VI) capture by amidoxime/carbon nitride composites: Evidence of EXAFS and modeling. *Chemosphere*, *274*, 129743. doi:10.1016/j.chemosphere.2021.129743 PMID:33540307

Huang, F., Singer, G., Zhou, Y., Sha, Z., Chen, J., Han, Z., Brown, S. A., Zhang, J., & Wang, C. H. (2021). Creating ionic pathways in solid-state polymer electrolyte by using PVA-coated carbon nanofibers. *Composites Science and Technology*, *207*, 108710. doi:10.1016/j.compscitech.2021.108710

Ibhadon, A. O., & Fitzpatrick, P. (2013). Heterogeneous Photocatalysis: Recent Advances and Applications. *Catalysts, 3*(1), 189–218. doi:10.3390/catal3010189

Ilankoon, N. (2014). *Use of iron oxide magnetic nanosorbe...* Retrieved February 9, 2023, from https://scholar.google.com/scholar?hl=en&as_sdt=0%2C5&q=Ilankoon%2C+N.+%282014%29.+Use+of+iron+oxide+magnetic+nanosorbents+for+Cr%28VI%29+removal+from+aqueous+solutions%3A+A+review.+J.+Eng.+Res+Appl.%2C+4%2C+55%E2%80%9363.&btnG=

Irshad, M. A., Nawaz, R., Wojciechowska, E., Mohsin, M., Nawrot, N., Nasim, I., & Hussain, F. (2023). Application of Nanomaterials for Cadmium Adsorption for Sustainable Treatment of Wastewater: A Review. *Water, Air, and Soil Pollution*, *234*(1), 1–17. doi:10.100711270-023-06064-7

Ju-Nam, Y., & Lead, J. (2016). Properties, Sources, Pathways, and Fate of Nanoparticles in the Environment. *Engineered Nanoparticles and the Environment: Biophysicochemical Processes and Toxicity*, 95–117. doi:10.1002/9781119275855.ch6

Kadam, A., Saratale, R. G., Shinde, S., Yang, J., Hwang, K., Mistry, B., Saratale, G. D., Lone, S., Kim, D. Y., Sung, J. S., & Ghodake, G. (2019). Adsorptive remediation of cobalt oxide nanoparticles by magnetized α-cellulose fibers from waste paper biomass. *Bioresource Technology*, *273*, 386–393. doi:10.1016/j.biortech.2018.11.041 PMID:30458408

Kalaitzidou, K., Zouboulis, A., & Mitrakas, M. (2020). Cost evaluation for Se(IV) removal, by applying common drinking water treatment processes: Coagulation/precipitation or adsorption. *Journal of Environmental Chemical Engineering*, *8*(5), 104209. doi:10.1016/j.jece.2020.104209

Kaur, R., Wani, S. P., Singh, A. K., & Lal, K. (n.d.). *Wastewater production, treatment and use in India*. Academic Press.

Khan, S. T., & Malik, A. (2019). Engineered nanomaterials for water decontamination and purification: From lab to products. *Journal of Hazardous Materials*, *363*, 295–308. doi:10.1016/j.jhazmat.2018.09.091 PMID:30312926

Kidd, J., Westerhoff, P., & Maynard, A. (2021). Survey of industrial perceptions for the use of nanomaterials for in-home drinking water purification devices. *NanoImpact*, *22*, 100320. doi:10.1016/j.impact.2021.100320 PMID:35559977

Kumar, V., Singh, S., Singh, J., & Upadhyay, N. (2015). Erratum: Potential of plant growth promoting traits by bacteria isolated from heavy metal contaminated soils. Bulletin of Environmental Contamination and Toxicology, 91(6). doi:10.100700128-015-1551-3

Lam, E., & Luong, J. H. T. (2014). Carbon materials as catalyst supports and catalysts in the transformation of biomass to fuels and chemicals. *ACS Catalysis*, *4*(10), 3393–3410. doi:10.1021/cs5008393

Larramendy, M. L., Soloneski, S., Larramendy, M. L., & Soloneski, S. (2015). *Emerging Pollutants in the Environment - Current and Further Implications*. Emerging Pollutants in the Environment - Current and Further Implications. doi:10.5772/59332

Li, J., Wang, X., Zhao, G., Chen, C., Chai, Z., Alsaedi, A., Hayat, T., & Wang, X. (2018). Metal–organic framework-based materials: Superior adsorbents for the capture of toxic and radioactive metal ions. *Chemical Society Reviews*, *47*(7), 2322–2356. doi:10.1039/C7CS00543A PMID:29498381

Li, M., Yin, J. J., Wamer, W. G., & Lo, Y. M. (2014). Mechanistic characterization of titanium dioxide nanoparticle-induced toxicity using electron spin resonance. *Yao Wu Shi Pin Fen Xi*, *22*(1), 76–85. doi:10.1016/j.jfda.2014.01.006 PMID:24673905

Li, Q., Mahendra, S., Lyon, D. Y., Brunet, L., Liga, M., Li, D., & Alvarez, P. J. J. (2008). Antimicrobial nanomaterials for water disinfection and microbial control: Potential applications and implications. *Water Research*, *42*(18), 4591–4602. doi:10.1016/j.watres.2008.08.015 PMID:18804836

Liu, F., Zhu, S., Li, D., Chen, G., & Ho, S. H. (2020). Detecting Ferric Iron by Microalgal Residue-Derived Fluorescent Nanosensor with an Advanced Kinetic Model. *iScience*, *23*(6), 101174. doi:10.1016/j.isci.2020.101174 PMID:32498017

Liu, X., Pang, H., Liu, X., Li, Q., Zhang, N., Mao, L., Qiu, M., Hu, B., Yang, H., & Wang, X. (2021). Orderly Porous Covalent Organic Frameworks-based Materials: Superior Adsorbents for Pollutants Removal from Aqueous Solutions. *The Innovation*, *2*(1), 100076. doi:10.1016/j.xinn.2021.100076 PMID:34557733

Lu, S. Y., Jin, M., Zhang, Y., Niu, Y. B., Gao, J. C., & Li, C. M. (2018). Chemically Exfoliating Biomass into a Graphene-like Porous Active Carbon with Rational Pore Structure, Good Conductivity, and Large Surface Area for High-Performance Supercapacitors. *Advanced Energy Materials*, *8*(11), 1702545. doi:10.1002/aenm.201702545

Madannejad, R., Shoaie, N., Jahanpeyma, F., Darvishi, M. H., Azimzadeh, M., & Javadi, H. (2019). Toxicity of carbon-based nanomaterials: Reviewing recent reports in medical and biological systems. *Chemico-Biological Interactions*, *307*, 206–222. doi:10.1016/j.cbi.2019.04.036 PMID:31054282

Man, P., He, B., Zhang, Q., Zhou, Z., Li, C., Li, Q., Wei, L., & Yao, Y. (2019). A one-dimensional channel self-standing MOF cathode for ultrahigh-energy-density flexible Ni–Zn batteries. *Journal of Materials Chemistry. A, Materials for Energy and Sustainability*, *7*(48), 27217–27224. doi:10.1039/C9TA11759H

Mohamed, M. B. (n.d.). *Low Cost Nanomaterials for Water Desalination and Purification*. Academic Press.

Moon, G. H., Kim, D. H., il Kim, H., Bokare, A. D., & Choi, W. (2014). Platinum-like Behavior of Reduced Graphene Oxide as a Cocatalyst on TiO 2 for the Efficient Photocatalytic Oxidation of Arsenite. *Environmental Science & Technology Letters*, *1*(2), 185–190. doi:10.1021/ez5000012

Morrison, E. S., Shields, M. R., Bianchi, T. S., Liu, Y., Newman, S., Tolic, N., & Chu, R. K. (2020). Multiple biomarkers highlight the importance of water column processes in treatment wetland organic matter cycling. *Water Research*, *168*, 115153. doi:10.1016/j.watres.2019.115153 PMID:31655441

Mudasir, P., & Naqash, J. A. (2021). The effect of water cement ratio on the characteristics of multi-walled carbon nanotube reinforced concrete. *Materials Today: Proceedings*, *43*, 3852–3855. doi:10.1016/j.matpr.2020.12.1176

Naidoo, S., & Olaniran, A. O. (2013). Treated Wastewater Effluent as a Source of Microbial Pollution of Surface Water Resources. *International Journal of Environmental Research and Public Health, 11*(1), 249–270. doi:10.3390/ijerph110100249

Narayan, R. (2010). Use of nanomaterials in water purification. *Materials Today*, *13*(6), 44–46. doi:10.1016/S1369-7021(10)70108-5

Nasrollahzadeh, M., Sajjadi, M., Iravani, S., & Varma, R. S. (2021a). Carbon-based sustainable nanomaterials for water treatment: State-of-art and future perspectives. *Chemosphere*, *263*, 128005. doi:10.1016/j.chemosphere.2020.128005 PMID:33297038

Nasrollahzadeh, M., Sajjadi, M., Iravani, S., & Varma, R. S. (2021b). Green-synthesized nanocatalysts and nanomaterials for water treatment: Current challenges and future perspectives. *Journal of Hazardous Materials*, *401*, 123401. doi:10.1016/j.jhazmat.2020.123401 PMID:32763697

Nowack, B., & Bucheli, T. D. (2007). Occurrence, behavior and effects of nanoparticles in the environment. *Environmental Pollution*, *150*(1), 5–22. doi:10.1016/j.envpol.2007.06.006 PMID:17658673

Oji, I. S., Chukwuma, N. P., Friday, N. K., & Philip, P.-E. O. (2018). Domestic Wastewater Treatment and Reuse in Awka Urban, Anambra State, Nigeria. *IIARD International Journal of Geography and Environmental Management, 4*(2). www.iiardpub.org

Oki, T., & Kanae, S. (2006). Global hydrological cycles and world water resources. Science, 313(5790), 1068–1072. doi:10.1126cience.1128845

Panahi, Y., Mellatyar, H., Farshbaf, M., Sabet, Z., Fattahi, T., & Akbarzadehe, A. (2018). Biotechnological applications of nanomaterials for air pollution and water/wastewater treatment. *Materials Today: Proceedings*, *5*(7), 15550–15558. doi:10.1016/j.matpr.2018.04.162

Pradeep, T., & Anshup. (2009). Noble metal nanoparticles for water purification: A critical review. *Thin Solid Films*, *517*(24), 6441–6478. doi:10.1016/j.tsf.2009.03.195

Prasad, R., Bhattacharyya, A., & Nguyen, Q. D. (2017). Nanotechnology in sustainable agriculture: Recent developments, challenges, and perspectives. *Frontiers in Microbiology*, *8*(JUN), 1014. doi:10.3389/fmicb.2017.01014 PMID:28676790

Prasad, R., Kumar, V., & Prasad, K. S. (2014). Nanotechnology in sustainable agriculture: Present concerns and future aspects. *African Journal of Biotechnology*, *13*(6), 705–713. doi:10.5897/AJBX2013.13554

Puay, N. Q., Qiu, G., & Ting, Y. P. (2015). Effect of Zinc oxide nanoparticles on biological wastewater treatment in a sequencing batch reactor. *Journal of Cleaner Production*, *88*, 139–145. doi:10.1016/j.jclepro.2014.03.081

Qamar, M. A., Javed, M., Shahid, S., Shariq, M., Fadhali, M. M., Ali, S. K., & Khan, M. S. (2023). Synthesis and applications of graphitic carbon nitride (g-C3N4) based membranes for wastewater treatment: A critical review. *Heliyon*, *9*(1), e12685. doi:10.1016/j.heliyon.2022.e12685 PMID:36660457

Qu, X., Alvarez, P. J. J., & Li, Q. (2013). Applications of nanotechnology in water and wastewater treatment. *Water Research*, *47*(12), 3931–3946. doi:10.1016/j.watres.2012.09.058 PMID:23571110

Rambabu, K., Bharath, G., Banat, F., & Show, P. L. (2021). Green synthesis of zinc oxide nanoparticles using Phoenix dactylifera waste as bioreductant for effective dye degradation and antibacterial performance in wastewater treatment. *Journal of Hazardous Materials*, *402*, 123560. doi:10.1016/j.jhazmat.2020.123560 PMID:32759001

Rana, N., Ghosh, K. S., Chand, S., & Gathania, A. K. (2018). Investigation of ZnO nanoparticles for their applications in wastewater treatment and antimicrobial activity. *IJPAP, 56*, 19–25. https://nopr.niscpr.res.in/handle/123456789/43515

Rasalingam, S., Peng, R., & Koodali, R. T. (2014). Removal of hazardous pollutants from wastewaters. *Journal of Nanomaterials*, *2014*, 1–42. Advance online publication. doi:10.1155/2014/617405

Rodrigues, S. M., Demokritou, P., Dokoozlian, N., Hendren, C. O., Karn, B., Mauter, M. S., Sadik, O. A., Safarpour, M., Unrine, J. M., Viers, J., Welle, P., White, J. C., Wiesner, M. R., & Lowry, G. (2017). Nanotechnology for sustainable food production: Promising opportunities and scientific challenges. *Environmental Science. Nano*, *4*(4), 767–781. doi:10.1039/C6EN00573J

Rudakiya, D. M., & Pawar, K. (2017). Bactericidal potential of silver nanoparticles synthesized using cell-free extract of Comamonas acidovorans: in vitro and in silico approaches. *Biotech, 7*(2). doi:10.1007/s13205-017-0728-3

Saharan, P., Chaudhary, G. R., Mehta, S. K., & Umar, A. (2014). Removal of water contaminants by iron oxide nanomaterials. *Journal of Nanoscience and Nanotechnology*, *14*(1), 627–643. doi:10.1166/jnn.2014.9053 PMID:24730287

Savage, N., & Diallo, M. S. (2005). Nanomaterials and water purification: Opportunities and challenges. *Journal of Nanoparticle Research*, *7*(4–5), 331–342. doi:10.100711051-005-7523-5

Shah, A. I., Din Dar, M. U., Bhat, R. A., Singh, J. P., Singh, K., & Bhat, S. A. (2020). Prospectives and challenges of wastewater treatment technologies to combat contaminants of emerging concerns. *Ecological Engineering*, *152*, 105882. doi:10.1016/j.ecoleng.2020.105882

Sharma, V. K., Filip, J., Zboril, R., & Varma, R. S. (2015). Natural inorganic nanoparticles – formation, fate, and toxicity in the environment. *Chemical Society Reviews*, *44*(23), 8410–8423. doi:10.1039/C5CS00236B PMID:26435358

Singh, K. K., Singh, A., & Rai, S. (2022). A study on nanomaterials for water purification. *Materials Today: Proceedings*, *51*, 1157–1163. doi:10.1016/j.matpr.2021.07.116

Sitko, R., Zawisza, B., & Malicka, E. (2013). Graphene as a new sorbent in analytical chemistry. *Trends in Analytical Chemistry*, *51*, 33–43. doi:10.1016/j.trac.2013.05.011

Solis, K. L. B., Kwon, Y. H., Kim, M. H., An, H. R., Jeon, C., & Hong, Y. (2020). Metal organic framework UiO-66 and activated carbon composite sorbent for the concurrent adsorption of cationic and anionic metals. *Chemosphere*, *238*, 124656. doi:10.1016/j.chemosphere.2019.124656 PMID:31472345

Tang, J., & Wang, J. (2020). Iron-copper bimetallic metal-organic frameworks for efficient Fenton-like degradation of sulfamethoxazole under mild conditions. *Chemosphere*, *241*, 125002. doi:10.1016/j.chemosphere.2019.125002 PMID:31590027

Tarekegn, M. M., Hiruy, A. M., & Dekebo, A. H. (2021). Nano zero valent iron (nZVI) particles for the removal of heavy metals (Cd2+, Cu2+ and Pb2+) from aqueous solutions. *RSC Advances*, *11*(30), 18539–18551. doi:10.1039/D1RA01427G PMID:35480950

Thines, R. K., Mubarak, N. M., Nizamuddin, S., Sahu, J. N., Abdullah, E. C., & Ganesan, P. (2017). Application potential of carbon nanomaterials in water and wastewater treatment: A review. *Journal of the Taiwan Institute of Chemical Engineers*, *72*, 116–133. doi:10.1016/j.jtice.2017.01.018

Thoniyot, P., Tan, M. J., Karim, A. A., Young, D. J., & Loh, X. J. (2015). Nanoparticle–Hydrogel Composites: Concept, Design, and Applications of These Promising, Multi-Functional Materials.*Advancement of Science*, *2*(1–2), 1400010. Advance online publication. doi:10.1002/advs.201400010 PMID:27980900

Thostenson, E. T., Ren, Z., & Chou, T. W. (2001). Advances in the science and technology of carbon nanotubes and their composites: A review. *Composites Science and Technology*, *61*(13), 1899–1912. doi:10.1016/S0266-3538(01)00094-X

Tiwari, J. N., Mahesh, K., Le, N. H., Kemp, K. C., Timilsina, R., Tiwari, R. N., & Kim, K. S. (2013). Reduced graphene oxide-based hydrogels for the efficient capture of dye pollutants from aqueous solutions. *Carbon*, *56*, 173–182. doi:10.1016/j.carbon.2013.01.001

Tsarev, S., Collins, R. N., Fahy, A., & Waite, T. D. (2016). Reduced Uranium Phases Produced from Anaerobic Reaction with Nanoscale Zerovalent Iron. *Environmental Science & Technology*, *50*(5), 2595–2601. doi:10.1021/acs.est.5b06160 PMID:26840619

Voisin, H., Bergström, L., Liu, P., & Mathew, A. P. (2017). Nanocellulose-Based Materials for Water Purification. *Nanomaterials*, *7*(3), 57. doi:10.3390/nano7030057

Wang, H., Wang, T., Zhang, B., Li, F., Toure, B., Omosa, I. B., Chiramba, T., Abdel-Monem, M., & Pradhan, M. (2014). Water and Wastewater Treatment in Africa - Current Practices and Challenges. *Clean - Soil, Air. Water (Basel)*, *42*(8), 1029–1035. doi:10.1002/clen.201300208

Wang, J., Tang, X., Xu, Y., Cheng, X., Li, G., & Liang, H. (2020). Hybrid UF/NF process treating secondary effluent of wastewater treatment plants for potable water reuse: Adsorption vs. coagulation for removal improvements and membrane fouling alleviation. *Environmental Research, 188*, 109833. doi:10.1016/j.envres.2020.109833 PMID:32798951

Wang, J., Zhang, S., Pan, B., Zhang, W., & Lv, L. (2011). Hydrous ferric oxide–resin nanocomposites of tunable structure for arsenite removal: Effect of the host pore structure. *Journal of Hazardous Materials, 198*, 241–246. doi:10.1016/j.jhazmat.2011.10.036 PMID:22047722

Wang, M. M., Wang, J., Cao, R., Wang, S. Y., & Du, H. (2017). Natural Transformation of Zinc Oxide Nanoparticles and Their Cytotoxicity and Mutagenicity. *Journal of Nanomaterials, 2017*, 1–12. Advance online publication. doi:10.1155/2017/8457960

Wang, S., Ng, C. W., Wang, W., Li, Q., & Li, L. (2012). A comparative study on the adsorption of acid and reactive dyes on multiwall carbon nanotubes in single and binary dye systems. *Journal of Chemical & Engineering Data, 57*(5), 1563–1569. doi:10.1021/je3001552

Wawa, A. I. (2021). Challenges facing Wastewater Management in Fast Growing Cities in Tanzania: A Case of Dodoma City Council. *Huria: Journal of the Open University of Tanzania, 27*(1). https://www.ajol.info/index.php/huria/article/view/204348

Xue, X., Cheng, R., Shi, L., Ma, Z., & Zheng, X. (2017). Nanomaterials for water pollution monitoring and remediation. *Environmental Chemistry Letters, 15*(1), 23–27. doi:10.100710311-016-0595-x

Yadav, A., Chowdhary, P., Kaithwas, G., & Bharagava, R. N. (2017). Toxic Metals in the Environment : Threats on Ecosystem and Bioremediation Approaches. Handbook of Metal-Microbe Interactions and Bioremediation, 128–141. doi:10.1201/9781315153353-11

Yan, W., Lien, H. L., Koel, B. E., & Zhang, W. X. (2012). Iron nanoparticles for environmental clean-up: Recent developments and future outlook. *Environmental Science. Processes & Impacts, 15*(1), 63–77. doi:10.1039/C2EM30691C PMID:24592428

Yang, X., Guo, N., Yu, Y., Li, H., Xia, H., & Yu, H. (2020). Synthesis of magnetic graphene oxide-titanate composites for efficient removal of Pb(II) from wastewater: Performance and mechanism. *Journal of Environmental Management, 256*, 109943. doi:10.1016/j.jenvman.2019.109943 PMID:31989978

Yang, Y., Xie, Y., Pang, L., Li, M., Song, X., Wen, J., & Zhao, H. (2013). Preparation of reduced graphene oxide/poly(acrylamide) nanocomposite and its adsorption of Pb(II) and methylene blue. *Langmuir, 29*(34), 10727–10736. doi:10.1021/la401940z PMID:23895359

Yaqoob, A. A., Parveen, T., Umar, K., & Ibrahim, M. N. M. (2020). Role of Nanomaterials in the Treatment of Wastewater: A Review. *Water, 12*(2), 495. doi:10.3390/w12020495

Yin, Z., Cui, C., Chen, H., Duoni, Yu., Yu, X., & Qian, W. (2020). The Application of Carbon Nanotube/Graphene-Based Nanomaterials in Wastewater Treatment. *Small, 16*(15), 1902301. doi:10.1002mll.201902301 PMID:31788946

Zhang, Y., Chi, H., Zhang, W., Sun, Y., Liang, Q., Gu, Y., & Jing, R. (2014). Highly Efficient Adsorption of Copper Ions by a PVP-Reduced Graphene Oxide Based On a New Adsorptions Mechanism. *Nano-Micro Letters*, *6*(1), 80–87. doi:10.1007/BF03353772 PMID:30464959

Zheng, X., Wu, R., & Chen, Y. (2011). Effects of ZnO nanoparticles on wastewater biological nitrogen and phosphorus removal. *Environmental Science & Technology*, *45*(7), 2826–2832. doi:10.1021/es2000744 PMID:21381661

Zhu, Y., Murali, S., Cai, W., Li, X., Suk, J. W., Potts, J. R., & Ruoff, R. S. (2010). Graphene and Graphene Oxide: Synthesis, Properties, and Applications. *Advanced Materials*, *22*(35), 3906–3924. doi:10.1002/adma.201001068 PMID:20706983

Chapter 3
Eco-Friendly Integration of Seawater Desalination With Carbon Mineralization of Brine Wastes

Luqman Abidoye
International Maritime College, Oman

Oyetunji B. Okedere
Osun State University, Nigeria

Kazeem O. Rabiu
Osun State University, Nigeria

Kehinde Oyewole
Osun State University, Nigeria

ABSTRACT

Challenges of insufficient freshwater for domestic and industrial uses in different parts of the world necessitate the development of desalination technology. However, brine wastes and CO_2 emitted from the desalination process are of detrimental effects to the environments and living beings, owing to high-level of alkalinity and presence of toxic levels of various chemicals. This study acknowledged efforts being made by scientists to integrate carbon mineralization of brine wastes with desalination system to generate various carbonates: nesquehonite, ragonite, calcite, dolomite, etc. However, the carbonation process behaves differently under the effects of impurities in the CO_2 stream. Studying and analyzing the process using PHREQC simulation software, the carbonation of brine using pure CO_2 shows high drop in pH of the solution, thereby increasing the level of the acidity. The presence of impurity in the CO_2 further worsens the acidity problem in the presence of N_2 and O_2.

DOI: 10.4018/978-1-6684-7303-0.ch003

DESALINATION PROCESSES AND CHALLENGES

Water is essential to the survival of man, yet a very significant proportion of the world's population is still experiencing scarcity of water (World Bank, 2017). Increase in world's population, perennial water scarcity, extreme conditions of weather and uncertainties about hydrological cycles (drought, precipitation and flood) are presently exerting tremendous pressures on water resources world-wide. The availability or non-availability of water is closely linked to quality of life, public health and the extent of burden of diseases world-wide (World Bank, 2017). It is for these reasons that several efforts are being undertaken by governments all over the world to ensure its availability and safety. The availability of safe drinking water is one of the principal pillars of the Sustainable Development Goals (SDGs); and this is captured in the 6th goal as Clean Water and Sanitation. A further dissection of the sixth pivotal goal of the SDGs shows that targets aimed at ensuring that the goal is attained, are set for attainment by 2030. The strengthening of international cooperation and capacity building support to developing countries in water and sanitation- related programmes stood out among the eight targets. Among other programmes, desalination is seen as one of the key activities to be pursued for the attainment of this mission as it is presently playing leading roles in clean water supply in Maldives, Malta, Saudi Arabia and the Bahamas (UNEP, 2019).

Simply put, desalination is a technique that is deployed to remove salt from water. While the idea of water desalination is not entirely new as it is dated back to the time of the 2nd World War, its potential in the solution to the global water supply has not been fully explored. The seas and the oceans around the world constitute inexhaustible endowments of water resources that could help in the attainment of the SDGs. The issues of drought, climate change and global population increase are currently putting pressure on the available water resources. In the same vein, poverty and arid nature of some nations especially in the developing countries have further deepened the need for investment in desalination of seawater for safe global water availability.

The approaches in the desalination of water are either by thermal or membrane separation although; there are sub-classifications of these methods. The thermal techniques could be by vapour compression, multiple effect distillation or multi-stage flash distillation while the popular membrane technologies include electro-dialysis, reverse electro-dialysis and reverse osmosis. The products of a desalination process for treatment of saline water are usually two; one part that is very low in salt concentration and the other with higher salt concentration than the original saline water (brine concentrate). According to Liyanaarachchi (2014), desalination efforts are presently tilting towards the development of membrane technologies which are believed to require lesser energy compared to the thermal process.

Notwithstanding, the major breakthroughs in the development of seawater desalination techniques, quite a number of challenges still exist and the attainment of the SDG in the area of clean water availability is closely associated with overcoming these drawbacks. One such drawback is the handling of the brine concentrate that is left after fresh water has been separated from the saline seawater feed. Brine effluents (wastes) have tremendous deleterious effects on the environment (Ariono et al., 2016; UNEP, 2019). A major feature of brine waste is its richness in salt concentration which makes it highly corrosive and detrimental to coastal structures and ecosystems. There could also be other pollutants depending on other chemicals used in the treatment process. As much as one and half litres of liquid effluents containing chlorine and copper are produced per litre yield of drinking water in majority of the existing desalination processes (UNEP, 2019). Most modern techniques employed for desalination now favour the recovery of salt from the brine concentrate in order to address the environmental challenges posed by its disposal. Another challenge is the high energy requirement of the process. This chapter therefore focuses on the prospects and challenges of various desalination techniques and the characteristics of the resulting brine wastes.

Seawater Characteristics

The nature of seawater is such that it consists of a mix of water, salts and several organic and inorganic substances which are received from riverine, mineral dissolution, inflows, air and lands. Generally, the chemical and physical properties of seawater are defined by major and trace ions. About 13 ions (18980 mg/L chloride, 10556 m/L sodium, 2649 m/L sulphate, 1262 mg/L magnesium, 400 mg/L calcium, 380 mg/L potassium, 140 mg/L bicarbonate, 1 mg/L fluoride, 13 mg/L strontium, 65 mg/L bromide, 26 mg/L borate, and less than 1 mg/L silicate and iodide) are present in concentrations in excess of 1 part per million (1 ppm) (Brown, 2016; AQUARIUS, 2022). Most of the ions exist as free ions although, a few (sulphate) may exist as paired ions, which is mostly in association with metallic ions like magnesium and calcium. Although, salinity varies slightly from one location to another and with depth as a result of differential mixing, evaporation and precipitation, the ratio of ions is generally uniform. Seawater also contains dissolved gases such as nitrogen, oxygen, carbon dioxide and argon which are associated with interfacial diffusion and tide motions.

The presence of several salts imparts salinity to seawater which is expressed as the gramme of salt per kilogram of seawater. The density and the pH are higher than those of fresh water. Compared with other water bodies, the salinity of the ocean water is very high usually in the order of 35,000 ppm. At a salinity of 35 part per thousand (ppt), the densities of seawater at 4 °C and 15 °C are 1.02781 and 1.0255 gcm^{-3} respectively. The pH of seawater which is a reflection of its acidity or alkalinity ranges between 7.5 and 8.4. Marked variations exist in the particulate compositions with seawater having particles from 1 μm to 1mm (Liao et al., 2019). Seawater exhibits properties such as higher heat capacity, latent heat of fusion, latent heat of vapourization, latent heat of melting, thermal expansion, turbidity and viscosity (AQUARIUS, 2022).

Seawater Desalination Techniques

As discussed in the preceding sections, seawater is an abundant resource that has the potential for supply of water to meet human needs as detailed in the SDGs, its saline nature makes desalination a vital and mandatory process in the exploitation of seawater for man's safe water need. Essentially, desalination is the removal of the salts and other dissolved substances from seawater and other water resources up to an acceptable level for human consumption. All desalination processes will yield water that is very low in salt (desired for consumption) and brine (highly saline solution than the original feed waters). A good desalination process must take into consideration from the onset what becomes of the brine after clean water has been extracted due to the various environmental pollution issues associated with handling of brine. Desalination methods may be broadly classified as thermal or membrane technologies. Figure 1 presents a general overview of the available desalination technologies some of which are subsequently discussed.

Figure 1. Desalination technology types
Source: Curto et al. (2021)

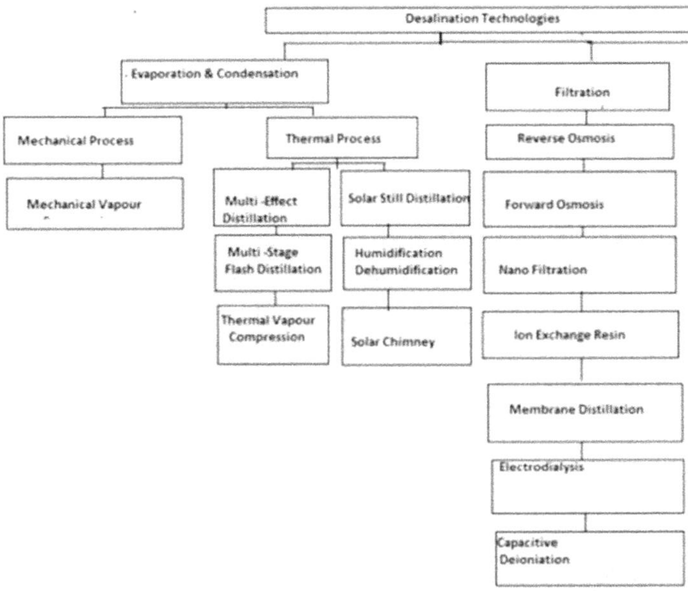

Thermal Desalination

This is about the first approach adopted for desalination of seawater. It entails the application of high temperature to vapourize the salty water. The vapour produced is subsequently condensed to fresh water and mineralized to produce the portable water.

Multi-Stage Flash Distillation

Multi-stage flash distillation presently accounts for bulk of the desalinated seawater consumed globally (Saadat et al. 2018). It involves a combination of evaporation and condensation processes with in-built capacity for recovery of latent heat of vapourization that is subsequently used to preheat up the incoming fresh feed water. The recovery of heat is essential for the economy of the process in the face of global rising cost of fuel and energy. The system operates at moderate temperature and pressure. The technique is advantageous in the sense that it can be used to produce water with as low as 10 mg/L total dissolved solids. It can be hybridized with other techniques and it is not usually affected by the salinity of the feed water. It however has some demerits associated which include high cost of installation and operation, low recovery ratio of freshwater from the feed water and high level of technical know-how to operate. Although, heat transfer problem due to scale formation is usually addressed with bulk boiling, scaling and fouling of the evaporator is still very significant (Saadat et al. 2018; Curto et al. 2021).

Multiple Effect Distillation

These have been in use as far back as 7 decades ago but were overtime replaced with flash distillation techniques essentially due to heat transfer and scaling issues associated with the heat transfer tubes

(Saadat et al., 2018). In this approach, the condensation of vapour takes place in each stage and the heat released is recycled for preheating of fresh feed. Evaporation takes place in the hot tubes in an operation that is achieved at low pressure and fairly low temperature to improve performance and minimize scaling and corrosion. It has a better thermal management compared to the multi flash distillation method.

Vapour Compression Distillation

In this method, feed water is vapourized and the resulting vapour is compressed under high pressure and subsequently used to heat up and evaporate the incoming feed water. Vapour compression is usually achieved mechanically or by steam ejectors. Mechanical compression is usually in a single stage while steam ejection can be multi stage. The plant is very compact; hence the cost is fairly reasonable. The operation is does not require specialized training or extreme technical know-how. Quality water can be obtained from the process but it is largely used on a small scale installations. High cost of compression facilities, scaling and fouling are its major drawbacks.

Membrane Technology

Essentially, membranes separate salty water by causing a selective passage or transport of components across the membrane. Concentration, temperature, pressure, or electrochemical potentials can serve as the gradient precursor and driving force.

Reverse osmosis and electrodialysis are the major classifications. In reverse osmosis, the feed water is pressurized and made to pass through a water permeable membrane which separates the salts from the water. The brine stays on the pressurized side of the membrane while water passes through with no phase change that is associated with the thermal treatment. This method removes dissolved solids, organics and colloidal materials as well as microbes. The energy requirement is the pressure applied to the feed water which ranges between 15-25 bars and 54-80 bars for brackish and seawater respectively (Buros, 2000; Saadat, 2018). Improvement in the technology has further reduced the osmotic pressure requirement with the availability of low pressure membranes capable of reducing the pressure by 50%. The method is versatile as it can handle marked range of flow rates up to 750,000 L/day and 450,000 L/day feed of brackish and seawater respectively with little or no chemical treatments. The capacity can be easily increased. Major drawbacks are the high cost of membrane with fairly short lifespan. There may also be need for pretreatment to reduce suspended particles as a result of stormy weather, sea tides and waves that could cause elevated levels of suspended particles. They can also be sensitive to pH, oxidizer and microbes. Pre-treatment of feed water is vital in the use of reverse osmosis.

In electrodialysis, direct current is used to create an electric field in the water to be desalinated. The electric field drives the ions through the membrane to the electrode of opposite polarity. Only ionic compounds are separated due to the nature of the driving force. Minerals are separated in the cell pairs which comprise a cation and anion transfer membranes, as well as two spacers. Just like in reverse osmosis, no phase change occurs during desalination and the system requires less pre-treatment of the water. The concern about osmotic pressure is eliminated.

Brine Wastes Characteristics

Brine effluents could be generated from quite a number processes including desalination of water, ion exchange regeneration and ballast water (Ariono et al., 2016). The negative effects of brine on the environment are associated with their very high salt concentration and the presence of other pollutants. Depending on the source, the characteristics of brine water may vary slightly, but they are generally high in salt concentration. Typical characteristics of brine wastes are presented in Table 1.

Table 1. Brine sources and characteristics

Types	Characteristics and Sources
Brine from Desalination Plant	It has elevated salt concentration and total dissolved solid (TDS) 40 g/L. Other properties include hardness, presence of heavy metal and other chemicals (Anti-scalant, cleaning agents etc.) used in the process as well as high temperature than ambient if any of the thermal approved is used.
Spent Brine Regenerant	High salinity ≥ 50 g/L, other properties include hardness and presence of organic compounds (Decolorization plants)
Blow-Down Water	High in salinity, total suspended solids (TSS), total organic compounds (TOC) and TDS between 50- 300 g/L
Rejected Bittern Brine from salt ponds	High salinity up to 400 g/L with sodium, Magnessium, potassium and bromide
Ballast Water	Seawater with aquatic species

Source: Ariono et al. (2016)

Environmental Issues and Eco- Friendly Handling of Brine Wastes

Desalination involves quite a number of stages with the aim to produce (1) water needed by man (ii) reject water with high salinity. The latter is of environmental concern because it is also high in total dissolved solids (70,000 mg/L) and has several chemical residues based on those used during pretreatment (NaOCl, $FeCl_3$, $AlCl_3$, H_2SO_4, HCl, $NaHSO_3$). Some of these pretreatment are applied to minimize alga growth, corrosion, scaling as well as for pH control (Mavukkandy et al., 2019). The environmental or ecological impacts of improperly disposed brine waste cannot be over-emphasized. Arising from its physico-chemical properties (salinity, temperature, density and other constituents), it can significantly change the physico-chemical features of the receiving water which in turns may affect the balance of marine ecosystem to a great extent. A change in aquatic eco-system salinity can cause a shift in osmotic pressure of the system with the attendant deleterious effects on organisms living in the habitat. There could also be acute or chronic poisoning of organisms associated with the release toxic substances like heavy metals and other constituents of the brine waste. Specific instances of the adverse effects and high toxicity of desalination effluents (Copper and Chlorine) have been reported (Robert et al., 2010; Hoepner and Lattemann, 2003. Ariono et al., 2016). Corrosion of coastal beaches and other coastal structures by desalination plants has also be reported (Al-Barwani and Purnama, 2007). Unlined ponds, soil and groundwater which received brine wastes from desalination plants have also been negatively impacted (Mohammed et al., 2005). Other reported impacts include eutrophication due to algae blooming and aquatic bio-invasion of some dangerous bacteria (Goncalves and Gagnon, 2012).

The methods currently being implemented for brine disposal include sea and land disposal, evaporation, membrane distillation, forward osmosis, deep electrodialysis, capacitive distillation, well injection, capacitive deionization (Leyland et al. 2019). More recently, trends in brine wastes management aimed at zero liquid discharge (ZLD) via harvesting of freshwater and solid salts have been advocated. Eco-friendly approaches in the management of brine could be viewed from the perspective of volume reduction or ZLD/crystallization (Mavukkandy et al., 2019). Poirier et al. (2022) demonstrated the recovery of freshwater and salts of high purity (calcite, anhydrite, sodium chloride and epsomite). Some of these salts are needed by other industries for production of a binder such as cement. In ZLD approach, an innovative heat recovery through a heat transfer fluid and a compressor for energy reduction is applied.

Steps and Reactions of Carbon Mineralization of Brine Wastes

Carbon dioxide (CO_2) sequestration using saline/brine has appeared as a positive possibility to reduce the harmful impacts of global warming due to its large storage capacity and its favourable conversion process. Consequently, studies on CO_2 solubility in brine water in relation to its reaction with the brine's cation content are required. Brines has the concentrations of Mg, Ca, Fe, Cl and Na ions, hence, they would dissolve with CO_2 to form $MgCO_3$(s), $CaCO_3$(s), Fe_2CO_3(s) and other useful materials.

The solubility of CO_2 in brine water would form mineral carbonate precipitates and have been reported by various researchers (Druckenmiller & Maroto-Valer, 2005; Tang et al., 2021); Soong et al 2004; Maries & Hills, 2013). The study on the improvement of carbonates using a buffer solution in the presence of brine and coal ash at room temperature as well as high pressure and temperature conditions has been carried out (Liu & Maroto-Valer, 2011). In the same vein, Grace et al. (2012) used the applied statistical method to study the factors affecting carbonation of fly ash impacted by brine. Jo et al. (2012) assessed the factors affecting mineral carbonation of fly ash in brine reservoir. Uliasz-Bochenczyk et al. (2009) utilized brine and fly ash as the CO_2 storage option using aqueous carbonation reaction. In addition, Dilmore et al. (2009) analysed the mineralization of CO_2 by reaction of saline waste water and by-product of caustic. Researches on the acceleration of mineral carbonation of waste materials using some catalysts have studied (Jacobs & Hitch, 2011; Pronost et al., 2011).

It is commonly known that mineralization is a popular type of CO_2 sequestration method, as it will form solid products that can sequester CO_2 permanently without compromising the atmosphere. A derivation of alkaline earth metals for carbonation are desalination brines, which is a type of concentrated seawater containing calcium, magnesium, sodium, iron, and chloride. The alkaline cations can aid in the formation of carbonates by injecting CO_2 directly into the brine.

The steps involve in storing CO_2 through mineralisation is based on a number of sequential chemical reactions. In step 1 (Equation 1), CO_2 dissolution takes place and the solubility is affected by different factors such as pressure, temperature, and brine salinity. The dissolution of CO_2 gas in brine is via the simplified reactions [1] to [5], as described below. Once the CO_2 gas (Reaction [1]) dissolved, there is the formation of carbonic acid (i.e. Reaction [2]). The carbonic acid then splits into bicarbonate (Reaction [3]) and carbonate ions (Reaction [5]). The dissolved CO_2 can also form a reaction with hydroxide ions (OH^-) to form bicarbonate ions (Reaction [4]). These reactions therefore substantiate that the increase in pH of an aqueous solution tends to increase the dissolution of CO_2 gas.

$$CO_2(g) \leftrightarrow CO_2(aq) \tag{1}$$

$$CO_2(aq) + H_2O \leftrightarrow H_2CO_3 \tag{2}$$

$$H_2CO_3 \leftrightarrow H^+ + HCO_3^- \tag{3}$$

$$CO_2(aq) + OH^- \leftrightarrow HCO_3^- \tag{4}$$

$$HCO_3^- \leftrightarrow H^+ + CO_3^{2-} \tag{5}$$

By combining Equations (1) – (5), The following equation is derived,

$$CO_2(g) + H_2O \leftrightarrow H^+ + HCO_3^- \tag{6}$$

Table 2. Factors affecting CO_2 dissolution in brine water

Factors	Explanation
Temperature	The solubility of CO_2 declines with a rise in temperature, under a constant salinity and pressure. For example, between the temperature of 25° C and 60° C, solubility of CO_2 is dominant. Nevertheless, when the temperature is beyond 100° C, solubility increases again with temperatures (Liu & Maroto-Valer, 2011; Rabiu et al., 2017). Temperature has a higher effect than pressure (Liu & Maroto-Valer, 2011).
Pressure	The solubility of CO_2 in solution increases with increases in pressure, under a constant salinity and temperature. Higher pressure results to the formation of higher carbonation. However, pressure have a lower impact on the carbonates formation when compare with brine pH (Liu & Maroto-Valer, 2011).
Brine salinity	The salinity effect is based on the compositions of salts, for example, monovalent salts solutions (KCl or NaCl) have fewer constrains on CO_2 dissolution in comparation to divalent salt solutions ($MgCl_2$ or $CaCl_2$).
Brine pH	The pH of brine is the most vital factor that affects the CO_2 dissolution. The precipitation of mineral carbonation is favoured by pH of 9.0 (Liu & Maroto-Valer, 2011). Since the conversion is somehow slow, it is recommended that some additives may be added to the brine formation to enhance brine pH so as to quicken or speed up the rate of carbonation reactions in the formations (Liu & Maroto-Valer, 2011).

Moreover, step 2 describes the formation of nesquehonite carbonates and can be shown in the following reactions:

$$HCO_3^- \leftrightarrow OH^- + CO_2(g) \tag{7}$$

$$Mg^{2+} + HCO_3^- + OH^- + 2H_2O \leftrightarrow Mg(OH)(HCO_3)\cdot 2H_2O(s) \tag{8}$$

From reaction [7], the pH solution will rise and CO_2 gas will escape from the mixture during the precipitation of nesquehonite. By substituting Equations [7] and [8], the complete reaction for nesquehonite precipitation can be acquired as described in Equation [9].

$$Mg^{2+} + 2HCO_3^- + 2H_2O \leftrightarrow Mg(OH)(HCO_3)\cdot 2H_2O(s) + CO_2(g) \tag{9}$$

Overall, the reaction of carbon mineralisation with brine is described in Reaction [10]. The high CO_2 pressure and low temperature enhanced the dissolution of CO_2 gas in Step 1, and will lead to a decline in pH. However, the precipitation of nesquehonite in Step 2 favours low CO_2 pressure and tolerably high temperature, resulting in a rise in pH solution. Combining Equations [6] and [9], the overall reaction for CO_2 mineralization is

$$Mg^{2+} + CO_2(g) + 4H_2O \leftrightarrow Mg(OH)(HCO_3)\cdot 2H_2O(s) + 2H^+ \tag{10}$$

As shown in Reaction [10], CO_2 mineralization will result in the unwanted acidizing of the aqueous solution (Ma, 2012).

Reactions Involved in Calcite Dissolution Rates

Calcite is the universal specie in CO_2 sequestration reservoirs and it is one of the important precipitate that forms from carbon mineral trapping mechanism. Hence, precipitation kinetics of calcite should be well studied. The following reactions are suggested by some researchers and the reactions appear simultaneously during dissolution of mineral under ambient conditions:

$$CaCO_3 + H^+ \leftrightarrow Ca^{2+} + HCO_3^- \tag{11}$$

$$CaCO_3 + H_2CO_3^* \leftrightarrow Ca^{2+} + 2HCO_3^- \tag{12}$$

$$CaCO_3 + H_2O \leftrightarrow Ca^{2+} + HCO_3^- + OH^- \tag{13}$$

Here, $H_2CO_3^*$ is the addition of the concentrations of carbonic acid H_2CO_3 together with dissolved $CO_2(aq)$. The equations 11 to 13 have the forward/backward constants rate and are specify as k_1/k_{-1}, k_2/k_{-2}, k_3/k_{-3}, respectively. The calcite precipitation $rate_+$ (in mol $m^{-2}s^{-1}$) is the sum of the three independent rates as shown in equation 14.

$$rate_+ = k_1 a_{H^+} + k_2 a_{H_2CO_3^*} + k_3 a_{H_2O} \tag{14}$$

The reactions given by equations 11, 12 and 13 are suggested to take place with the back-reaction. Studies on the dissolution rates of calcite by using microscopic techniques have been carried out. Shiraki et al. (2000) used Atomic Force Microscopic (AFM) to investigate the dissolution kinetic of calcite in 0.1M sodium chloride (NaCl) at room temperature and proposed an empirical dissolution rate law, as follows:

$$rate_{+} = k_1 m_{H^+} + k_2 m_{HCO_3^-} + k_3 m_{OH^-} \tag{15}$$

Where m_i are molarities of dissolved minerals and $k_1 = 10^{-0.66\pm0.13}$, $k_2 = 10^{+0.13\pm0.06}$ and $k_3 = 10^{-3.99\pm0.08}$ $m^{-2}s^{-1}$.

Phreeqc Modeling: Mineralization of Brine Waste and Effects CO_2 Impurities

The process of carbon mineralization of brine waste needs to be handled with caution as the system components between extreme acidity and alkalinity with the injection of CO_2. The process set-up comprises of CO_2 gas stream and brine stream. The gas stream requires purification of the flue gas emission through filtration of particulates and then compressed to concentrate the CO_2 in the stream. It is also known that the compression pressure raised the yield of the carbonated minerals (Abidoye and Das 2020). The pH of the brine stream needs to be adjusted to raise the efficiency of the process, since the CO2 is an acidic gas (Liu and Maroto-Valer, 2011). Any formation of suspension, mainly from salt production, should be filtered as soon as formed during pH adjustment. Thus, the filtration tank is required after the pH adjustment tank.

In a carbonation reactor, CO_2 is bubbled into the pH-adjusted brine stream at suitable conditions to precipitate out of the solution the carbonated products and the excess gas stream can then be recycled. The liquid stream can also be recycled to maximize the carbonates yield by reacting more with CO2. The liquid stream can be further purified into cleaner water for industrial use. The carbonated products can then be analysed using instruments like Scanning Electron Microscope (SEM), for microstructural imaging of the materials; X-ray diffractometer (XRD) for crystalline phase identifications; field emission scanning electron microscope (FESEM) equipped with an energy dispersive spectrometer (EDS), thermogravimetric analysis (TGA), Fourier transform infrared spectroscopy (FTIR), etc.

Many softwares are available to examine the performance of the brine mineralization process under virtual conditions before costly experiments are performed. The softwares include PHREEQC, ASPEN, COMSOL, etc. The program (PHREEQC) is a free software developed by U.S. Geological Survey. It can perform a wide variety of aqueous geochemical calculations. It implements chemical reactions and transport processes in natural or polluted water, in laboratory experiments, or in industrial processes. The program utilized the basis of equilibrium chemistry of aqueous solutions interacting with minerals, gases, solid solutions, exchangers, and with the capability to model kinetic reaction. ASPEN simulation software is also suitable to design and setup the process system and inspect the effect of various conditions on the optimization of the carbonate generation. The design can simulate purifier for flue gas followed by mixing of the CO_2 with recycled CO_2, emanating from the reactor outlet. The combined stream can be subjected to stagewise compressor, which can be conveniently simulated by ASPEN. The output of the compressor can be connected to react with brine after its pH adjustment. Centrifugal pump is known to be good for viscous flow like brine. Heater and heat-exchanger as well as cooler have to be strategically located to optimize the conversion of the brine minerals into carbonates. However, ambient operations will be economically and operationally efficient. The conditions of the process are crucial not only to the conversions process but also to the selectivity between calcite, aragonite magnesite, dolomite and other forms of carbonates.

SIMULATION AND OPTIMIZATION

The overwhelming abundance of non-fresh water in many parts of the world necessitates the process of desalination to take care of daily potable water needs for the people and the industries. Since World War II, the need for utilization of seawater for itinerary soldiers became imperative (Sajjad and Rasul, 2015). Since that time, the development of the desalination processes have ever been growing into complex industrial system

Typical brine composition used in this chapter was extracted from the literature sources. Main components were obtained from the work of Bang et al. (2017) as well as from the publication of Parkhurst and Appelo (2013) as shown in Table 3.

Table 3. Concentrations of the major cations and anions in the brine

Component	Concentration/Quantity	Unit
pH	8.22	
Density	1.023	g/cm^3
Ca	714	ppm
Mg	2226	ppm
Na	17987	ppm
K	858	ppm
Si	4.28	ppm
Cl	32220	Ppm
SO_4	4766.0	Ppm

The PHREEQC software was used to investigate the effects of impurities in CO_2 stream on the formation of the carbonation products and the pH of the system. In the program, thermodynamic equilibrium exists among dissolved species and determines the unknowns for each specie. These unknowns include, the activity, a_i, activity coefficient, γ_i molality, m_i and moles in solution, n_i The equations governing the activities of the aqueous species are expressed below,

$$a_i = \gamma_{1,i} \tag{16}$$

$$n_i = m_i W_{aq} \tag{17}$$

W_{aq} is the mass of solvent water in an aqueous solution.

The general mass-action is expressed as

$$K_i = a_i \prod_m^{M_{aq}} a_m^{-c_{m,i}} \tag{18}$$

K_i is a temperature-dependent equilibrium constant, $c_{m,i}$ is the stoichiometric coefficient of master species *m* and *i* is the total number of aqueous master species. The total moles of an aqueous species is also obtained from equation (18),

$$n_i = m_i W_{aq} = K_i W_{aq} \frac{\prod_m^{M_{aq}} a_m^{-c_{m,i}}}{\gamma_i} \tag{19}$$

Activity coefficients of aqueous species are defined with the Davies equation,

$$\log \gamma_i = -A z_i^2 \left(\frac{\sqrt{\mu}}{1+\sqrt{\mu}} - 0.3\mu \right) \tag{20}$$

z_i is the ionic charge of an aqueous specie *i*. "A" is the constant depending on temperature only.

These equations were then differentiated with respect to the master unknowns (the natural log of the activities of master species or components, the natural log of the activity of water, a_{H_2O}, the ionic strength, μ, and the mass of solvent water in an aqueous solution, W_{aq}). E.g.,

$$dn_i = n_i \left[dIn\left(W_{aq}\right) + \sum_m^{M_{aq}} c_{m,i} dIn\left(a_m\right) - \frac{\partial In\left(\gamma_i\right)\partial\mu}{\partial\mu} \right] \tag{21}$$

$$\frac{\partial In\gamma_i}{\partial\mu} = -In\left(10\right)\left[Az_i^2 \left(\frac{1}{2\sqrt{\mu}(\sqrt{\mu}+1)^2} - 0.3 \right) \right]$$

The results of the simulation are presented as follows

Pure CO_2 and Brine Mineralization

Figure 2 shows the effects of pure CO_2 on brine carbonation. The pH drops considerably upon slight injection of 0.01 mole of pure CO_2 per kg of the brine solution. The value

of pH drops from 8.22 to 4.27 (48%). This further dropped to 3.72 (54%) upon injection of 0.1 mole of pure CO_2. The steepness of the pH fall reduced at the injection of 1 mole of pure CO_2 leading the pH value fall to 3.22. The fall of pH further intensified upon injections of 10 and 20 moles of pure CO_2, but the steepness was less significant in relation to earlier injected moles of the gas and that of 20 moles injected gas culminated to the pH value of 2.66 (67%) compared to 54% fall in pH at 0.1 mole injected gas. This is in line with the findings of Castrillon-Munoz et al. (2022). The figure further shows that the carbonation of the brine leads to increase in the concentration of calcite, aragonite and dolomite in the solution. But, these minerals remain dissolved as they are undersaturated at ambient condition under the pure CO_2 injection. Out of all mineralization products, calcite remains the least-undersaturated which was closely overlain by aragonite level of undersaturation while the dolomite remains the most undersaturated, as shown by the Saturation Index in the figure. According to Rackley (2017), the pH of the medium determines the nature of trapping mechanism in operation in the formation. They found that water-mineral supersaturation with CO2 drives up the pH of the system.

Figure 2. Brine mineralization of pure of CO_2 under ambient conditions showing saturation indices of calcite, aragonite and dolomite in secondary y-axis

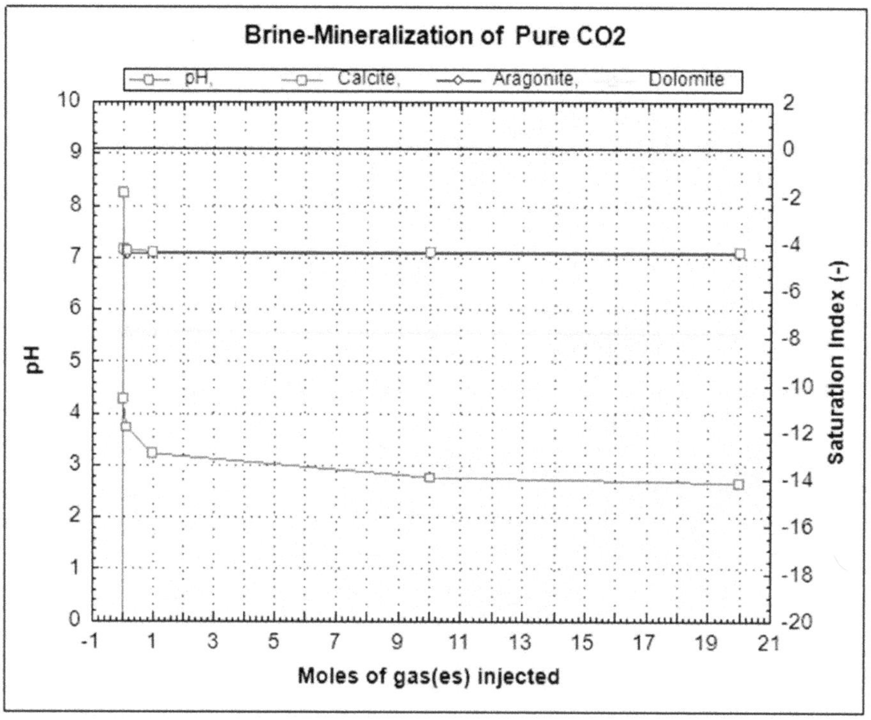

However, since CO_2 will hardly exist in pure state during carbonation of brine waste, further simulations, shown in Figures 3 and 4, show the pH and Saturation indices of the carbonation products under impure CO_2 injected into the brine for mineralization purposes, under ambient conditions. Figure 3 shows the profiles of these parameters for gas molar composition containing 80% CO2 with the remaining being N_2 and O_2, under atmospheric ratio. Under the impurities, the pH (measured by charge balance) of the system falls below 1 unlike the case with pure CO2, which remains above 2.6 at 20 mole of injected CO_2. This indicates rapid acidity generation in the presence of oxygen and nitrogen gases in the stream. The rapid fall of pH became more dramatic at 50% CO_2 (shown in Figure 4), where the pH was close to 0. The reason for this dramatic fall of pH was attributed to the presence of N_2, which partially ionizes to produce strong acid $N_2 + O_2 \rightarrow 2NO2NO + 4H_2O \rightarrow 2NO_3^- + 8H^+$

Therefore, the higher the percentage of the N_2 compositions, the lower the pH. It is pertinent to prevent impurity in CO_2, especially N_2, since it can react with CO_2 to produce string acid. But, the irony is that N_2 is very much abundant in the atmosphere. However, it is instructive to purify the CO_2 stream as much as possible to minimize the effect of acidity on the process equipment.

Figure 3. Brine mineralization of 80% CO_2 under ambient conditions showing saturation indices of calcite, aragonite and dolomite in the secondary y-axis

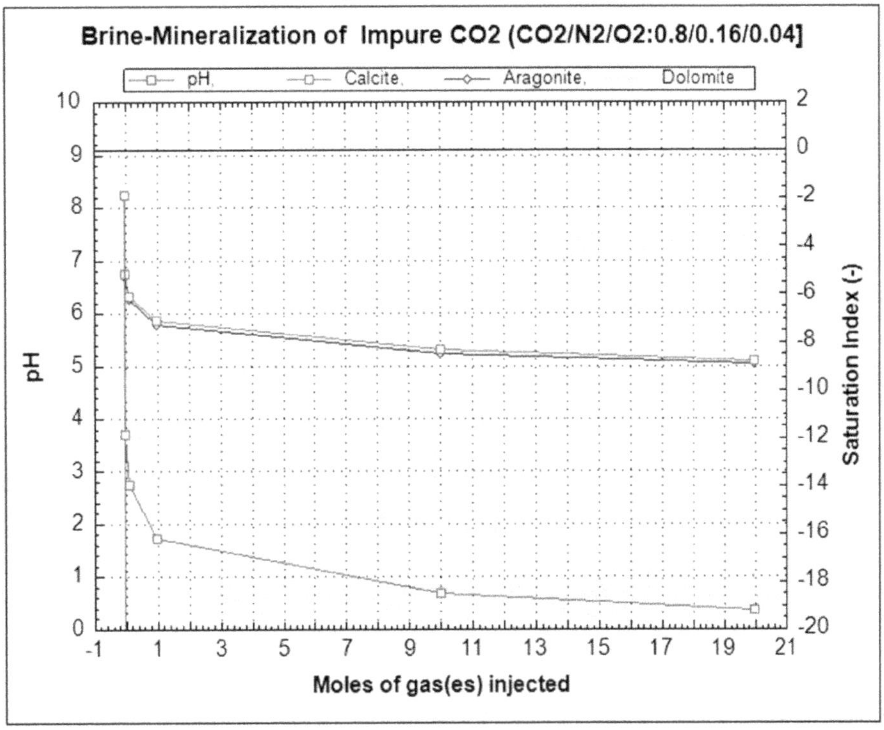

Figure 4. Brine mineralization of 50% CO_2 under ambient conditions showing saturation indices of calcite, aragonite and dolomite in the secondary y-axis

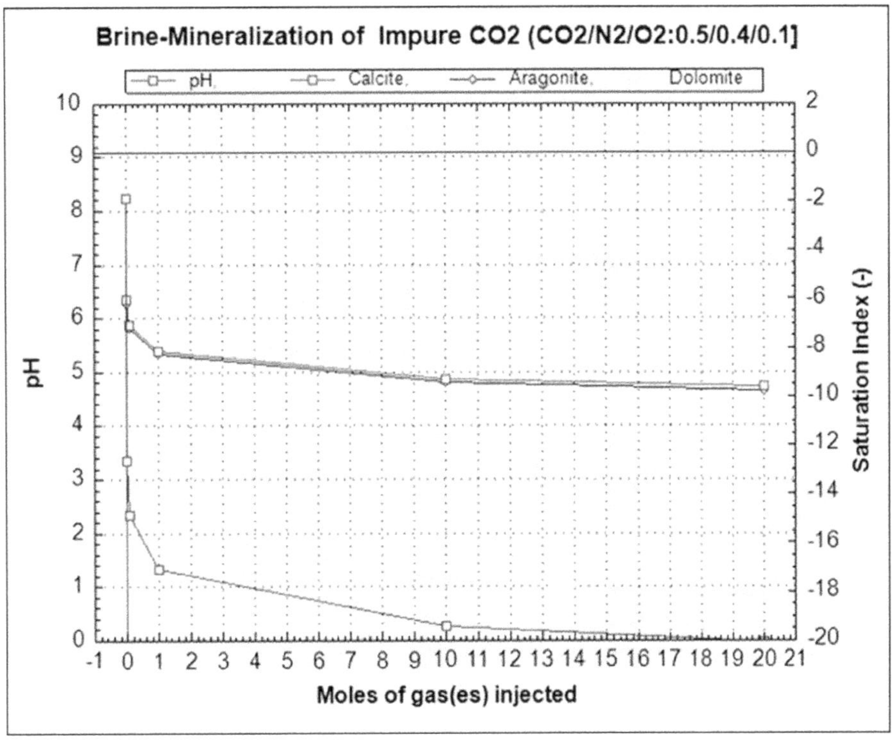

Another point to observe was the fact that the carbonation products- calcite, aragonite and dolomite remain undersaturated in the solution and therefore remained dissolved. This indicates that the product will be difficult to separate from the brine waste at the end of the process.

To counter the effect of the hyperacidity, alkaline solution (NaOH) was introduced into the simulation. This is shown in Figure 5. At 10% NaOH added, the pH was seen standing above 5 even at 20 moles of injected CO_2 stream. The presence of the alkaline solution stabilised the solution pH. It can also be observed in the figure that the saturation indices of the products show that most of the carbonation products had been or nearly fully saturated. Especially, dolomite had started precipitating after few moles of injection while others also show some degree of precipitation. Furthermore, Figure 6 shows that the presence of the alkaline stabilizes the system pH even in the presence N_2 and $O_{2,}$ which had earlier caused excessive pH decline (see Figures 3 and 4).

Also observed in Figure 6 was that more carbonation products have precipitated out of the solution as the pH increases in the presence of the alkaline medium.

Figure 5. Brine mineralization of pure CO_2 with NaOH (10 mol%) under ambient conditions showing saturation indices of calcite, aragonite and dolomite in secondary y-axis

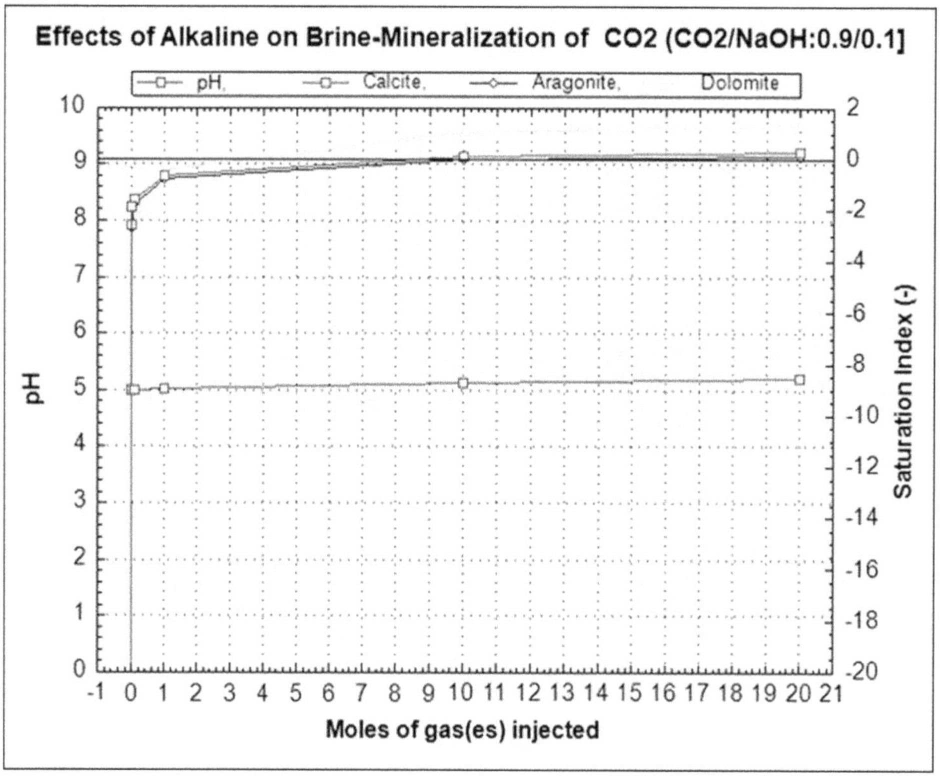

Figure 6. Brine mineralization of impure CO_2 with NaOH (10 mol%) under ambient conditions showing saturation indices of calcite, aragonite and dolomite in secondary y-axis

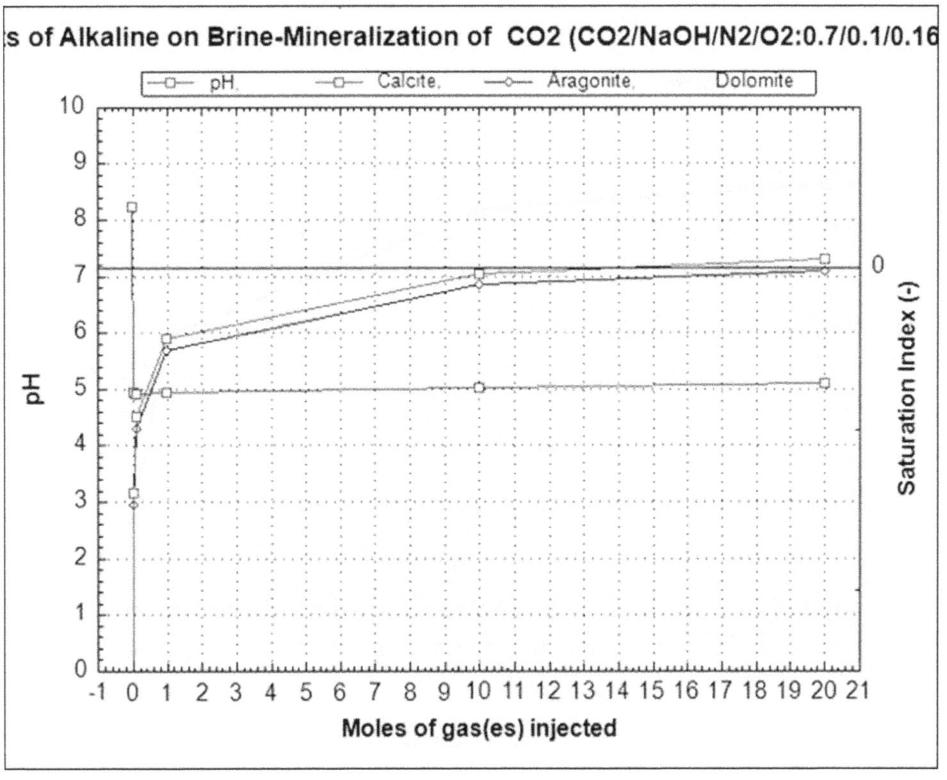

CONCLUSION

Desalination technology is essential for the attainment of SGG's goal 6: Clean Water and Sanitation. But, the desalination process comes with challenges to the environment and living beings as result of toxic brine discharge and greenhouse gases emissions. The carbonation of brine wastes with CO_2 is a feasible process that can safeguard the environment and also stabilized the environmentally-unfriendly components of the brine. Several desalination techniques are currently available- membrane, multi-stage flash distillation, multiple- effect and vapour-compression distillation, etc. The carbonation of brine using pure CO_2 leads to high drop in pH of the solution, thereby increasing the level of the acidity. The presence of impurity in the CO_2 further worsens the acidity problem in the presence of N_2 and O_2. The effect can be effectively corrected through introduction of alkaline medium. Addition of natural alkaline minerals are to catalyze the precipitation of the carbonation products- calcite, aragonite and dolomite in the system, which were previously dissolved and difficult to separate under high acidity, is recommended.

REFERENCES

Abidoye, L. K., & Das, D. B. (2021). Carbon storage in portland cement mortar: Influences of hydration stage, carbonation time and aggregate characteristics. *Cleanroom Technology*, *3*(3), 563–580. doi:10.3390/cleantechnol3030034

Al-Barwani, H. H., & Purnama, A. (2007). Re-assessing the Impact of Desalination Plants Brine Discharges on Eroding Beaches. *Desalination*, *204*(1-3), 94–101. doi:10.1016/j.desal.2006.03.536

AQUARIUS. (2022). *Properties of fresh Water and Seawater*. AQUARIUS Educational and Public Outreach. Retrieved November 24, 2022 from https://aquarius.umaine.edu/activities/prop_fresh_sea.pdf

Ariono, D., Purwasasmita, M., & Wenten, I. G. (2016). Brine Effluents: Characteristics, Environmental Impacts and their Hanling. *Journal of Engineering Technology and Science*, *48*(4), 367–387. doi:10.5614/j.eng.technol.sci.2016.48.4.1

Brown, W. (2016). *Physical Properties of Seawater*. Springer Handbook of Ocean Engineering. doi:10.1007/978-3-319-16649-0_5

Buros, O. K. (2000). *The ABCs of Desalting Topsfield*. International Desalination Association.

Castrillon-Munoz, F. J., Gibson, J. J., & Birks, S. J. (2022). Carbon dissolution effects on pH changes of RAMP lakes in northeastern Alberta, Canada. *Journal of Hydrology. Regional Studies*, *40*, 101045. doi:10.1016/j.ejrh.2022.101045

Curto, D. Franzitta, V., & Guercio, A. (2021) A Review of Water Desalination Techonologies. *Applied Science, 11*, 670. doi:10.3390/app11020670

Dilmore, R. M., Howard, B. H., Soong, Y., Griffith, C., Hedges, S. W., DeGalbo, A. D., Morreale, B., Baltrus, J. P., Allen, D. E., & Fu, J. K. (2009). Sequestration of CO2 in Mixtures of Caustic Byproduct and Saline Waste Water. *Environmental Engineering Science*, *26*(8), 1325–1333. doi:10.1089/ees.2008.0395

Druckenmiller, M. L., & Maroto-Valer, M. M. (2005). Carbon sequestration using brine of adjusted pH to form mineral carbonates. *Fuel Processing Technology*, *86*(14–15), 1599–1614. doi:10.1016/j.fuproc.2005.01.007

Goncalves, A. A., & Gagnon, G. A. (2012). Recent Technologies for Ballast Water Treatment. *Ozone Science and Engineering*, *34*(3), 174–195. doi:10.1080/01919512.2012.663708

Grace, M. N., Wilson, G. M., & Leslie, P. F. (2012). Statistical testing of input factors in the carbonation of brine impacted fly ash. *Journal of Environmental Science and Health. Part A, Toxic/Hazardous Substances & Environmental Engineering*, *47*(2), 245–259. doi:10.1080/10934529.2012.640898 PMID:22242877

Hashim, A., & Hajjaj, M. (2005). Impact of Desalination Plant of Alicante (SE Spain). *Desalination*, *182*, 373–393. doi:10.1016/j.desal.2005.04.020

Hoepner, T., & Lattemann, S. (2003). Chemical Impacts of Seawater Desalination Plants– A Case Study of Northern red Sea. *Desalination*, *152*(1-3), 133–140. doi:10.1016/S0011-9164(02)01056-1

Jacobs, A. D., & Hitch, M. (2011). Experimental mineral carbonation: Approaches to accelerate CO2 sequestration in mine waste materials. *International Journal of Mining, Reclamation and Environment*, *25*(4), 321–331. doi:10.1080/17480930.2011.608883

Jo, H. Y., Kim, J. H., Lee, Y. J., Lee, M., & Choh, S.-J. (2012). Evaluation of factors affecting mineral carbonation of CO2 using coal fly ash in aqueous solutions under ambient conditions. *Chemical Engineering Journal*, *183*, 77–87. doi:10.1016/j.cej.2011.12.023

Leyland, D., Chivavava, J., & Lewis, A. E. (2019). Investigations into Ice Scaling during Eutectic Freeze Crystallization of Brine Streams at Low Scraper Speeds and High Supersaturation. *Separation and Purification Technology*, *220*, 33–41. doi:10.1016/j.seppur.2019.03.025

Liao, R., Li, Q., & Mao, X. (2019). A Prototype for Detection of Particles in Seawater by using Polarized-Light Scattering. *Oceans*, 1–4. Advance online publication. doi:10.1109/OCEANSE.2019.8867414

Liu, Q., & Maroto-Valer, M. M. (2011). Parameters affecting mineral trapping of CO 2 sequestration in brines: Parameters affecting mineral trapping of CO 2 sequestration in brines. Greenhouse Gases. *Science and Technology*, *1*(3), 211–222. doi:10.1002/ghg.29

Liyanaarachchi, S., Shu, L., Muthukumaran, S., Jegatheesan, V., & Baskaran, K. (2014). *Problems in seawater industrial desalination processes and potential sustainable solutions: A review*. doi:10.1007/s11157-013-9326-y

Ma, J. (2012). *CO2 Mineralization Using Reactive Species* [MSc Thesis].

Maroto-Valer, M. M., Kuchta, M. E., Zhang, Y., Andrésen, J. M., & Fauth, D. J. (n.d.). *Comparison of Physical and Chemical Activation of Serpentine for Enhanced CO2 Sequestration*. Academic Press.

Mavukkandy, M.O., Chabib, C.M., Mustafa, I., Al Ghaferi, M., & AlMarzooqi, F. (2019). Brine Management in Desalination Industry: From Waste to Resources Generation. *Desalination, 472*. doi:10.1016/j.desal.2019.114187

Mohammed, A. M. O. (2005). Impact of Land Disposal of Rejects Brine from Desalination Plants on Soil and Groundwater. *Desalination*, *182*(1-3), 411–433. doi:10.1016/j.desal.2005.02.035

Parkhurst, D.L., & Appelo, C.A.J. (2013). *Description of input and examples for PHREEQC version 3—A computer program for speciation, batch-reaction, one-dimensional transport, and inverse*. Academic Press.

Poirier, K., Al Mhanna, N., & Patchiolla, K. (2022). Techno-Economic Analysis of Brine Treatment by Multi- Crystallization Separation Process for Zero Liquid Discharge. *Separations*, *9*(10), 295. doi:10.3390eparations9100295

Pronost, J., Beaudoin, G., Tremblay, J., Larachi, F., Duchesne, J., Hébert, R., & Constantin, M. (2011). Carbon Sequestration Kinetic and Storage Capacity of Ultramafic Mining Waste. *Environmental Science & Technology*, *45*(21), 9413–9420. doi:10.1021/es203063a PMID:21919443

Rabiu, K. O., Han, L., & Bhusan Das, D. (2017). CO2 Trapping in the Context of Geological Carbon Sequestration. In *Encyclopedia of Sustainable Technologies* (pp. 461–475). Elsevier. doi:10.1016/B978-0-12-409548-9.10124-1

Roberts, D. A., Johnson, E. L., & Knott, N. A. (2010). Impacts of Desalination Plants Discharges on Marine Environments: A Critical Review of Published Studies. *Water Research*, *44*(18), 5117–5128. doi:10.1016/j.watres.2010.04.036 PMID:20633919

Saadat, A. H. M., Islam, M. S., Islam, M. S., Parvin, F., & Sultana, A. (2018). Desalination Techniques for Developing Countries: A Review. *Journal of Scientific Research*, *10*(1), 77–97. doi:10.3329/jsr.v10i1.33179

Sajjad, M., & Rasul, M. G. (2015). Simulation and optimization of solar desalination plant using Aspen Plus simulation software. *Procedia Engineering*, *105*, 739–750. doi:10.1016/j.proeng.2015.05.065

Shiraki, R., Rock, P. A., & Casey, W. H. (2000). Dissolution Kinetics of Calcite in 0.1 M NaCl Solution at Room Temperature: An Atomic Force Microscopic (AFM) Study. *Aquatic Geochemistry*, *6*(1), 87–108. doi:10.1023/A:1009656318574

Tang, Y., Hu, S., He, Y., Wang, Y., Wan, X., Cui, S., & Long, K. (2021). Experiment on CO2-brine-rock interaction during CO2 injection and storage in gas reservoirs with aquifer. *Chemical Engineering Journal*, *413*, 127567. doi:10.1016/j.cej.2020.127567

Uliasz-Bocheńczyk, A., Mokrzycki, E., Piotrowski, Z., & Pomykała, R. (2009). Estimation of CO2 sequestration potential via mineral carbonation in fly ash from lignite combustion in Poland. *Energy Procedia*, *1*(1), 4873–4879. doi:10.1016/j.egypro.2009.02.316

UNEP. (2019). *Towards Sustainable Desalination*. Retrieved August 7, 2022 from https://www.unep.org/news-and-stories/story/towards0sustainable-desalination

World Bank. (2017). *Water Resources Management*. Retrieved August 6, 2022 from https://www.worldbank.org/en/topic/wwaterresourcesmanagement

Chapter 4
A High-Performance Grid-Integrated Wind Energy Conversion System (WECS) for Sustainable Energy Development

Abdul Hameed Kalifullah
National University of Science and Technology, Oman

Thomas Thangam Jeyaraj
National University of Science and Technology, Oman

ABSTRACT

The environmental problems such as global warming and contamination have prompted the usage of renewable energy sources (RES) as well as storage units in the modern power systems. Among several RES, the WECS is considered as a highly efficient system due to its numerous benefits such as ease of availability as well as pollution free. In this chapter, a switched-reluctance generator-based wind energy conversion system is proposed that reduces switching burdens in the grid and improves the average output power with reduced error. The system utilizes a DC-DC converter that generates improved efficiency with reduced switching stress thereby solving the problem of reverse voltage polarity. Further, an improved grey wolf optimization (GWO)-based PI controller is employed to reduce the error of mean power. The proposed approach is simulated as well as validated generating an enhanced efficiency of 98% and reduced THD of 2.2%. The obtained results clearly indicate a dynamic response with reduced harmonic content and low power factor that assures sustainable energy.

DOI: 10.4018/978-1-6684-7303-0.ch004

INTRODUCTION

Recently, all countries worldwide are concerned about the impacts caused by prolonged utilization of fossil fuels in environment. Subsequently, various schemes are initiated for promoting the utilization of clean energy resources. Adoption of these schemes insists the countries to reduce the emissions of below a specific limit. The European Union 2020 suggests that a particular renewable energy quantity has to be integrated with grid. Renewable energy is often viewed as the unavoidable future trend in global electric power development. The proportion of power grid renewable energy generation has shown a steady increase as the cost of renewable energy generation has decreased and power electronics technology has advanced rapidly. These improvements in generation of renewable energy has started replacing thermal power plants and thus emerged as an efficient solution. Among renewable energy resources, wind energy faces a rapid growth and presently contributes a considerable part of the power generation. Due to the nature of wind resource, unsteady patterns generate in the wind power production. Added to intermittency, the patterns of wind output power generates rapid ramps in both negative as well as positive directions. Incorporating these intermittent sources with grid lead to intense dilemmas in grid stability. Apart from the surge in generation of wind power, the stability of grid must be maintained (Jun Hashimoto et al., 1988) (Moghaddam et al., 2019; Sattar et al., 2020).

Figure 1. The switched reluctance generator (SRG) in wind energy conversion system (WECS)

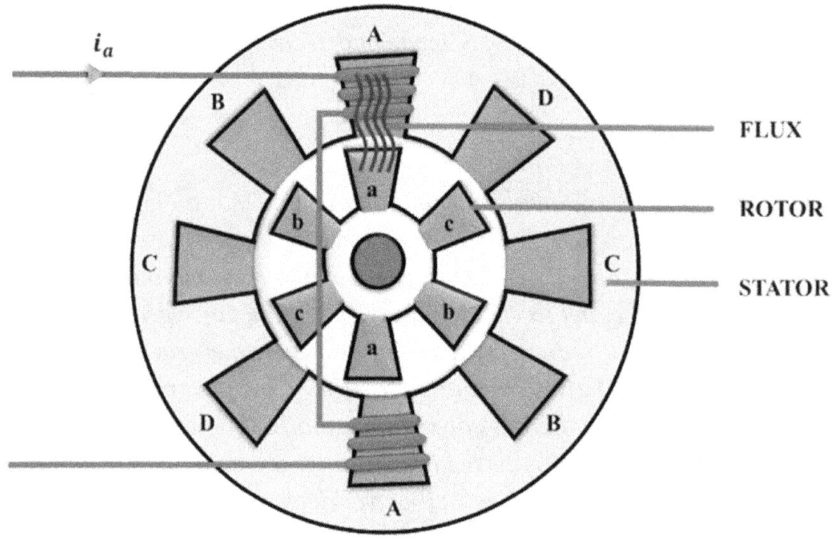

is widely utilized owing to its advantages like simple design, improved efficiency, absence of windings or magnets in rotor, fault tolerance and robustness (Neto et al., 2018; Bartolo et al., 2017). Hence, heat losses generally occur in stator which in turn provides easy cooling as well as increased speed of operation. Added to this, SRG has captivated the interest of many researchers due to its less repair costs and maintenance (Mendez et al., 2014; Gan et al., 2018; Sikder et al., 2014). The connection of SRG with the utility grid is either in direct manner or in interfaced manner. In direct manner, the voltage source converter (VSC) is exploited for the regulation of DC-bus voltage and the maximum power point tracking is carried out (Rahmanian et al., 2017; Barros et al., 2017; Mossad, 2020). Considering the interfaced manner of connection, a DC-DC converter is generally deployed for enhancing the resultant voltage from SRG thereby generating an optimal performance.

Generally, the DC-DC converters fall under two categories: isolated as well as non-isolated converters. The former converters include a transformer and generates improved voltage gain by changing the transformer turns ratio. Fly back (Gangavarapu et al., 2019; Gangavarapu et al., 2019; Kolincio et al., 2019; Fu et al., 2019), push-pull (Hassan et al., 2020; Tarzamani et al., 2019; Khatun et al., 2020), and forward type converters (Abbas et al., 2021; Qian et al., 2017; Zhang et al., 2021) are the types of isolated converters. These converters have a disadvantage of increased voltage stress owing to leakage inductance in transformer which further minimizes the efficiency. Taking non-isolated DC-DC converters into account, classical boost converter (Valdez-Resendiz et al., 2018; Zahng et al., 2017; Sferlazza et al., 2019) utilizes large duty cycle generating increased step-up voltage gain. This in turn leads to increased switching voltage stress with reverse recovery problems thereby reduces step up conversion. An improved step-up converter efficiency is obtained by utilizing coupled inductor (Fan et al., 2020; Pourjafar et al., 2019). It generates an increased voltage gain by adjusting the turns ratio of the coupled inductor. Anyway, high voltage spike and voltage stress occurs which in turn affects the main switch. Added to this, active as well as passive clamping methods (Salvador et al., 2017) are adopted along with coupled inductors resulting in increased converter size and cost. Hence, improved step-up converters are introduced by integrating two converters for achieving increased voltage gain. This paper utilizes an interleaved single-ended primary inductor converter (SEPIC) which regulates and boosts the output voltage to the required level. It generates a step-up conversion ratio with increased efficiency and non-inverting output voltage when compared to other interleaved topologies (Slah et al., 2017; Alzahrani et al., 2019; Granados-Luna et al., 2019; Azar et al., 2020).

Traditional PI controllers (Afghoul et al., 2018; Ganthia et al., 2020; Mosaad et al., 2019) are widely utilized for the achievement of dynamic performance of the SRG based WECS. The appropriate selection of control gains is mandatory for the effective functioning of PI controllers. Basic trial and error methods are exploited for the tuning of PI parameters but it is substantial and time consuming. Henceforth, recently intelligent optimization algorithms are adopted for PI parameter tuning and this approach exploits grey wolf optimizer (GWO). It is straightforward, robust as well as simple and solves the complicated optimization issues easily in the absence of local optima trapping. In addition, a PWM-3ϕ voltage source inverter (VSI) (Guimaraes et al., 2018) adopting theory is utilized for the conversion of DC power to AC power. The modulation index of the inverter is changed with relevant to the DC link voltage and the wind speed to maintain a constant grid voltage.

Within this context, an SRG based WECS is proposed in this approach which utilizes an interleaved SEPIC DC-DC converter. It generates an enhanced buck-boost conversion ratio along with non-inverting output voltage and provides improved energy efficiency with the utilization of reduced components. A GWO-PI controller inspired by the hunting behaviour of grey wolves is adopted for optimizing the weight parameters of the controller. It is a population based approach and identifies the fittest solution from the complete population. Finally, 3ϕ VSI adopting theory is utilized for the generation of sinusoidal currents with increased power factor and reduced harmonic contents.

PROPOSED METHODOLOGY

The environmental problems such as global warming and contamination has prompted the usage of RES as well as storage units in the modern power systems. Among several RES, the WECS is considered as a highly efficient system due to its numerous benefits such as ease of availability as well as pollution free. In variable speed applications, the SRG based WECS is commonly employed in the structure of turbine. This work integrates the wind energy with the grid along with interleaved SEPIC and the corresponding illustration of block is displayed in Figure 2.

Figure 2. Illustration of interleaved SEPIC

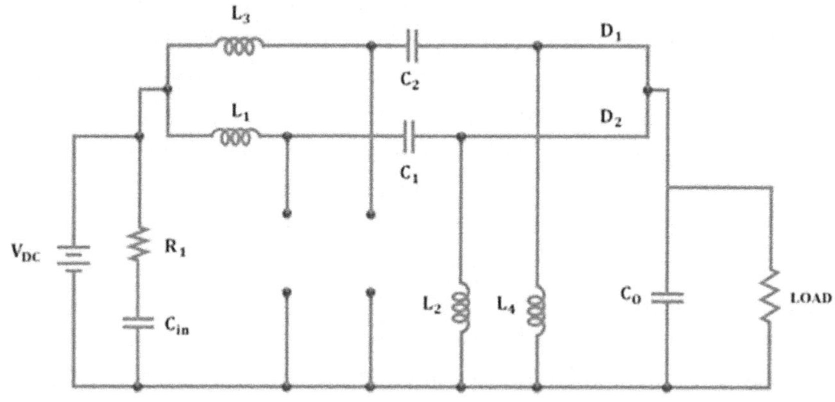

The proposed methodology comprises a SRG based WECS, which is integrated with the grid through interleaved SEPIC and 3Φ VSI. The wind energy attained from wind turbine (WT) is converted by utilizing SRG, which is then transformed into a regulated DC with the utilization of a PWM rectifier. Now, this output is improved with the aid of an interleaved SEPIC, which is given to the grid through a 3Φ VSI. A GWO based PI controller is utilized, which effectively controls the switches of interleaved SEPIC. The link voltage is retained as constant through this technique, which is given to the grid through3Φ VSI. By utilizing *d–q* reference frame, the power injected in the grid is controlled. Thus the grid synchronization and reactive power compensation are accomplished with less THD.

MODELING OF PROPOSED METHODOLOGY

SRG-Based WECS

The SRG based WECS is instigated, which increases the utilization of wind energy, high torque and less maintenance when analogized to conventional generators. The WT is a machine, which transforms the kinetic energy into mechanical energy. The mechanical torque (T_m) is given to the shaft of WT, which is expressed as,

$$T_m = \frac{1}{2}\rho AR\frac{C_p(\lambda)}{\lambda}V_w^2 \quad (1)$$

Where the air density is specified as ρ, the area swept by the blades is specified as A, the power coefficient is specified as $C_p(\lambda)$, the tip speed ratio is specified as λ, the wind speed is specified as V_w and the radius of turbine blade is specified as R. Theoretically the maximum value of C_p is 0.593, the value of air density is 1.225kg/m3. The ratio of tip speed is expressed as,

$$\lambda = \frac{R\omega_m}{V_w} \quad (2)$$

The dynamic equation of generator, which is expressed as,

$$J\frac{d\omega_m}{dt} = T_m - T_e \quad (3)$$

Figure 3. Schematic representation of SRG based WECS

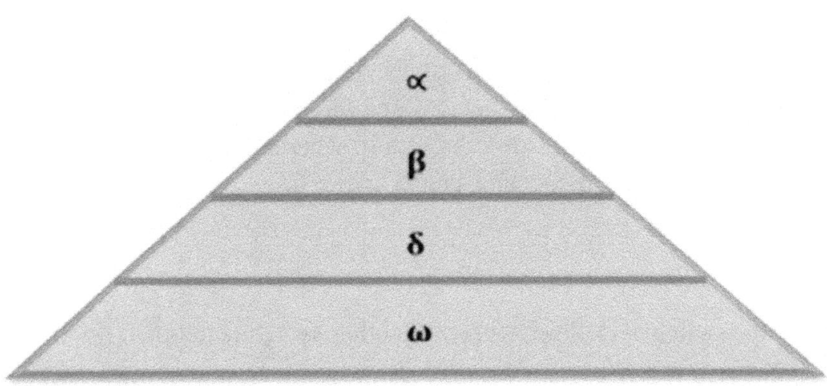

Where the moment of inertia is specified as J, shaft rotational speed is specified as ωmand the electromagnetic torque is specified as T_e.

When the SRG's stator as well as rotor poles are aligned in its location, the phase inductance increases and the reluctance decreases. When the rotor pole removes from its aligned location, the current flows through the stator windings lowers, whereas the inductance achieves the value of minimum. Due to magnetic field, the torque produced in SRG keeps the poles to its location of minimum reluctance. The SRG generates torque by synchronizing excitation with rotor position. When the poles are divided, the reluctance starts rises through which the negative torque is created. The SRG computes the torque produced by a phase under linear operating conditions. It is expressed as,

$$T_e = \frac{1}{2} i^2 \frac{dL}{d\theta} \tag{4}$$

Where the electromagnetic torque is specified as T_e, the phase current is specified as i, the phase inductance is specified as L and the position of rotor is specified as θ.

The torque generated by SRG is negative, since it works in the range of $(dL/d\theta<0)$. The overall torque produced by 4-phase generator is equal to the sum of the torques produced by individual phases. It is expressed as,

$$T_e = T_a(\theta, i_a) + T_b(\theta, i_b) + T_c(\theta, i_c) + T_d(\theta, i_d) \tag{5}$$

The winding of every phase is energized serially in SRG, whereas during the working of SRG, an electronic driver circuit is required, which enables the current passes through the phases serially. An asymmetric half bridge (AHB) converter is utilized that operates the SRG, whereas this converter comprises two switches as well as two diodes per phase. The circuit layout of AHB converter is displayed in

The switching circuit of phase *A* for three different modes such as exciting mode, freewheeling mode and generating mode are displayed in Figure 5.

Exciting Mode

In this mode, the SRG works as a motor, whereas the switches S_1 and S_2 are in excitation. When supply is given to the windings, the phase winding gets excited through the switches S_1 and S_2. As a result of pulling force in rotor poles, the current flows through it, whereas the rotor pole tends to rotates in its fixed location. After that, the phase inductance begins increasing.

Freewheeling Mode

In this mode, one of the switch is in close, whereas another switch is in open.

Generating Mode

In this mode, the switches S_1 and S_2 are in open, whereas the energy stored in the winding is given to the source through diodes. Because of $dL/dh<0$, the phase current gets increased due to back emf(e). The phase voltage equation is expressed as,

$$V = R.i + \frac{d\psi}{dt} \tag{6}$$

Where the voltage applied to a phase winding is specified as V, the phase current is specified as i, the phase resistance is specified as R and the magnetic flux is specified as ψ.

Figure 4. Torque produced by four-phase generator

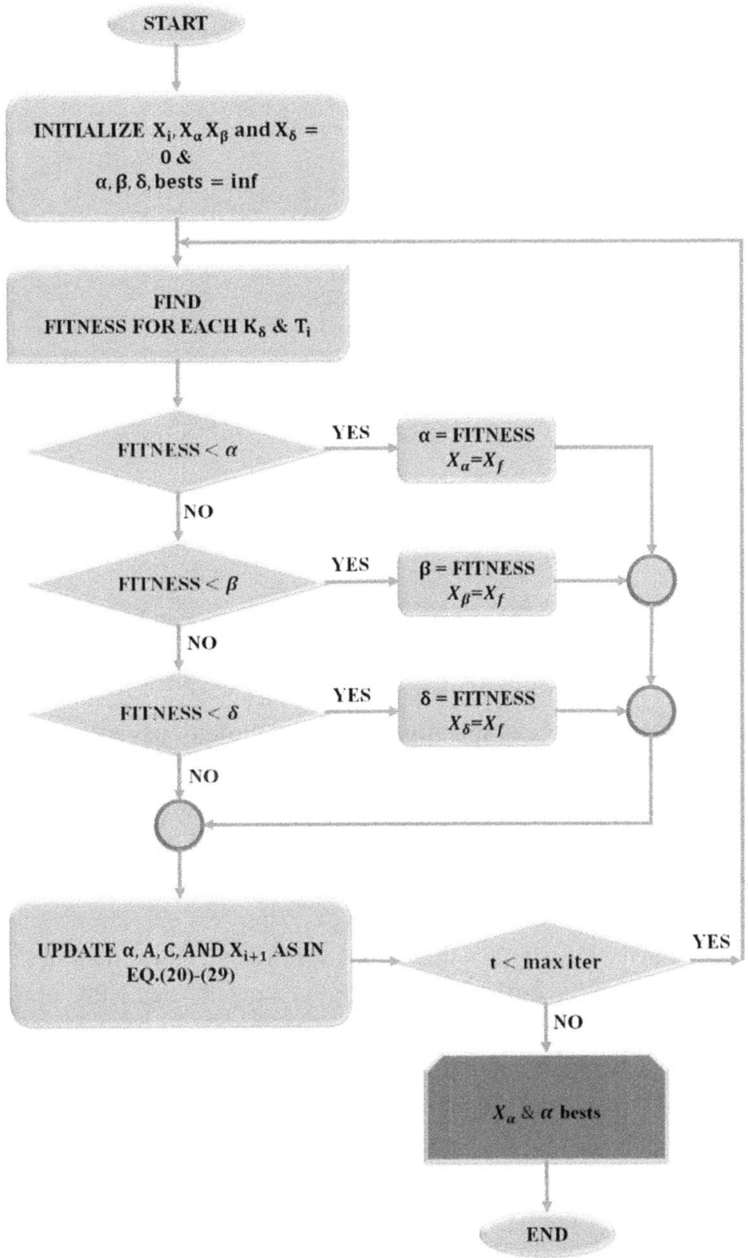

Figure 5. Freewheeling and generating mode

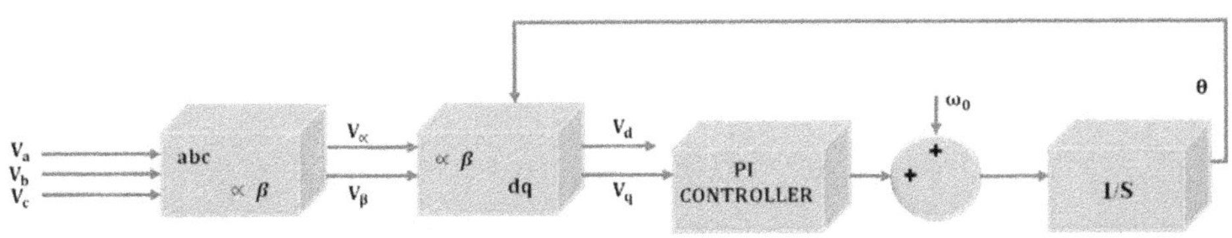

$$\psi = L(i, \theta).i \tag{7}$$

$$V = R.i + L\frac{di}{dt} + i\frac{d\theta}{dt}\frac{dL}{d\theta} \tag{8}$$

$$\omega = \frac{d\theta}{dt} \tag{9}$$

The voltage equation is rearranged, which is expressed as,

$$V = R.i + L\frac{di}{dt} + i.\omega\frac{dL}{d\theta} \tag{10}$$

$$e = \omega.i\frac{dL}{d\theta} \tag{11}$$

$$V = R.i + L\frac{di}{dt} + e \tag{12}$$

Where the back electromotive force is specified as e and the rotational speed is specified as ω. Thus the wind energy attained from the WT is transformed by utilizing SRG, which is then converted into DC with the utilization of PWM rectifier and this output is given to the interleaved SEPIC.

Interleaved SEPIC Converter

The SEPIC is a DC-DC converter, whereas the output voltage is greater than, lesser than or equal to the input voltage that depends on the ratio of duty cycle. As a result, the SEPIC performs the operation buck boost without polarity reversal.

To further minimize the size of filter and maximize the efficiency, the interleaving technique is utilized in the proposed SEPIC. For parallel connected n SEPIC cells, phase shift the gate signals by$2\pi/n$ in the interleaved arrangement. As a result, some ripples are eliminated as well as the output capacitor's RMS current is also lowered due to this phase shift.

The interleaved SEPIC's circuit comprises two components such as converter cell 1 as well as converter cell 2 in which the converter cell 1 contains L_1, S_1, C_1, L_2 and D_1, whereas converter cell 2 contains L_3, S_2, C_2, L_4 and D_2. The switches S_1 and S_2 are operated in interleaved manner i.e. with a phase shift of 180° and the inductive currents, IL_1, IL_2, IL_3, and IL_4 works under continuous conduction mode (CCM). When the duty cycle ratio >0.5, the converter operates in boost mode, whereas when duty cycle ratio <0.5, the converter operates in discontinuous conduction mode (DCM).

Figure 6. Interleaved SEPIC's circuit layout

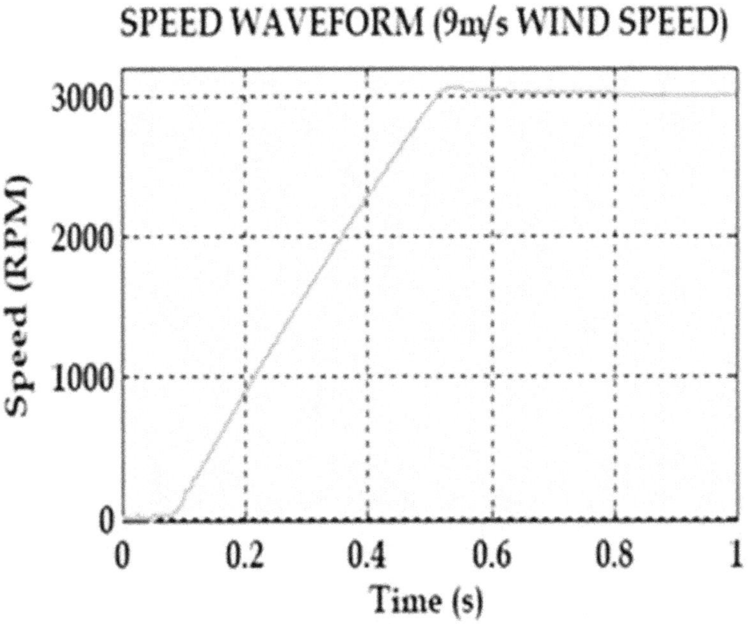

Operating Modes

The interleaved SEPIC works in the same way as traditional SEPIC. The firing pulses of both converters are overlap boost mode, because the duty ratio is >0.5. As a result, at-least one switch is turned *ON* during the complete working cycle. However, there is no overlap between the firing pulses in buck mode. As a result, one switch is turned *OFF* for a portion of working cycle.

All the capacitor values in the converter are chosen to be sufficiently big to block the direct current (DC). As result, the average current passes through them is zero. The load current is supplied by the inductances L_3 and L_4. As a result, the average current flows through these inductances is same as that of current flowing through the load.

If the current flows through the inductances L_1 and L_2 never drops to zero, which is referred as SEPIC's CCM. The average voltage across the capacitances C_1 and C_2 during the working of steady-state in SEPIC is equal to the input voltage (V_m). Because the capacitances C_1, C_2 blocks DC, the average current flows through it is equal to zero, which makes the inductances L_1 and L_2 as the source of load current. As a result, the average current flows through the inductance L_2 is the same as that of average load current.

Figure 7. Working interleaved SEPIC in boost mode

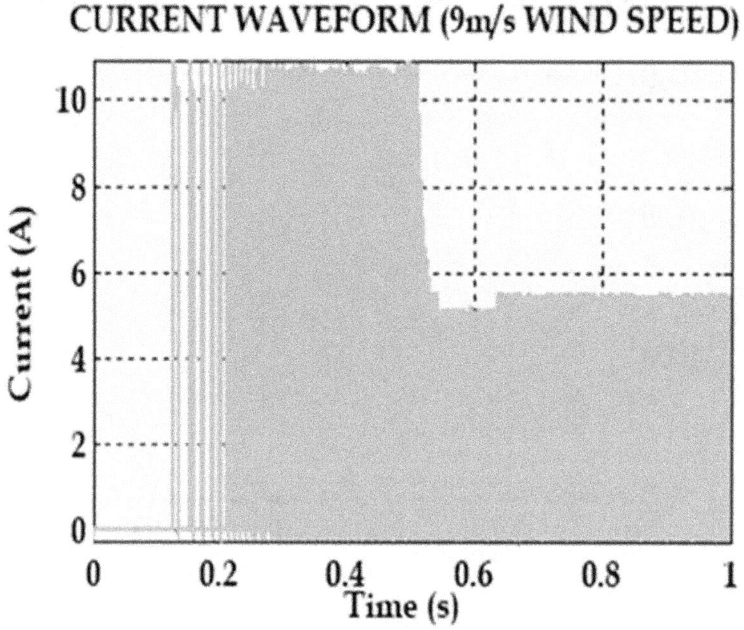

Mode 1

The switch S_1 is open and switch S_2 is closed in this mode. The capacitances C_1 and C_2 are charged with the positive terminal of input voltage. Through the switch S_2 and inductance L_4, the inductance L_3 charges as well as the capacitance C_2 discharges. As a result of this the diode D_2 becomes reverse biased. The diode D_1 is forward biased because the switch S_1 is open, whereas the inductances L_1 and L_2 are discharged via the load.

Mode 2

Both the switches S_1 and S_2 are closed in this state. So all the inductances L_1, L_2, L_3, and L_4 gets charged during this state. The mode 2 circuit layout is displayed in

Mode 3

The switch S_1 is closed and switch S_2 is open in this state. Hence the current flows through the inductances L_1 and L_2 increases, whereas the current through inductances L_3 and L_4 decreases.

Figure 8. Mode 1 circuit layout

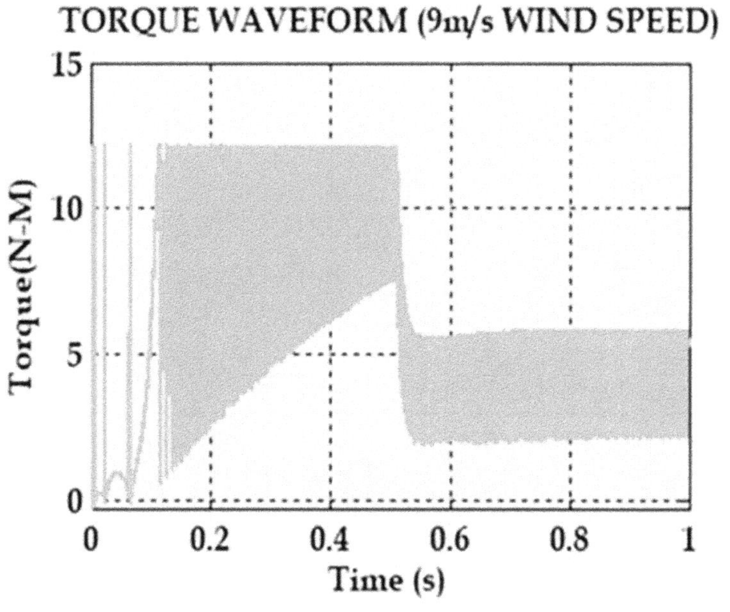

Figure 9. Mode 2 circuit layout

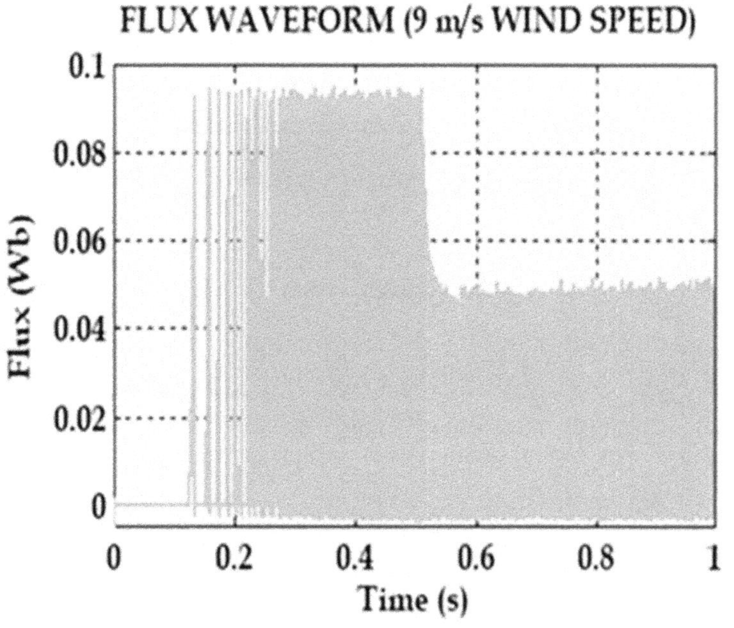

Figure 10. Mode 3 circuit layout

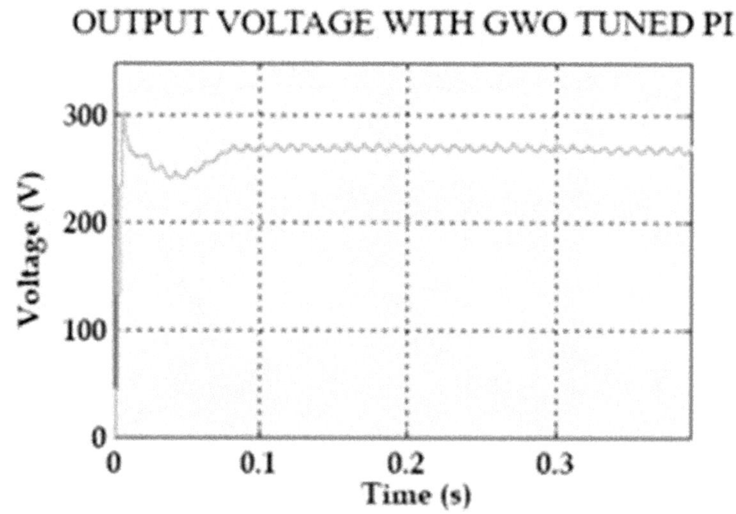

Mode 4

Both the switches S_1 and S_2 are open in this state. So all the inductances L_1, L_2, L_3, and L_4 gets discharged during this state.

Figure 11. Mode 4 circuit layout

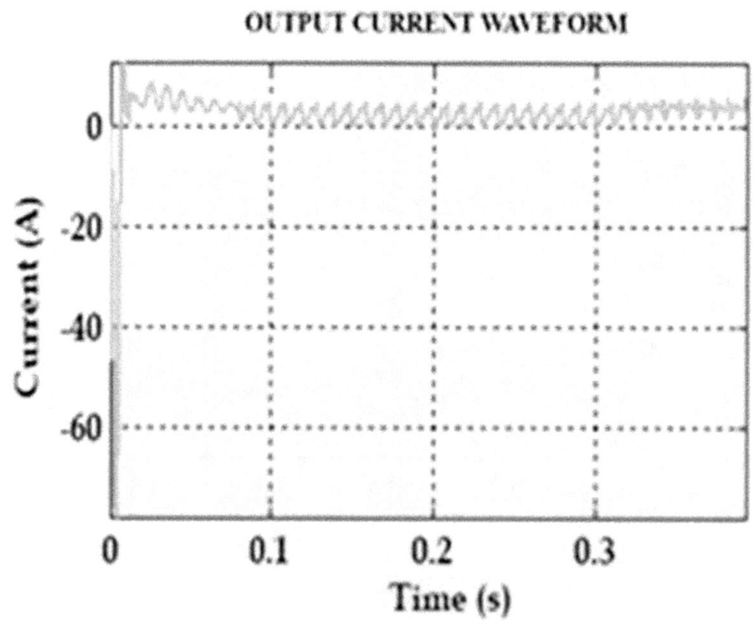

Design Equations

The SEPIC's operates on the basis of constant current in the intermediate ripple-filter the inductances(L_3 and L_4). The boost inductances are chosen as (L_1 and L_2), whereas on the basis of allowable ripple current and voltage, the capacitances are chosen as (C_1, C_2, and C_0).

Output voltage, $V_0 = DV_{in}/(1-D)$ (13)

Boost inductances, $L_1,L_2 = DV_{in}/[f_s(\Delta IL_i)]$ (14)

Intermediate capacitances, $C_1,C_2 = DV_{dc}/[(Rf_s)(\Delta VC_i)]$ (15)

Output inductive filter, $L_3,L_4 = (1-D)V_{dc}/[f_s(\Delta IL_0)]$ (16)

Output capacitive filter, $C_0 = I_{av}/(2\omega\Delta V_{dc})$ (17)

Where the switching frequency is specified as f_s, the SEPIC's average input voltage is specified as V_{in}, the duty cycle is specified as D, the ripple current IL_i's peak to peak value is specified as ΔIL_i, the ripple current IL_0's peak to peak value is specified as ΔIL_0, the inductive filter is specified as L_0, the boost inductance is specified as L_i, the capacitive filter is specified as C_0, the boost capacitance is specified as C_i and the DC output current's average value is specified as I_{av}.

GWO-Based PI Controller

The GWO based PI controller is utilized to tune the PI parameters, through which the interleaved SEPIC is controlled by utilizing this optimization technique. It is a meta-heuristic algorithm, which imitates the behaviour of a grey wolves and it lives in a group of 5-12 members. This group has a leader referred as alpha (α), the secondary one is referred as beta (β), who assists (α) in decision-making.

The grey wolves search the food, surround it, hunt it and attack it. The following is the arithmetic model of food's surroundings,

$$\vec{D} = \left|\vec{D}.\vec{X}_{pi} - \vec{X}_i\right| \tag{18}$$

$$\vec{X}_{i+1} = \vec{X}_{pi} - \vec{A}.\vec{D} \tag{19}$$

Where the grey wolf's place is specified as $\overrightarrow{X_i}$, the food's place is specified as $\vec{X}_{pi}$ and the distance is specified as D. The vectors A as well as C are computed by the expressions,

$$\vec{a} = 2 - 2 \times t / Max\_iter \tag{20}$$

$$\vec{A} = 2\vec{a}.\vec{r}_1 - \vec{a} \tag{21}$$

Figure 12. Remaining members of this group referred as δ and ω

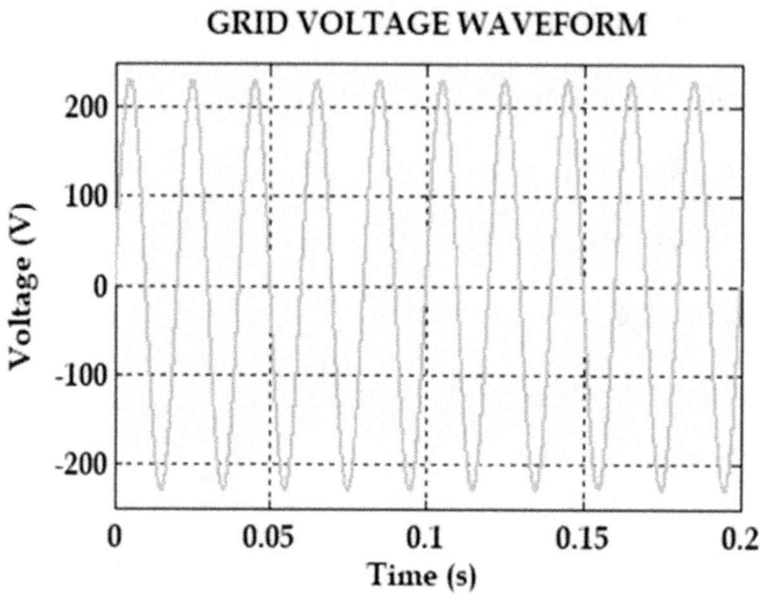

$$\vec{C} = 2.\vec{r}_2 \tag{22}$$

Where the random numbers between [0,1] are specified as r_1 and r_2. While the iterations increased, the variable a linearly decreased from 2 to 0. When $|A|>1$, the place of food search is achieved by diverging the search entities, which is mentioned as exploration. When $|A|<1$, the food attaining i.e. exploitation, which is achieved through the convergence of the search entities. The hunting is achieved by α entities with β and δ entities as shown in the equations (11) to (13).

$$\vec{D}_\alpha = \left|\vec{C}_1.\vec{X}_{\alpha i} - \vec{X}_i\right| \tag{23}$$

$$\vec{D}_\beta = \left|\vec{C}_2.\vec{X}_{\beta i} - \vec{X}_i\right| \tag{24}$$

$$\vec{D}_\delta = \left|\vec{C}_3.\vec{X}_{\delta i} - \vec{X}_i\right| \tag{25}$$

$$\vec{X}_1 = \vec{X}_{\alpha i} - \vec{A}_1.\vec{D}_\alpha \tag{26}$$

$$\vec{X}_2 = \vec{X}_{\beta i} - \vec{A}_2.\vec{D}_\beta \tag{27}$$

$$\vec{X}_3 = \vec{X}_{\delta i} - \vec{A}_3.\vec{D}_\delta \tag{28}$$

Figure 13. The GWO algorithm's flowchart

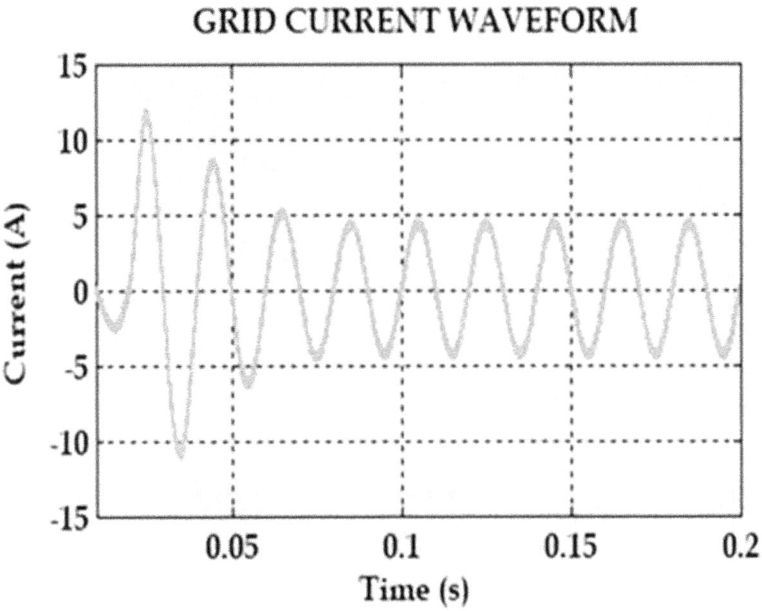

$$\vec{X}_{i+1} = \frac{\vec{X}_1 + \vec{X}_2 + \vec{X}_3}{3} \quad (29)$$

Through this optimized algorithm, the parameters of PI such as K_p and K_i are tuned, which provides best output and retains the link voltage as constant with the aid of GWO based PI.

Three Phase Grid Connected VSI Using *dq* Reference Frame

The *d–q* reference is utilized to synchronize the grid power with the wind power, which is achieved by controlling the 3Φ VSI. By utilizing Clarke's transformation is applied on the 3Φ voltage vector in the synchronous frame PLL to convert the values of *abc* into $\alpha\beta$ stationary reference frame, whereas by utilizing Park's transformation converts the value of $\alpha\beta$ to *d–q* rotating frame, which is displayed in Figure 14.

The *d–q* reference frames' feedback loop regulate the angular position which makes the *q*-axis component is in the steady state, whereas during the condition of steady state, the voltage amplitude is in the component of *d*-axis.

Under condition of balanced, the components of d and q axis are expressed as,

$$\begin{bmatrix} V_d \\ V_q \end{bmatrix} = \begin{bmatrix} cos\hat{\theta} & sin\hat{\theta} \\ -sin\hat{\theta} & cos\hat{\theta} \end{bmatrix} \begin{bmatrix} Ucos\theta \\ Usin\theta \end{bmatrix} = \begin{bmatrix} Ucos\left(\theta - \hat{\theta}\right) \\ Usin\left(\theta - \hat{\theta}\right) \end{bmatrix} \quad (30)$$

Figure 14. Park's transformation converts the value of αβ to d–q rotating frame

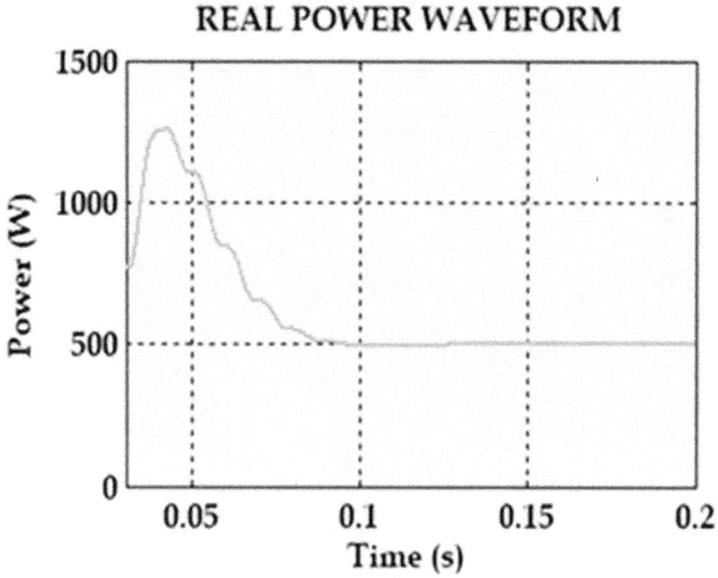

Where the input signal's amplitude is specified as U, the input signal's phase is specified as θ, the output of PLL is specified as $\hat{\theta}$, the component of d-axis is specified as V_d, and the component of q axis is specified as V_q.

The component q-axis is specified as phase as well as the component d-axis is specified as amplitude in the steady state. Under the condition of unbalanced utility, the generalized voltage vector (no harmonics on voltage) is expressed as,

$$V = V_+ + V_- + V_0 \tag{31}$$

Where the subscript + specifies the positive component, the subscript – specifies the negative component and the subscript 0 specifies the zero component. The Clarke's transformation's $\alpha\beta$ component is expressed as,

$$V_{\alpha\beta\gamma} = \begin{bmatrix} V_\alpha \\ V_\beta \\ V_\gamma \end{bmatrix} = T_{\alpha\beta/abc} \begin{bmatrix} V_a \\ V_b \\ V_c \end{bmatrix} \tag{32}$$

$$T_{\alpha\beta/abc} = \frac{2}{3}\begin{bmatrix} 1 & -\frac{1}{2} & -\frac{1}{2} \\ 0 & \frac{\sqrt{3}}{2} & \frac{\sqrt{3}}{2} \\ \frac{1}{2} & \frac{1}{2} & \frac{1}{2} \end{bmatrix} \tag{33}$$

The component of zero-sequence is ignored because it is on the γ-axis. The voltage vector on $\alpha\beta$-plane is expressed as,

$$V_{\alpha\beta} = T_{\alpha\beta/abc}\left(V_{+} + V_{-}\right) = \begin{bmatrix} U_{+}cos\theta_{+} & U_{-}cos\theta_{-} \\ U_{+}sin\theta_{+} & U_{-}sin\theta_{-} \end{bmatrix} \tag{34}$$

By utilizing Park's transformation, the $\alpha\beta$ frame is transformed to the component of d–q frame. It is expressed as,

$$V_{dq} = T_{dq/\alpha\beta}V_{\alpha\beta} = \begin{bmatrix} U_{+}\cos(\theta_{+} - \hat{\theta}) + U_{-}\cos(\theta_{-} - \hat{\theta}) \\ U_{+}\sin(\theta_{+} - \hat{\theta}) + U_{-}\sin(\theta_{-} - \hat{\theta}) \end{bmatrix} = \begin{bmatrix} U_{+} + U_{-}cos\left(2\omega t\right) \\ U_{+} + U_{-}sin\left(2\omega t\right) \end{bmatrix} \tag{35}$$

$$T_{dq/\alpha\beta} = \begin{bmatrix} cos\hat{\theta} & sin\hat{\theta} \\ -sin\hat{\theta} & cos\hat{\theta} \end{bmatrix} \tag{36}$$

Where the voltage vector's angular frequency is specified as ω and $\hat{\theta} = \theta_{+} = \theta_{-} = \omega t$. Through this technique, the 3Φ VSI is controlled by applying proper gating pulses and thus the grid synchronization and reactive power compensation are accomplished with the less THD.

RESULTS AND DISCUSSION

The performance of the proposed system is validated and analyzed by simulation and hardware implementation. The designing, modelling and simulation of the proposed system is carried out under varying wind speed. The reference power with variable speed of wind and the output power by SRG are obtained. A DC motor is utilized for rotating the SRG and a DC source is exploited for powering the driver circuit. The operation of the interleaved SEPIC DC-DC converter is controlled by the GWO-PI controller which in turn generates the efficient gain parameter values. The parameters of the SRG and the interleaved converter utilized are illustrated in Table 1 and Table 2 respectively.

Table 1. SRG parameters

Parameters	Value
Nominal power	1.13KW
Nominal voltage	48V
Number of rotor poles	6
Number of stator poles	8
One phase winding resistance	0.5Ω
Minimum inductance	0.1mH
Maximum inductance	0.75mH
Moment of inertia	0.0017Kg.m^2
Coefficient of friction	0.001NM/(*rad/s*)

Table 2. Interleaved SEPIC converter parameters

Components	Values
Input voltage	20V
Output current	1A
Output voltage	280V
Resistance	110Ω
Duty cycle	0.84
Switching frequency	50KHz
Capacitors C_1,C_2	1.9μF
Capacitor C_0	470μF
Inductors L_1,L_2, L_3,L_4	366μH

SIMULATION RESULTS

The simulation of the proposed SRG based WECS utilizing an interleaved SEPIC DC-DC converter is carried out in MATLAB/SIMULINK. The obtained waveforms are given as follows.

Figure 15. Waveform for speed

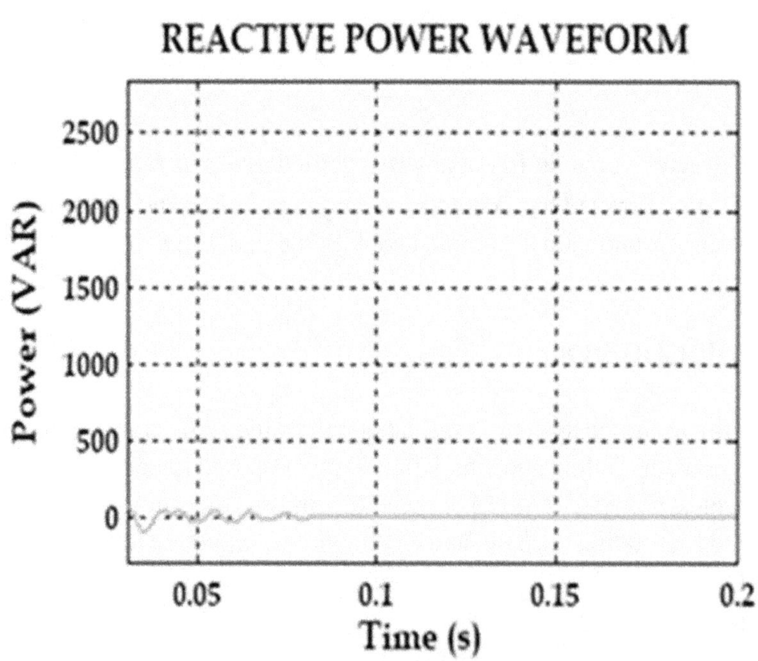

The simulated waveform indicates that the speed of the wind is maintained at 9m/s after 0.5s. In relevant to the wind speed, the generated current, torque and flux are shown below.

Figure 16. Generated current in relevance to wind speed

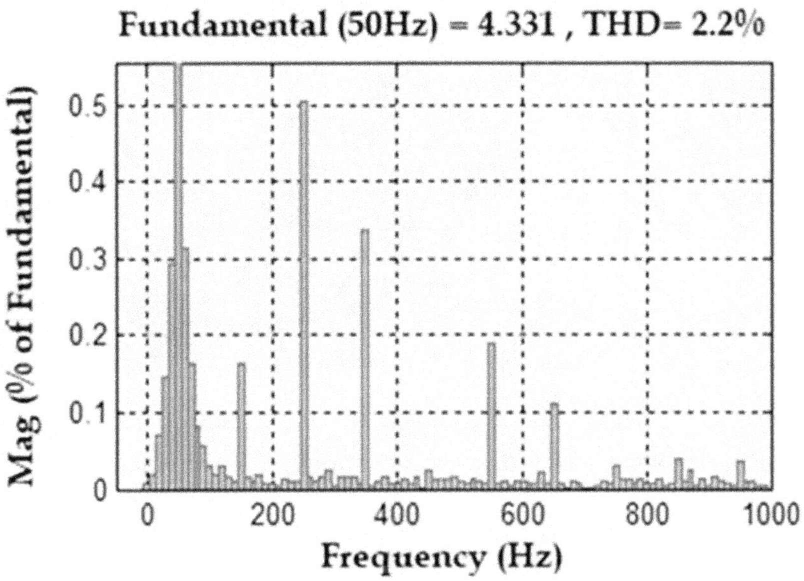

Figure 17. Generated flux in relevance to wind speed

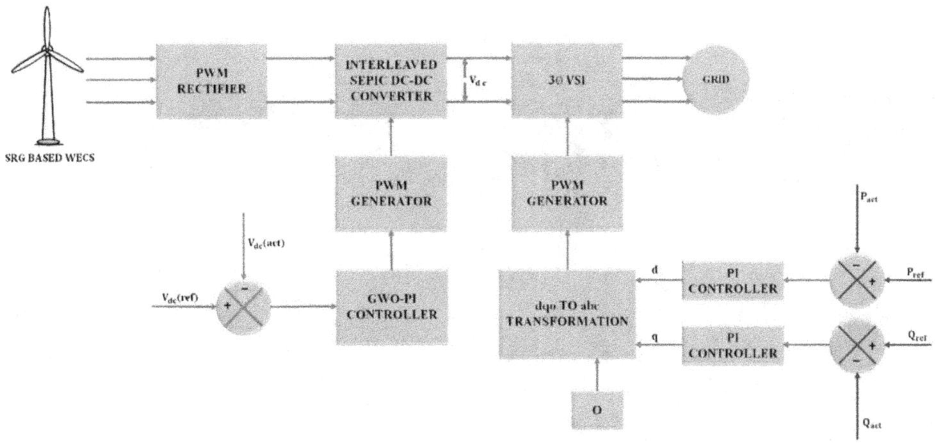

Figure 18. Generated torque in relevance to wind speed

Due to the change in wind speed, there occurs variation in the obtained current, torque and flux values after 0.5s.

Figure 19. Output waveforms of the voltage and current

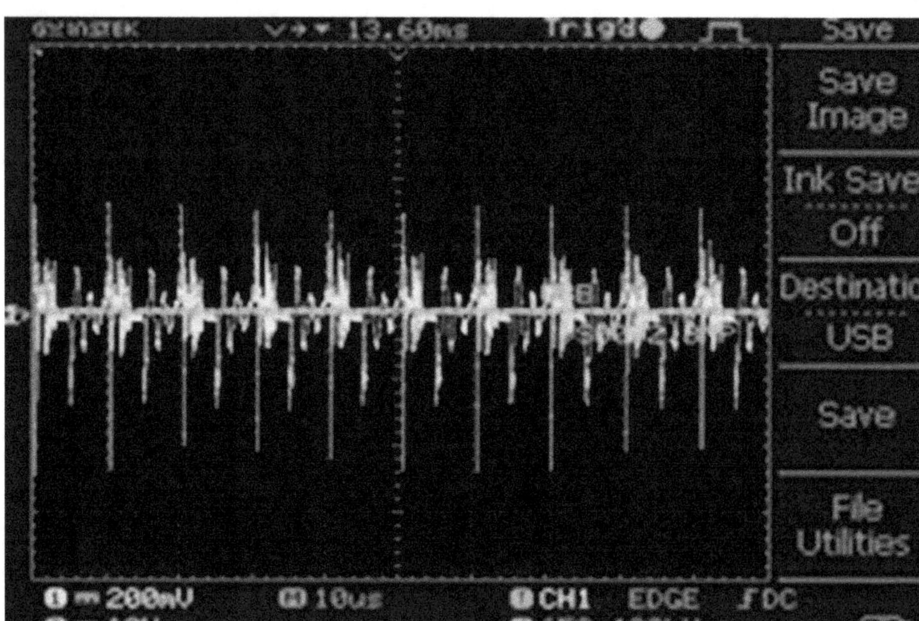

An input voltage of 20V is boosted to 270V as shown in figure 17. The proposed approach exploits closed loop PI controller which generates the gain parameters. These parameters are utilized for reducing the error in speed. Due to the variation in current, the flux as well as torque which are closely coupled also exhibit variations. These variations are maintained to a constant value by the PI controller after 0.1s.

Figure 20. The grid waveforms for voltage and current which are sinusoidal in nature with reduced harmonics

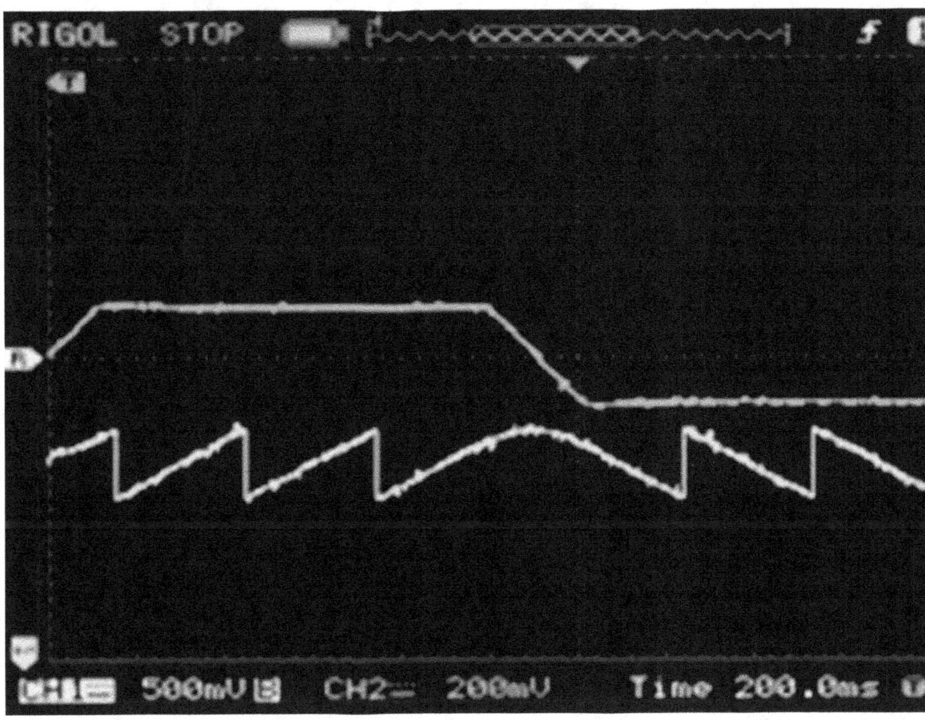

This denotes a dynamic flow of power into the grid.

The d axis component regulates the real power and the q axis component regulates the reactive power. These values are subjected to dq transformation and the corresponding outputs are applied to the PWM generator which in turn generates the pulses required for 3ϕ VSI. The corresponding waveforms obtained for the real power and reactive power are shown in denotes that the obtained THD value of the proposed system is 2.2% which compensates the IEEE 929-2000 standard. This generates a distortion less waveform with reduced harmonics to be injected into the grid.

HARDWARE RESULTS

It includes a SRG based WECS, an interleaved DC-DC SEPIC converter, 3ϕ VSI and an FPGA controller for performing switching operation of both converter as well as the inverter. The performance of the proposed system in real time is discussed as follows.

Figure 21. The hardware setup of the proposed SRG-based WECS

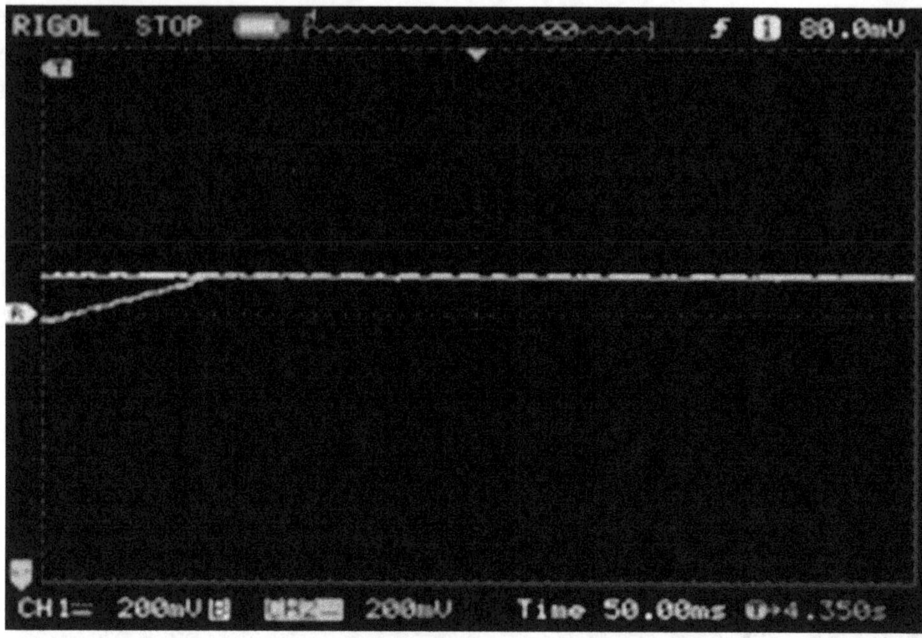

Figure 22. Obtained current and magnetic flux waveforms for SRG at the wind speed of 9m/s

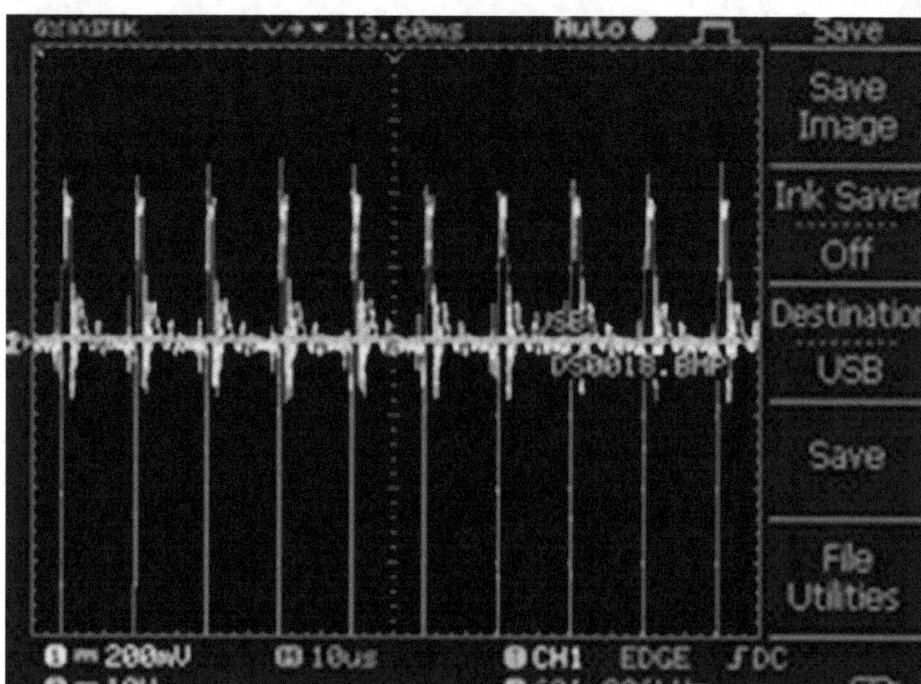

Due to the presence of uncertainties in wind power and environment, the parameters get distorted. This in turn results in nonlinearities of current further affecting the magnetic flux.

Figure 23. Obtained speed and torque for the SRG at the wind speed of 9m/s

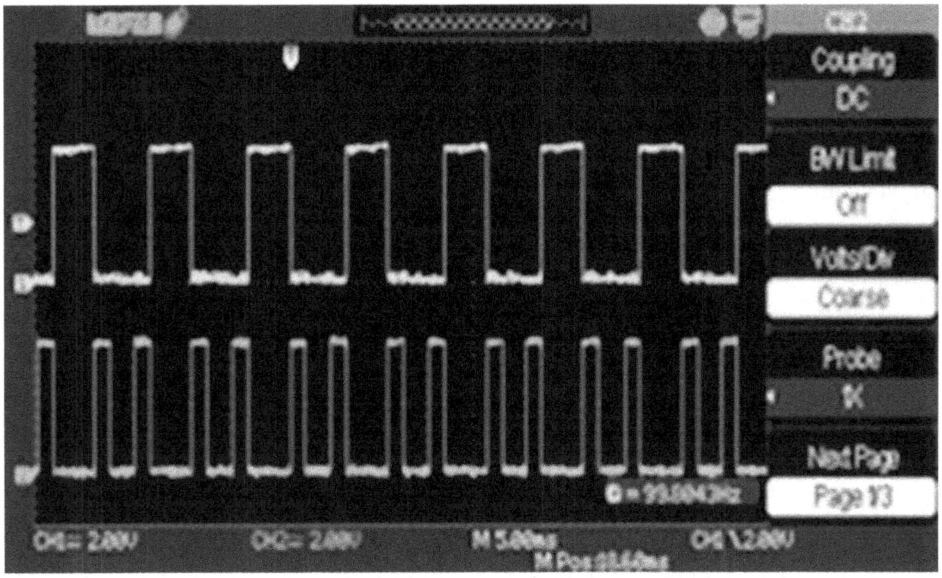

The obtained waveform for torque indicates more distortions due to the difference in processor running time.

Figure 24. Waveforms for PWM pulse and the DC link voltage with PI controller

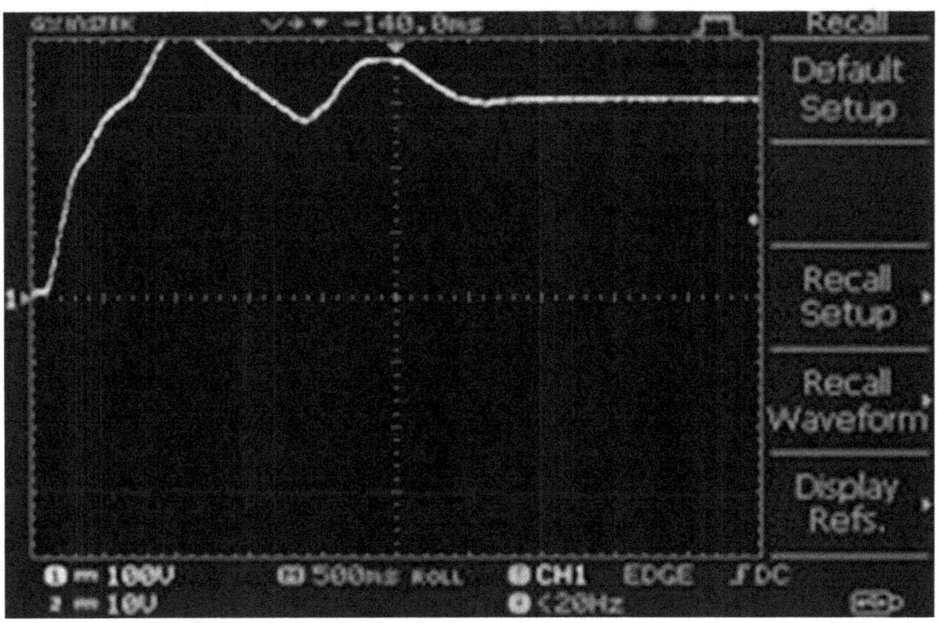

The DC link voltage has to be maintained constant irrespective of the variations in the speed of the wind. Hence, the gain parameters of the PI controller are varied so that the speed of SRG is maintained constant. The utilized GWO-PI controller effectively performs the control operation which in turn generates optimized outputs with improved efficiency.

Table 3. Comparison of gain parameters

Parameters	GA		PSO		GWO	
	K_p	K_i	K_p	K_i	K_p	K_i
Minimum value	298.96	596.01	398.55	731.75	479.51	958.40
Maximum value	496.18	1002.6	500.00	1002.50	500.00	1002.28
Mean	396.87	793.79	445.76	871.41	480.57	963.97
Standard deviation	38.83	77.68	53.68	93.59	6.32	160.86

The GWO approach is compared with the genetic algorithm (GA) and particle swarm optimization (PSO) approaches as shown in Table 3. The GWO exhibited a rapid convergence due to less number of iterations when compared to GA and PSO. The GWO-PI controller generates an increased K_p gain of 480.57 and K_i gain of 963.97 which indicates an improved performance. These results confirm that utilizing GWO for tuning the parameters of PI controller is more efficient when compared to other approaches.

Figure 25. Comparison of rise time and settling time values of the GWO tuned PI controller with GA and PSO approaches

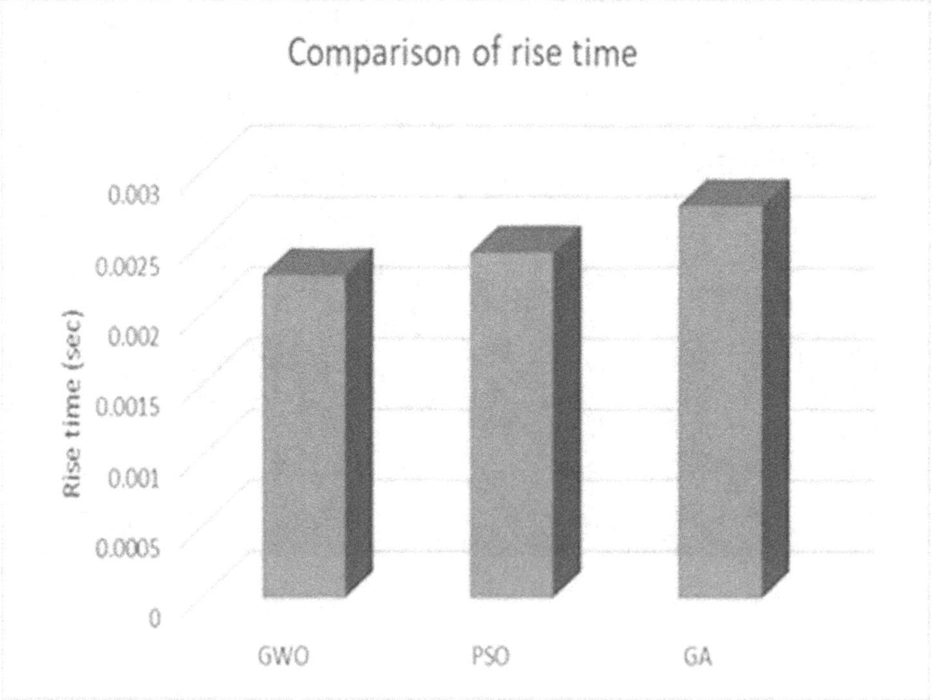

The GWO approach generates efficient values of 0.00228s rise time and 0.00406s settling time which is less when compared to other approaches.

The proposed system generates an improved efficiency of 98%.

Figure 26. Comparison of the proposed system with grid-connected SRG interfaced with bidirectional DC-DC converter in WECS using PI controller and direct power control of SRG-based WECSs using optimised fractional-order PI controller (Mosaad et al., 2020) in terms of efficiency

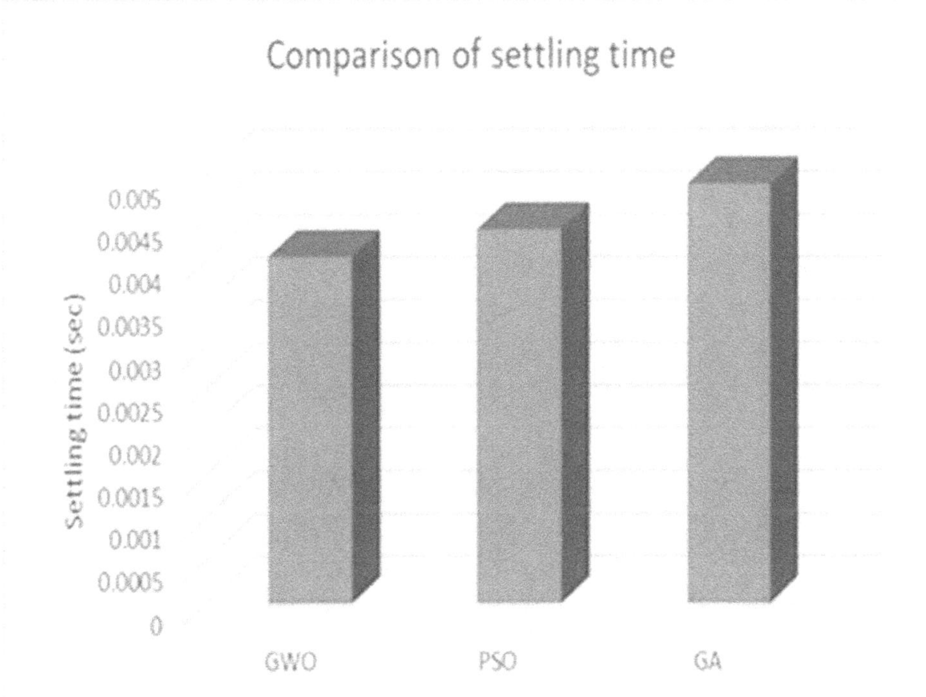

CONCLUSION

This paper introduces an SRG based high performance WECS that operates at wide speed range with an operating range greater than induction and synchronous generators. It comprises of an interleaved SEPIC DC-DC converter for regulating an optimal SRG excitation voltage value. Although, the duty cycle, supply voltage and load resistance vary, the desired output voltage is obtained from the interleaved converter. The utilization of GWO for PI parameter tuning generates increased gain values which in turn indicates an improved control performance. It provides low standard deviation and rapid convergence into global optimum. Subsequently, 3ϕ VSI adopting *dq* theory is exploited through which sinusoidal currents that are in phase with voltage across the grid are injected with minimized harmonic contents. The obtained values for efficiency is 98% and THD is 2.2%. The experimental results prove that the extraction of maximum wind power is carried out over wide speed range indicating improved performance.

Figure 27.

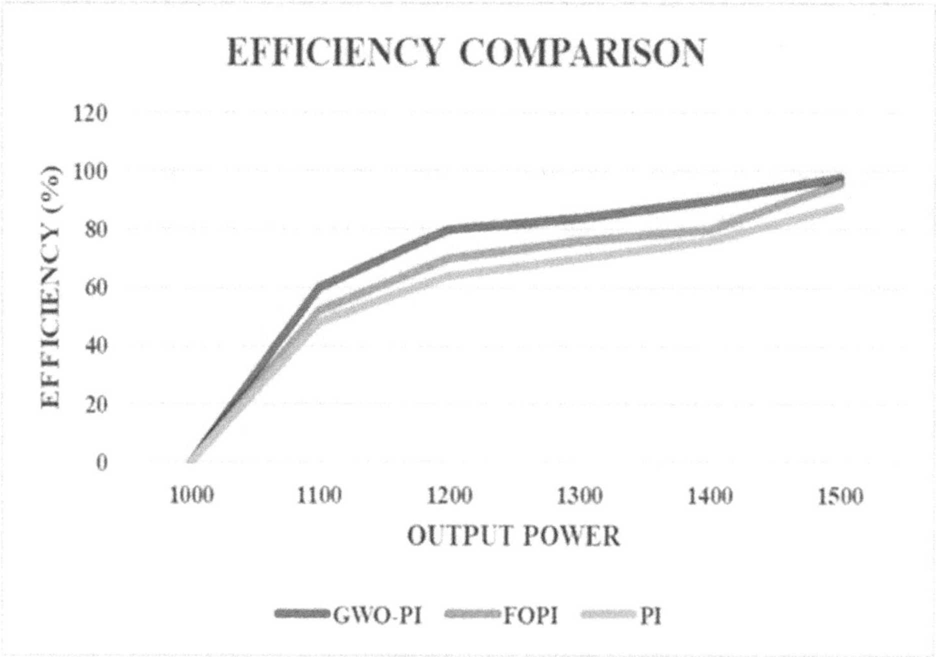

Figure 28.

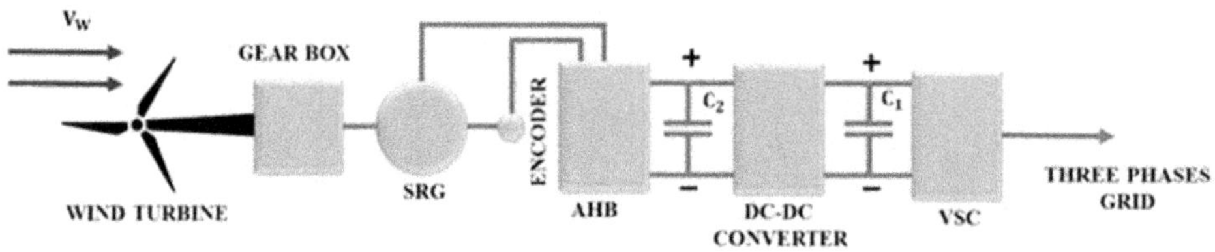

Figure 29.

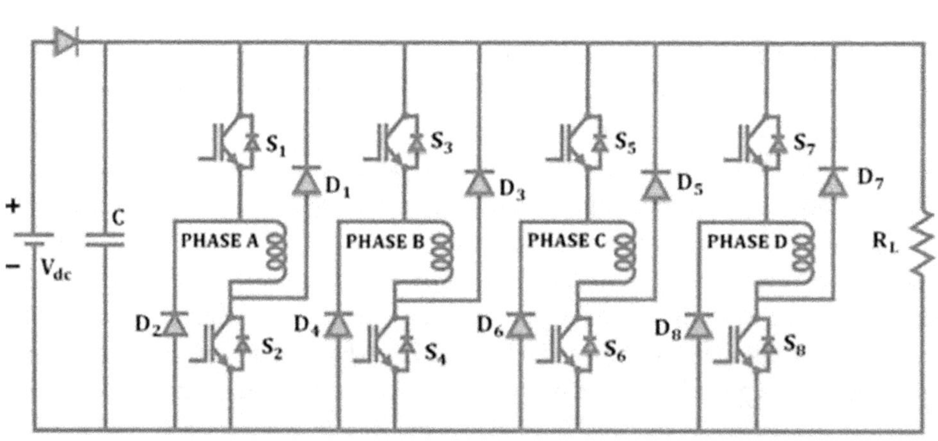

Figure 30.

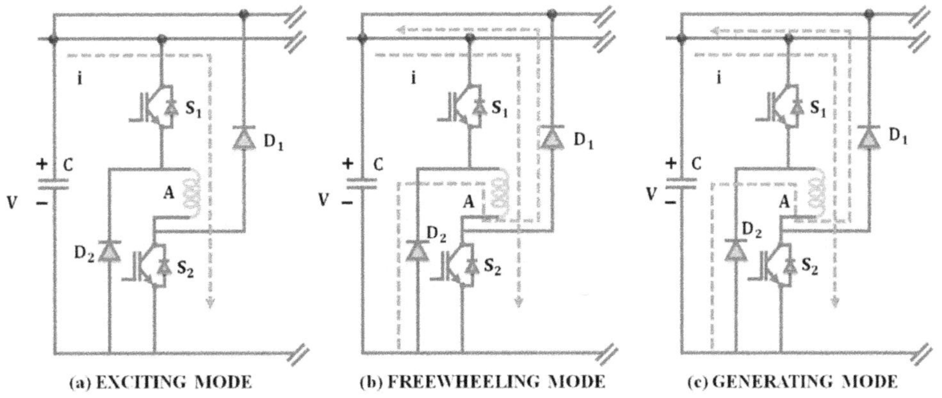

Figure 31.

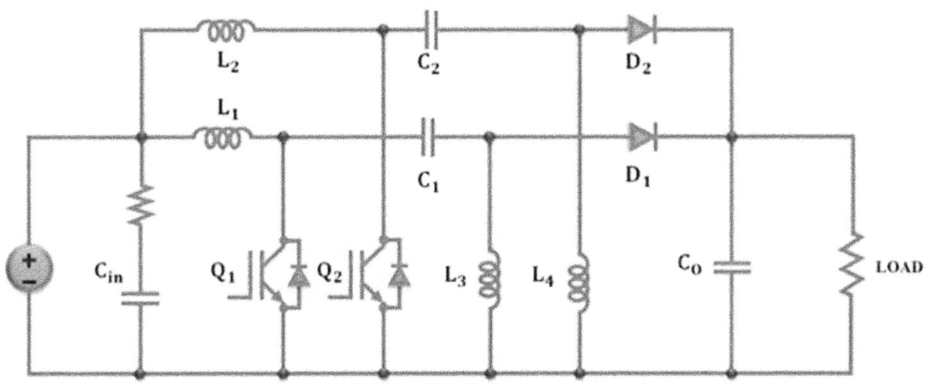

Figure 32.

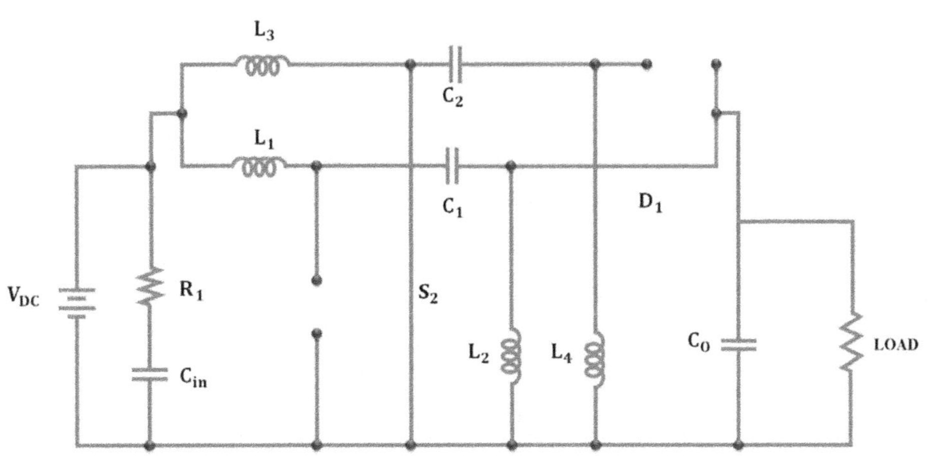

Figure 33.

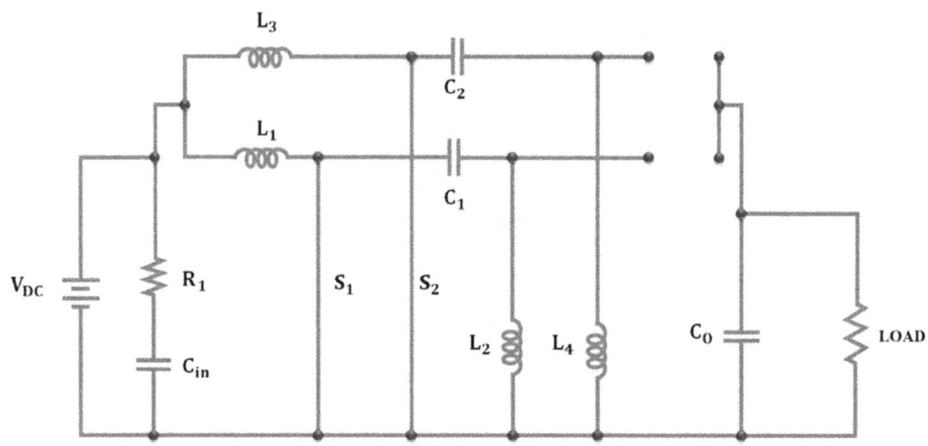

Figure 34.

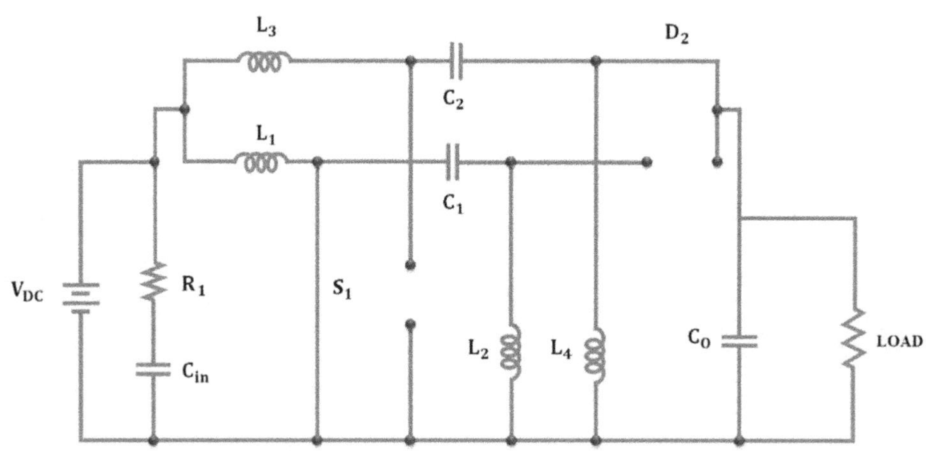

REFERENCES

Abbas, G., Asad, M. U., Gu, J., Alelyani, S., Balas, V. E., Hussain, M. R., Farooq, U., Awan, A. B., Raza, A., & Chang, C. (2021). Multivariable unconstrained pattern search method for optimizing digital PID controllers applied to isolated forward converter. *Energies, 14*(1), 77.

Afghoul, H., Krim, F., Babes, B., Beddar, A., & Kihel, A. (2018). Design and real time implementation of sliding mode supervised fractional controller for wind energy conversion system under sever working conditions. *Energy Conversion and Management, 167*, 91–101.

Alzahrani, A., Ferdowsi, M., & Shamsi, P. (2019). A family of scalable non-isolated interleaved DC-DC boost converters with voltage multiplier cells. *IEEE Access, 7*, 11707–11721.

Azer, P., & Emadi, A. (2020). Generalized state space average model for multi-phase interleaved buck, boost and buck-boost DC-DC converters: Transient, steady-state and switching dynamics. *IEEE Access, 8*, 77735–77745.

Barros, T. A. S., dos Santos Neto, P. J., Nascimento Filho, P. S., Moreira, A. B., & Ruppert Filho, E. (2017). An Approach for Switched Reluctance Generator in a Wind Generation System with a Wide Range of Operation Speed. *IEEE Transactions on Power Electronics, 32*(11), 8277–8292.

Bartolo, J.B., Degano, M., & Espina, J. (2017). Design and initial testing of high speed 45-kw switched reluctance drive for aerospace application. *IEEE Trans. Ind. Electron, 64*(2), 988-997.

Fan, X., Sun, H., Yuan, Z., Li, Z., Shi, R., & Ghadimi, N. (2020). High voltage gain DC/DC converter using coupled inductor and VM techniques. *IEEE Access, Vol, 8*, 131975–131987.

Fu, K., & Chen, W. (2019). Evaluation method of flyback converter behaviors on common-mode noise. *IEEE Access, Vol, 7*, 28019–28030.

Gan, C., Jin, N., & Sun, Q. (2018). Multiport bidirectional SRM drives for solar-assisted hybrid electric bus powertrain with flexible driving and self-charging functions. *IEEE Trans. Power Electron, 33*(10), 8231–8245.

Gangavarapu, S., & Akshay, K. R. G. (2019). A three-phase single-stage isolated flyback-based PFC converter with leakage energy recovery clamping circuit. *IEEE Transactions on Transportation Electrification, 5*(4), 1155–1168.

Gangavarapu, S., Rathore, A. K., & Khadkikar, V. (2019). High-Efficiency Three-Phase Single-Stage Isolated Flyback-Based PFC Converter With a Novel Clamping Circuit. *IEEE Transactions on Industry Applications, 56*(1), 718–729.

Ganthia & Barik. (2020). Steady-state and dynamic comparative analysis of PI and fuzzy logic controller in stator voltage oriented controlled DFIG fed wind energy conversion system. *Journal of The Institution of Engineers (India): Series B, 101*(3), 273-286.

Granados-Luna, T.-R., Araujo-Vargas, I., Forsyth, A. J., Cano-Pulido, K., Velázquez-Elizondo, P.-E., Cervantes, I., Gómez-Olguín, F., & Villarruel-Parra, A. (2019). Two-Phase, Dual Interleaved Buck–Boost DC–DC Converter for Automotive Applications. *IEEE Transactions on Industry Applications, Vol, 56*(1), 390–402.

Guimaraes, J. S., Ricardo de Almeida, B., Tofoli, F. L., & Demercil de Souza, O. (2018). Three-phase grid-connected wecs with mechanical power control. *IEEE Transactions on Sustainable Energy, Vol, 9*(4), 1508–1517.

Hashimoto, J., Ustun, T. S., Suzuki, M., Sugahara, S., Hasegawa, M., & Otani, K. (2021). Advanced Grid Integration Test Platform for Increased Distributed Renewable Energy Penetration in Smart Grids. *IEEE Access, Vol, 9*, 34040–34053.

Hassan, Abbassi, Jerbi, Mehmood, Tahir, Cheema, Elavarasan, Ali, & Khan. (2020). A novel algorithm for MPPT of an isolated PV system using push pull converter with fuzzy logic controller. *Energies, 13*(15), 4007.

Khatun, K., Vakacharla, V. R., Kizhakkan, A. R., & Rathore, A. K. (2020). Small-signal analysis and control of snubberless naturally clamped soft-switching current-fed push–pull DC/DC converter. *IEEE Transactions on Industry Applications, Vol, 56*(4), 4299–4308.

Kolincio, M., Chrzan, P. J., & Musznicki, P. (2019). Multitransformer primary-side regulated flyback converter for supplying isolated IGBT and MOSFET drivers. *IEEE Transactions on Industrial Electronics, Vol, 67*(2), 1005–1012.

Mendez, S., Martinez, A., & Millan, W. (2014). Design, characterization, validation of a 1-kw AC self-excited switched reluctance generator. *IEEE Trans. Ind. Electron, 61*(2), 846-855.

Moghaddam, I. N., Chowdhury, B., & Doostan, M. (2019). Optimal sizing and operation of battery energy storage systems connected to wind farms participating in electricity markets. *IEEE Trans. Sustain. Energy, 10*(3), 1184–1193.

Mosaad, M. I. (2020). Direct power control of SRG-based WECSs using optimised fractional-order PI controller. *IET Electric Power Applications, 14*(3), 409–417.

Mosaad, M. I., Abu-Siada, A., & El-Naggar, M. F. (2019). Application of superconductors to improve the performance of DFIG-based WECS. *IEEE Access, 7*, 103760–103769.

Neto, D. S., Jose, P., Andre dos Santos Barros, T., Hancco Catata, E., & Ruppert Filho, E. (2021). Grid-connected SRG interfaced with bidirectional DC-DC converter in WECS. *IEEE Transactions on Energy Conversion.*

Neto, Barros, & Paula. (2018). Design of computational experiment for performance optimization of a switched reluctance generator in wind systems. *IEEE Trans. Energy Convers., 33*(1), 406-419.

Pourjafar, S., Sedaghati, F., Shayeghi, H., & Maalandish, M. (2019). High step-up DC–DC converter with coupled inductor suitable for renewable applications. *IET Power Electronics, 12*(1), 92–101.

Qian, T., & Wu, Q. (2017). A scheme of a resonant forward-flyback converter with suppressed frequency variation. *IEEE Transactions on Power Electronics, 33*(5), 3711–3716.

Rahmanian, E., Akbari, H., & Sheisi, G. H. (2017). Maximum Power Point Tracking in Grid Connected Wind Plant by Using Intelligent Controller and Switched Reluctance Generator. *IEEE Transactions on Sustainable Energy, 8*(3), 1313–1320.

Salvador, M. A., Lazzarin, T. B., & Coelho, R. F. (2017). High step-up DC–DC converter with active switched-inductor and passive switched-capacitor networks. *IEEE Transactions on Industrial Electronics, 65*(7), 5644–5654.

Sattar, Al-Durra, Caruana, Debouza, & Muyeen. (2020). Testing the Performance of Battery Energy Storage in a Wind Energy Conversion System. *IEEE Transactions on Industry Applications, 56*(3), 3196 – 3206.

Sferlazza, A., Albea-Sanchez, C., Martínez-Salamero, L., Garcia, G., & Alonso, C. (2019). Min-type control strategy of a DC–DC synchronous boost converter. *IEEE Transactions on Industrial Electronics, 67*(4), 3167–3179.

Sikder, C., Husain, I., & Sozer, Y. (2014). Switched reluctance generator control for optimal power generation with current regulation. *IEEE Trans. Ind. Appl., 50*(1), 307-316.

Slah, Mansour, Hajer, & Faouzi. (2017). Analysis, modeling and implementation of an interleaved boost DC-DC converter for fuel cell used in electric vehicle. *International Journal of Hydrogen Energy, 42*(48), 28852-28864.

Tarzamni, H., Babaei, E., Esmaeelnia, F. P., Dehghanian, P., Tohidi, S., & Mohammad, B. B. S. (2019). Analysis and reliability evaluation of a high step-up soft switching push–pull dc–dc converter. *IEEE Transactions on Reliability, 69*(4), 1376–1386.

Valdez-Resendiz, J. E., Rosas-Caro, J. C., Mayo-Maldonado, J. C., & Llamas-Terres, A. (2018). Quadratic boost converter based on stackable switching stages. *IET Power Electronics, 11*(8), 1373–1381.

Zhang, G., Iu, H. H.-C., Zhang, B., Li, Z., Fernando, T., Chen, S.-Z., & Zhang, Y. (2017). An impedance network boost converter with a high-voltage gain. *IEEE Transactions on Power Electronics, 32*(9), 6661–6665.

Zhang, Z., Liao, M., Jiang, D., Yang, X., & Li, S. (2021). High step-up isolated forward-flyback DC/DC converter based on resonance with pulse frequency modulation. *Journal of Power Electronics, 21*(2), 483–493.

Chapter 5
Offshore Wind Turbine Site Selection in the Northern Aegean Sea by Utilizing GIS-Based Maritime Spatial Planning and Exergy Efficiency

Arife Tugsan Isiacik Colak
National University of Science and Technology, Oman

Hamid Reza Soltani Motlagh
National University of Science and Technology, Oman

Gizem Senel
Istanbul Technical University, Turkey

Cigdem Goksel
Istanbul Technical University, Turkey

ABSTRACT

Maritime spatial planning (MSP) is a mechanism of supporting adaptive decision making in response to possible conflicts over offshore wind turbine locations. The main purpose of this study is to determine the location of potential offshore wind turbine sites in the Northern Aegean Sea in Turkey by creating a geographical information systems-based marine spatial planning system for the region. A decision model has been proposed in which GIS and multi-criteria decision-making techniques are used to determine alternative areas. While the analytic hierarchy process (AHP) was utilized to identify viable sites based on several essential physical factors in this study, the Fuzzy TOPSIS approach was used to compare the Gökçeada, Bozcaada, and Enez study regions and find the optimal area. In the final stage, the areas where offshore wind turbine's locations are calculated are energy and exergy efficiency.

DOI: 10.4018/978-1-6684-7303-0.ch005

INTRODUCTION

According to U.S. Energy Information Administration (2013) the use of fossil fuels such as oil, natural gas and coal, which meet a large part of the energy demand, has increased significantly in the last century (Ritchie et al., 2020). However, using these fuels increase the average world temperature, reduces the ozone layer and pollutes nature and the environment (Rosati, 2017). The climate crisis caused by these impacts has created significant environmental problems in the world (Shivanna, 2022). Increases in the accumulation of greenhouse gases in the atmosphere due to human activities cause global warming. Since greenhouse gases come predominantly from energy production, where energy comes from and how it is consumed has become a critical issue in climate policy (IPCC, 2018). The energy sector, which causes two-thirds of global greenhouse gas emissions, plays a crucial role in achieving climate protection targets. Sengul et al. (2014) have stated that the environmental impact and exhaustibility of fossil fuels have led people to move to renewable energy sources that are environmentally friendly and do not have a limited lifespan. These are natural sources of energy, the source of which does not depend on any substance, and most of them are formed by the deformation of solar energy (Holechek et al., 2022).

British Petroleum (2020) reported that the most significant increase in the desire to use renewable energy globally, which has grown significantly, includes wind energy, hydropower, solar energy, geothermal energy, hydrogen energy, wave energy, and biomass energy. Wind energy, which stands out from the European Union (EU) perspective, is an essential resource with its environmentally friendly nature and sustainability. In recent years, due to the high potential of wind energy systems, it has been the subject of many research studies in wind turbine energy, exergy, and economic analysis.

Regarding the determinations for the current status, potential and future availability of wind energy worldwide, 2020 has been the best year for the global wind industry, with 53% annual growth and 93 GW of new capacity installed. As of today, 743 GW of wind energy capacity is being used worldwide. New annual installations are expected to exceed 20 GW in 2025 and 30 GW in 2030, with the global offshore wind market growing at an average annual growth rate of 18.6% by 2024 and 8.2% by the end of the decade. Following "World Forum Offshore Wind" (2020) in 2019, with offshore wind turbines accounting for 10% of global new wind power installations, over 205 GW of new offshore wind capacity was expected to be added to the global offshore wind sector over the next decade.

In Europe, the world's largest regional offshore wind market, 450 GW is foreseen in the 2050 plan for offshore wind turbines, and this capacity is expected to meet 30% of Europe's electricity. 85% of the 450 GW target in 2050 belongs to the North Sea. In 2050, the EU is expected to maintain steady growth. However, new construction outside of Europe, mainly from China and North America, are anticipated to outnumber those in Europe by 2030. The forecast is that in the short term (2020-2024), China and Taiwan will account for the majority of growth outside of Europe; by 2024, the United States is expected to participate in this growth with the introduction of the Offshore Project (WFO, 2021).International Renewable Energy Agency (IRENA) Year 2020 Renewable Energy Statistics Report indicated that the total renewable energy production capacity in Turkey in 2010, was 17369 megawatts (MW), whereas 44587 MW of this value in 2019. According to the "World Energy Statistics Report" (2020), Turkey met 45.3 terawatt-hours (TWh) of renewable energy sources in 2019. The growth rate in electricity generation has reached 20%. In 2019, 21.7 TWh of electricity was produced by wind energy in Turkey. In this direction, when we look at the use of renewable energy sources; wind energy is in the first place, geothermal energy and biomass energy are in the second place with 12.7 TWh of electricity production, and solar energy is in the third place with a total output of 10.9 TWh (British Petroleum, 2020).

Turkey's wind energy potential brings the need for an unlimited and sustainable wind energy supply to reduce foreign dependency on energy and is a candidate for EU membership. Wind energy turbines installed and operated in recent years are accepted as an alternative energy source in many countries. Ilkiliç and Aydin (2015) acknowledged that wind energy has great potential among all renewable energy sources due to Turkey's geographical location. In terms of total installed capacity, Turkey is one of the leading six countries in Europe and among the top twenty countries worldwide (Cali et al., 2018). According to Argin et al. (2019), the goal is to reach 20 GW of installed wind generating capacity by 2023 (TUREB, 2021).

In light of the rapid growth in the use of renewable energy worldwide, wind energy is showing significant growth. From the perspective of preserving the environment, wind energy is an essential resource that is environmentally friendly. (Jabber, S., 2014). Considering the Aegean Sea's geopolitical significance, spatial analysis is thought to be the key to selecting a wind farm site. In this study, the primary objective is to assess the suitability of potential wind turbine sites in the northern Aegean Sea area of Turkey and to determine the suitability areas to develop a marine spatial planning system using Geographic Information Systems (GIS). The use of GIS and multi-criteria decision analysis (MCDA) techniques together has been proposed as a decision model. In this study, while the analytical hierarchy process (AHP) has been used to evaluate the weight of the criteria related to the appropriate locations for wind turbine sites, the Fuzzy TOPSIS method has been employed to rank the studied areas including Gökçeada, Bozcaada, and Enez in order to determine which area is suitable for wind turbine development. Bozcaada and Gökçeada were selected due to their high wind speed capacity, close proximity to the mainland, and compliance with the study's various criteria. It is, however, a costly process to deploy offshore wind turbines and put the system into operation.

Consequently, in order to establish wind power plants in the off-shore area, it is necessary to ensure the efficiency of the energy generated. A final analysis of the energy and exergy efficiency of the selected sample turbine has been performed. Meteorological parameters derived from the five-year period and data from the determined stations were used for calculating efficiencies.

Description of the Problem

Turkey is covered by sea on three sides and therefore has a great advantage for offshore wind farms. The identification of areas with high potential for offshore wind turbines is a necessity. A multi-faceted spatial information system is a suitable method that has been used in this research to evaluate and study for proper location and find the best result. Currently, there is no offshore wind farm project realized in Turkey.

As a result, in light of the current environmental conditions, it is necessary to discuss, study, and design offshore wind energy extraction projects to maximize energy efficiency. Furthermore, it is critical to ensure that the project does not harm communities, potential harm to marine life, restrictions on environmental protection laws, and considerations of political and legal tensions in interactions with the country's neighbours when locating and installing offshore wind farms. This research aims to use a GIS-based Marine Spatial Planning System to identify regions for Offshore Wind Power Plants in the North Aegean Sea, where average wind potential is high, to gain more from renewable energy Turkey and to assess energy efficiency.

The project's main purpose is to find potential wind turbine locations in the North Aegean Sea area by excluding restricted regions from the study area and constructing a North Aegean Marine Spatial

Information System to achieve this goal; to set a decision model in which GIS and multi-criteria decision-making techniques are used together for identifying potential regions to establish wind farms in the offshore area, by evaluating the energy efficiency obtained from the selected area.

In this context, the secondary purpose of this study is to reveal the energy and exergy potential of the study area using the ideal wind turbine model and key variables, including wind speed, temperature and relative humidity.

Within the scope of the study, the criteria and constraints that are the basis for the evaluation of the wind power plants located in the North Aegean Sea have been utilized and applied in a multi-faceted manner in the context of marine spatial planning. Bathymetry, wind speed, distance to shipping routes, distance to underwater cables and pipelines, distance to residential areas, distance to shore, proximity to ports, distance to transformer centres, anchorage area, distance to a fault line, distance to shipwrecks and mine sites, training and the military regions, diving prohibited areas and Protection Area were determined as evaluation and restriction criteria. AHP was used to determine the criterion weights, and the Fuzzy TOPSIS method was used to rank the potential areas obtained regarding AHP results. The system's energy efficiency analysis was carried out using the meteorological parameters of the relevant stations on the turbine model selected for the potential wind turbine areas.

WIND ENERGY AND ENERGY/EXERGY EFFICIENCY

Definition of Energy and Exergy

Energy is often defined as the ability to do work. Max Planck describes energy as "the ability of a system to produce activity outside of itself". Energy can occur or be used in different forms. Electrical, thermal, mechanical, chemical, kinetic, magnetic, and nuclear energy are the most common. While doing work, a system may use some or all of these energy types. All these energies contained in the system are called total energy (Karakoc et al., 2011). The maximum available energy supplied from the system in each state is defined as exergy. The ultimate work obtained from a system, which is considered exergy, happens if the system can move to the equilibrium condition with the surrounding state (dead state) through a reversible process from a particular initial state. In the reversible process, a state change can be performed in the opposite direction after occurring in one direction, without leaving any environmental traces. The resulting work is called usability or exergy. This definition sets the upper limit of the work that a machine can do without violating the laws of thermodynamics.

According to Kurtaran (2019), the total exergy equals the sum of its kinetic, potential, thermal and chemical exergy. Ikbal (2010) argued that it causes warming in the earth's atmosphere with the rays coming from the sun with wind energy. When the air is heated, the air whose density decreases rises and is replaced by cold air. This air current causes the formation of air movements that carry kinetic energy to a large extent as the earth rotates around itself. Wind energy is the motion (kinetic) energy of the airflow that creates the wind. Wind turbines are systems that convert the kinetic energy of the air in motion into mechanical energy. Electricity is produced in the generator, which is connected to the turbine's blade arrangement. The turbine blade function converts the kinetic energy in the wind current into mechanical energy and rotates the generator. Average wind and hourly wind speeds for the places where wind farms will be established are mainly obtained from meteorological stations.

Energy and Exergy Analysis

Karakoc et al. (2011) stated that this analysis is based on thermodynamics' first and second laws to determine how energy and exergy are distributed. Based on thermodynamic equations to express the concept of equilibrium in a process in quantitative terms, the sum of the difference between the net input and output, along with the difference produced and consumed within the system, must be calculated to obtain the accumulation rate a process. Analysis of energy and exergy balance through the relevant equations can lead to a clear picture of potential capacities, which improves the performance of wind turbines.

Kareem (2018) examined wind energy in Iraq based on the second law of thermodynamics. Bernoulli's equation was applied to determine the relationship between air pressure at different altitudes and different regions. According to the results of the research, air pressure and wind speed have inverse relationships, and pressure decreases with increased height, which affects wind speed.

According to Shahin et al. (2006), Dincer and Marc (2007), the dynamic pressure of wind when impacting a wind turbine is greater than the static pressure, and the speed decreases. To calculate the dynamic pressure and kinetic energy of wind flow at a constant height, the Bernoulli equation can be used. Moreover, this study demonstrated that Bernoulli's equation can be applied to calculate entropy. This is because it is dependent on variables such as the system's input and output pressures as well as temperature.

Due to the positive thrust generated by the wind turbine, downstream wind speeds are lower than upstream wind speeds. The Bernoulli equation can be utilized to calculate the control volumes on both sides of the turbine rotor since work is not performed on both sides. It is assumed that the upstream and downstream pressures will be equal, and the wind speed across the turbine rotor remains constant. According to Bernoulli's equation, the sum of dynamic and static pressures is constant if friction is not considered (James et al., 2006).

In another study that extended Bernoulli's theorem to wind energy systems and assumed an isentropic flow, the reduction in cut-in speed was demonstrated by reducing the cross-sectional area. It is therefore possible for the turbine to generate energy at lower wind speeds, thereby increasing the annual energy generated by it (Luca et al., 2012).

Energy and Exergy Efficiencies

The application of the first law of thermodynamics to determine the maximum accessible energy is insufficient. To determine the maximum accessible energy in any system, an exergy analysis based on the second law of thermodynamics should be done. As a result, the system's potential to produce more efficient work in terms of energy availability is also necessary for system improvements.

It is appropriate to examine both energy and exergy efficiency to see how system performance changes during a process. Energy efficiency is the ratio of output to total input energy:

$$\eta = W\text{out} / W\text{wind} \; [14] \tag{2.1}$$

The exergy efficiency is the ratio of the total generated energy to the total input exergy:

$$\varepsilon = W\text{output}/ \; E\,\text{flow} \; [14] \tag{2.2}$$

Numerous new and renewable energy sources, including as wind, solar, hydraulic, and wave energy, are associated with meteorological variables, according to "Enerji Portali" (2020). A significant gap in understanding regarding energy investments may result if meteorological parameters of renewable energy sources are not evaluated and/or considered..

Offshore Wind Turbine Energy and Exergy Analysis

Exergy analysis of wind turbines gives an idea of increased performance rather than energy research.

The output of a wind turbine is calculated as the ratio of electricity generated to wind potential energy in the region swept by the wind turbine. As weather factors vary, the association between wind turbine efficiencies and meteorological variables produces the output of wind turbine efficiency models.

Based on the interaction between meteorological and wind turbine characteristics, meteorological parameters such as wind speed, pressure, temperature, and relative humidity have a substantial influence on wind turbine energy and efficiency. For the purpose of analyzing such efficiencies, monthly average wind, temperature, pressure and relative humidity values for Bozcaada, Gökçeada and Enez meteorology stations were collected from the Turkish State Meteorological Service. Bozcaada Damlacık (2015-2019 years), Gökçeada Uğurlu Port (2015-2019 years), Enez Breakwater (2015-2020 years) meteorological data stations; monthly average relative humidity, monthly average wind speed and monthly average temperature data were compiled.

Using meteorological data collected in each area, wind frequency diagrams were constructed using annual average monthly wind values. From 2015-2019, average wind power and speed were computed based on monthly wind power density using Equation 2.3.

$$P = \frac{1}{2}\rho V^3 d \tag{2.3}$$

The Weibull shape, scale, and Weilbull numbers are calculated using the gamma function based on the energy pattern factor. The Weibull shape and scale factor table and diagram calculate the wind probability density and cumulative function.

The energy and exergy amount of the wind was calculated based on the HTW5.0-126 wind turbine model. The technical specifications of the turbine are shown in Table 1. In summary, the turbine is selected according to the following (Hitachi, 2017).

- Turbines are suitable for offshore use where specifications, parameters and data are available for calculation.
- The turbine rated power is vital to achieve maximum capacity at different ranges of wind speed and the pattern or probability of the wind.
- The turbine diameter is an essential factor in the blade swept area, affecting power generation, especially at low wind speed.

Today, most wind turbines can generate power from the low wind speed (cut-in speed) to the very high possible wind speed (cut-out speed). The correct choice of a turbine is a function of wind quality and the project's capital cost.

Choosing a wind turbine with a low power generation can lower the project's initial cost. Still, it also necessitates the use of more turbines, which increases the project's operating and maintenance costs compared to total power generation, potentially leading to an increase in the price per kilowatt-hour for electricity produced.

Table 1. Turbine technical specifications

Parameter	Features
Output Power	5000Kw
Rotor radius	126m
Height	90m and above
Number of Wings	3
rotor direction	in the direction of the wind
Turbine output voltage	33,000 V
Cut-in speed	4 m/s
Cut-out speed	25m/s

Wind power at different speeds can be found with the following equation:

$$P = \frac{1}{2}\rho A V^3 \tag{2.4}$$

where A is the blade, Swept Area, and ρ is the air density at a given height. The following equation calculates the variation of the wind speed at different stages:

$$C_p = 4a(1 - \mathrm{a})^2 \tag{2.5}$$

The relationship between the axial induction factor and the wind speed is defined as:

$$a = V_1 - \frac{V_2}{V_1} \tag{2.6}$$

Where V1 and V2 are the wind speed at altitude and the wind speed before the turbine blade. The wind speed before and after the wind blade is equal. V2 = V3. Downstream wind speed V4 is equal to:

$$\mathrm{V4} = \mathrm{V1}(1 - 2\mathrm{a}) \tag{2.7}$$

The exergy of the wind can be found as:

$$\textit{Total exergy} = ex(\textit{Kenetic}) + ex(\textit{Potential}) + ex(\textit{Physical}) + ex(\textit{Chemical}) \tag{2.8}$$

The potential exergy in a wind turbine is negligible. The physical exergy, chemical exergy and kinetic energy are calculated to analyse the wind turbine exergy. Bernoulli's equation is applied to calculate the pressure changes upstream, before the turbine blade, after the turbine blade, and downstream. The variation of pressure can be found using the Bernoulli equation as follows:

$$p_1 + \frac{1}{2}\rho V_1^2 = p_2 + \frac{1}{2}\rho V_2^2 \quad (2.9)$$

$$p_3 + \frac{1}{2}\rho V_3^2 = p_4 + \frac{1}{2}\rho V_4^2 \quad (2.10)$$

Here, p1, p2, p3, and p4 are the wind pressures upstream, before the turbine, after the turbine, and downstream.

Wind energy and exergy efficiencies are as follows:

Energy efficiency η = E out/W (2.1)

Exergy efficiency ψ = E out/Ex (2.2)

W and Ex are total input energy and exergy, respectively, and Eout is generated energy or electricity output.

MARINE SPATIAL PLANNING

MSP is a public process that analyses the spatial and temporal distribution of human activities in marine regions and prepares to implement it in a coordinated manner to fulfil ecological, economic, and social goals and objectives set by a political process (Ehler & Douvere, 2007). Ehler and Douvere (2007) mentioned that ecosystem-based, integrated, place or area-based, flexible, participative, strategic, and predictive are the main characteristics of maritime spatial planning. To fulfil the social and economic goals for maritime regions in a clear and planned manner, the MSP process balances development demands with the principle of ecosystem protection. An effective MSP process should ensure integration across sectors and government agencies, be strategic and forward-looking, and involve active stakeholder engagement throughout the entire process.

Many states reserved marine areas for a variety of human activities, including shipping, oil and gas production, offshore energy, offshore aquaculture, and waste disposal. Nevertheless, once this planning is carried out on a sectoral or case-by-case basis, with little regard for other human activities or their impacts on the marine environment, there are two basic types of conflict. It is the presence or potential of issues such as (1) conflicts between human uses of marine areas (user conflicts), (2) human uses of marine areas jeopardizing the sustainability of ecosystem services on the environment in which they are located, and biodiversity degradation (user-user conflicts, environmental conflicts).Identifying and mapping spatial and temporal distributions, as according Lawrence et al. (2002), are critical elements of effective marine spatial planning. The ultimate outcomes of marine spatial planning are the development of harmonious utilization and the reduction of conflicts between uses. Pressures from human activities have often led to initiatives to improve the management of marine areas. A Blue Growth strategy was developed by the European Commission in September 2012. One of the potential drivers of Blue Growth is renewable energy. The authors of Ehler and Douvere (2010) emphasize that it is important to evaluate the spatial analysis and decision-making criteria prior to the installation of wind turbines on the sea.

The Importance of Geographic Information Systems in Marine Spatial Planning

In the opinion of Ehler and Douvere (2009) to realize MSP, decision-makers and/or planners need Geographical Information Systems to access data, use, archive, manage and analyze related data. In addition, there is the inclusion and processing of new data in the system at different stages of Marine Spatial Planning, which is an iterative process. Snickars and Pitkänen (2007) defined that MSP aims to minimize conflicts between other marine uses and their adverse effects by allocating space and zoning for different uses. The primary problem in MSP is balancing different ecological, economic, and social objectives. On significant ecological scales, sustainable spatial planning must address the cumulative consequences of all anthropogenic activities on the maritime environment. GIS, a system designed to gather, store, update, analyze, manage, and show various forms of geographic data, might be included in the MSP approach in this context. Heywood et al. (2006) stated that GIS is an effective tool for describing and interrogating these data and analyzing results in their spatial distribution or visualizing them in the form of thematic maps. Therefore, GIS can also evaluate management performance and obtain indicators to measure impacts on certain biophysical, socio-economic and management factors. Also, GIS allows the combining of spatial data taken and produced from different sources at different scales. According to Stelzenmüller et al. (2013) and Noble et al. (2019) advances in GIS technology have contributed to the advancement of many areas of spatial planning, including MSP, to solve environmental problems and integrate large amounts of data from different sources. Spyridonidou and Vagiona (2020) and Christidis and Law (2012) agreed that GIS is an effective tool in determining suitable areas for installing wind turbines on land or offshore. The development and application of GIS-based models can assist in resolving relevant difficulties in marine regions. As a consequence, the efficiency and profitability of offshore wind are boosted while detrimental effects are minimized.

Multi-Criteria Decision Analysis (MCDA)

Malczewski (2006) has argued that GIS-based MCDA is the most effective method for spatial positioning of wind turbines. MCDA is a technique used in making decisions for spatial problems with many criteria and data layers and is widely used in spatial planning and positioning of wind turbines. Mahdy and Bahaj (2018) stated that at its most basic level, GIS-MCDA can be expressed as a process that transforms and combines geographic data and value judgments (decision maker's preferences) to obtain information by decision. In this context, Malczewski (1999) concluded that GIS is used to view, analyze and process different geographic data in separate layers, generate new data layers and/or identify suitable lands. In contrast, MCDA is used to evaluate suitability by comparing the developed criteria of alternatives. Site selection for renewable energy is a complex multi-criterion decision making problem. The GIS-MCDA process generally defines the problem, normalizes the criteria layers, and determines the criteria weights. After these steps, regional methods.

In this way, the most suitable location for the construction of renewable energy plants will be determined according to the following steps (Malczewski & Rinner, 2015).

- **Step1: Defining the Problem**

In accordance with Shao et al. (2020) all decision-making methods begin with the definition of the problem. At this stage, alternatives are formulated, and the problem and study area are defined. This phase includes the search for conditions that require decisions.

- **Step 2: Evaluation Criteria**

Noble et al. (2019) explained that after the decision problem is determined, the evaluation criteria (targets and attributes) are determined. The second stage, the definition of evaluation criteria, is a complex and vital process involving the description and grouping of exclusion criteria that have a particular impact in assessing the suitability of a specific region for the installation of renewable power plants.

Relationships between evaluation criteria and geographical assets are expressed with maps. There are two types of benchmark maps, evaluation and restriction. The evaluation criteria map is used to evaluate the performance of alternatives. Evaluation benchmark maps are also called feature maps, thematic maps, or data layers. On the other hand, Constraint maps show the limitations on the value that attributes and decision variables can take (Shao et al. 2020).

- **Step 3: Normalization of Criteria Layers**

Malczewski and Rinner (2015) mentioned that the third stage is the normalization of the layers of the criteria. The data obtained are obtained from the actual numerical values and/or the opinions of the decision-maker. They are then normalized so that they can be used in quantitative calculations. For this purpose, metrics are classified based on the types of information available to create the maps. Several approaches such as Linear Scale Transformation, Value/Benefit function, and Fuzzy Logic can make criteria layers comparable.

- **Step 4: Determination of Criterion Weights**

Saaty (1980) defined that in the fourth stage, the weight of the criteria is determined and evaluated. The weight of each criterion is an assigned value that shows its importance compared to other considered criteria. From the perspective of the CBS-MCDA, weighting criteria are examined under two headings, global and local. International techniques include ranking, gradation, pairwise comparison and entropy approaches based on the assumption of spatial homogeneity of preferences. These methods require decision-makers to state their preferences according to evaluation criteria.

MCDA offers methods such as Analytical Hierarchy Process Method (AHP), Preference Ranking Technique by Similarity to Ideal Solution (TOPSIS) to structure decision problems and design, evaluate and priorities alternative decisions (Mahdy & Bahaj, 2018).

Analytic Hierarchy Process Method (AHP)

To cope with uncertainty in decision-making, the MCDA technique paired with fuzzy set theory is commonly utilized. Fuzzy TOPSIS is a hybrid of fuzzy set theory and the classic TOPSIS method, in which the value of the measures is represented by fuzzy triangular numbers. In general, decision-makers employ linguistic variables to evaluate divided weights of criteria and evaluations of options in the fuzzy TOPSIS technique.

The Fuzzy TOPSIS approach was used to rank the sites after the relevant areas were found in this investigation. Enez, Gökçeada South, Gökçeada East, Gökçeada West (Uurlu Harbor), and Bozcaada are the alternative areas. First, AHP can be integrated into GIS to estimate weights associated with criteria map layers. After that, the weights are combined with the criteria map layers using the weighted linear combination (WLC) rule. This approach is important for problems involving many alternatives, where pairwise comparison of other options is difficult. Second, the AHP method is used to consolidate priorities for all levels of the hierarchy structure, including the level that represents the alternatives. Relatively few alternatives can be evaluated with this approach (Malczewski, 1999, Malczewski & Rinner, 2015, Saaty, 1980).

Preference Ranking Technique According to Similarity to Fuzzy Ideal Solution (Fuzzy TOPSIS)

The MCDA technique is often combined with fuzzy set theory to deal with uncertainty in decision-making. Vagiaona et al. (2018) applied an integrated methodology to evaluate and prioritize suitable sites in Greece for sustainable offshore wind turbine development at the regional level. The study included the application of various positioning criteria (technical, spatial, economic, social and environmental) and GIS, AHP and TOPSIS were used. As a result of the entire method, a regional decision-making process can be used to plan offshore wind turbines.

Daneshvar et al., (2019) investigated wind power plant location selection in Turkey in their study. For this purpose, alternative wind farm locations have been determined for Turkey.. To achieve this aim, the TOPSIS method was used together with the intuitive fuzzy set.

Hwang & Yoon (1981) expressed that the Ranking of Preference Technique by Similarity (Relative Closeness) to the Ideal Solution (TOPSIS), is one of the widely used MCDA methods. This method is based on the principle that the chosen alternative should have a Positive Ideal Solution (PIS) from the shortest distance and a Negative Ideal Solution (NIS) from the furthest distance. Chen (2000) extended the TOPSIS method with fuzzy triangular numbers and introduced the fuzzy TOPSIS method.

STUDY AREA AND DATA

Study Area

When the Wind Energy Potential Atlas (REPA) produced for an altitude of 50 m is examined, the coastal part of the Aegean, the Marmara Sea and the southern and eastern parts of the Mediterranean have the high potential (YEGM, 2018). From the western most point of the southern boundary to the Meriç River in the north, the Aegean Sea follows the Greek coast in the west. It follows the Turkish coasts connecting the mouth of the Meriç River, which also serves as a land border between Turkey and Greece, and its eastern border is defined by the Turkish Straits Maritime Traffic Regulations as the southern border of the Dardanelles Strait, with the line connecting Turkey's Mehmetçik Cape and Kumkale Cape. It continues till it reaches the Aegean Sea's southern boundary on the eastern side. Figure 1 depicts the work area as well as the work area's limits. The overall research area is 407,060,7 hectares.

Figure 1. The study area

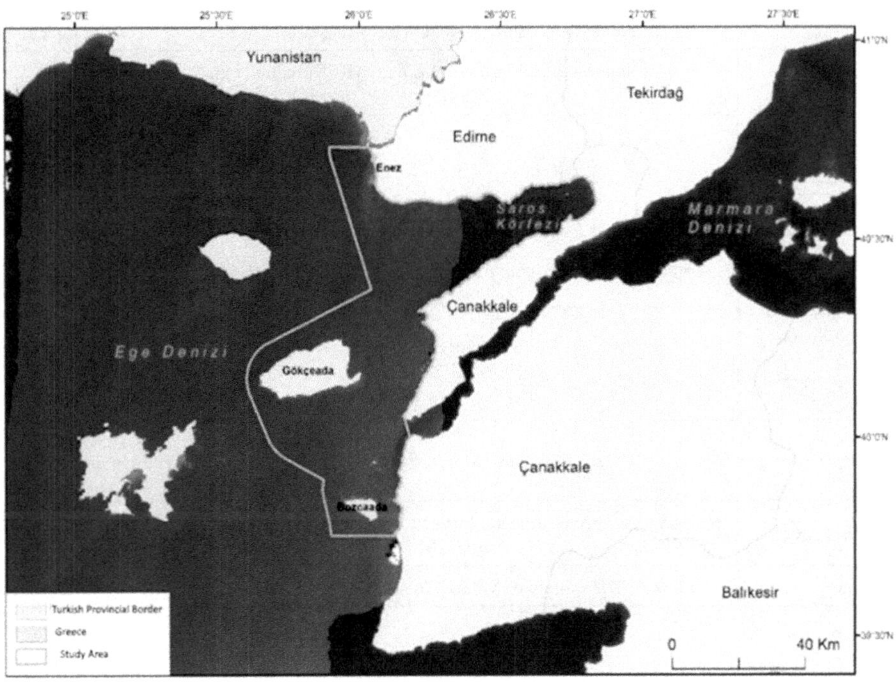

Establishment of the North Aegean Geographic Information System

Many critical studies have been carried out on the location selection of offshore wind power plants in the world and Turkey. In general, the criteria and constraints used to differ in these studies depending on the geographical location and strategic importance of the region to be studied. According to the extent and priority, the same differences appear in the analysis and the selection of value ranges. When previous studies on related subjects are examined, it is seen that most of the categorization and value assignment processes are based on expert opinions. All criteria and constraints used in this study were entered into the created database and analyzed with GIS, reaching conclusions.

In determining the criteria to select suitable locations for offshore wind turbines in the North Aegean Sea and the areas to be excluded from the analysis, current literature, legal regulations, ecological, economic and social restrictions have been evaluated. The criteria assessed in this study are bathymetry, wind speed, ship density, distance to underwater cables, distance to residential areas, distance to shore, distance to port and anchorage areas, and distance to transformer centers. Restriction criteria are determined as military training and shooting areas, protection zones, dive-prohibited areas, mined areas, shipwrecks, and anchorage areas (Table 2).

Table 2. Data included in the North Aegean Sea information system

Data	Source	MPS Coverage
Bathymetry	EMODnet Central Portal (EMODnet, 2020)	Ecological/Geographic
Ship Routes		Socio-Economic-Political
Wind speed	Global Wind Atlas[(Global Wind Atlas, 2021)_	Ecological/Geographic
Substations	Turkey Renewable Energy Wind Energy (RES) Sahin et al., (2005)	Socio-economic
Shoreline	Department of Navigation Hydrography and Oceanography	Ecological/Geographic
Residential Areas		Socio-economic
Harbor and Anchorage Areas		Socio-economic
Diving Areas		Socio-Economic/Ecological
Underwater Cables		Socio-economic
Military Training and Shooting Fields		Political
Anchorage Areas		Socio-economic
Mine Fields	Notice to Mariners (2018)	Socio-Economic/Geographical
Sunken Ships		Socio-economic
Saros Conservation Area	Saros Special Environmental Protection Area Management Plan (Saros, 2018)	Ecological
Fault Lines	General Directorate of Mineral Research and Exploration	Geographical

A GIS database was created by collecting the necessary data for the criteria and using ArcMap 10.6 software. After cutting all the obtained data to remain within the study area, the Universal Transverse Mercator (UTM) projection was transferred to the World Geodetic System (WGS) 1984 Datum, Zone 34 N coordinate system integrity data and to be able to analyze it. In this study, the Marine Spatial Planning North Aegean Sea Information System is structured under four main headings: ecological, socio-economic, geographical and political. Table 4.3 lists the data and sources included in the North Aegean Information System created with the Marine Spatial Planning perspective within the scope of this study. As the base data of this study, the mosaic Aegean image created from Landsat 8 OLI satellite images with a spatial resolution of 30 meters from 2018 was used. In addition, maps of Edirne and Çanakkale in 1/25 000 and 1/50 000 scales and Aegean maps in various scales obtained from the Department of Navigation, Hydrography and Oceanography were used.

Determining Criteria for Multi-Criteria Decision Analysis

The criteria to be considered in the stage of a suitable location for offshore wind turbines may vary according to the study area and purpose (Tortumluoğlu & Doğan, 2021; Colak et al., 2021). The related research evaluated the criteria, limits, and constraints used for offshore wind power plants, and a summary is presented in Appendix-A Table 4.4. When the studies carried out in the same sea areas were examined, it was determined that there was no standard criterion evaluation and differences in the upper and/or lower limits of the parameters. However, it is observed that the criteria excluded in the assessment, in general, are similar in many studies. Considering the relevant studies and the data obtained, the requirements are explained in the following sections. The criteria used in this study are summarized in Table 3. In the criteria selected for Bozcaada, Gökçeada and Enez regions, the characteristics of the chosen locations were considered, and the literature on the positioning of wind turbines in the North Aegean Sea was used (Argin et al., 2019; Spyridonidou & Vagiona, 2020; Colak et al., 2021; Vagiona & Karanikolas, 2012; Vasileiou et al.,2017; Spyridonidou & Dimitra, 2020; Tercan et al., 2020; Vagiona & Kamilakis, 2018; Aydin et al., 2010, Bili & Vagiona, 2018; Latinopoulos & Kechagia, 2015; Spyridonidou et al., 2020).

Table 3. Limits of the criteria used in the study and external criteria

	Criteria	Restriction Limit	Ideal Value	Situation
P1	Bathymetry	10 < x > 60 m	30m	Constraint/Criteria
P2	Wind speed	0 - 6 m/s	8 m/s	Constraint/Criteria
P3	Distance to ship routes	0 - 3000 meters	5 km	Constraint/Criteria
P4	Distance to underwater cables and pipelines	0 - 500 meters	1.5 km	Constraint/Criteria
P5	Distance to residential areas	0 - 1500 meters	3 km	Constraint/Criteria
P6	Distance to shore	0 - 1500 meters	3 km	Constraint/Criteria
P7	Proximity to ports	0 - 1500 meters	3 km	Constraint/Criteria
P8	Distance to substations	-	1.5-2 km	Criterion
P9	Anchoring area	all	-	Restriction
P10	Distance to the fault line	0 - 1500 meters	-	Restriction
P11	Distance to shipwrecks	0-500m	-	Restriction
P12	Training and shooting ranges	all	-	Restriction
P13	Diving areas	all	-	Restriction
P14	Saros Conservation zone	all	-	Restriction

Creation of the Suitability Map of the North Aegean Sea With AHP

Locating and charting optimal sites for offshore wind turbine installation, the problem definition was first made, the North Aegean Information System was created, and then the evaluation and restriction criteria described in Section 4.3 were determined. As given in Table 3, while bathymetry, wind speed, distance to ship routes, distance to residential areas, distance to shore, proximity to ports and distance to transformer centres are determined as evaluation criteria, distance to underwater cables and pipelines, anchorage area, distance to the fault line, shipwrecks are determined. Distance, training and shooting areas, no-diving areas and the area defined as Saros Protection area are included in the study as restriction areas. In addition, unsuitable values for bathymetry, wind speed, ship routes, distance to residential areas, distance to shore and proximity to ports were also defined as restriction areas.

Preparation and Classification of Data

Pixel values within the constraint values for bathymetry and wind speed criteria were extracted from these data with the feature extraction (spatial analysis) tool. Firstly, a raster data layer was created by using the Euclidean distance (spatial analysis) tool to extract the restriction limits determined for the criteria of distance to ship routes, distance to underwater cables and pipelines, distance to residential areas, distance to shore, proximity to ports, distance to fault lines and shipwrecks. Then, the relevant constraint limits were extracted from these data with the feature extraction (spatial analysis) tool.

All the restraint data obtained were converted to vector data and then *combined* with the data of mooring areas, training and shooting ranges, no-dive areas and Saros Protected Area data using the *merge* tool to form a single restriction data layer. During the classification of the data, values in the range of 1 (least suitable) – 4 (most convenient) were assigned for each criterion, based on the ideal values given in Table 4.5 and the minimum/maximum values of the data (Appendix-A Table 4). All data were reclassified and made ready for analysis using the *reclassification (spatial analysis)* tool, based on Appendix-A Table 5.1 for the classification of the data to create the suitability map.

Table 4. Values determined for grading criteria

	Min-Max	1 (Least Suitable)	2 (Less Suitable)	3 (Available)	4 (Most Suitable)
Bathymetry (m)	-10 - -60	-60 – -50	-50 - -40	-40 - -30	-30- -10
Wind speed (m/s)	6.0011 - 9.88	6-6.5	6.5-7.5	7.5-8	over 8
Distance to ship routes (km)	3 - 68,105	3 - 15	15 - 30	30 - 45	45 - 68.105
Distance to residential areas (km)	1.5 - 27.5	1.5 - 2	2 - 2.5	2.5 - 3	3 - 27.5
Distance to shore (km)	1.5 - 20,342	1.5 - 2	2 - 2.5	2.5 - 3	3 - 20,342
Proximity to ports (km)	1.5 - 35.002	1.5 - 2	2 - 2.5	2.5 - 3	3 - 35.002
Distance to substations (km)	0.948 - 53.243	53,243 - 30	20-30	11- 20	0.948 - 11

Determination of Weights with the Analytical Hierarchy Method

AHP, one of the multi-criteria decision-making methods, was used to determine the weights of the criteria. After determining the requirements, pairwise comparisons are made in this method, and the consistency ratio is calculated. If the consistency ratio is within the acceptable limit, the criterion weights are determined. Table 5.3 shows the results of the calculated weight for each criterion.

Table 5. Criterion weights

Criteria	Weight
Bathymetry	0.27
Wind speed	0.415
Distance to ship routes	0.120
Distance to residential areas	0.027
Distance to shore	0.090
Proximity to ports	0.041
Distance to substations	0.050

Creation of Suitability Map

The thematic maps of the classified evaluation criteria and the weights generated by the AHP approach were combined to create an appropriate map of the sites where offshore wind turbines can be built. As a result of this procedure, a single data layer was created that could be used to rate potential offshore wind turbine sites. Then, using the Natural Range (Jenk's) data classification approach, the raster data layer was divided into four classes: least suitable, less suitable, appropriate, and highly suitable (Figure 2).

Sequencing of Eligible Fields With Fuzzy TOPSIS

The difficulty of choosing a wind power plant site that takes into account complex criteria is frequently solved using MCDA techniques. Fuzzy TOPSIS is the approach that is most frequently used to rank alternatives according to criteria. The option selected is the one that is the furthest from the negative ideal solution and the one that is the most similar to the positive ideal solution.

Based on Saaty (1977,1980), Khan et al. (2021) explained that the Consistency Ratio (CR) estimates probability value by dividing Consistency Index(CI) by Random Index (RI). According to Saaty (1980), the consistency index depends on the lambda max from the pairwise criteria comparison matrix, as well as the number of criteria in this study. Lambda can be determined for each of the criteria by dividing the weight of the criteria by the sum of the weighted criteria in each row of the matrix. Lambda max can be derived by dividing the sum of the lambda by the number of criteria. According to Ammarpala et al. (2018), Saaty (1980) provided the Random Index (RI) value in the form of a table depending on the size of the matrix. Consequently, the consistency ratio can be calculated by calculating the consistency index and selecting the random index based on the number of criteria in the random index table.

To cope with uncertainty in decision-making, the MCDA technique paired with fuzzy set theory is commonly utilized. Fuzzy TOPSIS is a hybrid of fuzzy set theory and the classic TOPSIS method, in which the value of the measures is represented by fuzzy triangular numbers. In general, decision-makers employ linguistic variables to evaluate divided weights of criteria and evaluations of options in the fuzzy TOPSIS technique.

Figure 2. Suitability map

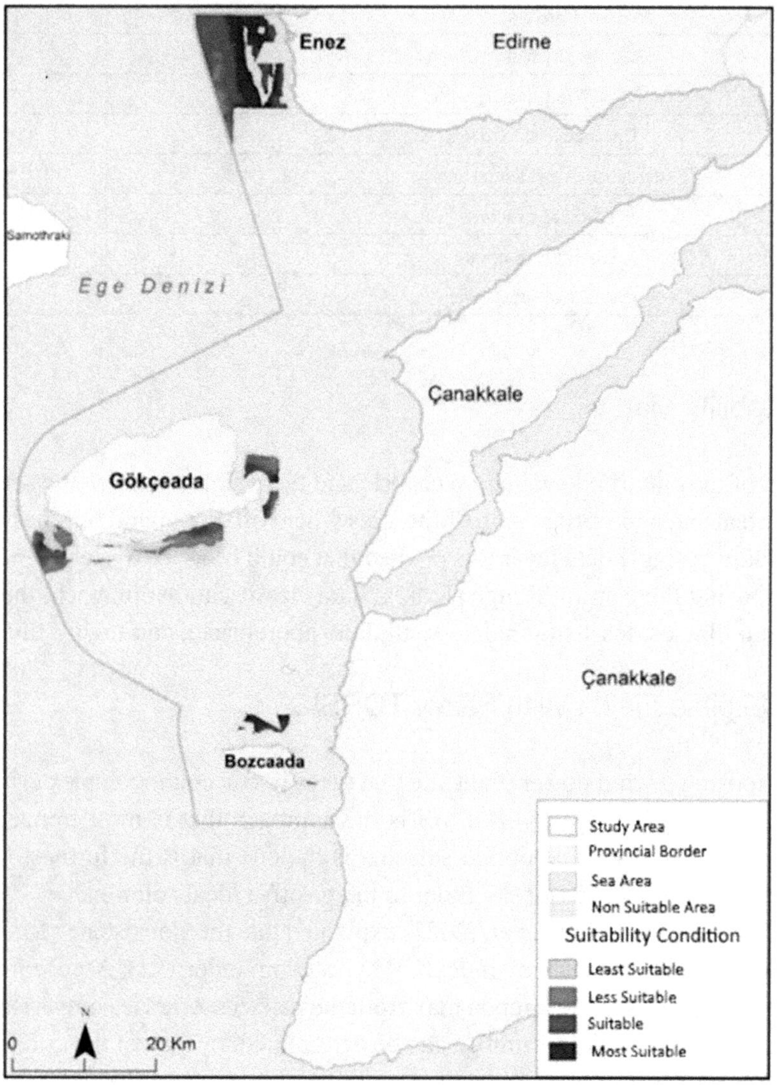

The Fuzzy TOPSIS approach was used to rank the sites after the relevant areas were found in this investigation. Enez, Gökçeada South, Gökçeada East, Gökçeada West (Ugurlu Harbor), and Bozcaada are the alternative areas.The criteria were defined as the seven criteria used in the AHP analysis. The Fuzzy TOPSIS method was applied to calculate the closeness coefficient (CC_i) for each alternative and to rank the other options (Table 6).

Table 6. The closeness coefficient values calculated for the alternatives and their rankings

	CC_i	Normalise CC_i	Ranking
Enez	0.62	0.206	2
Gökçeada Uğurlu Port	0.49	0.163	4
Gökçeada South	0.50	0.166	3
Gökçeada East	0.48	0.162	5
Bozcaada	0.90	0.30	1

Energy and Exergy Efficiency

Offshore Wind Turbine Energy and Exergy Analysis Application Results

The data required for this study for the Bozcaada, Gökçeada and Enez regions were collected from the Turkish State Meteorological Service. Mean monthly relative humidity, mean monthly wind speed and mean monthly temperature were used as baseline data to analyze energy and exergy efficiency. Wind frequency values were calculated based on the annual average of monthly wind values of the data to determine the graph of wind frequency changes in each region. Then, calculations related to Weibull distribution were performed, and the values of Weibull shape factor, Weibull scale factor, wind probability density function and cumulative function were determined. After calculating the wind blade speeds, wind exergy was calculated. Wind energy and exergy efficiencies were found in the last step by applying the efficiency formulas (2.1) and (2.2) explained in Section 2.

Location 1: Bozcaada

Available meteorological data and selected characteristics of wind turbines are used to calculate energy and exergy efficiencies. In Appendix-B, Figure 3, the wind frequency diagram for Bozcaada was calculated based on the average monthly wind values for the years 2015-2019. Monthly average wind power and speed are produced between 2015-2019 and are shown in Appendix-B Figure 4. The results of the energy efficiency and exergy efficiency analysis are shown in Appendix-B, Figure 5. The best energy and exergy efficiency occur in the wind speed range between 5 m/s and 9 m/s. The amount of the generated energy at the wind speed of 9 m/s, shows a decreasing trend with a slight slope (Appendix-B Figure 6). The total energy and exergy results calculated in this area show that the total (potential) kinetic energy is 111.71 GWh, the total energy (considering rated power) is 19.62 GWh, the total generated energy is 16.24 GWh, and the total exergy is 75.66 GWh.

Location 2: Gokceada

Appendix C Figure 7 shows the wind frequency diagram for Gökçeada, calculated based on the monthly average of wind values in 2015-2019. Between 2015-2019, monthly average wind power and speed were established and shown in Appendix-C Figure 8. The energy efficiency and exergy efficiency analysis results are shown in Appendix-C Figure 9, where the best energy efficiency and exergy efficiency occurs in the wind speed range between 5 m/s and 10 m/s. The amount of energy produced, 10 m/s, shows a decreasing trend with a slight slope. Appendix C Figure 10 shows these changes. The total energy and exergy results calculated in this area show that the total (potential) kinetic energy is 6.55 GWh. The total energy (considering rated power) is 5.09 GWh. The total generated energy is 4.21 GWh, and the total exergy is 5.47 GWh.

Location 3: Enez

Appendix-D Figure 11 illustrates the wind frequency diagram achieved for Enez based on the average monthly wind values from 2015-2020. Between 2015-2020, monthly average wind power and speed were created and shown in Appendix-D Figure 12. The energy efficiency and exergy efficiency analysis results are shown in Appendix-D Figure 13, where the best energy efficiency and exergy efficiency occurs in the wind speed range between 4.5 m/s and 10 m/s. The amount of energy produced, 10 m/s, shows a decreasing trend with a slight slope. Appendix-D Figure 14 illustrates these changes. The total energy and exergy results calculated in this area show that the total (potential) kinetic energy is 19.22 GWh, the total energy (considering rated power) is 18.95 GWh, the total generated energy is 15.68 GWh, and the total exergy is 17.44 GWh.

Energy calculations show that the current wind energy in Bozcaada, Gökçeada and Enez are 19.62, 5.09 and 18.95 GW, respectively. However, due to all losses in the system, including technical and transportation, the amount of the generated energy in Bozcaada, Gökçeada and Enez are 16.24, 4.21 and 15.68 GWh, respectively. In addition, total exergy in Bozcaada is 75.66 GWh, in Gökçeada is 5.47 GWh and in Enez is 17.44 GWh.

Comparative results for exergy efficiency at all three sites indicate a higher energy efficiency than exergy efficiency at wind speeds of 5 to about 12 m/s in Bozcaada. In the same situation, the exergy efficiency is the opposite of the energy efficiency in the other two stations, and the exergy efficiency is higher than the energy efficiency. Exergy efficiency is more elevated at wind speeds above 12 m/s in all positions.

CONCLUSION

In this study, to determine the location and energy-exergy efficiency for establishing Offshore Wind Turbines in the North Aegean, the information system was created by considering the ecological, socio-economic, political and geographical features. This perspective provides a holistic and versatile assessment with a multidisciplinary perspective, regardless of a criterion, constraint or any sector and/or professional discipline.

A total of fourteen criteria were used in this study. Of these criteria, Bathymetry, wind speed, distance to shipping routes, distance to underwater cables and pipelines, distance to residential areas, distance to shore, proximity to ports and distance to transformer centres were used as criteria, anchorage area, distance to the fault line, distance to shipwrecks, training and shooting locations, diving prohibited areas and Saros Conservation zone are considered as restrictions. Fisheries and tourism data were not used as a criterion in the study since reliable data could not be provided. After the literature study, the differences in the criteria and limits used in offshore turbine positioning in different geographical areas were determined, and the criteria suitable for the study were evaluated.

As a result, this study was carried out in four stages, the regions suitable for the installation of offshore wind power plants for the North Aegean Sea were defined as Enez, Gökçeada South region, Gökçeada East region, Gökçeada West region and Bozcaada. A suitability map was created with AHP, and suitable areas were determined. The Fuzzy TOPSIS method was applied to determine that the relevant areas are superior to the others and to evaluate the weights of the criteria and the ratings of the alternatives. According to the closeness coefficient values calculated for the alternatives, the alternatives are respectively; It was obtained as Bozcaada, Enez, Gökçeada South, Gökçeada Uğurlu Harbor and Gökçeada East.

To compare the superiority of these areas to each other, the meteorological parameters and wind power in the area where the offshore wind turbines will be located and the energy and exergy efficiency of the system has been calculated. This calculation shows how much more work can be obtained from the system and how efficient the system is according to the given energy. The relationships between the efficiency of the wind turbine and the meteorological variables are shown using the five-year average monthly meteorological variables. The efficiency performance of wind energy systems has been evaluated if the HTW5.0-126 wind turbine model study areas are offshore. When the total energy and exergy and the total energy produced in the turbine are calculated; For Bozcaada, Gökçeada and Enez, energy and exergy efficiency have been revealed in a specific wind speed range. The most crucial point to be emphasized in this calculation is the relationship between wind turbine efficiencies and meteorological variables. Since the most critical parameter, wind power, is proportional to the cubic wind speed, the variance in energy and exergy efficiency more closely reflects the variability in wind speed than the other three meteorological variables. In addition, the height above sea level where the turbine blades will be located is critical for wind power to be obtained. The diameter of the turbine is also an essential factor in terms of the swept area of the turbine blade, which affects power generation, especially at low wind speed. This factor is the most significant advantage of offshore wind turbines over onshore wind turbines. The blade diameter of turbines located on land cannot be as wide as the blade diameter of offshore turbines due to environmental and habitat restrictions. In addition, since the roughness of the turbines placed in the open sea areas is close to zero, the wind reaching the turbines directly hits the blades, increasing the resulting wind power. According to the results of the total energy and exergy of the study areas and the total generated energy of the turbine, in order of efficiency, Bozcaada, Enez and Gökçeada.

Bozcaada study area is in a very advantageous position. It is the most vital area in terms of wind power, close to the mainland, ideal distance to the port, far from the earthquake fault line, and close to the Ezine transformer centre. However, the suitable area of Bozcaada was found only in the northern sea area of Bozcaada, mainly due to the Çanakkale Strait traffic in the western and eastern parts and the fact that the south, west and northwestern parts of Bozcaada are the anchorage areas. Nevertheless, Bozcaada is the ideal area in this study since it is a sea area that receives a lot of wind in the northern region of Bozcaada, and the wind parameter is the highest weight criterion in the criteria. Bathymetry, being close to the mainland and the port, are other advantages of Bozcaada.

Gokceada wind power; compared to Bozcaada's wind power, it can be easily stated that Gökçeada is much weaker than Bozcaada, according to the five-year meteorological data obtained and long-term measurements in the literature study. However, according to the energy and exergy calculations made, it is seen that efficiency will be achieved when offshore wind turbines are installed. Criteria for the comparison of suitable areas in the three regions of Gökçeada; bathymetry is the distance to the ship routes, the distance to the urban areas, the proximity to the ports and the distance to the transformer centres. The fault passing through the island's northern coast is the continuation of the North Anatolian Fault in the Aegean Sea. The fault line in the northern part of Gökçeada has been observed. Gökçeada's southern coast is not eroded by severe waves due to meteorological-oceanographic features, while there is a slightly sloping, flat coastal shape. This makes the south of Gökçeada more suitable than other areas. Another important factor is the existence of Gökçeada Marine Park. Therefore, the southeast direction of Gökçeada is partially excluded from the study. The suitability map in Gökçeada was examined as three regions. The south of Gökçeada is the ideal region of the island. The western part of Gökçeada is close to the ship traffic towards the Dardanelles Strait and is far from the mainland transformer centre, which has created a disadvantageous situation. There is also an anchorage area, albeit a small one, but the presence of Uğurlu Port in its immediate vicinity has created an advantageous situation. In the east of Gökçeada, while proximity to the mainland transformer centre is an advantage, being close to the mainland - island ferry line very close to the port is a disadvantage. In addition, there are underwater cable routes from the mainland towards the eastern application area of Gökçeada. It was concluded that the southern area of Gökçeada is the most suitable. It is away from ship routes and anchorage areas. It is close to the mainland transformer centre compared to the west of Gökçeada. The distance to the coast and living areas is also within ideal limits. The appropriate area, which emerges as a long line parallel to the shore compared to other areas of the island, is important in the placement of the turbines. However, it should be noted that Gökçeada is the area with the weakest wind power among. Choosing wide-bladed models in turbine selection can eliminate this disadvantage.

The Enez region came in second in terms of efficiency and convenience. The main reason for this is wind power. There are many restrictive criteria in the Enez region. Enez is an important region in terms of sea tourism and fishing. It is also within the scope of "land military forbidden zones" and first and second degree "marine military forbidden zones". Gala Lake National Park, declared as Turkey's 36th National Park is located here. Located in the Meriç Delta, it has the characteristics of a wetland that is included in the lists of Turkey's important natural areas, important bird areas and important plant areas and meets the Ramsar criteria. After excluding the above protected and prohibited areas from the study area, which is not very close to the Enez port, was considered a table area. Enez is more advantageous than Gökçeada in terms of wind power and bathymetry. However, being close to the ship routes, Enez port and living areas and being far from the transformer centre are the disadvantageous criteria of Enez.

This study, which aims to reveal the location selection of offshore wind turbines and the current wind power potential to and from offshore wind turbines with the North Aegean Sea Information System, was created with a marine spatial planning perspective and a multidisciplinary perspective. In today's world, where data production speed is breakneck, data repetition should be minimal so that high data standards and quality. In this regard, the management, updating, storage and analysis of data in the context of space and time concepts of GIS, which was created with the principles of marine spatial planning, is of great importance for the effective use of offshore wind energy studies.

When evaluated from this point of view, it is imperative strategically, economically and ecologically that Turkey prepares a comprehensive, integrative and future-oriented marine spatial planning that brings together all stakeholders and sectors. It should be stated that it is possible to benefit from the resources in our seas in a safe, sustainable, environmentally friendly and effective manner, and revealing the current wind power potential of our country is beneficial for the use of local and clean energy resources.

REFERENCES

Ammarapala, V., Chinda, T., Pongsayaporn, P., Ratanachot, W., Punthutaecha, K., & Janmonta, K. (2018). Cross-border shipment route selection utilizing analytic hierarchy process (AHP) method. *Songklanakarin Journal of Science and Technology*, *40*. Advance online publication. doi:10.14456jst-psu.2018.3

Argin, M., Yerci, V., Erdogan, N., Kucuksari, S., & Cali, U. (2019). Exploring the offshore wind energy potential of Turkey based on multi-criteria site selection. *Energy Strategy Rev.*, *23*, 33–46. doi:10.1016/j.esr.2018.12.005

Aydin, N. Y., Kentel, E., & Duzgun, S. (2010). GIS-based environmental assessment of wind energy systems for spatial planning: A case study from Western Turkey. *Renewable & Sustainable Energy Reviews*, *14*(1), 364–373. doi:10.1016/j.rser.2009.07.023

Bili, A., & Vagiona, D. G. (2018). Use of multi-criteria analysis and GIS for selecting sites for onshore wind farms: The case of Andros Island (Greece). *Eur. J. Environ. Sci.*, *2018*(8), 5–13.

British Petroleum Company. (2020). *BP statistical review of world energy*. British Petroleum Co.

Cali, U., Erdogan, N., Kucuksari, E., & Argin, M. (2018). Techno-economic analysis of high potential offshore wind farm locations in Turkey. *Energy Strategy Reviews*, *22*, 325–336. doi:10.1016/j.esr.2018.10.007

Chen, C. T. (2000). Extension of the TOPSIS for group decision-making under fuzzy environment. *Fuzzy Sets and Systems*, *114*(1), 1–9. doi:10.1016/S0165-0114(97)00377-1

Christidis, T., & Law, J. (2012). Review: The use of geographic information systems in wind turbine and wind energy research. *Journal of Renewable and Sustainable Energy*, *4*(1), 012701. Advance online publication. doi:10.1063/1.3673565

Colak, T. A. I., Senel, G., & Goksel, C. (2021). GIS-based maritime spatial planning for site selection of offshore wind farms with exergy efficiency analysis: A case study. *International Journal of Exergy*, *34*(2), 255–268. doi:10.1504/IJEX.2021.113008

Daneshvar, B., Yildizbasi, A., & Arikan, Ü. (2018). *Using ıntuitionistic fuzzy TOPSIS in site selection of wind power plants in Turkey* (Vol. 2018). Advances in Fuzzy Systems.

De Luca, R., & Desideri, P. (2012). Wind energy: An application of Bernoulli's theorem generalized to isentropic flow of ideal gases. *European Journal of Physics*, *34*(1), 189–197. doi:10.1088/0143-0807/34/1/189

Dincer, I., & Rosen, M. (2007). *Exergy Analysis of Renewable Energy Systems.*, doi:10.1016/B978-008044529-8.50013-6

Ehler, C. (2018). In Offshore Energy and Marine Spatial Planning. In Marine Spatial Planning: an idea whose time has come. Routledge. doi:10.4324/9781315666877

Ehler, C., & Douvere, F. (2007). V*isions for a sea change: Report of the first international workshop on marine spatial planning*. UNESCO.

Ehler, C., & Douvere, F. (2010). An International Perspective on Marine Spatial Planning Initiatives, Environments. *Journal Volume Environments, 37*(3), 9–20.

EU Commission Maritime Affairs and Fisheries. (2018). *Annual economic report on EU blue economy*. Directorate-General for Maritime Affairs and Fisheries European Commission.

HannahR.MaxR.RosadoP. (2022). *Energy*. Retrieved from: https://ourworldindata.org/energy

Heywood, I., Cornelius, S., & Carver, S. (2006). *An Introduction to Geographical Information Systems* (3rd ed.). Pearson Education Limited.

Holechek, J. L., Geli, H. M. E., Sawalhah, M. N., & Valdez, R. (2022). A Global Assessment: Can Renewable Energy Replace Fossil Fuels by 2050? *Sustainability, 14*(8), 4792. doi:10.3390u14084792

Hwang, C. L., & Yoon, K. P. (1981). *Multiple attribute decision making: methods and applications, a state-of-the-art survey*. SpringerVerlang. doi:10.1007/978-3-642-48318-9

İkbal, F. (2010). *Rüzgâr Enerjisi Sistemlerinde Optimizasyon*. İstanbul Üniversitesi Fen Bilimleri Enstitüsü Elektrik-Elektronik Mühendisliği Anabilim Dalı.

Ilkiliç, C., & Aydin, H. (2015). Wind power potential and usage in the coastal regions of Turkey. *Renewable & Sustainable Energy Reviews, 44*, 78–86. doi:10.1016/j.rser.2014.12.010

IPCC. (2018). Global Warming of 1.5°C. An IPCC Special Report on the impacts of global warming of 1.5°C above pre-industrial levels and related global greenhouse gas emission pathways, in the context of strengthening the global response to the threat of climate change, sustainable development, and efforts to eradicate poverty. IPCC.

Kareem, F. (2018). Applied second law of thermodynamics on the wind energy over Iraq. *Journal of Engineering and Sustainable Development., 22*(02), 75–92. Advance online publication. doi:10.31272/jeasd.2018.2.20

Khan, A., Chatterjee, S., & Weng, Y. (2021). *3-UHI drivers and mapping the urban thermal environment. Urban Heat Island Modeling for Tropical Climates*. Elsevier.

Kurtaran, A. (2019). İsdemir Kuvvet Santrali Enerji ve Ekserji Analizi. Batman Üniversitesi Fen Bilimleri Enstitüsü Otomotiv Mühendisliği Anabilim Dalı, Batman.

Latinopoulos, D., & Kechagia, K. (2015). GIS-based multi-criteria evaluation for wind farm site selection. A regional scale application in Greece. *Renewable Energy, 78*, 550–560. doi:10.1016/j.renene.2015.01.041

Lawrence, D., Kenchington, R., & Woodley, S. (2002). *The Great Barrier Reef: Finding the Right Balance*. Melbourne University Press.

Mahdy, M., & Bahaj, A. S. (2018). Multi-criteria decision analysis for offshore wind energy potential in Egypt. *Renewable Energy, 118*, 278–289. doi:10.1016/j.renene.2017.11.021

Malczewski, J. (1999). *GIS and multi-criteria decision analysis*. John Wiley & Sons.

Malczewski, J. (2006). GIS-based multi-criteria decision analysis: A survey of the literature. *International Journal of Geographical Information Science, 20*(7), 703–726. doi:10.1080/13658810600661508

Malczewski, J., & Rinner, C. (2015). *Multicriteria decision analysis in geographic information science* (1st ed.). Springer-Verlag Berlin Heidelberg. doi:10.1007/978-3-540-74757-4

Manwell, J., Mcgowan, J., & Rogers, A. (2006). *Wind Energy Explained: Theory* (2nd ed.). Design and Application., doi:10.1260/030952406778055054

Noble, M. M., Harasti, D., Pittock, J., & Doran, B. (2019). Linking the social to the ecological using GIS methods in marine spatial planning and management to support resilience: A review. *Marine Policy, 108*, 103657. Advance online publication. doi:10.1016/j.marpol.2019.103657

Rossati, A. (2017, January). Global Warming and Its Health Impact. *The International Journal of Occupational and Environmental Medicine, 8*(1), 7–20. doi:10.15171/ijoem.2017.963 PMID:28051192

Saaty, T. L. (1977). Scaling Method for Priorities in Hierarchical Structures. *Journal of Mathematical Psychology, 15*(3), 234–281. doi:10.1016/0022-2496(77)90033-5

Saaty, T. L. (1980). *The analytic hierarchy process*. McGraw-Hill.

Şahin, A., Dincer, I., & Rosen, M. (2006). New Spatio-Temporal Wind Exergy Maps. Journal of Energy Resources Technology-transactions of The Asme - J Energ. *Resources and Technology, 128*(3), 194–202. Advance online publication. doi:10.1115/1.2213271

Sahin, M., Bilgili, H., & Akilli, H. (2005). The wind power potential of the eastern Mediterranean region of Turkey. *Journal of Wind Engineering and Industrial Aerodynamics, 93*(2), 171–183. doi:10.1016/j.jweia.2004.11.005

Şengül, Ü., Sibel, T., Şermin, A., & Şengül, A. (2014). Türkiye Gökçeada'da yenilenebilir enerji kaynaklarinin potansiyeli. *Akademik Araştırmalar ve Çalışmalar Dergisi, 6*(11), 41–55.

Shao, M., Han, Z., Sun, J., Xiao, C., Zhang, S., & Zhao, Y. (2020). A review of multi-criteria decision-making applications for renewable energy site selection. *Renewable Energy, 157*, 377–403. doi:10.1016/j.renene.2020.04.137

Shivanna, K. R. (2022). Climate change and its impact on biodiversity and human welfare. *Proc.Indian Natl. Sci. Acad., 88*(2), 160–171. doi:10.100743538-022-00073-6

Snickars, M., & Pitkänen, T. (2007). *GIS tools for marine planning and management: BALANCE Interim (Report No. 28)*. Interreg Baltic Sea Region. Retrieved from https://balance-eu.org/xpdf/balance-interim-report-no-28.pdf

Spyridonidou, S., & Vagiona, D. G. (2020). Systematic Review of Site-Selection Processes in Onshore and Offshore Wind Energy Research. *Energies, 13*(22), 5906. Advance online publication. doi:10.3390/en13225906

Spyridonidou, S., Vagiona, D. G., & Loukogeorgaki, E. (2020). Strategic Planning of OffshoreWind Farms in Greece. *Sustainability, 12*(3), 905. doi:10.3390u12030905

Stelzenmüller, V., Lee, J., South, A., Foden, J., & Rogers, S. I. (2013). Practical tools to support marine spatial planning: A review and some prototype tools. *Marine Policy, 38*, 214–227. doi:10.1016/j.marpol.2012.05.038

Tercan, E., Tapkın, S., Latinopoulos, D., Dereli, M., Tsiropoulos, A., & Ak, A. (2020). A GIS-based multi-criteria model for offshore wind energy power plants site selection in both sides of the Aegean Sea. *Environmental Monitoring and Assessment, 192*(10), 652. doi:10.100710661-020-08603-9 PMID:32964332

TUREB. (2021). *Türkiye Rüzgar Enerjisi İstatistik Raporu.* Türkiye Rüzgar Enerjisi Birliği.

U.S. Energy Information Administration. (2013). *International Energy Outlook 2013 with Projections to 2040.* U.S. Energy Information Administration Office of Energy Analysis, U.S. Department of Energy.

Vagiona, D. G., & Kamilakis, M. (2018). Sustainable site selection for offshore wind farms in the South Aegean— Greece. *Sustainability, 10*(3), 749. doi:10.3390u10030749

Vasileiou, M., Loukogeorgaki, E., & Vagiona, D. G. (2017). GIS-based multi-criteria decision analysis for site selection of hybrid offshore wind and wave energy systems in Greece. *Renewable & Sustainable Energy Reviews, 73*, 745–757. doi:10.1016/j.rser.2017.01.161

WFO. (2020). *Global offshore wind report 2019.* World Forum Offshore Wind.

WFO. (2021). *Global offshore wind report 2020.* World Forum Offshore Wind.

YEGM. (2018). *Türkiye Rüzgâr Enerjisi Potansiyel Atlası.* http://www.yegm.gov.tr/YEKrepa/REPA-duyuru_01.html

APPENDIX A: VALUES DETERMINED FOR GRADING CRITERIA

Table 7. Values determined for grading criteria

	Min-Max	1 (Least Suitable)	2 (Less Suitable)	3 (Available)	4 (Most Suitable)
Bathymetry (m)	-10 - -60	-60 – -50	-50 - -40	-40 - -30	-30- -10
Wind speed (m/s)	6.0011 - 9.88	6-6.5	6.5-7.5	7.5-8	over 8
Distance to ship routes (km)	3 - 68,105	3 - 15	15 - 30	30 - 45	45 - 68.105
Distance to residential areas (km)	1.5 - 27.5	1.5 - 2	2 - 2.5	2.5 - 3	3 - 27.5
Distance to shore (km)	1.5 - 20,342	1.5 - 2	2 - 2.5	2.5 - 3	3 - 20,342
Proximity to ports (km)	1.5 - 35.002	1.5 - 2	2 - 2.5	2.5 - 3	3 - 35.002
Distance to substations (km)	0.948 - 53.243	53,243 - 30	20-30	11- 20	0.948 - 11

APPENDIX B: BOZCAADA WIND ENERGY AND EXERGY GRAPHS

Figure 3. Wind frequency diagram of Bozcaada

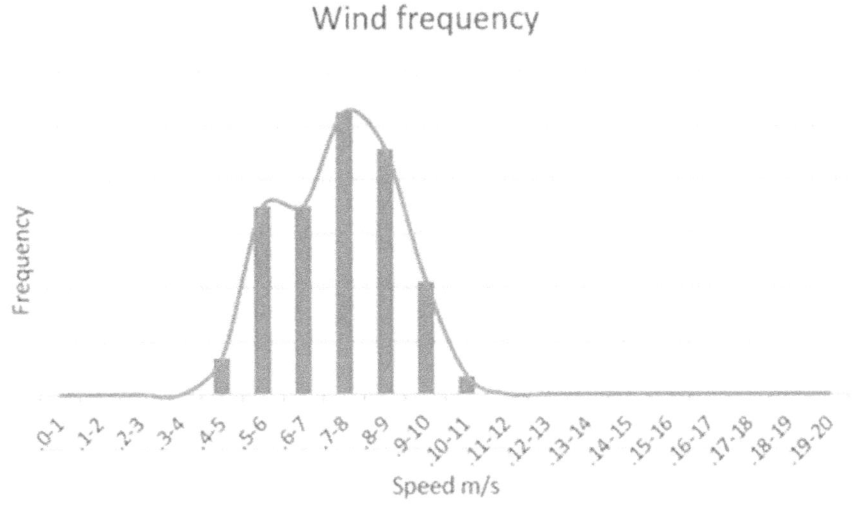

Figure 4. Bozcaada average power and speed diagram

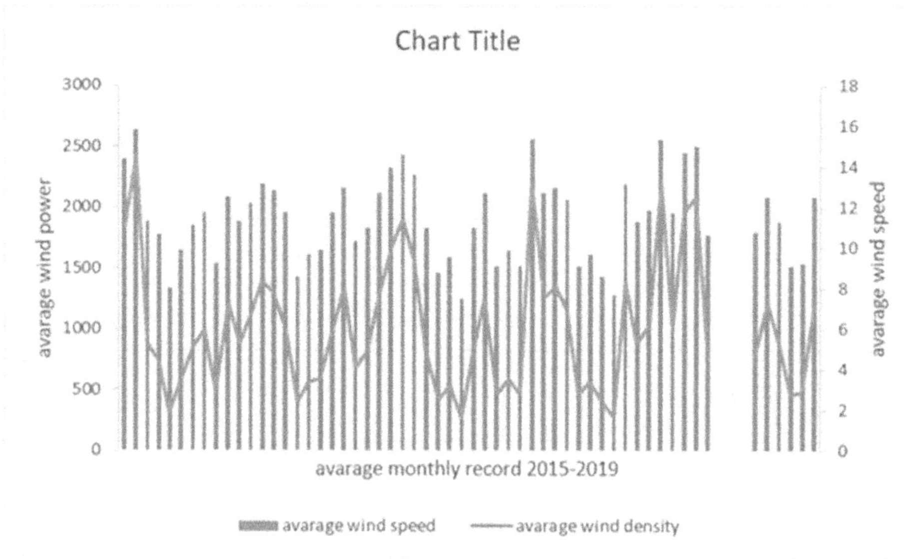

Figure 5. Energy and exergy efficiency distribution in Bozcaada

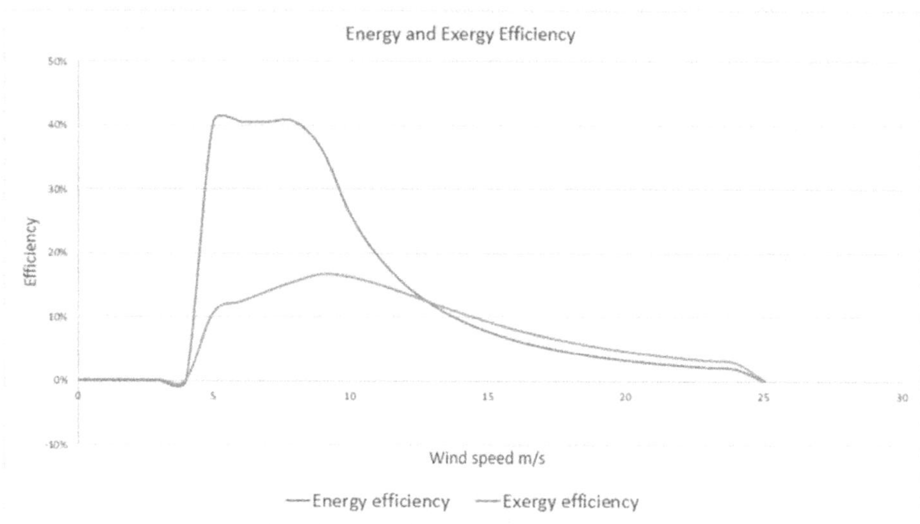

Figure 6. Wind exergy and generated energy from wind at different wind speeds in Bozcaada

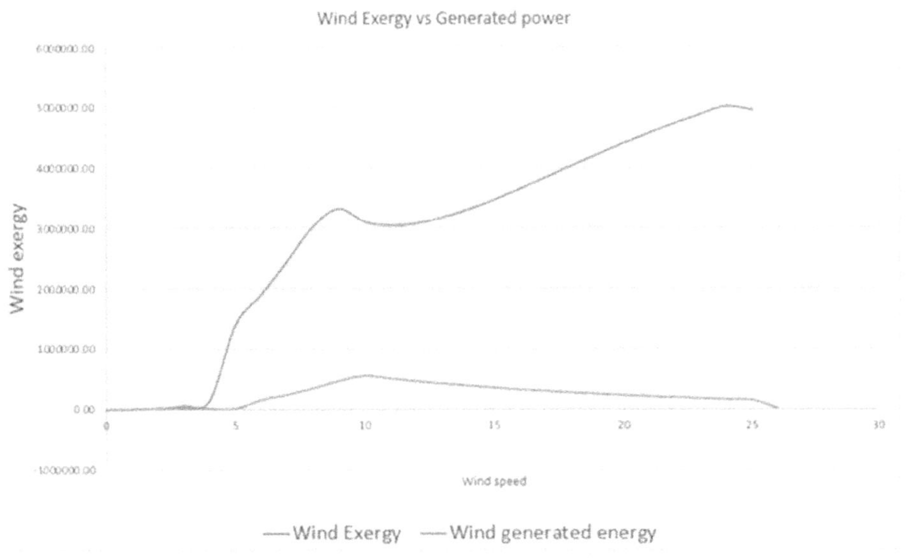

APPENDIX C: GÖKÇEADA WIND ENERGY AND EXERGY GRAPHS

Figure 7. Gökçeada wind frequency diagram

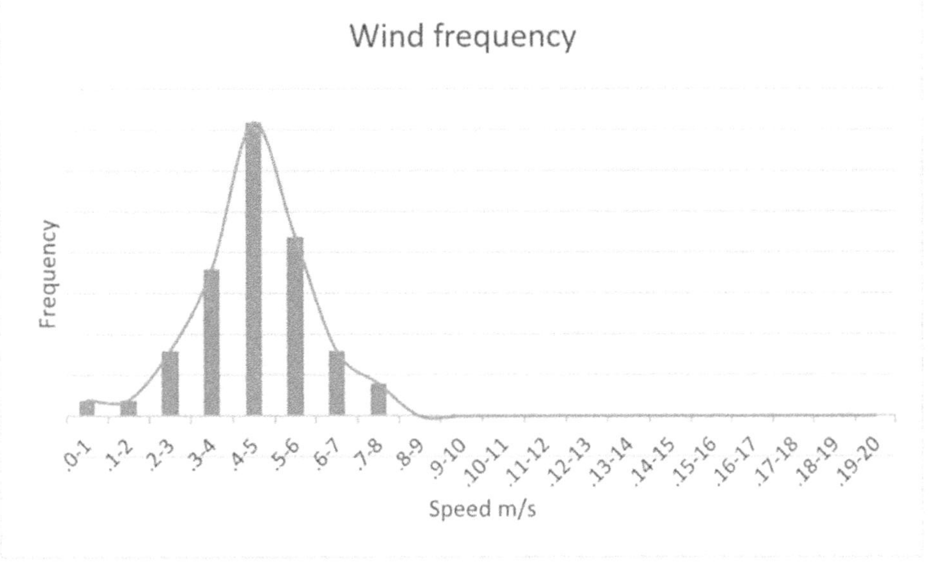

Figure 8. Gokceada average power and velocity diagram

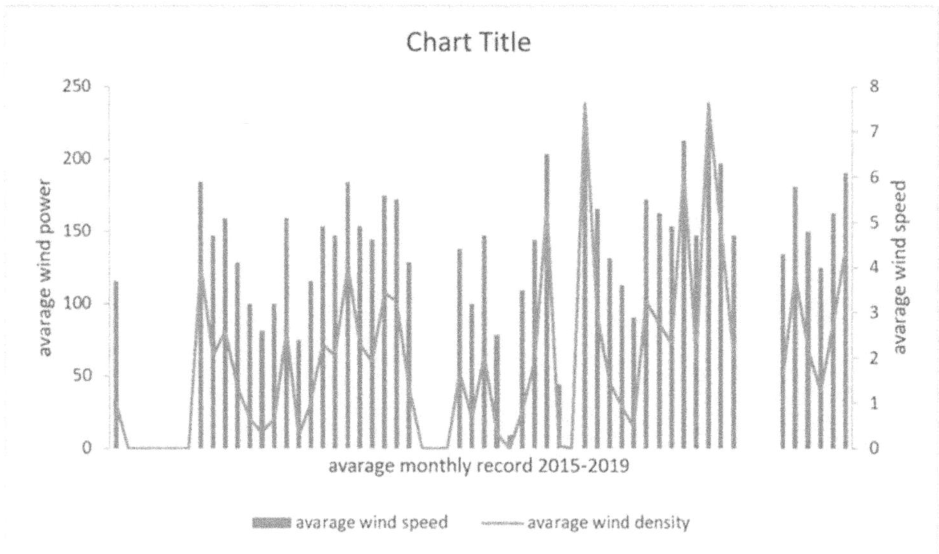

Figure 9. Energy and exergy efficiency distribution in Gökçeada

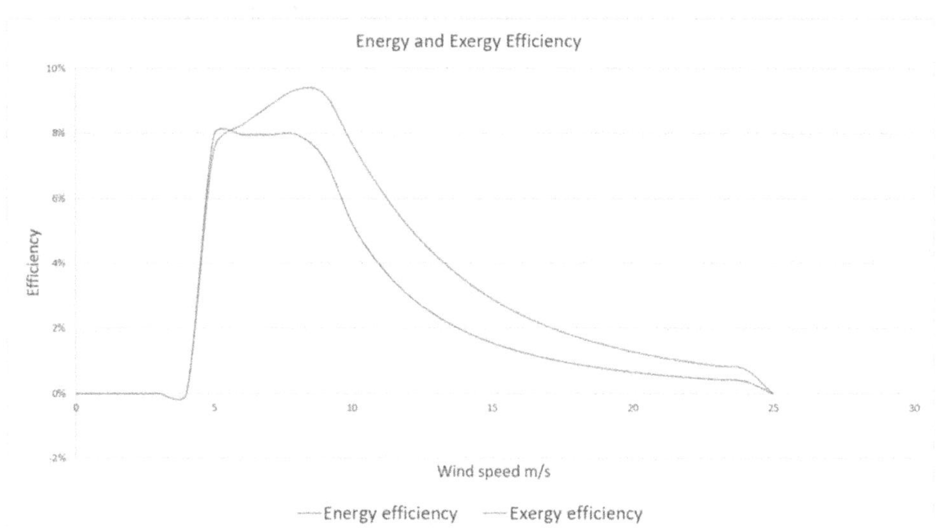

Figure 10. Wind exergy in Gökçeada and the generated energy from the wind at different wind speeds

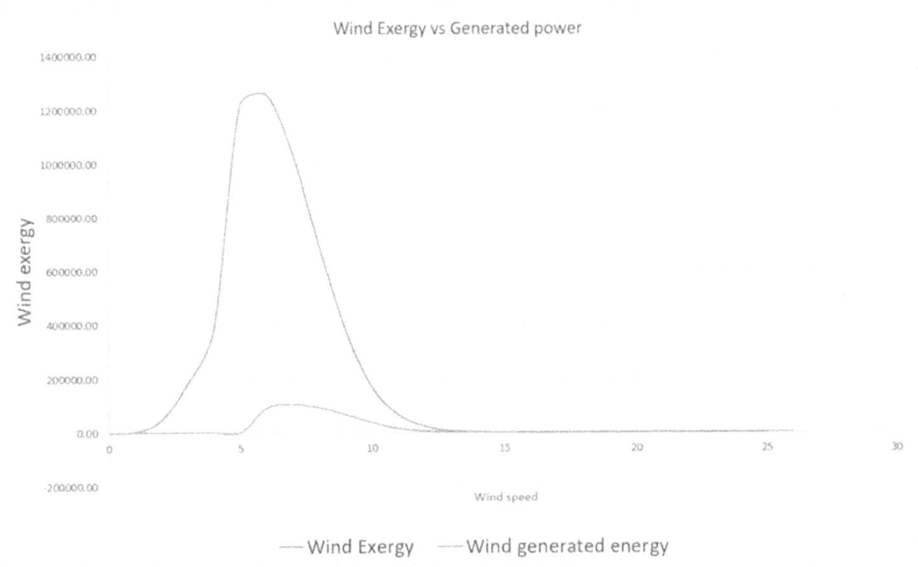

APPENDIX D: ENEZ WIND ENERGY AND EXERGY GRAPHS

Figure 11. Enez wind frequency diagram

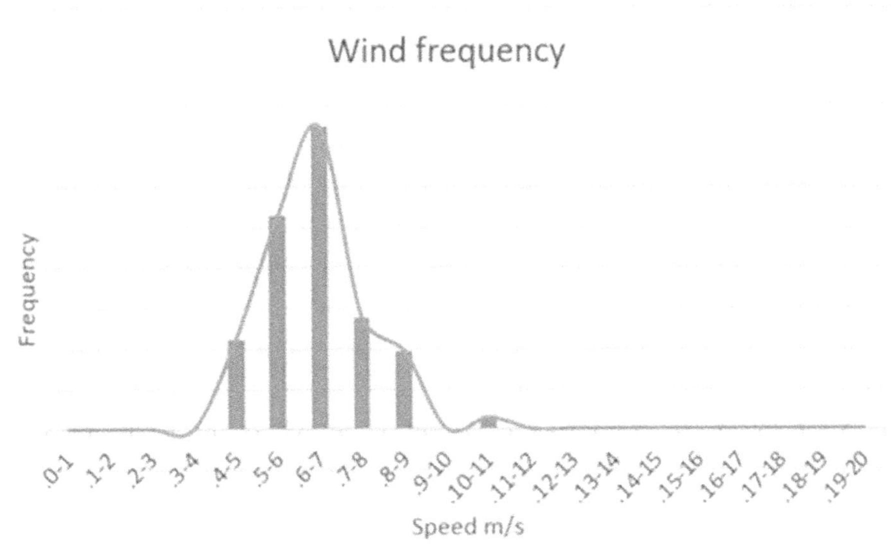

Figure 12. Enez average power and speed diagram

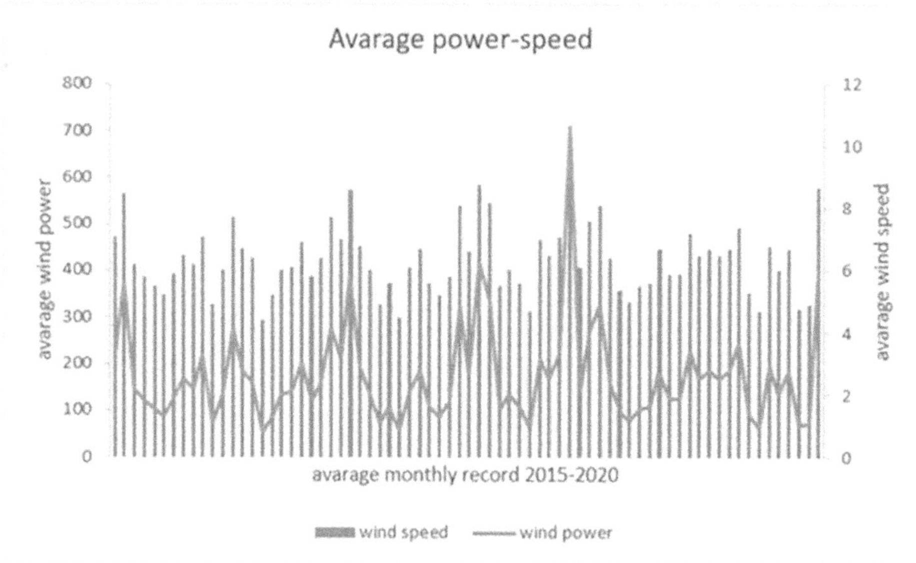

Figure 13. Enez energy and exergy efficiency distribution

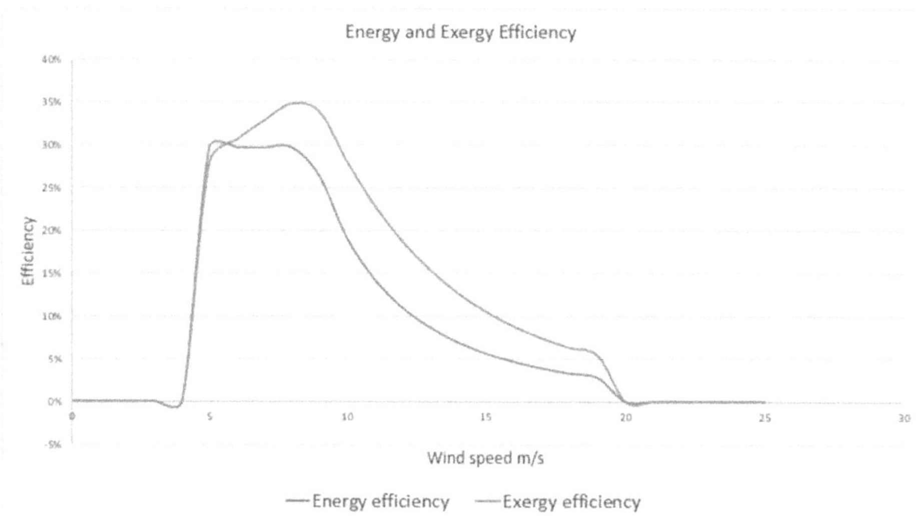

Figure 14. Enez wind exergy and generated energy from wind at different wind speeds

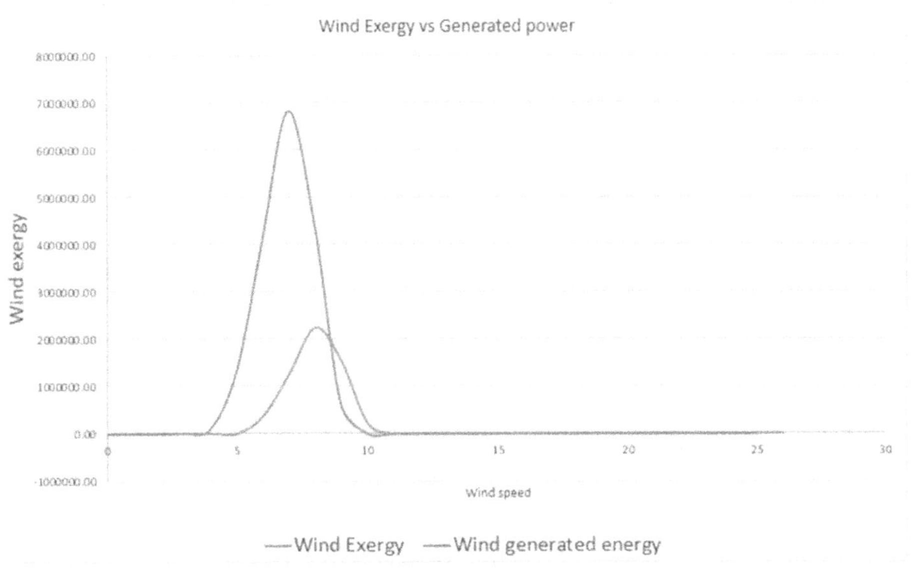

Chapter 6
Factors Affecting Farm Income and Climate Adaptation Strategies in the Farming Sector of Gujarat State of India

Ajay Kumar Singh
DIT University, Dehradun, India

Shah Nawaz Ashraf
Mohammed VI Polytechnic University, Morocco

Sandeep Kumar Sharma
DIT University, Dehradun, India

Sanjeev Kumar
Hansraj College, University of Delhi, India

ABSTRACT

This study observed the effect of climatic factors, social-economic profile of agriculturists, appropriate technology, climate adaptation strategy (CAS), and development institutions on farm income Ha^{-1} in Gujarat using a log-linear regression model. Euler's theorem was used to observe the expected farm income Ha^{-1} by the years of 2040s, 2060s, 2080s, and 2100s. Logit and probit regression models were also employed to assess the CAS affecting factors. The results indicated that farm income Ha^{-1} decreases as marginal increase in climatic variables. The results based on Euler's theorem showed that farm income Ha^{-1} may decline by 12.01%, 17.22%, 22.80%, and 28.66% by the years of 2040s, 2060s, 2080s, and 2100s, respectively. It highlighted that farmers' CAS were significantly associated with family size, yearly income, total arable land, income from cash crops, financial support from government, agricultural development institutions, collaboration with different stakeholders, and skills and technical support from technology developers.

DOI: 10.4018/978-1-6684-7303-0.ch006

INTRODUCTION

The production of agricultural sector is not possible without climatic and agro-ecological factors, geographical location and availability of ecosystem services (Mondal et al., 2015; Chandio et al., 2021; Mondejar et al., 2021; Singh & Jyoti, 2021; Musafiri et al., 2022). Maximum temperature, minimum temperature, rainfall, precipitation and relative humidity are vital climatic factors which also work as inputs for the agricultural sector. Since, climatic factors cannot control by farmers. Therefore, extreme variation in climatic factors during various seasons may be caused to reduce production and yields of food-grain and crops (Kumar et al., 2014; Singh et al., 2020a). Accordingly, food, nutritional, health and livelihood security; farmer's income, jobs, rural development and agro-industrial production decline due to climate change (Singh & Issac, 2018; Mligo et al., 2022; Singh et al., 2022c). Earlier evidence indicates that most climatic factors like maximum temperature, minimum temperature, precipitation, rainfall and relative humidity are instable due to rising greenhouse gases (GHGs) emissions in the atmosphere (Gadgil, 1995; Akinnagbe & Irohibe, 2014; Kumar et al., 2016; Jat et al., 2016; Zilberman et al., 2018; Balasha et al., 2023). Scientific literature also claimed that GHGs emissions from human or natural activities are responsible for climate change and extreme variability in climatic factors (Kumar & Sharma, 2014; Singh et al., 2016; Singh, 2017). Agricultural and its allied sector has large contribution in GHGs emissions in atmosphere. This sector, therefore, can be considered as a cause and consequences for variability in climatic factors (Kumar & Sharma, 2013; Kumar et al., 2015; Mondejar et al., 2021).

Furthermore, plant growth, production and yield of a specific crop can tolerate the impact of a climatic factor up to a certain extent (Singh, 2017). Prior evidence also found a non-linear association of productivity of individual crop with climatic variables (Singh & Sharma, 2018). Thus, farming activities can be negatively affected due to extreme variability in climatic variables (Akinnagbe & Irohibe, 2014; Asrat & Simane, 2018; Biswas et al., 2020). It is, therefore, seemed that global countries cannot achieve sustainable food security (SFS) and sustainable agricultural development (SAD) in era of climate change. SFS is a state in which agricultural sector can meet the food requirement of all people, and fodder requirement of all livestock as sustaining the common property of natural resources and ecosystem services (Singh et al., 2022b). SFS is an integration of all components which ensure the food security of all (Kumar et al., 2017). SAD also helps to meet the sustainability in economic, social and environmental aspects in the agricultural sector (Singh et al., 2022e). Hence, any decline in agricultural production due to climate change will reduce the SFS and SAD (Singh et al., 2022b; Balasha et al., 2023; Akpo et al., 2023). Thereupon, production scale of agro-industrial sector may decline due to low production of commercial crops like sugarcane, cotton, mustard, potato, soyabean, sesame, groundnut, etc. (Weldesilassie et al., 2015; Singh & Jyoti, 2019). For instance, production activities of sugar and textile industries are expected to be declined due to low production of sugarcane and cotton (Kumar et al., 2015; Jyoti & Singh, 2020). Subsequently, employment opportunities of workers in the agricultural and its allied sector to be decreased (Zilberman et al., 2018). It would also reduce the income of growers of commercial crops (Singh et al., 2019; Weldesilassie et al., 2015).

High variability in climatic factors propagate several bacteria and viruses which increase or germinate new diseases which hamper the human health (Singh & Singh, 2020). Also, climate change may be answerable for rising food prices which reduce demand and supply of food-grain products in the market (Singh et al., 2020a). Consumption pattern and income of farming community may be inclined due to climate change (Singh & Sharma, 2018). Henceforth, livelihood security of producers and consumers will be vulnerable due to existence of climate change (Singh et al., 2022c). Climate change has a negative impact on implementation of ongoing government policies of development (Kumar et al., 2015). As land and water are essential inputs for growth of agricultural sector. While, soil fertility and quality, water quality and air quality are contaminating due to overuse of fertilizer and chemicals, inappropriate practices of technologies and unprotective land management in this sector. Climate change also may be accountable to increase labour migration or labour displacement (Zilberman et al., 2018). Hence, climate change has an indispensable influence on ecosystem services, natural resources, human resources, human livelihood and socio-economic development (Akpo et al., 2023). Agricultural sector, therefore, needs digital technology, environmental technology, green technology, appropriate technology (AT), climate adaptation strategy (CAS), suitable land management practices, appropriate institutional support and proper credit accessibility to minimize GHGs emissions from various sources to reduce the negative consequences of climatic factors in the growth of this sector (Sharma & Singh, 2017; Singh et al., 2020b; Msomba et al., 2021; Singh & Kumar, 2022; Musafiri et al., 2022; Balasha et al., 2023).

Adaptation strategy is a specific practice or action that can be used to avoid the risk or negative consequences of exogeneous and endogenous variables in production activities (Malhi et al., 2021; Msomba et al., 2021; Singh et al., 2022a). In agricultural sector, climatic factors are exogeneous variables which cannot be controlled by farmers easily. However, the impact of most climatic factors in the agricultural sector can be minimized using different practices which is known as climate adaptation strategy (CAS) (Akinnagbe & Irohibe, 2014; Asrat & Simane, 2018; Luu et al., 2019; Msomba et al., 2021; Musafiri et al., 2022; Biswas et al., 2020; Akpo et al., 2023). Detection of adaptation strategies, therefore, would be helpful for agricultural scientists and farmers to make climate change related policies and climate action plan for better growth of agricultural and its allied sectors (Smit & Skinner, 2002; Asrat & Simane, 2018; Fierros-González & López-Feldman, 2021; Kangogo et al., 2021). Earlier studies have noticed that farmers are applying various technologies, drought-resistant and disease-resistant varieties of seeds, planted trees, organic fertilizer, mixed and change in cropping pattern, crop diversification, change in plating time, heat tolerant crops, advance irrigation facilities, electricity for irrigation, water conservation, land management practices, social communication, green farming and mixed cropping pattern as a CAS in the agricultural sector (Luu et al., 2019; Biswas et al., 2020; Sargani et al., 2022; Mligo et al., 2022; Swami & Parthasarathy, 2022). For instance, climate smart agricultural, digital technologies and appropriate technologies would reduce the stress of climate change in farming (Zilberman et al., 2018; Mondejar et al., 2021). Sustainable land management practices would avoid the adverse impact of climate change in cultivation (Akpo et al., 2023). Land intensification, growing tree on field borders, nutrient management, crop rotation, subsidy for farmers, crop insurance policies, appropriate agricultural marketing, better pricing of products, livestock rearing business, income of farmers, skills and technical support, livelihood assets of farmers, financial accessibility and agricultural development institutions also regulate CAS of farmers (Jamal et al., 2021; Singh et al., 2022a; Sargani et al., 2022; Akpo et al., 2023). Entrepreneurial skills of farmers may be positive to increase their consciousness towards CAS in the cultivation (Kangogo et al., 2021). CAS may be a useful driver to increase crop yield, food security, income and welfare of farming households in the presence of climate change (Falco & Veronesi 2013; Kumar et al., 2014; Pakmehr et al., 2020; Singh et al., 2022b). Hence, it also seemed that CAS would be supportive to increase sustainable agricultural development. CAS of farmers also depend on economic capacity, credit and market accessibility, income from livestock rearing occupation, family size, gender, age, education level and off-farms activities of farmers; farmers' membership in agricultural co-operative societies, access to information, technical support for farmers, availability of technologies, farmer's association with universities and research institutions, agricultural extension services and agricultural R&D expenditure (Ndamani & Watanabe, 2016; Balasha et al., 2023).

Technological development and innovation may bring several opportunities to reduce GHGs emissions and improve the productivity of most resources including environmental sustainability (Singh et al., 2020b). Hence, it may be useful to address the issue of climate change effectively in the farming activities (Singh and Singh, 2021). Applications of technologies would be supportive to eradicate the pressure of anthropogenetic activities and reduce negative impact of waste materials on ecosystem services (Abdullahi et al., 2015). Subsequently, applications of technological development and innovation can be used as climate adaptation strategy (CAS) in the agricultural sector to diminish the intensity of climate change (Singh & Jyoti, 2021). Hence, technological change and innovation has a significant impact on agricultural productivity (Abdullahi et al., 2015; Ashraf & Singh, 2021). For instance, use of digital technologies may be conducive to avoid the climate change risk in farming activities (Mondejar et al., 2021). Digital technologies and devices such as social media, mobile apps, climate information, information and communication technologies (ICT), weather forecasting instruments and GIS would be helpful for farmers to apply CAS in cultivation. Furthermore, applications of appropriate technology (AT) are useful to meets economic, social and environmental viability of production sector (Swain, 2016; Moon & Hwang, 2018). AT reduces waste materials and abate GHGs emissions in production activities (Moon & Hwang, 2018). Hence, use of AT would reduce possibility of climate change in future. AT may be beneficial to increase land pattern, mechanization of farm, cropping pattern, crop diversification, seed germination, soil fertility and quality, land and labour productivity, crop diversification, efficiency of inputs, proper marketing of Agri-products, water conservation and human health (Shabbir & Yaqoob, 2019; Siddick, 2019; Singh et al., 2022a; Swami & Parthasarathy, 2022).

The agricultural sector is a sole sector to meet food security of people and provide fodder to livestock and animals in India. The implications of agricultural sector cannot be denied in India due to high dependency of large population on this sector to maintain their livelihood security and rural development (Singh et al., 2022d). Furthermore, extreme poverty, high income inequality, high population growth, overwhelming urbanization, limited access of capital, low capital formation, poor technological accessibility of farmers, inappropriate infrastructural development in rural area, low economic capacity and literacy of farmers, low technological skills of people, low credit accessibility of farmers, and low per capita income are the basic characteristics of Indian economy (Sharma and Singh, 2017; Singh & Jyoti, 2021). Therefore, food security, nutritional security, health security, income of people, food prices stability, and agricultural and rural development policies will be in more vulnerable position in presence of climate change in India (Singh et al., 2022c).

Existing studies have observed that agricultural production activities would be adversely affected due to incidence of climate change in India (Sharma and Singh, 2017). Numerous researchers and scientists have detected that productivity and production of most crops are likely to be declined in India due to climate change (Gadgil, 1995; Yadav et al., 2015; Mondal et al., 2015; Singh and Singh, 2021; Chandio et al., 2021). Food security of farming and non-farming community would be declined due to climate change (Singh et al., 2022b). Thereupon, climate change has the negative impact on sustainable livelihood security (Singh et al., 2022c). It is, therefore, appeared that climate change has severe adverse impact on most activities of production activities which further hamper the social-economic development, agricultural sustainability, sustainable agricultural development and sustainable development in India (Singh and Singh, 2021; Singh et al., 2022d). Most studies have reported a positive contribution of technological development in the agricultural sector (Swain, 2016). Applications of technological development in farming activities would be positive to increase the land productivity, production, income of farmers, soil quality, soil fertility and environmental sustainability. For example, few researchers reported a positive contribution of technological development in cultivation (Ashraf & Singh, 2021).

Prior literature has also argued that cropped area, production and productivity of many crops could improve due to technological advancement in the agricultural sector (Kumar et al., 2017; Ashraf & Singh, 2021). Accordingly, technological development would be helpful to mitigate the negative consequences of climate change in most sectors (Singh et al., 2022d). Total factor productivity of agricultural sector was also improved due to technological advancement in India (Shabbir & Yaqoob, 2019). Furthermore, various adaptation practices and strategies may be positive to avoid risk of climate change in farming sector (Jat et al., 2016; Biswas et al., 2020; Swami & Parthasarathy, 2022). Therefore, India needs to implement climate action policy, develop climate adaptation strategy (CAS), and appropriate technology, digital technology and innovation, and digitalization to minimize climate change impact in the agricultural sector (Chandio et al., 2021; Singh et al., 2022a). Consequently, it would assist to meet food and health security of population, increase farmer's income and maintain sustainability in ecosystem services; create extensive jobs for agricultural workers, maintain price stability of food-grain products, reduce poverty and income inequality, and increase agricultural sustainability in India. Afore-mentioned goals can be achieved through application of AT, technological development, CAS and suitable institutional support in the agricultural sector. Hence, this study assessed the role of AT, technological development, CAS of farmers and agricultural development institutions in the agricultural sector using farm level information. Accordingly, the present study made a significant contribute to explore the answer on below questions:

- How has CAS had an effective contribution to alleviate the negative impact of climatic factors in the cultivation?
- How AT, CAS, and institutional support can be applied to avoid the risk of climate change in the cultivation?
- Can social-economic factors of farmers and institutional support be used as CAS in the agricultural sector?

As per the stated research questions, the present study is realized below objectives:

- To investigate the effect of climatic factors, AT and CAS on farm income Ha^{-1} using a log-linear regression model.
- To observe the expected farm income Ha^{-1} in diverse climate change perspectives using a Euler's theorem.
- To assess the impact of social-economic characteristics of farmers on CAS using Logit and Probit regression equations.

APPLIED RESEARCH METHODOLOGY

Summary of Study Area

The proposed study is used primary information of 400 farmers who were randomly selected from 8 districts of Gujarat. These 8 districts were considered as per their significant contribution in gross cropped area (Figure 1), agricultural workers (Figure 2), agricultural district domestic products (Figure 3) and gross irrigated area (Figure 4) in the state (CMIE, 2018 – 19). Furthermore, these districts have significant share in cropped area of rice (43.90%), wheat (34.92), maize (22.87%), millet (39.69%), sorghum (61.83%), arhar (47.88%); and commercial crops like castor seed (28.41%), rapeseed & mustard (57.66%), sesamum (28.88%), cotton (26.26%), groundnut (23.07%), sugarcane (84.12%), onion (54.67%) and potato (62.55%) in Gujarat (CMIE, 2018 – 19).

Data Collection Technique

Two villages from each district were selected for field visits. Total 25 farmers from each village were identified based on their socio-economic features, yearly income, family and land size, irrigated area, occupation, cultivation of multiple-crops, technical and education literacy, membership in agricultural co-operative societies, financial supports, and use of various technologies in cultivation. The required information was collected through face to face interview of selected farmers. The interview of randomly selected respondents was conducted using a well-structured questionnaire during December 01, 2019 to February 28, 2020. Survey team make an approach to obtain desire information from planned 400 farmers. As few farmers could not provide their opinion for various questions such as concept of appropriate technology and its components, agricultural technology, financial support from the Government and banking sector, farmer's association with Agri-entrepreneurs, crop insurance, skilled supports from agricultural development institutions, information technologies, and CAS, etc. Only 240 farmers could provide the complete information on most desired variables which were essential to include in descriptive and empirical investigation. Accordingly, information of few observations was dropped. For this, statistical property (i.e., mean, standard deviation and skewness, etc.) of all variables was estimated before and after dropping the information of 160 observations. As the statistical property of all relevant variables were not significantly diverse before and after missing the observations. Therefore, it was rational to dropped the information of those sample which do not have the related information. Otherwise, it was difficult to run the regression analysis. Further, the occurrences of missing observations were also random. After dropping the samples, this study applied Cronbach's alpha test to check the scale reliability of undertaken variables. The statistical values of Cronbach's alpha test were more than 0.7 for all variables. Hence, estimates infer the sample size of selected variables have the adequacy. Information of stated climatic factors for selected districts were derived from the online website of International Crop Research Institute for Semi-Arid Tropics (ICRISAT).

Rationality of Dependent and Explanatory Variables

In this study, farm income was calculated as aggregate economic value of all crops (which were cultivated by farmers during survey period) at farm harvest price (FHP). Thereafter, farm income Ha^{-1} was evaluated as ratio of aggregate economic value with gross cropped area. It is also known as land productivity which reflect the overall soil quality and fertility as well. Accordingly, it was used as a dependent variable in a standardize form. Furthermore, CO_2 concentration, average maximum temperature and temperature, average rainfall and precipitation, actual evapotranspiration and relative humidity were used as climatic factors by previous researchers. Highlighted variables were also used as independent variables in the regression equations to capture the impact of climatic factors in the agricultural sector. As the information of mentioned climatic factors were available during 1991-2015 for selected districts.

Coefficient of variation (CV) in these variables were included to capture the influence of climate change on land productivity.

Land management practices increase as age of farmers increases (Akpo et al., 2023). Most family member of farming community generally engaged in cultivation. Therefore, agricultural productivity may be increased as involvement of family members increase in cultivation (Jamal et al., 2021). Thus, farmer's age and family size can be used as CAS in farming (Fierros-González & López-Feldman, 2021). Educated farmers have a better understanding on applications of AT, CAS and other inputs to avoid the risk of climate change (Kumar et al., 2016; Akpo et al., 2023). Moreover, male farmers can be associated with various stakeholders, and be a member of agricultural co-operative societies and agricultural extension centres. Thus, productivity of male and female farmers may be different (Singh & Singh, 2019; Singh et al., 2022a).

Agricultural production to be increased as yearly income of the farmer increases. Accordingly, farmers can use different CAS and technologies to get better return in farming (Jamal et al., 2021). Furthermore, agricultural production is a product of arable land and irrigated land. Irrigated area can tolerate the impact of high temperature, relative humidity and sun intensity on crop yield. Human resources are also supportive to increase agricultural productivity (Singh, 2020; Sargani et al., 2022). Use of agricultural labour Ha^{-1} was also used as independent variable. Soil fertility and quality, and land productivity to be improved as application of fertilizer Ha^{-1} increases (Kumar et al., 2016; Msomba et al., 2021).

Cash crop farming provide the better return to the farmers and it may be useful to increase farm income Ha^{-1} (Singh et al., 2022a). Cash crop growers can also increase their collaboration with sugar, textile, oil processing and Agri-based industries. Agri-based industries also provide the regular training to the farmers. Crop diversification is also useful to increase agricultural production and farm income Ha^{-1} (Singh, 2020). Hence, income from cash crops and crop diversification were also included as control variables in the log-linear regression equation.

Technological development also has a positive impact in the agricultural sector (Ashraf & Singh, 2021). Cost of technologies Ha^{-1} was used to identify the impact of technological development on farm income Ha^{-1} (Singh et al., 2022a). AT is an integration of social, economic and environmental viability of technologies (Musunuri, 2014; Siddick, 2019; Singh et al., 2022a). Hence, use of AT in cultivation has a positive impact on SAD. Therefore, farmer's view on social, economic and environmental viability of technologies in binary format were considered to examine the impact of AT on farm income Ha^{-1}. Poor and small farmers cannot afford the high cost of agricultural inputs and they have low production Ha^{-1}. Therefore, financial support for small farmers from financial organizations (i.e., government and banking sector) would be useful for them to purchase new seeds, fertilizer, pesticides, technologies, instruments of irrigation and other inputs. Moreover, CAS are also incorporated to examine its impact on farm income Ha^{-1}. Agricultural development institutions provide skilled and technical support to the farmers (Jamal et al., 2021). Thus, the role of agricultural development institutions in the agricultural sector is remarkable (Zilberman et al., 2018; Singh et al., 2022b). Technology developers' industries also provide the skilled training to the farmers to operate new technologies. Hence, abovementioned variables were also included as independent variables in the regression models.

Figure 1. Share of districts in gross cropped area (in %) of Gujarat (CMIE 2018-19)

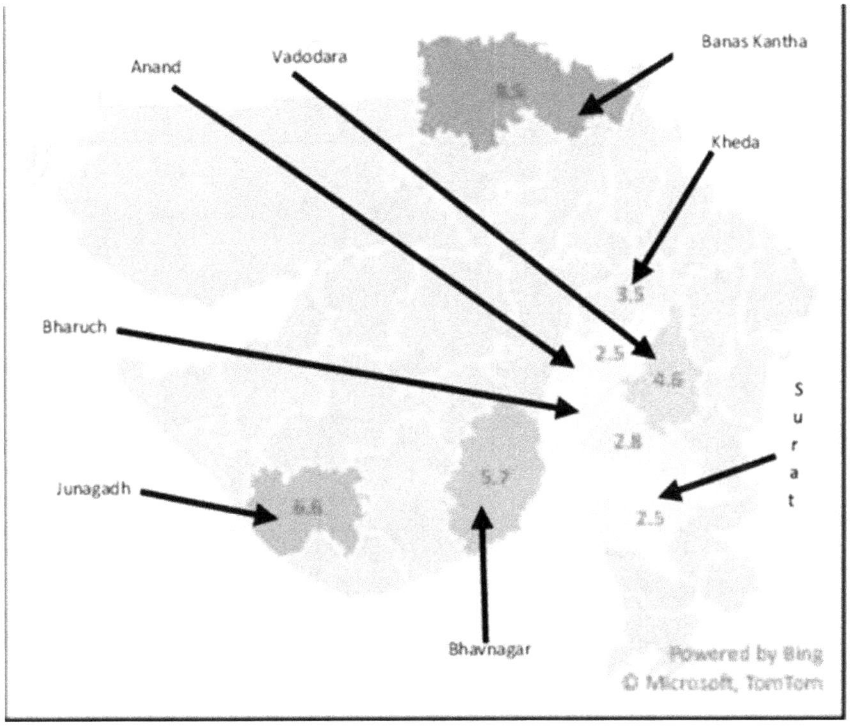

Figure 2. Share of districts in agricultural labours (in %) of Gujarat (CMIE 2018-19)

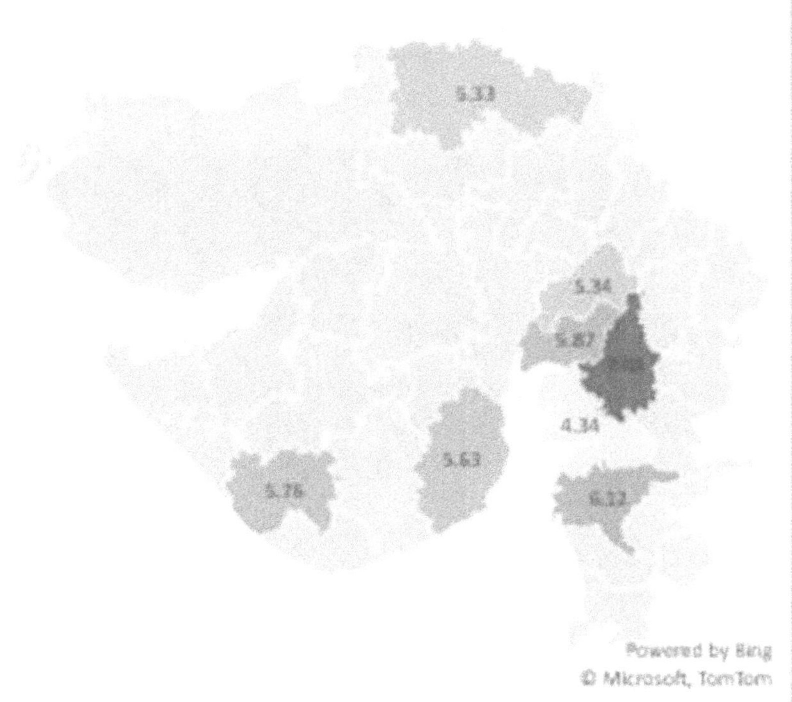

Figure 3. Share of districts in agriculture DDP (in %) in Gujarat (CMIE 2018-19)

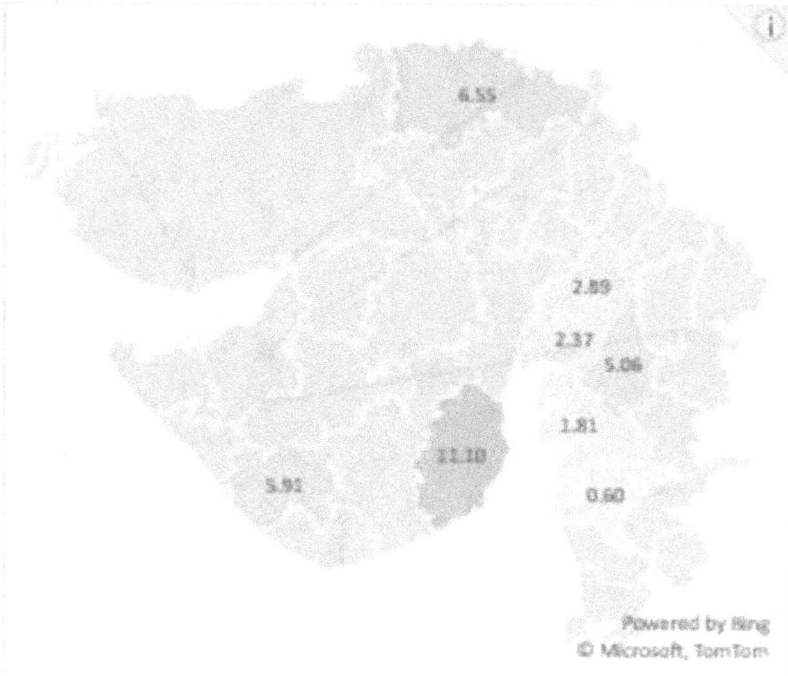

Figure 4. Share of districts in gross irrigated area (in %) in Gujarat (CMIE 2018-19)

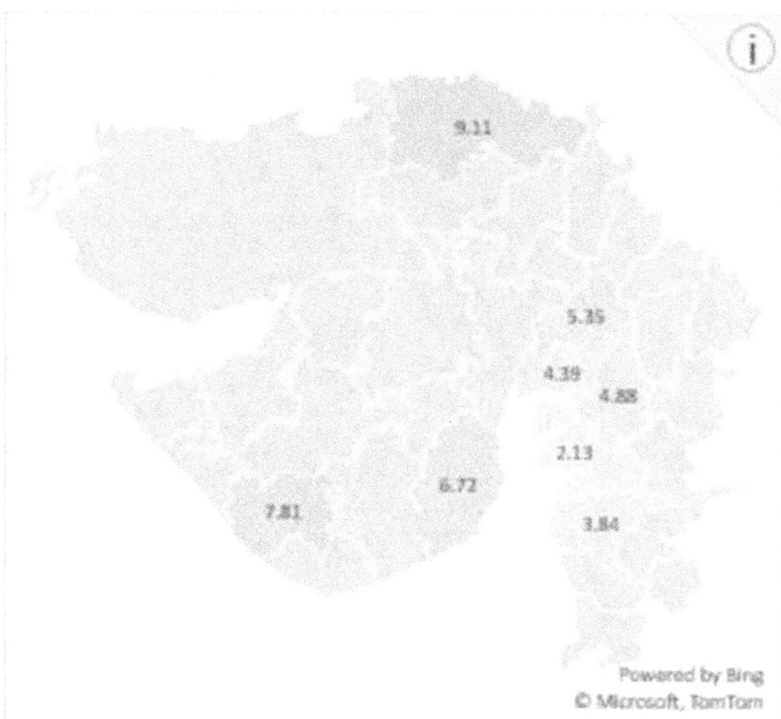

EMPIRICAL ANALYSIS

Regression Model for Association of Farm Income With Explanatory Variables

Prior studies argued that functional form of Cobb-Douglas production function (log-linear regression model) produce consistent coefficient of explanatory variables with dependent variable in the agricultural production analysis (Singh et al., 2017). Also, the statistical inference of results estimated through log-linear regression model are easy and simple. Thereupon, the model produces the elasticity of output with respect to a specific input. In general, elasticity measures the percentage variation in output due to 1% variation in input (Kumar et al., 2017). Log-linear regression model, therefore, was employed to observe the impact of climatic and socio-economic factors on farm income Ha^{-1} (land productivity) in this study. Previous studies have preferred this model in agricultural production analysis (Singh et al., 2022a). It hypothesized that farm income Ha^{-1} (land productivity) depend on climatic factors and farmer's social–economic characteristics which is specified as:

$$(fiph) = f\{(cvaaea), (cvaamaxtem), (cvaamintem), (cvaapre), (cvaarf), (famsizres), (edulevres), (genres), (annincfam), (tagla), (irrare), (ualph), (faph), (ficcph), (cdi), (cotepeha), (apptec), (finsupgov), (adstfa)\} \quad (1)$$

Here, *fiph* is farm income Ha^{-1} or land productivity; *cvaaea* is coefficient of variation in actual annual average evapotranspiration, *cvaamaxtem* is coefficient of variation (CV) in annual average maximum temperature, *cvaamintem* is CV in annual average minimum temperature, *cvaapre* is CV in annual average precipitation, *cvaarf* is CV in annual actual rainfall, *famsizres* is family size, *edulevres* is education level, *genres* is gender; *annincfam* is yearly income, *tagla* is total agricultural land, *irrare* is gross irrigated land, *ualph* is applications of agricultural labour Ha^{-1}, *faph* is fertilizer applications Ha^{-1}, *ficcph* is farm income from cash crops Ha^{-1}, *cdi* is crop diversification index, *cotepeha* is cost of technologies Ha^{-1}, *apptec* is appropriate technology (AT), *finsupgov* is financial accessibility of farmers from government and banks, and *adstfa* is farmers' climate adaptation strategy (CAS) in equation (1). After adopting the log-linear regression equation, the above-mentioned function was transformed as:

$$ln(fiph)_i = \alpha_0 + \alpha_1 ln(tagla*cvaaea)_i + \alpha_2 ln(tagla*cvaamaxtem)_i + \alpha_3 ln(tagla*cvaamintem)_i + \alpha_4 ln(tagla*cvaapre)_i + \alpha_5 ln(tagla*cvaarf)_i + \alpha_6 ln(famsizres)_i + \alpha_7 ln(edulevres)_i + \alpha_8 ln(annincfam)_i + \alpha_9 ln(tagla)_i + \alpha_{10} ln(irrare)_i + \alpha_{11} ln(ualph)_i + \alpha_{12} ln(faph)_i + \alpha_{13} ln(ficcph)_i + \alpha_{14} ln(cdi)_i + \alpha_{15} ln(cotepeha)_i + \beta_{16}(apptec)_i + \beta_{17}(finsupgov)_i + \beta_{18}(adstfa)_i + \beta_{19}(genres)_i + \lambda_i \quad (2)$$

Here, *ln* is logarithm of associated factors; α_0 is intercept term; $\alpha_1, \alpha_2, \ldots, \alpha_{15}$ are elasticities of land productivity with respect to corresponding explanatory variables; $\beta_{16}, \ldots, \beta_{19}$ are the coefficient of binary variables with farm income Ha^{-1}; λ_i is the residual term; and *i* is i^{th} respondents in the equation (2). The specification of other variables is given in Table 1.

Table 1. Explanation of dependent and explanatory variables

Variables	Symbol	Unit/Measurement
Farm income Ha^{-1} [Dependent Variable]	*fiph*	Rs $Ha.^{-1}$
Coefficient of variation (CV) in actual average evapotranspiration	*cvaaea*	mm
CV in annual average maximum temperature	*cvaamaxtem*	^0C
CV in Annual average minimum temperature	*cvaamintem*	^0C
CV in Annual average precipitation	*cvaapre*	mm
CV in Annual actual rainfall	*cvaarf*	mm
Members in family	*famsizres*	Number
Education level of respondents	*edulevres*	Years
Gender of respondents	*genres*	Male =1; Female=0
Annual income of the respondents	*annincfam*	Rs.
Gross agricultural land	*tagla*	Ha.
Gross irrigated area	*irrare*	Ha.
Application of human resource Ha^{-1}	*ualph*	Number
Application of fertilizer Ha^{-1}	*faph*	Kg $Ha.^{-1}$
Farm income from cash crops Ha^{-1}	*ficcph*	Rs
Crop diversification index	*cdi*	%
Cost of technology Ha^{-1}	*cotepeha*	Rs $Ha.^{-1}$
Appropriate technology	*apptec*	Number
Financial support for farmers from financial institutions	*finsupgov*	Yes =1; No =0
Climate adaptation strategy (CAS)	*adstfa*	1 = Yes; 0 = No

Applications of Euler's Theorem

Euler's theorem is useful to examine the nature of a production function (Vîlcu and Vîlcu, 2013). The theorem is helpful to assess the marginal impact of individual variable on output. Thereupon, integrated impact of all explanatory variables on output can be evaluated using this theorem. Therefore, this theorem was used to examine the nature of log-linear production function model. As per the Euler's theorem, a new production function was established as:

$$N\times(fiph)= cvaaea\times[\alpha_1(fiph)/(cvaaea)] + cvaamaxtem\times[\alpha_2(fiph)/(cvaamaxtem)] + cvaamintem\times[\alpha_3(fiph)/(cvaamintem] + cvaapre \times [\alpha_4(fiph)/(cvaapre)] + cvaarf \times [\alpha_5(fiph)/(cvaarf)] \quad (3)$$

Here, $N\times(fiph)$ is new production function; N is the coefficient of farm income Ha^{-1} (N is quantity that measure the change in farm income due to 1% increase in the values of all climatic factors); $[\alpha_1(fiph)/(cvaaea)]$, $[\alpha_2(fiph)/(cvaamaxtem)]$, $[\alpha_3(fiph)/(cvaamintem]$, $[\alpha_4(fiph)/(cvaapre)]$ and $[\alpha_5(fiph)/(cvaarf)]$ are the 1st differentiation of equation (2) with respect to corresponding climatic variables in equation (3). If $N = 1$, $N < 1$ and $N > 1$, then the production function would be constant return to scale, decreasing return to scale and increasing return to scale, respectively (Vîlcu and Vîlcu, 2013). The value of 1st differentiation of equation (2) with respect to a specific climatic variable measure marginal change in farm income Ha^{-1} (Huong et al., 2019). In this study, therefore, the values of 1st differentiation of equation (2) with respect to all climatic factors were aggregated to examine their integrated impact on farm income. Accordingly, projected farm income was also examined in different climate change perspectives. For above-said investigations, it assumed that actual evapotranspiration will be increased by 2, 2.5, 3 and 3.5 mm by 2040, 2050, 2080 and 2100, respectively. Maximum temperature and minimum temperature are projected to rise by 0.5, 0.75, 1.0 and 1.5^{0}C by the said years (Kumar et al., 2016). The rainfall and precipitation may be varied by 4, 5, 6 and 7 mm in the mentioned years (Singh et al., 2017; Singh et al., 2021). Following equation was employed to observe the expected farm income Ha^{-1} in stated years:

$$[\Delta(fiph)]= [(\delta(fiph)/\delta(cvaaea)]\times\Delta(aaea)+ [(\delta(fiph)/\delta(cvaamaxtem)]\times\Delta(aamaxtem)+ [\delta(fiph)/\delta(cvaammintem]\times\Delta(aamintem)+ [\delta(fiph)/\delta(cvaapre)]\times\Delta(aapre)+ [\delta(fiph)/\delta(cvaarf)]\times\Delta(aarf)\} \quad (4)$$

Here, $\Delta(fiph)$ is projected farm income; $\Delta(aaea)$, $\Delta(aamaxtem)$, $\Delta(aamintem)$, $\Delta(aapre)$ and $\Delta(aarf)$ are presumed change in respective climatic factors; $[(\delta(fiph)/\delta(cvaaea)]$, $[(\delta(fiph)/\delta(cvaamaxtem)]$, $[\delta(fiph)/\delta(cvaammintem]$, $[\delta(fiph)/\delta(cvaapre)]$ and $[\delta(fiph)/\delta(cvaarf)]$ are the 1st differentiation of log-linear production function with respect to corresponding variables in equation (4).

Empirical Model for Climate Adaptation Strategies Influencing Factors

Previous studies have applied different models to analysis the climate adaptation strategy (CAS) influencing factors using primary data. Falco and Veronesi (2013); Ndamani and Watanabe (2016) applied multinomial endogenous switching and binary logistic regression model to assess the impact of social–economic profile of farmers on CAS. Some studies used logit regression model to observe factors affecting CAS (Schattman et al., 2020; Swami & Parthasarathy, 2022; Musafiri et al., 2022; Balasha et al., 2023). Mligo et al. (2022) considered linear regression model to investigate impact of CAS and other explanatory variables on food security. This study used combine binary and actual forms of dependent and independent variables in regression equation.

Logit and Probit regression models produce the probability of variable that may enhance the farmer's intention to apply various CAS to address the issue of climate change in cultivation (Singh, 2020). Hence, above-mentioned models were employed to analyses the factors affecting CAS. The form of both the models were formulated as:

$$ln\{Pr/(1 - Pr) =\theta_0 +\gamma_1 (ageres)_i +\theta_2 (famsizres)_i +\theta_3 (typfamres)_i +\theta_4 (annincfam)_i +\theta_5 (tagla)_i +\theta_6 (irrare)_i +\theta_7 (ualph)_i +\theta_8 (ficcph)_i +\theta_9 (faph)_i +\theta_{10} (cotepeha)_i +\theta_{11} (cdi)_i +\theta_{12} (finpro)_i +\theta_{13} (finsupgov)_i +\theta_{14} (farassstahol)_i +\theta_{15} (skitecsupfar)_i \quad (5)$$

Here, Pr is a function of probability of adaptation strategy adopted by cultivators to reduce the risk of climate change; clarification of remaining explanatory variables is given in Table 2; θ_0 is the constant coefficient, and $\theta_1 \ldots \theta_{15}$ are the coefficients of corresponding variables in equation (5).

Table 2. Factor affecting CAS

Variables	Symbol	Measurement/Unit
Farmers' climate adaptation strategy [Dependent Variable]	*adstfa*	1 = Yes; 0 = No
Farmer's age	*ageres*	Years
Family size	*famsizres*	Number
Types of family	*typfamres*	1 = Joint; 0 = Single
Yearly income	*annincfam*	Rs
Gross agricultural land	*tagla*	Ha
Gross irrigated land	*irrare*	Ha
Applications of human resources Ha^{-1}	*ualph*	Number
Farm income from cash crops Ha^{-1}	*ficcph*	Rs
Fertilizer applications Ha^{-1}	*faph*	Kg Ha^{-1}
Technological cost Ha^{-1}	*cotepeha*	Rs Ha^{-1}
Crop diversification index	*cdi*	%
Financial problem of farmers	*finpro*	1 = Yes; 0 = No
Financial support from financial institutions	*finsupgov*	1 = Yes; 0 = No
Farmer's association with agricultural development institutions	*farassstahol*	1 = Yes; 0 = No
Skills and technical support from technology developers	*skitecsupfar*	1 = Yes; 0 = No

Process for Selection of Rational Empirical Model

Normality Test: If a variable is not in normal form, then it may increase fallacy in the statistical results. Thus, statistical values of skewness and kurtosis were calculated to check the normality of all variables (Singh et al., 2022a).

Cronbach's Alpha Test: It examines the coefficient reliability of individual variable and set of variables (Singh et al., 2022a). It also measures the internal consistency of composite score of multiple variables. A variable has viability when the value of Cronbach's alpha test for this variable is detected 0.7 (Singh & Kumar, 2022).

Functional Form of Empirical Model: Well-defined functional form of an empirical model produces rational results. Singh and Jyoti (2021) used *Ramsay RESET* test to recognize the suitability of a regression equation. This test, therefore, was applied to observe the validity of log-linear regression equation.

Multi-Collinearity: Existence of multi-collinearity among the explanatory variables may be inappropriate to provide clarification on regression coefficient in a model (Msomba et al., 2021). Variance inflation factor (*VIF*) was estimated to observe the existence of multi-collinearity among the variables.

Heteroskedasticity: Presence of heteroskedasticity in the cross-sectional data produce unreliable coefficient of independent variables with output. Cameron & Trivedi decomposition of IM-test and Breusch-Pagan/Cook-Weisberg test were used to identify the existence of heteroskedasticity (Msomba et al., 2021).

Akaike Information Criterion (AIC) and Bayesian Information Criterion (BIC): Logit and Probit regression models were used to estimate the CAS affecting factors. Hence, statistical values of *AIC* and *BIC* were used to observe the validity of both the models. Regression equations were run through SPSS and STATA statistical software.

RESULTS

Statistical Property of Variables

The statistical properties of dependent and independent variables are given in Table 3. Statistical values of skewness infer that most variables were in normal form. Statistical values of *Cronbach's Alpha Test* for most variables were also found greater than 0.80 and scale reliability coefficient was 0.92. These variables, therefore, have reliability. Hence, all variables can be used for further empirical investigations.

Table 3. Statistical properties of explanatory variables

Variables/Statistics	Mean	SD	Skewness	Kurtosis	Alpha
logfiph	9.421	0.312	6.365	45.678	0.88
*logtagla*cvaaea*	0.006	0.756	-0.197	2.298	0.854
*logtagla*cvaamaxtem*	-2.493	0.700	-0.465	2.547	0.851
*logtagla*cvaamintem*	-1.990	0.701	-0.489	2.590	0.851
*logtagla*cvaapre*	0.787	0.686	-0.394	2.528	0.851
*logtagla*cvaarf*	1.036	0.704	-0.438	2.529	0.85
logfamsizres	1.715	0.314	-0.164	3.257	0.88
logedulevres	2.501	0.259	-0.373	1.846	0.871
genres	0.979	0.143	-0.423	1.723	0.884
loganninc fam	13.132	0.338	-0.911	3.964	0.87
logtagla	2.019	0.688	-0.485	2.612	0.85
logirrare	1.564	0.767	-0.491	2.681	0.852
logualph	3.977	0.193	-7.700	95.830	0.881
logfaph	5.086	0.371	2.009	12.552	0.873
logficcph	0.692	0.463	-0.830	1.689	0.88
logcdi	-1.769	0.784	-0.429	2.766	0.853
logcotepeha	7.821	0.213	-8.859	112.153	0.881
apptec	0.722	0.295	-0.512	1.909	0.873
finsupgov	0.438	0.497	0.252	1.063	0.88
adstfa	-0.124	0.694	-0.436	2.460	0.851

Source: Author's estimation based on field-based information

Regression Coefficient of Explanatory Variables with Farm Income Ha^{-1}

The coefficient of farm income Ha^{-1} with climatic, socio-economic factors, AT and CAS is given in Table 4. Statistically significant *F-value* demonstrate that applied model was best fit. The Chi^2 values under *Ramsey RESET* test for output and explanatory variables were also statistically insignificant. The estimates infer that proposed form of log-linear regression model was suitable and has competency to produce consistent coefficients. As per the R^2 value, 84% variation in farm income Ha^{-1} depend on included variables in the model. The impact of actual annual average evapotranspiration and annual average maximum temperature on farm income Ha^{-1} were reported negative. Farm income Ha^{-1} may decrease by 0.68% and 6.98% due to 1% increase in evapotranspiration and maximum temperature, respectively. The coefficient of minimum temperature and annual rainfall with farm income Ha^{-1} were also reported negative. The estimates imply that farm income Ha^{-1} may decline by 4.80% and 0.96% due to marginal change in minimum temperature and rainfall, respectively. The plant growth of most crops depends upon specific quantity of rainfall, maximum and minimum temperature, and evapotranspiration. The findings of previous studies have also noticed that productivity of most crops decline due to change in rainfall pattern in sowing, growing and harvesting time (Kumar et al., 2017). Consequently, any variability in mentioned variables have a negative impact on land productivity. The impact of precipitation on farm income Ha^{-1} was also reported positive. The results claimed that farm income Ha^{-1} may increase by 1.41% due to 1% increase in precipitation.

The coefficient of family size, education level, gender of respondents and yearly income with farm income Ha^{-1} were found positive. The estimates reveal that farm income Ha^{-1} may increase by 0.13% and 0.025% as 1% change in education level and annual income of farmers, respectively. As most members of farming household can be involved in cultivation. Subsequently, the crop production increases as involvement of family member in cultivation increase. Thus, family size showed a positive impact on farm income Ha^{-1}. Furthermore, literate farmers have better knowledge and understanding on appropriate time to use irrigation, fertilizer and other inputs in cultivation. Farm income Ha^{-1} is expected to be improved as education level of farmers increases. The estimates are consistent with previous studies such as Kumar et al. (2016); Biswas et al. (2020); Singh (2020); Swami and Parthasarathy (2022). Economically strong farmers can use different techniques, modern seeds, organic and green fertilizer, pesticides, irrigation, advance technologies and other inputs in farming. Hence, yearly income has a positive impact on farm income Ha^{-1}. Singh et al. (2022a) have also recognized the positive contribution of annual income of farmers in the agricultural production activities.

The coefficients of gender of farmers with farm income Ha^{-1} was observed positive. The estimate recommended that male farmers have better association with other farmers, be the member of agricultural cooperative society and other institutional bodies. Family size, education level and annual income were seemed crucial determinants to increase CAS. Mligo et al. (2022) have also found that off-farm income is positive to increase CAS. The agriculture sector cannot grow without arable land and irrigated area. Thus, farm income Ha^{-1} increases as arable and irrigated land increase. Further, farm income Ha^{-1} be enlarged by 2.50% as arable land increase by 1%.

Human resources (i.e., agricultural work force) showed a positive impact on farm income Ha^{-1}. The impact of fertilizer and cost of technology on farm income Ha^{-1} were appeared negative. The estimates specify the farm income Ha^{-1} declines by 0.10% and 0.01% as 1% increase in applications of fertilizer and cost of technology, respectively. Crop production may increase as use of recommended quantity of fertilizer Ha^{-1} increases (Kumar et al., 2016). In contrary, extensive use of fertilizer in the cultivation may reduce soil quality and fertility in long-run. Thus, farm income Ha^{-1} may decrease due to unnecessary use of fertilizer in cultivation. Cash crop farming and crop diversification provides the better return to the farmers. The coefficient of farm income from cash crop and crop diversification index with farm income Ha^{-1} were observed positive. Farm income Ha^{-1} increases by 0.062% and 3.38% as in cash crop income and crop diversification index, respectively. AT helps to increase water conservation, soil fertility and quality, ecosystem services, seed germination, crop productivity, profit of farmers, land productivity, marketing process; and reduce transportation cost and waste materials (Jat et al., 2016; Siddick, 2019). Thus, the impact of AT on farm income Ha^{-1} was found positive. Farm income Ha^{-1} may increase by

Table 4. Impact of explanatory variables on farm income Ha^{-1}

Number of observations	240	*Ramsey RESET* test for (*DV*) [*Prob > F* = 0.0197]			3.36	
F - Value [*Prob > F* =0.000]	0.77	*Ramsey RESET* test for (*IVs*) [*Prob > F* = 0.9038]			0.72	
R-squared	0.8434	*Chi²* [*hettest*] [*Prob > Chi²* = 0.0000]			215.09	
Adj. R-squared	0.8298	*Chi²* [*estat imtest*] [*Prob > Chi²* = 0.9382			137.07	
Root MSE	0.31534	*AIC*			146.2321	
Mean VIF	19980.08	*BIC*			215.8448	
logfiph [Dependent Variable]	*Reg. Coef.*	*Std. Err.*	*t – value*	*P>|t|*	[95% Conf. Interval]	
*logtagla*cvaaea*	-0.6810	1.377	-0.490	0.022	-3.394	2.033
*logtagla*cvaamaxtem*	-6.9810	12.473	-0.560	0.046	-31.562	17.600
*logtagla*cvaamintem*	-4.8050	5.568	-0.860	0.039	-15.779	6.169
*logtagla*cvaapre*	1.4120	2.986	0.470	0.037	-4.473	7.297
*logtagla*cvaarf*	-0.9570	1.018	-0.940	0.048	-2.964	1.049
logfamsizres	0.0170	0.078	0.210	0.131	-0.138	0.171
logedulevres	0.1370	0.144	0.950	0.043	-0.147	0.421
genres	0.0770	0.154	0.500	0.617	-0.226	0.380
logannincfam	0.0250	0.071	0.350	0.029	-0.115	0.165
logtagla	2.5010	2.366	1.060	0.092	-2.163	7.164
logirrare	0.0090	0.096	-0.090	0.127	-0.197	0.179
logualph	0.0270	0.109	0.250	0.151	-0.187	0.241
logfaph	-0.1040	0.062	-1.680	0.034	-0.226	0.018
logficcph	0.0620	0.048	1.300	0.016	-0.032	0.157
logcdi	3.3800	4.897	0.690	0.031	-6.271	13.031
logcotepeha	-0.0090	0.100	-0.090	0.028	-0.206	0.188
apptec	0.0170	0.117	0.150	0.012	-0.213	0.247
finsupgov	0.0480	0.044	1.080	0.040	-0.039	0.134
adstfa	6.1610	9.471	0.650	0.016	-12.505	24.827
Con. Coef.	-16.310	38.029	-0.430	0.068	-91.258	58.637

Source: Author's estimation

0.017% as 1% increase in AT in cultivation. However, most farmers cannot buy agricultural inputs during sowing time due to their financial restrictions. Therefore, the estimate also indicates that farm income Ha^{-1} may increase by 0.048% as financial support from government and banking sector for farmers increases by 1%. Also, farm income Ha^{-1} may increase by 6.16% due to 1% increase in CAS by farmers.

Projected Farm Income in Different Climate Change Perspectives

The integrated impact of all climatic factors on farm income Ha^{-1} was appeared negative. The estimate showed that farm income is expected to be decline by 12.01% due to 1% increase in the values of included climatic factors in the model. While, the integrated impact of non-climatic factors and CAS on farm income was found positive. Also, farm income may be increased by 12.35% when contribution of non-climatic factors increases by 1% in cultivation (Figure 5). Henceforth, the estimates claimed that some non-climatic factors can be effective to address the issue of climatic change in the cultivation. Furthermore, farm income Ha^{-1} is expected to be declined by 17.32%, 20.05%, 22.89% and 28.66% by the years 2040s, 2060s, 2080s and 2100s, respectively. The marginal impact of gross cropped area, farmer's education level and gender, annual income of farmers, irrigated area, farm income from cash crop farming, crop diversification, appropriate technology, financial support for farmers from the government and CAS on farm income were observed positive and statistically significant. Thus, these variables may reduce the negative consequences of climate variability in the agricultural production system in near future.

Figure 5. Projected farm income Ha^{-1} or land productivity (estimated by authors)

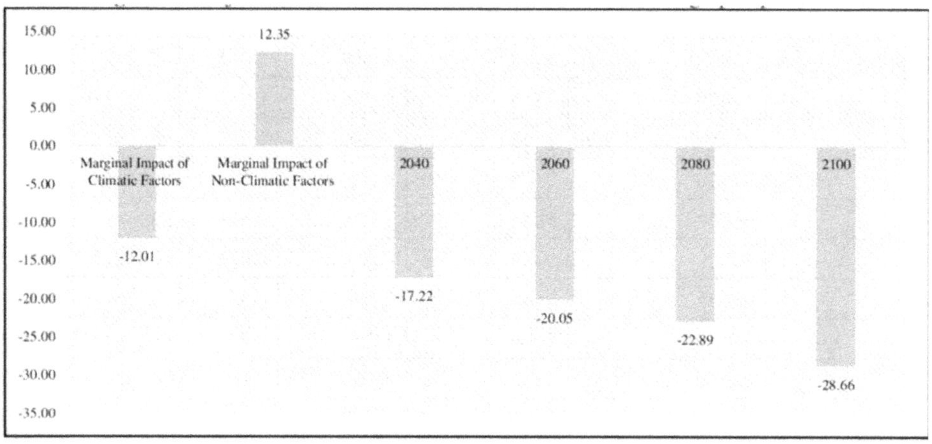

Regression Coefficient of Independent Variables with CAS

The coefficients of CAS with socio-economic factors were estimated using Logit and Probit regression models. (Table 5). Both the models produced similar values of *LR* Chi^2, *Log likelihood, Pseudo* R^2, *Log-Lik Intercept Only, Maximum Likelihood* R^2, *AIC* and *BIC*. These models produced same sign and magnitudes of coefficients of CAS with independent variables. The value of *likelihood ratio Chi-square* was also observed statistically significant which showed that these models produce reliable results.

The coefficients of family size, yearly income, gross arable land, applications of human resources, farm income from cash crops, use of fertilizer, use of technologies Ha^{-1} land, crop diversification, financial problems, financial support for farmers; farmer's collaboration with agricultural developmental institutions, and skilled and technical support for farmers were seemed useful to increase CAS of farmers. The estimates were seemed similar with previous studies such as Biswas et al. (2020); Singh (2020); Swami and Parthasarathy (2022); Singh et al. (2022a). The coefficient of age of respondents with CAS was found negative and statistically significant. It means that matured farmers have a better experience to use effective techniques in cultivation to get better return. Therefore, impact of farmer's age on CAS was seemed negative. Family size was appeared useful to increase CAS to manage the climate change impact in farming. Therefore, coefficient of family size with CAS was detected positive. Annual income of farmers would assist to increase CAS of farmers. The coefficient of arable land with CAS was also noticed positive. The result showed that income of farmers is expected to be increased as arable land increases. Accordingly, farmers can used different CAS to increase crop productivity. Human skills can be a best CAS in the agricultural sector. Cash crop farming will enhance income and purchasing power of farmers to buy new inputs and technologies. Subsequently, it would assist to the farmers to apply various technologies and CAS in farming. Therefore, coefficient of farm income from cash crops with land productivity was observed positive. Yield of specific crops may increase as application of fertilizer. Thus, fertilizer applications may be used as a CAS for cultivation of certain crops only. For instance, Singh (2017) have reported diverse impact of fertilizer application on yield of food-grain and cash crops in India.

The impact of crop diversification index on farm income Ha^{-1} was seemed positive. Crop diversification may be positive to increase the CAS of farmers. Any farmers cannot implement CAS without financial availability and accessibility. Thus, financial unavailability of farmers reduces their intention to apply CAS in cultivation. Also, poor and small farmer cannot use advance technologies, high yielding of seeds, fertilizer, pesticides, irrigation machine, electric engine and other inputs in cultivation. Hence, the estimate exhibited a negative impact of financial problem of farmers on CAS. The result indicates that financial support for farmers would be positive to increase CAS.

The coefficients of family's type, gross irrigated area, cost of technology Ha^{-1} and farmer's association with different stakeholders with CAS were not found statistically significant. Crops which grow on non-irrigated land needs more CAS to increase land productivity in presence of climate change. Moreover, the coefficient of gross irrigated area with CAS was seemed negative but statistically insignificant. Crops that grow on irrigated area can tolerate the adverse impact of high temperature and humidity effectively. Farmers can learn specific method to cultivate a particular crop from agricultural cooperative societies, agricultural extension offices and agricultural development agencies. Hence, agricultural development institutions should suggest scientific methods to reduce the negative return of climate change in the agricultural sector. Also, CAS may decrease as cost of technology Ha^{-1}increases. Farmer's association with different stakeholders also shown a positive impact on CAS.

Table 5. Regression coefficient of CAS with independent variables

Regression Model	Logit Model			Probit Model		
Number of obs.	240			240		
LR Chi^2(15)	230.66			231.15		
Prob > Chi^2	0.000			0.000		
Pseudo R^2	0.6961			0.6976		
Log-likelihood	-50.3521			-50.1029		
Adstfa [Dependent Variable]	*Log-Lik Intercept Only* = - 165.681	*Log-Lik Full Model* = - 50.352		*Log-Lik Intercept Only* = -165.680	*Log-Lik Full Model* = -50.103	
	D (224) = 100.704	*LR* (15) = 230.655		*D* (224) = 100.206	*LR* (15) = 231.154	
		Prob > LR = 0.000			*Prob > LR* = 0.000	
	McFadden's R^2 = 0.696	*McFadden's Adj. R^2* = 0.600		*McFadden's R^2* = 0.698	*McFadden's Adj R^2* = 0.601	
	Maximum Likelihood R^2 = 0.618	*Cragg & Uhler's R^2* = 0.825		*Maximum Likelihood R^2* = 0.618	*Cragg & Uhler's R^2* = 0.826	
	McKelvey and Zavoina's R^2 = 0.879	*Efron's R^2* = 0.739		*McKelvey and Zavoina's R^2* = 0.879	*Efron's R^2* = 0.735	
	*Variance of y** = 27.095	*Variance of error* = 3.290		*Variance of y** = 8.262	*Variance of error* = 1.000	
	Count R^2 = 0.904	*Adj. Count R^2* = 0.793		*Count R^2* = 0.900	*Adj. Count R^2* = 0.784	
	AIC = 0.553	*AIC*n* = 132.704		*AIC* = 0.551	*AIC*n* = 132.206	
	BIC = -1126.959	*BIC'* = - 148.446		*BIC* = - 1127.457	*BIC'* = - 148.944	
	Reg. Coef.	*Std. Err.*	*P>|z|*	*Reg. Coef.*	*Std. Err.*	*P>|z|*
ageres	-0.1057	0.0370	0.004	-0.0547	0.0180	0.002
famsizres	0.2290	0.1710	0.081	0.1136	0.0950	0.033
typfamres	-8.0788	1.5070	0.145	-4.5155	0.8140	0.123
annincfam	0.1290	1.2760	0.003	0.1130	0.9350	0.012
tagla	0.1103	0.2130	0.005	0.0418	0.0130	0.011
irrare	-0.2767	0.2440	0.256	-0.1371	0.1290	0.287
ualph	0.0755	0.0360	0.137	0.0452	0.0200	0.125
ficcph	0.4520	0.0130	0.007	0.3250	0.0230	0.042
faph	-0.0023	0.0020	0.061	-0.0013	0.0010	0.050
cotepeha	-0.0018	0.0010	0.143	-0.0010	0.0000	0.138
cdi	3.0495	3.3310	0.060	1.9537	1.7510	0.065
finpro	-1.5923	0.6100	0.009	-0.8531	0.3380	0.012
finsupgov	1.0607	0.7710	0.069	0.5682	0.4080	0.063
farassstahol	0.0267	0.5570	0.162	-0.0307	0.3140	0.122
skitecsupfar	0.8000	0.7140	0.042	0.3774	0.3790	0.020
Con. Coef.	6.5474	3.4490	0.058	3.5645	1.8930	0.060

Source: Author's estimation.

SUMMARY AND CONCLUDING REMARKS

The empirical results indicate that farm income Ha^{-1} is likely to be decreased due to change in marginal values of included climatic variables in regression model. Further, the estimates infer that farm income Ha^{-1} is expected to be declined by 0.68%, 6.98%, 4.80% and 0.96% in different years due to 1% increase in the climatic factors. The results based on Euler's theorem showed that integrated impact of all climatic factors was reported negative. Farm productivity is likely to be decreased by 12.01% due to 1% increase in actual values of undertaken climatic factors. While, the integrated marginal impact of non-climatic factors on farm income Ha^{-1} was reported positive. The estimates highlighted that non-climatic factors, institutional support and appropriate technology can be applied as CAS in cultivation. The projected results also focused that farm income Ha^{-1} is expected to be declined in near future. Thus, the estimates demonstrated that agricultural sector of Gujarat may be in more susceptible position due to climate change. Hence, farmers, agricultural scientists, government and rural developmental agencies should develop climate action plan to avoid the risk of climate change in this sector.

Moreover, the impact of several social-economic and institutional factors on farm income were found positive. Thus, these variables can be introduced as CAS in farming. For instance, the marginal impact of male farmers, human resource, arable land, income from cash crops, crop diversification, appropriate technology, financial accessibility of farmers and CSA on farm income Ha^{-1} were seemed positive. In future, education level and skills of famers may be a best CAS. The results also highlighted that farm income Ha^{-1} may be improved by 0.13%, 2.5%, 3.38% and 6.16% due to 1% increase in the education level of farmers, arable land, crop diversification and CAS, respectively in cultivation. Thus, agricultural development agencies and agricultural universities should organize more training programmes for farmers to increase their understating on climate change and reduce its impacts in farming. Farmers should cultivate cash crops to increase their income to buy new technologies and apply various CAS in cultivation.

The estimates also imply that social–economic profile of farmers has a important impact on land productivity. Gender of respondents, gross irrigated area and use of agricultural labour have a positive impact on farm income Ha^{-1}. The role of appropriate technology, financial accessibility of farmers and CSA were also found suitable to reduce negative consequences of climate change. Irrigated area, family size, education level, annual income, AT and financial support from government and banking sector, farm income from cash crops and crop diversification were also appeared positive to increase the purpose of cultivators to apply various CAS. Farmers can apply mixed cropping pattern, heat tolerance crops, less climate sensitive crops, green fertilizer, less chemical and pesticides and less water consuming crops as adaptation practices to mitigate the climate change impacts. Farmers should apply natural-based-adaptation practices such as protection of forest area, water management and conservation, sustainable land management practices, green technologies and green fertilizer to increase soil fertility and quality (Zilberman et al., 2018).

Appropriate technology, education level, income of farmers and financial availability were perceived most influencing factors of CAS. Hence, extensive application of appropriate technology, green technology, digital technology, information and communication technology (ICT), eco-adaptation-based technology, and eco-adaptation-based strategies in the agriculture sector, industrial and other sectors may be a general solution to abate the GHGs emissions. Above-mentioned initiations would minimize the possibility of climate change and extreme variability in other weather variables in near future (Singh et al., 2020b; Malhi et al., 2021). Use of appropriate technology would be supportive to sustain the common property of soil, land, water, air and other natural resources. Appropriate technology, therefore, may be conducive to increase the sustainability in economic, social and environmental development. It would build a path of sustainable development for long-run (Singh et al., 2021). Subsequently, Government and banking sector should provide green fund to the farmers for purchasing of green and environmental technology to protect ecosystem services and to mitigate the climate change impact in cultivation. ICT, digitalization and digital platform can be used as an appropriate technology to increase the communication of farmers with various stakeholders. ICT will work as a key driver to increase digitalization which enhance sustainable agricultural development (Mondejar et al., 2021). Other practices of appropriate technologies would be helpful to save water, increase efficiency of water, reduce waste materials, use of waste materials in non-agricultural activities, sustain soil fertility and quality, improve land productivity, increase seed germination, reduce high dependency of farmers on fertilizers and pesticides, and increase plant growth in farming (Zilberman et al., 2018; Singh et al., 2022a). Therefore, scientific research community should develop and discover environment friendly technologies, green technologies, appropriate technologies, digital technologies and digital devices for the agricultural sector.

REFERENCES

Abdullahi, H. S., Mahieddine, F., & Sheriff, R. E. (2015). *Technology impact on agricultural productivity: A review of precision agriculture using unmanned Aerial vehicles. Wireless and Satellite Systems.* Springer International Publishing.

Akinnagbe, O. M., & Irohibe, I. J. (2014). Agricultural adaptation strategies to climate change impacts in Africa: A review. *Bangladesh Journal of Agricultural Research, 39*(3), 407–418. doi:10.3329/bjar.v39i3.21984

Akpo, I. F., Zakari, F. T., Agani, F. O., Issaka, K., & Yabi, J. A. (2023). Effects of sustainable land management practices on food security on beneficiary households in context of adaptation to climate change in the communes of Malanville and Kouandé in Northern Benin. *International Journal of Agricultural Policy and Research, 11*(1), 16–26.

Ashraf, S. N., & Singh, A. K. (2021). Impact of technological change on growth and agricultural sector in Gujarat state of India: A time series data study. *Asian Development Policy Review, 9*(3), 144–160. doi:10.18488/journal.107.2021.93.144.160

Asrat, P., & Simane, B. (2018). Farmers' perception of climate change and adaptation strategies in the Dabus watershed, North-West Ethiopia. *Ecological Processes, 7*(7), 1–13. doi:10.118613717-018-0118-8

Balasha, A. M., Munyahali, W., Kulumbu, J. T., Okwe, A. N., Fyama, J. N. M., Lenge, E. K., & Tabbwe, A. N. (2023). Understanding farmers' perception of climate change and adaptation practices in the marshland of South Kivu, Democratic Republic of Congo. *Climate Risk Management, 39*(100469), 1–15. doi:10.1016/j.crm.2022.100469

Biswas, S., Chatterjee, S., & Roy, D. C. (2020). Understanding of farmers' perception of climate change and adaptation strategies: A case study in Jhargram district of West Bengal, India. *Journal of Applied and Natural Science, 12*(2), 207–212. doi:10.31018/jans.vi.2241

Chandio, A. A., Jiang, Y., Amin, A., Ozturk, I., Sinha, A., & Ahmad, F. (2021). Modeling the impact of climatic and non-climatic factors on cereal production: Evidence from Indian agricultural sector. *Environmental Science and Pollution Research International*. Advance online publication. doi:10.100711356-021-16751-9 PMID:34617217

Falco, S. D., & Veronesi, M. (2013). How can African agriculture adapt to climate change? A counterfactual analysis from Ethiopia. *Land Economics*, *89*(4), 743–766. doi:10.3368/le.89.4.743

Fierros-González, I., & López-Feldman, A. (2021). Farmers' perception of climate change: A review of the literature for Latin America. *Frontiers in Environmental Science*, *9*(672399), 1–7. doi:10.3389/fenvs.2021.672399

Gadgil, S. (1995). Climate change and agriculture-an Indian perspective. *Current Science*, *69*(8), 649–659.

Huong, N. T. L., Bo, Y. S., & Fahad, S. (2019). Economic impact of climate change on agriculture using Ricardian approach: A case of northwest Vietnam. *Journal of the Saudi Society of Agricultural Sciences*, *18*(4), 449–457. doi:10.1016/j.jssas.2018.02.006

Jamal, A. M., Antwi-Agyei, P., Baffour-Ata, F., Nkiaka, E., Antwi, K., & Gbordzor, A. (2021). Gendered perceptions and adaptation practices of smallholder cocoa farmers to climate variability in the Central Region of Ghana. *Environmental Challenges*, *5*(100293), 1–7. doi:10.1016/j.envc.2021.100293

Jat, M. L., Dagar, J. C., Singh, Y., Govaerts, B., Ridaura, S. L., Saharawat, Y. S., Sharma, R. K., Tetarwal, J. P., Jat, R. K., Hobbs, H., & Stirling, C. (2016). Climate change and agriculture: Adaptation strategies and mitigation opportunities for food security in South Asia and Latin America. *Advances in Agronomy*, *137*(1), 127–235. doi:10.1016/bs.agron.2015.12.005

Jyoti, B., & Singh, A. K. (2020). Projected sugarcane yield in different climate change scenarios in Indian states: A state-wise panel data exploration. *International Journal of Food and Agricultural Economics*, *8*(4), 343–365.

Kangogo, D., Dentoni, D., & Bijman, J. (2021). Adoption of climate-smart agricultural among smallholder farmers: Does farmers entrepreneurship matter? *Land Use Policy*, *109*(1), 1–13.

Kumar, A., Ahmad, M. M., & Sharma, P. (2017). Influence of climatic and non-climatic factors on sustainable food security in India: A statistical investigation. *International Journal of Sustainable Agricultural Management and Informatics*, *3*(1), 1–30. doi:10.1504/IJSAMI.2017.082917

Kumar, A., & Sharma, P. (2014). Climate change and sugarcane productivity in India: An econometric analysis. *Journal of Social and Development Sciences*, *5*(2), 111–122. doi:10.22610/jsds.v5i2.811

Kumar, A., Sharma, P., & Ambrammal, S. K. (2014). Climatic effects on food grain productivity in India: A crop wise analysis. *Journal of Studies in Dynamics and Change*, *1*(1), 38–48.

Kumar, A., Sharma, P., & Ambrammal, S. K. (2015). Climatic effects on sugarcane productivity in India: A stochastic production function application. *International Journal of Economics and Business Research*, *10*(2), 179–203. doi:10.1504/IJEBR.2015.070984

Kumar, A., Sharma, P., & Joshi, S. (2016). Assessing the impacts of climate change on land productivity in Indian crop agriculture: An evidence from panel data analysis. *Journal of Agricultural Science and Technology*, *18*(1), 1–13.

Luu, A. T., Hguyen, A. T., Trinh, Q. A., Pham, V. T., Le, B. B., Nguyen, D. T., Hoang, Q. N., Pham, H. T. T., Nguyen, T. K., Luu, V. N., & Hens, L. (2019). Farmers' intention to climate change adaptation in agriculture in the red river Delta Biosphere Reserve (Vietnam): A combination of structural equation modelling (SEM) and protection motivation theory (PMT). *Sustainability*, *11*(2993), 1–17. doi:10.3390u11102993

Malhi, G. S., Kaur, M., & Kaushik, P. (2021). Impact of climate change on agriculture and its mitigation strategies. *Sustainability*, *13*(1318), 1–21. doi:10.3390u13031318

Mligo, I., Misana, S., & Pauline, N. M. (2020). The effectiveness of adaptation strategies to climate change and variability in enhancing rural smallholder farmers' food security in Mvomero district, Tanzania. *Journal of the Geographical Association of Tanzania*, *42*(1), 45–63. doi:10.56279/jgat.v42i1.216

Mondal, P., Jain, M., DeFries, R. S., Galford, G. L., & Small, C. (2015). Sensitivity of crop cover to climate variability: Insights from two Indian agro-ecoregions. *Journal of Environmental Management*, *148*(1), 21–30. doi:10.1016/j.jenvman.2014.02.026 PMID:24680541

Mondejar, M. E., Avtar, R., Diaz, H. L. B., Dubey, R. K., Esteban, J., Gómez-Morales, A., Hallam, B., Mbungu, N. T., Okolo, C. C., Prasad, K. A., She, Q., & Garcia-Segura, S. (2021). Digitalization to achieve sustainable development goals: Steps towards a smart green planet. *The Science of the Total Environment*, *794*(148539), 1–20. doi:10.1016/j.scitotenv.2021.148539 PMID:34323742

Moon, Y., & Hwang, J. (2018). Crowdfunding as an alternative means for funding sustainable appropriate technology: Acceptance determinants of backers. *Sustainability*, *10*(1456), 1–18. doi:10.3390u10051456

Msomba, B. H., Ndaki, P. M., & Nassary, E. K. (2021). Assessing adaptation measures to climate change applied by out-growers for the sustainable production of sugarcane in Tanzania. *Journal of Global Agriculture and Ecology*, *12*(1), 18–29.

Musafiri, C. M., Kiboi, M., Macharia, J., Ng'etich, O. K., Kosgei, D. K., Mulianga, B., Okoti, M., & Ngetich, F. K. (2022). Smallholders' adaptation to climate change in Western Kenya: Considering socioeconomic, institutional and biophysical determinants. *Environmental Challenges*, *7*(100489), 1–11. doi:10.1016/j.envc.2022.100489

Musunuri, D. (2014). Appropriate technology and economic development of emerging economies— A myth or a reality. *Journal of Technology Management for Growing Economies*, *5*(1), 51–62. doi:10.15415/jtmge.2014.51003

Ndamani, F., & Watanabe, T. (2016). Determinants of farmers' adaptation to climate change: A micro level analysis in Ghana. *Scientia Agrícola*, *73*(3), 201–208. doi:10.1590/0103-9016-2015-0163

Pakmehr, S., Yazdanpanah, M., & Baradaran, M. (2020). Explaining farmers' response to climate change-induced water stress through cognitive theory of stress: An Iranian perspective. *Environment, Development and Sustainability*, *23*(1), 5776–5793.

Sargani, G. R., Jiang, Y., Chandio, A. A., Shen, Y., Ding, Z., & Ali, A. (2022). Impacts of livelihood assets on adaptation strategies in response to climate change: Evidence from Pakistan. *Environment, Development and Sustainability*. Advance online publication. doi:10.100710668-022-02296-5

Schattman, R. E., Hurley, S. E., Greenleaf, H. L., Niles, M. T., & Caswell, M. (2020). Visualizing climate change adaptation: An effective tool for agricultural outreach? *Weather, Climate, and Society*, *12*(1), 47–61. doi:10.1175/WCAS-D-19-0049.1

Shabbir, M. M., & Yaqoob, N. (2019). The impact of technological advancement on total factor productivity of cotton: A comparative analysis between Pakistan and India. *Journal of Economic Structures*, *4*(1), 1–24. doi:10.118640008-019-0160-4

Sharma, P., & Singh, A. K. (2017). Association of state-wise food security index with climatic factors in India: Evidence from state-wise panel data. *Journal of Global Agriculture and Ecology*, *6*(3), 196–205.

Siddick, S. A. (2019). Appropriate technologies for improving yield and income of small holders growing rice paddy in rainfed low lands of agro-biodiversity hotspots in India. *Agricultural Sciences*, *10*(1), 1497–1505. doi:10.4236/as.2019.1011110

Singh, A. K. (2017). *Climate Change and Productivity of Food-grain and Cash Crops in India*. LAP LAMBERT Academic Publishing.

Singh, A. K., & Issac, J. (2018). Impact of climatic and non-climatic factors on sustainable livelihood security in Gujarat state of India: A statistical exploration. *Agriculture and Food Sciences Research*, *5*(1), 30–46. doi:10.20448/journal.512.2018.51.30.46

Singh, A. K., & Jyoti, B. (2019). Measuring the climate variability impact on cash crops farming in India: An Empirical Investigation. *Agriculture and Food Sciences Research*, *6*(2), 155–164. doi:10.20448/journal.512.2019.62.155.165

Singh, A. K., & Jyoti, B. (2021). Projected food-grain production and yield in India: An evidence from state-wise panel data investigation during 1977-2014. *The Journal of Agricultural Sciences – Sri Lanka*, *16*(1), 108-125.

Singh, A. K., Jyoti, B., Kumar, S., & Lenka, S. K. (2021). Assessment of global sustainable development, environmental sustainability, economic development and social development index in selected economies. *International Journal of Sustainable Development and Planning*, *16*(1), 123–138. doi:10.18280/ijsdp.160113

Singh, A. K., & Kumar, S. (2022). Exploring the impact of sustainable development on social-economic, and science and technological development in selected countries: A panel data analysis. *Society & Sustainability*, *4*(1), 55–83. doi:10.38157s.v4i1.405

Singh, A. K., Kumar, S., Ashraf, S. N., & Jyoti, B. (2022a). Implications of farmer's adaptation strategies to climate change in agricultural sector of Gujarat: Experience from farm level data. *Research on World Agricultural Economy*, *3*(1), 42–57. doi:10.36956/rwae.v3i1.498

Singh, A. K., Kumar, S., & Jyoti, B. (2022b). Association of sustainable food security with climatic and non-climatic factors in Gujarat state of India: A district-wise panel data investigation. In Sustainable Agriculture and Food Security. World Sustainability Series. Springer Nature Switzerland. doi:10.1007/978-3-030-98617-9_16

Singh, A. K., Kumar, S., & Jyoti, B. (2022c). Does sustainable livelihood security have a relationship with climatic and geographical factors? evidence from a state-wise panel data investigation in India. *The IUP Journal of Applied Economics*, *21*(1), 7–35.

Singh, A. K., Kumar, S., & Jyoti, B. (2022d). Influence of climate change on agricultural sustainability in India: A state-wise panel data analysis. *Asian Journal of Agriculture*, *6*(1), 15–27. doi:10.13057/asianjagric/g060103

Singh, A. K., Kumar, S., & Jyoti, B. (2022e). Impact of the COVID-19 on food security and sustainable development goals in India: Evidence from existing literature. *GNOSI: An Interdisciplinary Journal of Human Theory and Praxis*, *5*(2), 94–109.

Singh, A. K., Narayanan, K. G. S., & Sharma, P. (2017). Effect of climatic factors on cash crop farming in India: An application of Cobb-Douglas production function model. *International Journal of Agricultural Resources, Governance and Ecology*, *13*(2), 175–210. doi:10.1504/IJARGE.2017.086452

Singh, A. K., Narayanan, K. G. S., & Sharma, P. (2019). Measurement of technical efficiency of climatic and non-climatic factors in sugarcane farming in Indian states: Use of stochastic frontier production function approach. *Climatic Change*, *5*(19), 150–166.

Singh, A. K., & Sharma, P. (2018). Implications of climatic and non-climatic variables on food security in developing economies: A conceptual review. *MOJ Food Processing & Technology*, *6*(1), 1–12. doi:10.15406/mojfpt.2018.06.00138

Singh, A. K., Sharma, P., & Singh, D. K. (2016). Measuring the influence of weather variables on productivity of food-grain crops in India: An application of Just & Pope's production technique. *ABBS Management Business and Entrepreneurship Review*, *7*(2), 29–46. doi:10.23874/amber/2016/v7/i2/121780

Singh, A. K., & Singh, B. J. (2020). Assessing the infectious diseases of students in different weather seasons in DIT University Dehradun, Uttarakhand (India). *Asian Journal of Multidimensional Research*, *9*(3), 34–48. doi:10.5958/2278-4853.2020.00055.5

Singh, A. K., & Singh, B. J. (2021). Projected productivity of cash crops in different climate change scenarios in India: Use of marginal impact analysis technique. *Finance & Economics Review*, *3*(1), 63–87.

Singh, A. K., Singh, B. J., & Negi, V. (2020b). Does sustainable development have a causal relationship with environmental development? Evidence from a country-wise panel data analysis. *International Journal of Technology Management & Sustainable Development*, *19*(2), 147–171. doi:10.1386/tmsd_00020_1

Singh, A. K., Singh, J. B., & Sharma, P. (2020a). Assessing the climate change impact on rice and wheat production in Uttar Pradesh and Haryana states of India. *Climatic Change*, *6*(21), 74–93.

Singh, S. (2020). Farmers' perception of climate change and adaptation decisions: A micro-level evidence from Bundelkhand region, India. *Ecological Indicators*, *116*(106475), 1–13. doi:10.1016/j.ecolind.2020.106475

Singh, S., & Singh, A. (2019). Farmers' perception of climate change and livelihood vulnerability in rainfall regions of India: A gender-environment perspective. *International Journal of Environment and Climate Change*, *9*(12), 878–889. doi:10.9734/ijecc/2019/v9i1230168

Smit, B., & Skinner, M. W. (2002). Adaptation options in agriculture to climate change: A typology. *Mitigation and Adaptation Strategies for Global Change*, *7*(1), 85–114. doi:10.1023/A:1015862228270

Swain, B. B. (2016). Does technological linkage in contract farming increase farm productivity and efficiency? The case of hybrid paddy seed cultivation in undivided Andhra Pradesh. *Agricultural Economics Research Review*, *29*(2), 211–224. doi:10.5958/0974-0279.2016.00049.5

Swami, D., & Parthasarathy, D. (2022). Assessing effectiveness of agricultural adaptation strategies in context of crop loss: A case study of the Indian subcontinent. *Regional Environmental Change*, *22*(71), 23–35. doi:10.100710113-022-01921-3

Vîlcu, A. D., & Vîlcu, G. E. (2013). On homogeneous production functions with proportional marginal rate of substitution. *Mathematical Problems in Engineering*, *2013*(732643), 1–5. doi:10.1155/2013/732643

Weldesilassie, A. B., Assefa, B., & Hagos, A. (2015). Productivity and welfare impact of climate change in sugarcane and cotton producing regions of Ethiopia. EDRI Research Report 23, Addis Ababa: Ethiopian Development Research Institute.

Yadav, M. K., Singh, R. S., Singh, K. K., Mall, R. K., Patel, C. B., Yadav, S. K., & Singh, M. K. (2015). Assessment of climate change impact on productivity of different cereal crops in Varanasi, India. *Journal of Agrometeorology*, *17*(2), 179–184. doi:10.54386/jam.v17i2.1000

Zilberman, D., Lipper, L., McCarthy, N., & Gordon, B. (2018). Innovation in response to climate change. Natural Resource Management and Policy, 52(1), 49-74.

Chapter 7
Advanced Fuels for Engine Emission Control

Kannan Gopal Radhakrishnan
PSNA College of Engineering and Technology, India

ABSTRACT

The importance of biodiesel increases day by day due to the depletion of petroleum reserves and increasing environmental concerns. The performance, emission, and combustion characteristics of diesel engine fuelled with waste cooking palm oil-based biodiesel were improved by varying the engine parameters like fuel injection pressure injection timing and thermal barrier-coated engine components. Further, simultaneous improvement of biodiesel fuel properties, as well as engine characteristics, were achieved by implementing fuel modification such as the use of a fuel-borne catalyst added biodiesel, ternary blends, and water micro-emulsion. Modified fuels such as biodiesel-diesel-ethanol or diestrol and diestrol-microemulsion fuel engine modification on the diesel engine. Experiments were carried out on a Kirloskar make, naturally aspirated, single-cylinder, four-stroke, water-cooled, and direct injection (DI) diesel engine developing 5.2 kW at a constant speed of 1500 rpm.

INTRODUCTION

The proposal of the book chapter under advanced fuels for engine emission control included the diesel engine characteristics using biodiesel generated from waste cooking palm oil, fuel additives added biodiesel, ternary fuels, and also the simultaneous effect of engine modification such as varying compression ratio, injection pressure, and injection timing.

The usage of biodiesel as an automotive fuel ensures energy security, and improvised air quality and is also a suitable alternative for petroleum reserves. Numerous feed stocks either from edible or non-edible vegetable oils could be used for biodiesel production which will result in the high price of biodiesel. Reusing used cooking oil as a feedstock for biodiesel production lowers the cost of making the fuel while also avoiding the expense of disposing and treating used cooking oil. Additionally, compared to diesel, it showed a number of technical limitations, including a lower calorific value, a higher viscosity, and poor low-temperature qualities.

DOI: 10.4018/978-1-6684-7303-0.ch007

The proposal of the book chapter under advanced fuels for engine emission control included the diesel engine characteristics using biodiesel generated from waste cooking palm oil, fuel additives added biodiesel, ternary fuels, and also the simultaneous effect of engine modification such as varying compression ratio, injection pressure, and injection timing.

The chapter has five subdivisions to highlight emission reduction by using fuel and engine modification strategies.

1. The Impact of Varying Injection Pressure and Timing on Biodiesel-Powered Diesel Engines
2. Impact of Fuel Borne Catalyst on Diesel Engine Using Biodiesel
3. Influence of Oxygenate Additives on Diesel Engine With Biodiesel
   3.1 Using Ternary Blends by the Addition of 1-Propanol and 1-Hexanol
   3.2 Using Diestrol (Biodiesel-Diesel-Ethanol)
4. Engine Study Using Diestrol-Microemulsion
5. Effect of Varying Compression Ratios and Thermal Barrier-Coated Piston on a Biodiesel-Fuelled Diesel Engine

A Kirloskar-brand, naturally aspirated, single-cylinder, four-stroke, water-cooled, and direct injection (DI) diesel engine producing 5.2 kW at a constant speed of 1500 rpm was used for the experimental evaluation of subdivisions from 1 to 4. The schematic diagram of the experimental setup is shown in Figure 1. A data acquisition system is a physical hardware component that is usually interfaced between the signal and the computer. It collects signals from the charge amplifier and TDC marker and processes them to obtain desired information. A voltage signal is produced by the charge amplifier that is proportionate to the cylinder pressure. Normally, each crank angle affects how much power is stored in the charge amplifier. The resolution of the TDC mark effectively determines the data gathering rate. An average of 20 data cycles with a resolution of 0.5°CA were recorded using data capturing equipment.

Figure 1. Schematic diagram of the engine setup

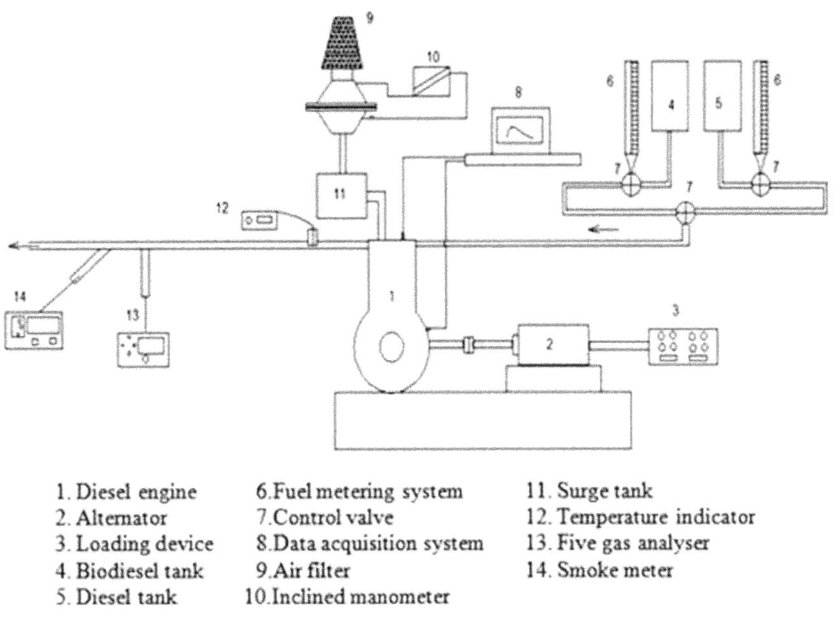

1. IMPACT OF VARYING INJECTION PRESSURE AND INJECTION TIMING ON DIESEL ENGINE WITH BIODIESEL

Figure 2 displays the findings achieved 220 bar injection pressure and 23°bTDC injection timing. A minimum BSEC of 12.88 MJ/kW h was recorded at an injection pressure of 280 bar and timing of 25.5obTDC, which is 4.2% higher than that of diesel. The result of biodiesel's increased viscosity and slow vaporisation, which come from inappropriate fuel and air mixing, is lessened by advance injection timing (Jindal et al., 2010; Hoang, 2021).

Figure 2. Change of BSEC of biodiesel

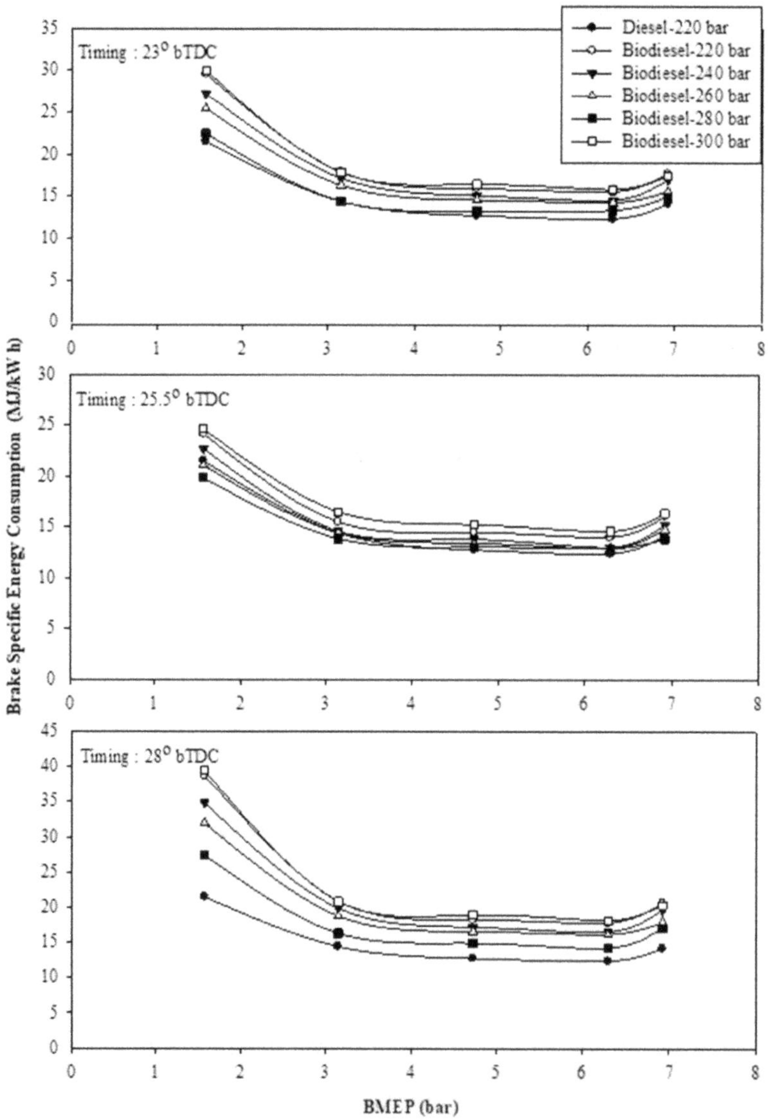

Figure 3 demonstrates that the engine operates at a maximum brake thermal efficiency of 30.5% at 280 bar and 25.5 ° bTDC, which is 1.4% higher than the diesel fuel at 100% load state.

Figure 3. Change of biodiesel's brake thermal efficiency

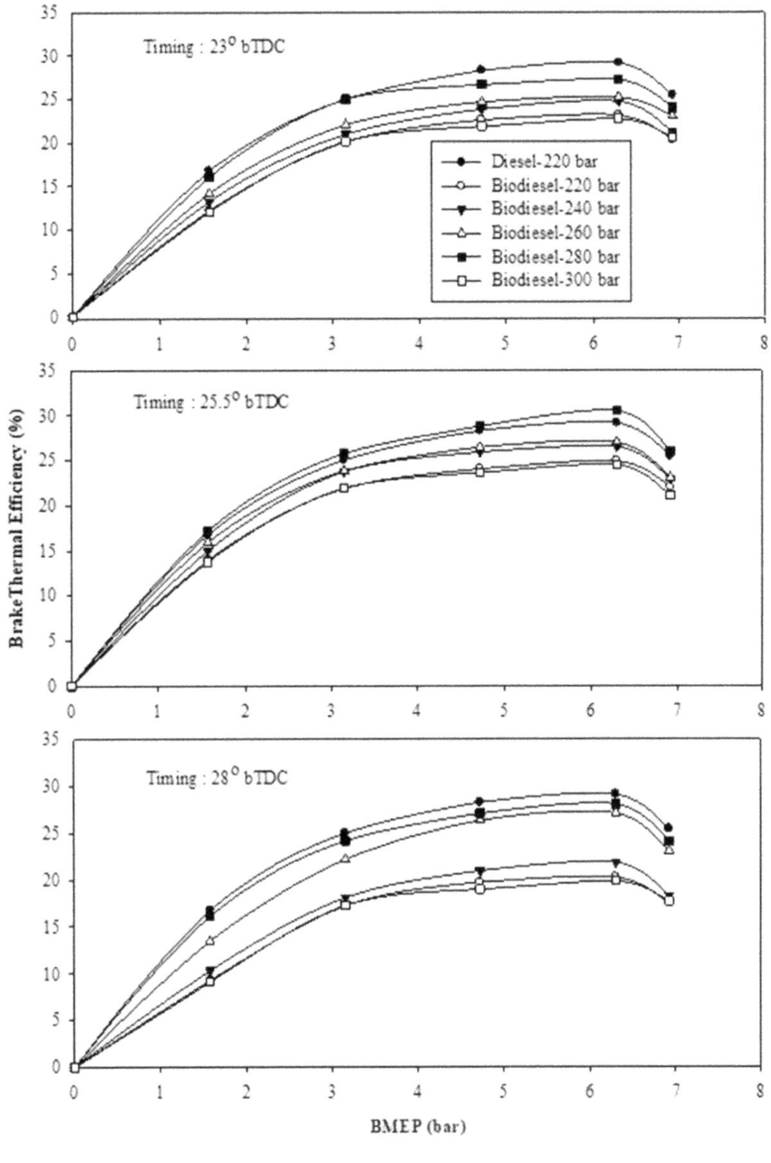

Figure 4 compares nitric oxide (NO) emissions for different injection pressures and test fuel durations. For all engine modifications, biodiesel NO emissions are less than those of diesel fuel. This is due to the lower iodine value of biodiesel (McGormick et al., 2005; Selvakumar Ramalingam and Mahalakshmi, 2020). The amount of accessible double bonds or the degree of unsaturation in the combination of fatty acids is indicated most accurately by the iodine number (Mittelbach and Remschmidt, 2004; Lapuerta et al., 2009). The amount of iodine in the biodiesel was discovered to be 57.3; the lower iodine value assured that the biodiesel include more saturated fatty acids (Chang and Van Gerpan, 1997).

Figure 4. Biodiesel no variance

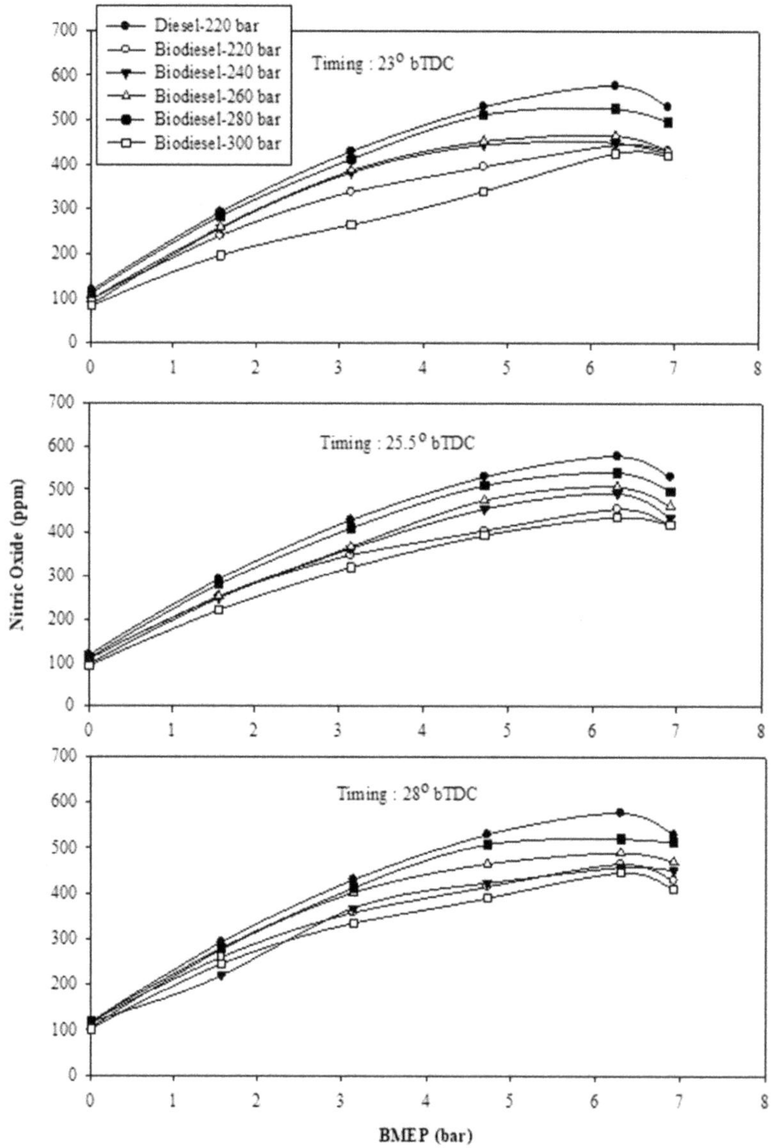

The lowest smoke opacity, measured in Figure 5 at 34.9% and attained at 280 pressure and 25.5 bTDC, is 17.9% lower than that of diesel. The key factors contributing to the decrease in smoke level are the oxygen content of biodiesel and the fuel's small particle diameter when injected at a high injection pressure. Also, an improvement in injection timing revealed longer length and higher temperatures during the expansion stroke, which causes the soot particles to oxidise (Cenk Sayin et al., 2010; Fayad, 2019).

Figure 5. Change of smoke emission for various IPs and ITs

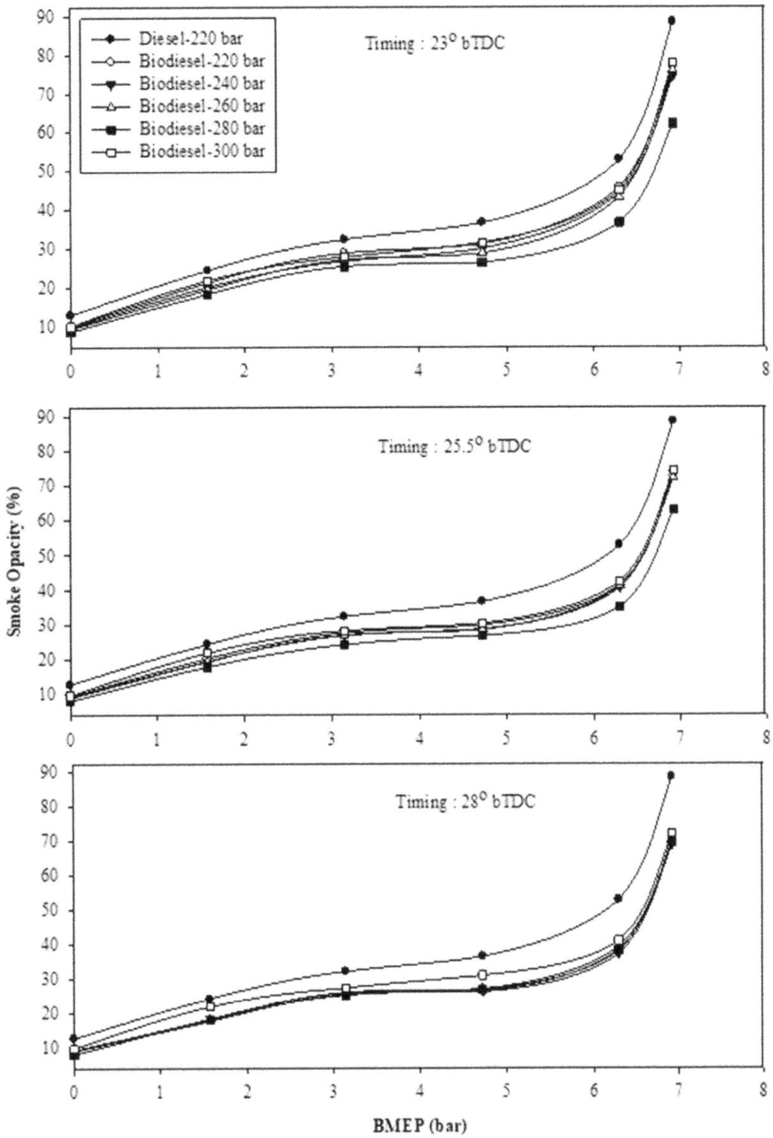

Figure 6 illustrates the main factors reducing the CO_2 emissions of biodiesel fuel at all injection pressures and timings. These factors include the lower carbon content of biodiesel in an equivalent volume of fuel consumed for the same load. The advanced 25.5°bTDC and higher 280 bar produced superior combustion as compared to other conditions, which resulted in the observation of the highest CO_2 emission.

Figure 6. Biodiesel emissions of carbon dioxide vary with varying injection pressures and timings

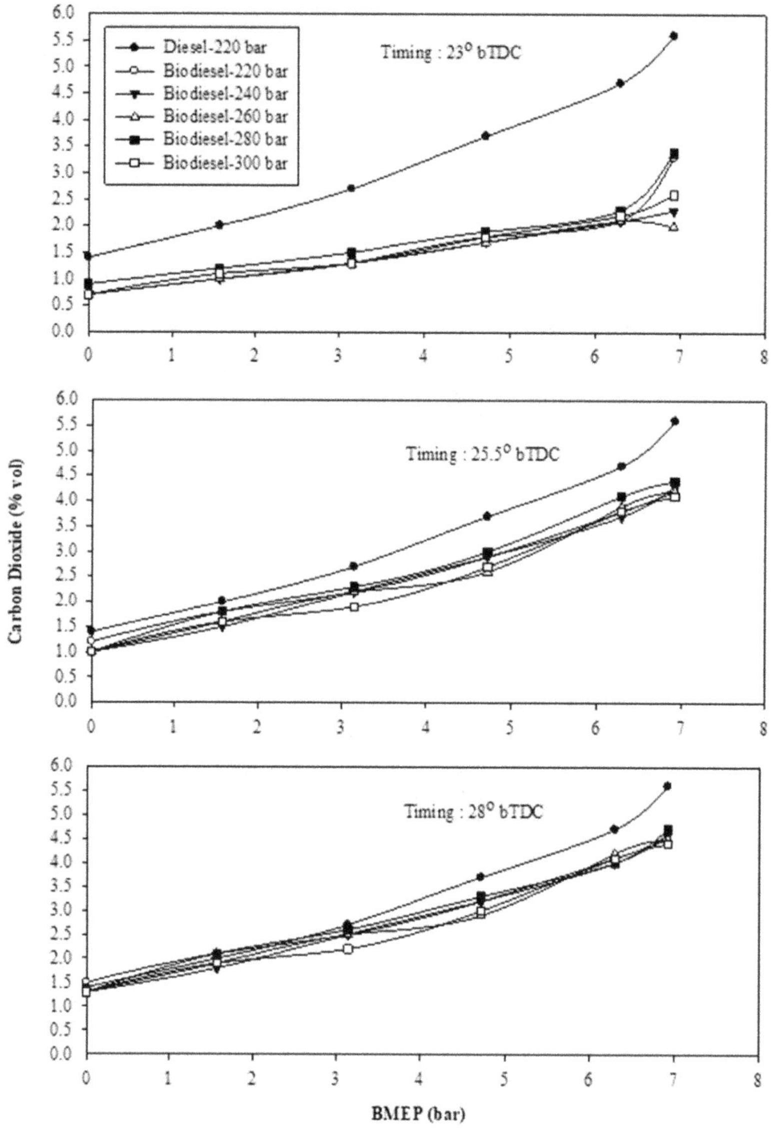

Figure 7 shows the fluctuation in cylinder gas pressure for test fuels under a 100% load condition. During the premixed burning phase, the fuel fraction that is burned determines the cylinder gas pressure. The fuel's aptitude for combining with air and burning is characterised by the cylinder gas pressure. As more fuel is consumed during the premixed combustion stage, the peak pressure and maximum rate of pressure rise increase.

Figure 7. Cylinder gas pressure of biodiesel under 100% load condition

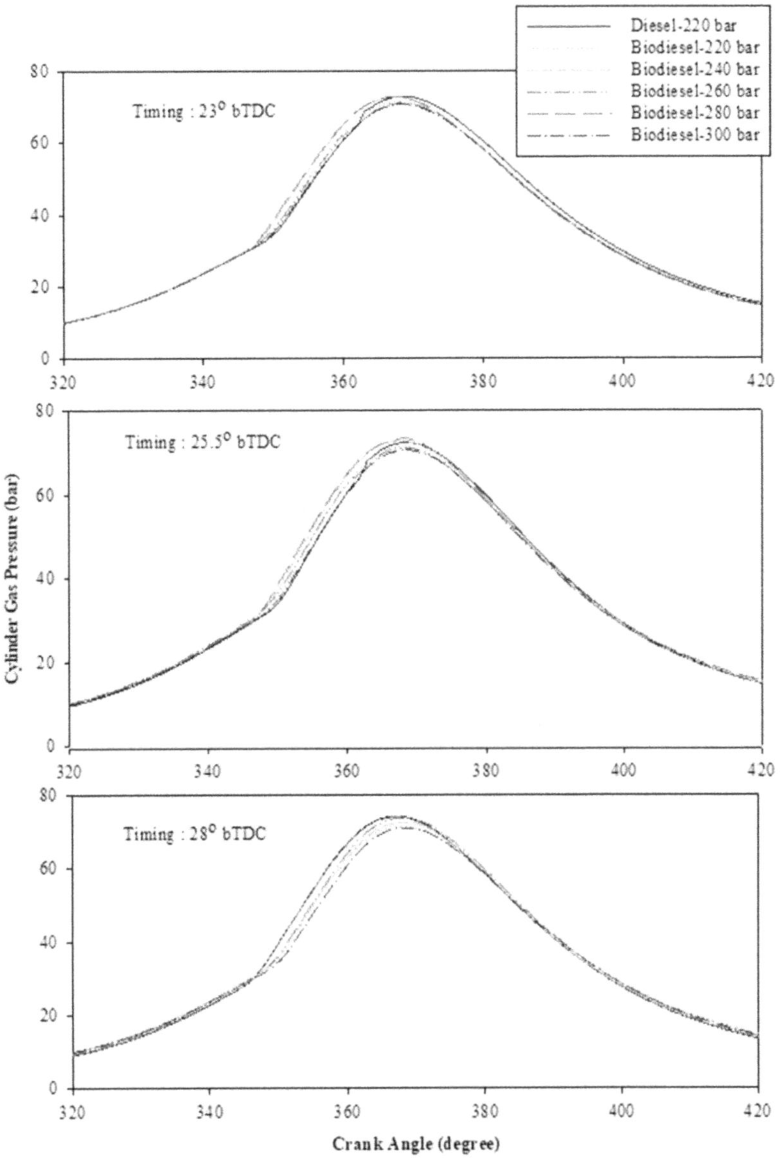

Figure 8 shows the variance in biodiesel heat release rate. Biodiesel has a maximum heat release rate of 27.8 J/ °CA at 280 bar at 25.5 °bTDC, which is 1.1% higher than that of diesel fuel. This is because the fuel builds up more due to the ignition delay caused by earlier fuel injection and the maximum heat release rate caused by a higher injection pressure during the premixed combustion phase (Puhan et al., 2009; Cavalcante et al., 2021; Simpson and Depcik, 2022).

Figure 8. HRR of biodiesel

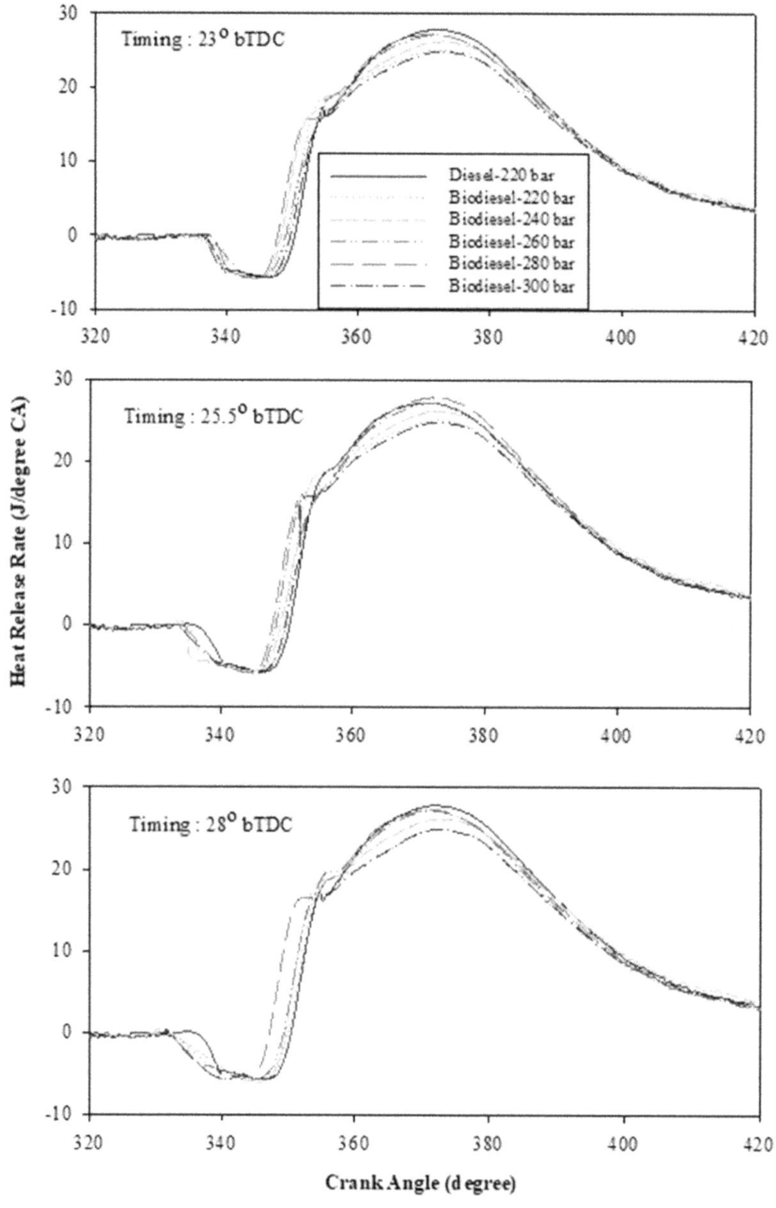

2. IMPACT OF FUEL-BORNE CATALYST (FBC) ON BIODIESEL-FUELED DIESEL ENGINE

This section elaborates on the effects of $FeCl_3$ as a fuel-borne catalyst on the characteristics of biodiesel-fueled engines. The amount of ferric chloride added to the biodiesel was 20 mol/L based on the acceptable fuel characteristics. The results obtained were compared with the optimal condition of pure biodiesel at 280 bar and 25.5° bTDC and diesel at 220 bar and 23° bTDC.

Figure 9 shows how the BSEC of the test fuels changed. At 6.2 BMEP, a slightly reduced BSEC of 12.5 MJ/kW hr was observed with biodiesel added per FBC at 280 bar and 25.5° bTDC . At the same operating and normal operating conditions, respectively, this BSEC value for biodiesel with FBC added is 3.2% and 19.9% lower than that of biodiesel without FBC.

Figure 9. BSEC of test fuels with BMEP

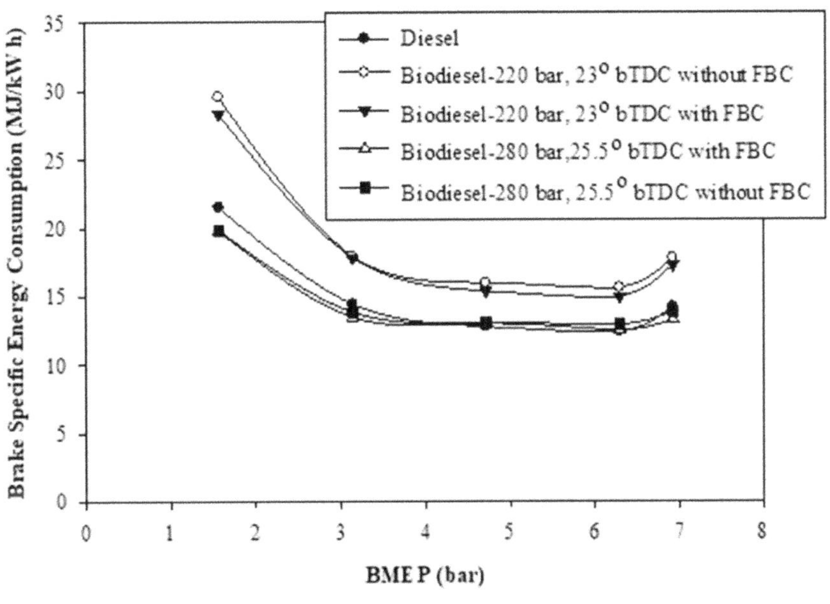

Figure 10 shows the variation in brake thermal efficiency for test fuels with BMEP. The decrease in biodiesel BTE is due to the low heating value of biodiesel (Keskin et al., 2007). At typical operating circumstances, the thermal efficiency of biodiesel compared to diesel has significantly improved at 280 bar with 25.5° bTDC. Moreover, the presence of air in the combustion chamber caused high valence (Fe(V)=O) iron oxo species, which in turn contributed to the catalytic impact of ferric chloride. To enhance combustion, high-value iron oxidising species have frequently been developed as primary oxidizers (Xiaopeng Shan and Lawrence Que 2006). Moreover, metals can act as oxidation catalysts to speed up the oxidation of hydrocarbons (Yang et al., 1998).

The variation in nitric oxide (NO) emissions is seen in Figure 11. The emission of biodiesel NO under various operating conditions was lower than that of diesel across the load range because of the drop in the iodine number of biodiesel. (Gürü et al., 2010; Chang and Van Gerpan, 1997). This study found that biodiesel made from leftover cooking oil had a Cetane rating of 61.

Thus, it can be argued that higher saturation levels (lower iodine numbers) and longer chain fatty acids reduce NO emissions (Ban-Weiss et al., 2007). The sources of the increased NO emission are faster combustion, hotter cylinder gas temperatures, and peak pressure that occurs at an earlier crank angle (Narayana Reddy and Ramesh, 2006; Rakopoulos et al., 2018; Simpson and Depcik, 2022).

Figure 10. BTE variations with of fuels

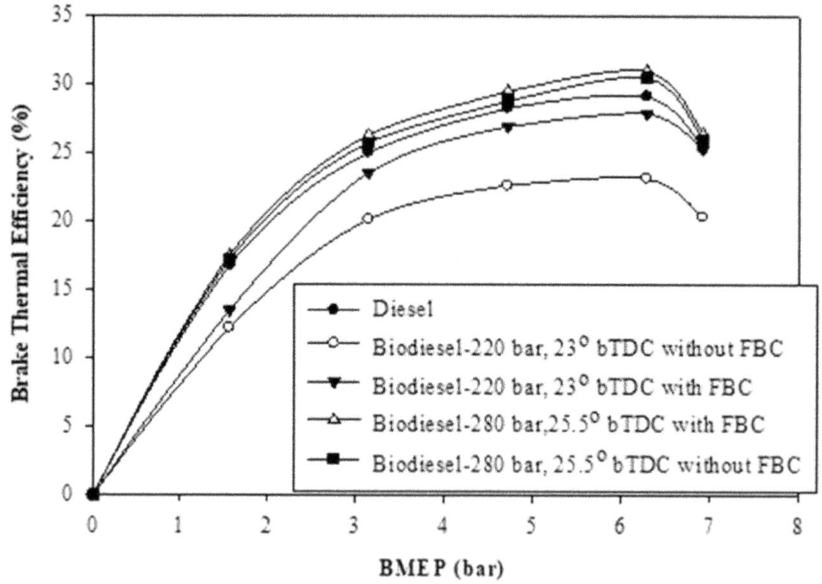

Figure 11. Nitric oxide of test fuels with BMEP

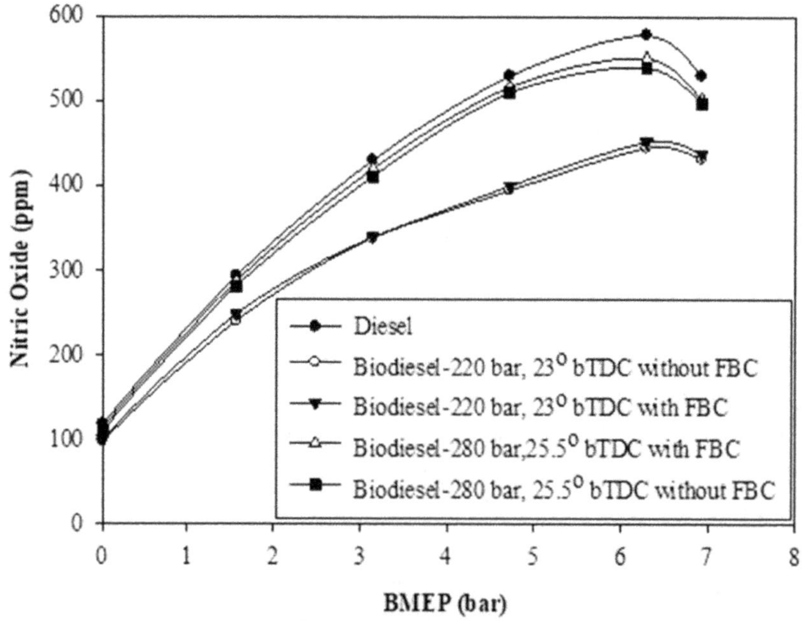

Figure 12 displays the smoke emissions. Because biodiesel contains more oxygen than diesel, biodiesel smoke emissions were lower at all loads. (Schmidt & Van Garpen, 1996). In comparison to conventional biodiesel operating circumstances, biodiesel at 280 bar and an 25.5 °bTDC showed the lowest smoke opacity, at 34.9%. The main factors contributing to the decrease in smoke level are the biodiesel's oxygen content and the fuel's small particle size when injected at a greater pressure. (Liang et al., 2019). Under the same operating conditions, smoke emissions with added biodiesel FBC were 3.9% lower than those with non-FBC biodiesel. The presence of $FeCl_3$ was responsible for the decrease in smoke emissions.

Figure 12. Smoke emissions with BMEP

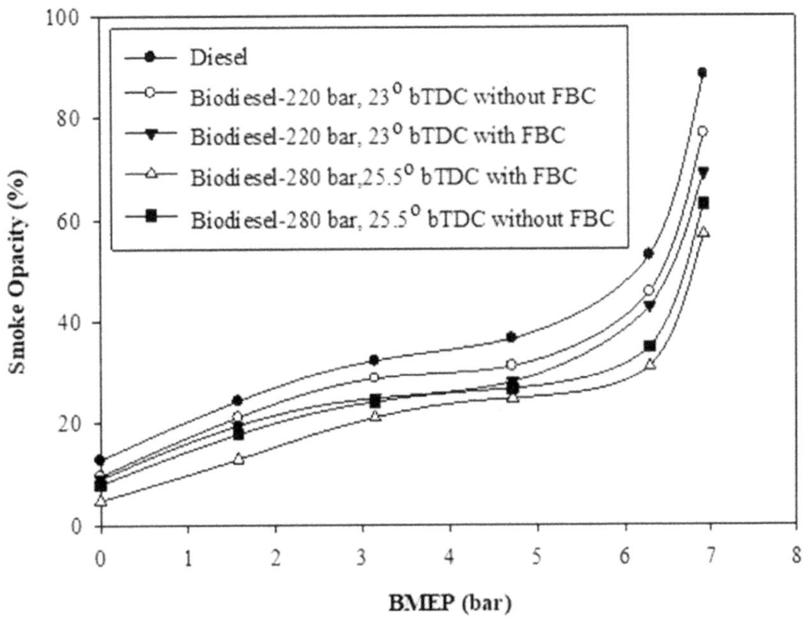

Figure 13 compares the un-burnt hydrocarbon (UHC) emissions of test fuels. The UHC emission of biodiesel was reduced to 13 ppm under optimized operation. This is due to improving spray characteristics, which led to better fuel and air mixing, as a result of higher injection pressure and timing (Rakopoulos et al., 2018; Liang et al., 2019; Simpson and Depcik, 2022). FBC included through the use of FeCl3, which catalyses the conversion of unburned hydrocarbons to CO_2 and water vapour, biodiesel has better combustion (Sadhik Basha et al., 2022).

Figure 14 displays the cylinder gas pressure from several test fuels operating at full load. Advancements in fuel injection timing cause combustion to begin earlier, whilst improvements in injection pressure cause biodiesel's ignition delay period to be shorter (Rakopoulos et al., 2018). The cylinder gas pressure for biodiesel with FBC was measured to be 75.8 bar (366.1 °CA) and for biodiesel without FBC to be 74.3 bar (367.2 °CA) under optimal operating conditions.

Figure 13. Unburnt hydro carbon emissions of test fuels

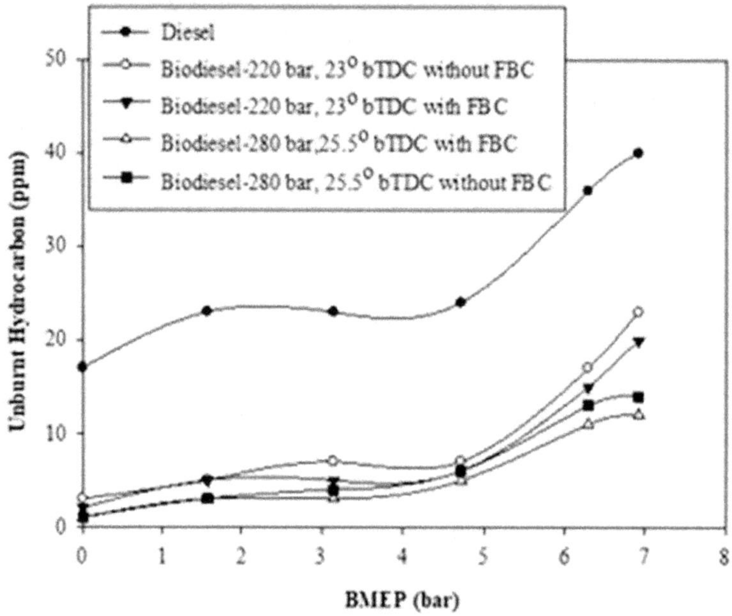

Figure 14. Cylinder gas pressure at 100% load

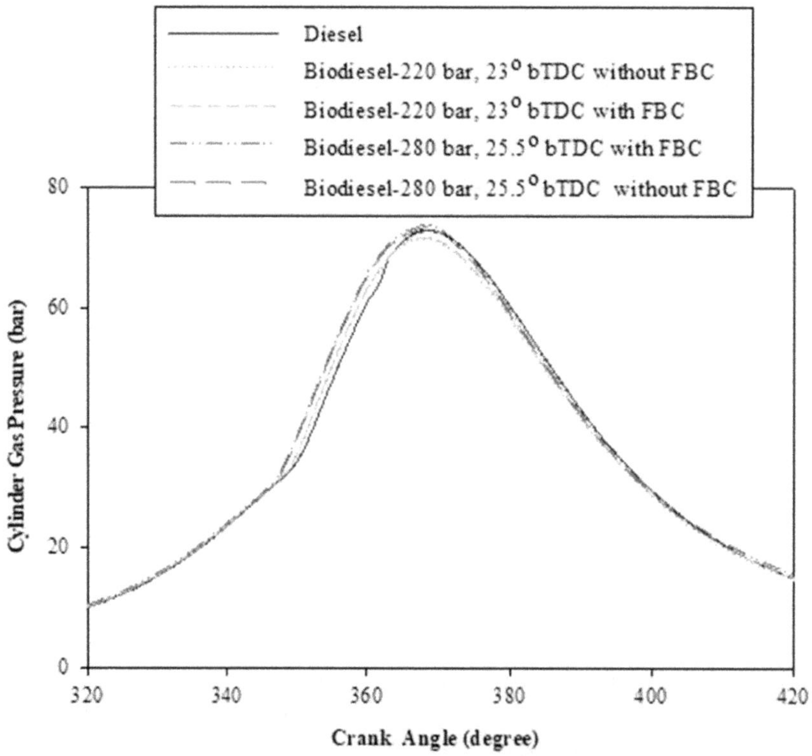

Figure 15 shows a comparison of the test fuels' rates of heat release. The cooling effect brought on by fuel vaporisation and heat losses from the engine cylinder walls accounts for the test fuels' negative heat release rate readings after ignition (Heywood, 1988). When fuel is injected sooner in the compression process, there is a shorter ignition delay, which results in a higher buildup of fuel and a faster heat release rate during the premixed combustion phase due to higher injection pressure. (Sukumar Puhan et al., 2009; Rakopoulos et al., 2018; Raghvendra Gautam and Saket Kumar, 2020). At optimal operation, biodiesel with FBC was found to have a maximum heat release rate of 29.2 J/ °CA, which is 4.6% higher than biodiesel without FBC. Biodiesel with FBC added had a 6.2% greater maximum heat release rate than diesel. This can be explained by the biodiesel's increased oxygen content, the earlier combustion that results from a higher cetane number that reduces the ignition delay, and the catalytic impact of FBC that provides better combustion and higher cylinder temperature (Simpson and Depcik, 2022; Sadhik Basha et al., 2022).

Figure 15. Variation of HRR

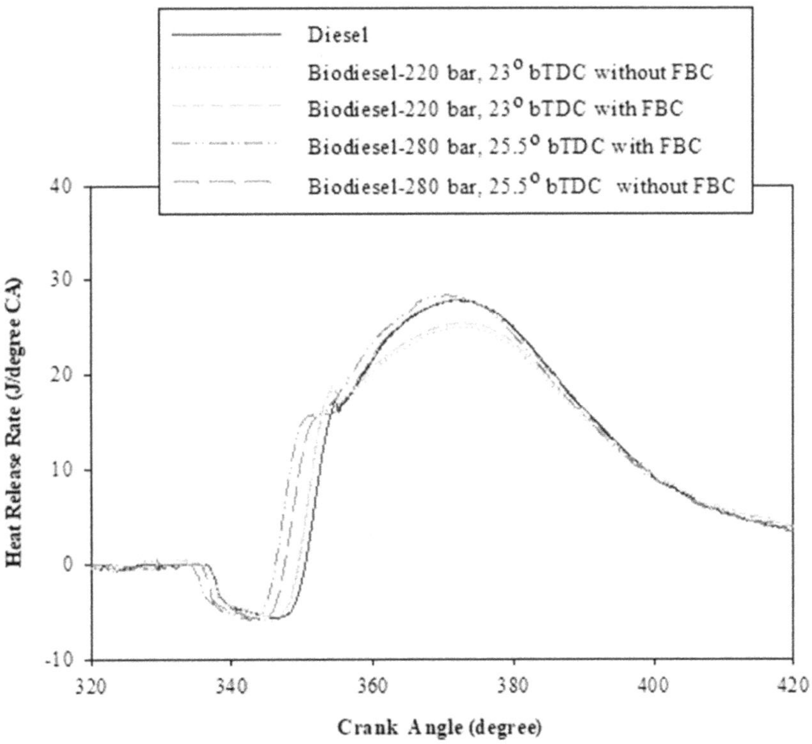

3. INFLUENCE OF OXYGENATE ADDITIVES ON DIESEL ENGINE FUELLED WITH BIODIESEL

3.1. Using Ternary Blends by the Addition of 1-Propanol and 1-Hexanol

Using 1-propanol and 1-hexanol, a biodiesel derived from cooking oil was utilised to compare the effects of alcohol on several aspects of diesel engine performance. Two ternary blends containing 20% by volume of higher alcohols (propanol and hexanol), 30% by volume of biodiesel, and 50% by volume of diesel were used for the engine studies. The outcomes were then contrasted with a binary mixture of pure diesel (D100), biodiesel, and a mixture of 50% diesel and 50% biodiesel (D50B50) (B100).

BTE is a measure of how well fuel's chemical energy is transferred into accessible crankshaft power (Ashok et al., 2017; Raghvendra Gautam and Saket Kumar, 2020). Figure 16 shows the BTE variance for several fuels at varied BMEP. The test sample B100 had the lowest BTE of the test samples due to its low calorific value and poor atomization properties brought on by its higher density and viscosity. The D50B50 pairing had a higher BTE. This is explained by the existence of diesel, which balances out biodiesel's poor calorific value (Ashok et al. 2019). HX20 blend among the ternary mixes showed better BTE than Pr20 blend, which was higher than D50B50 blend, due to its superior low heating value, atomization features, and higher oxygen content, which accelerate combustion and lead to higher BTE (Sadhik Basha, 2018). According to Nanthagopal et al. (2018), 1-butanol has a lower calorific value and lower engine BTE due to its higher oxygen concentration than 1-pentanol.

Figure 16. Brake thermal efficiency of test fuels at various BMEP

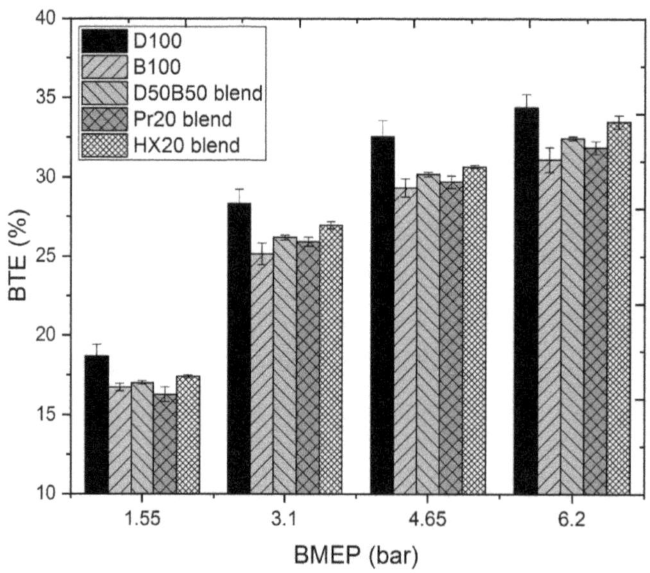

Fuel-borne oxygen plays a critical role in increasing engine temperature by supplying more oxygen for combustion processes because residence time is minimal for constant-speed engines (Ashok et al., 2019b; Raghvendra Gautam and Saket Kumar, 2020). NOx emissions for test samples at various BMPEs are shown in Figure 17.

Figure 17. NOx emissions of binary and ternary blends at various BMEP

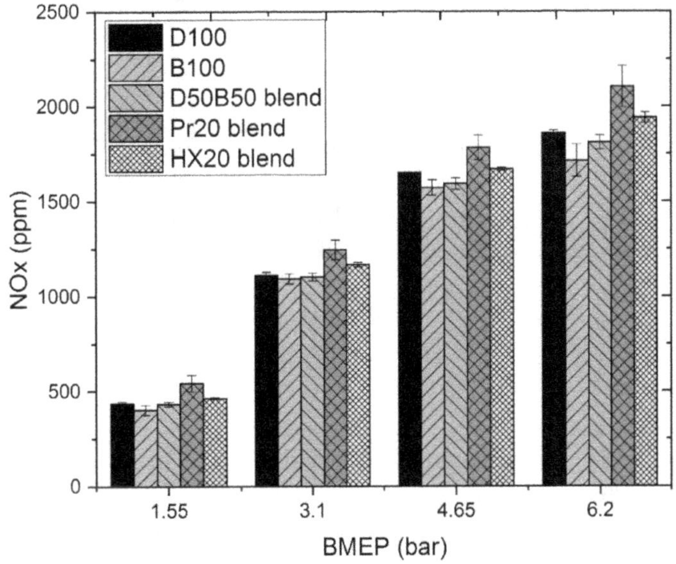

Due to the richer fuel-air ratio brought on by the higher BMEP, which allows a substantial amount of fuel to burn, rising gas temperature and hence NOx emission. Increased alcohol consumption and elevated NOx levels were reported by Damodharan et al. (2018). Due to its higher oxygen concentration and lower cetane number, Pr20 emits more NOx than HX20. Similar findings were found from Kumar et al. (2016), who observed that n-propanol mixes had higher NOx emissions than n-butanol and n-pentanol mixtures. This was partly because n-propanol has a lower cetane percentage and a higher oxygen content. The atomization of the fuel, the mixing of the fuel and air, the lack of oxygen, and the timing of the fuel injection are other factors that affect smoke emission. (Rakopoulos et al., 2018; Ghadikolaei et al., 2018; Liang et al., 2019; Rajasekaran et al., 2020; Sadhik Basha et al., 2020).

Figure 18 displays the smoke opacity of test fuels. As more fuel is injected, more fuel-rich zones are created, making it more difficult for the fuel to burn and increasing the amount of smoke that is produced as a result (Liang et al., 2019). The neat biodiesel operation creates the most smoke among the test samples when more fuel with a wider droplet width is delivered into the engine cylinder, mostly due of its poor atomization qualities. This negatively impacts vaporisation and favours soot formation. The addition of diesel to the D50B50 blend resulted in fewer smoke emissions when compared to B100.

Figure 18. Smoke opacity of test fuels

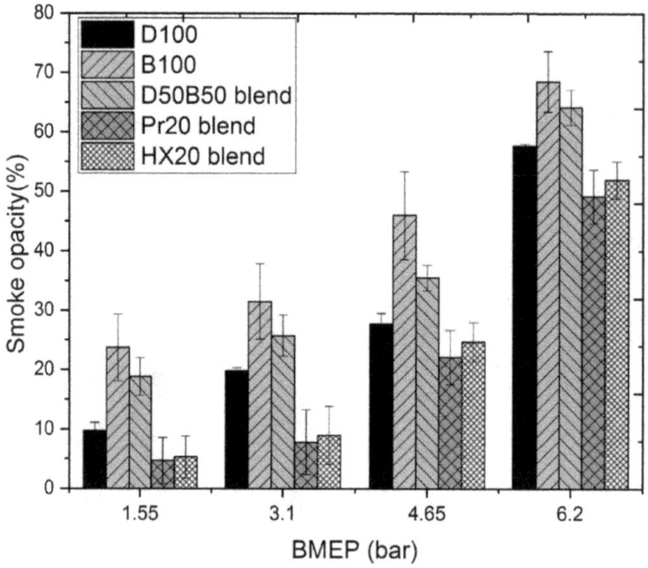

These findings were in good agreement with Rajesh et al. (2020)'s observation that low carbon chain alcohols had a reduced carbon/hydrogen content and inhibited smoke generation. The presence of oxygen atoms in propanol and hexanol, in contrast to the standard diesel operation, decreased smoke density, according to another observation from De Poures et al. (2020). Due to the increased amount of oxygen coupled to the fuel in the propanol mixture compared to the hexanol mixture, the smoke reduction was at its maximum level.

The creation of several fuel-rich zones, insufficient air-fuel mixing, and poor air entrainment are the causes of the unburned hydrocarbon (UHC) coming out the exhaust (Raghvendra Gautam and Saket Kumar 2020). The UHC emission of test samples is shown in Figure 19. Test sample B100 had the highest UHC emission of all the test samples because of its poor atomization characteristics, which caused larger sauter diameter droplets inside the combustion chamber to end up in unburned or partially burned hydrocarbons. The D50B50 mixture showed better UHC emission compared to B100 due to increased spray characteristics (Sadhik Basha, 2018).

The engine characteristics are depending only on the combustion properties (Ashok et al., 2018; Liang et al., 2019). Figure 20 displays the gas cylinder pressure at design power for several test samples. Diesel performed better than the other test samples in terms of self-ignition and heating value, resulting in a higher gas pressure. Due to ineffective atomization, the peak pressure of the D50B50 mixture was lower.

Variations in HRR is seen in Figure 21. B100 produced less heat release than the other studied fuels, mostly because of its subpar atomization characteristics.

Figure 19. Unburnt hydrocarbon emission of test fuels at various BMEP

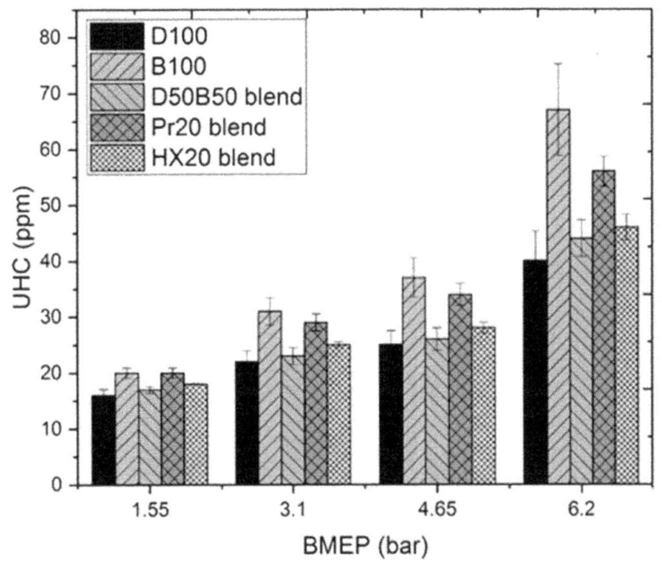

Figure 20. Cylinder gas pressures at 100% load

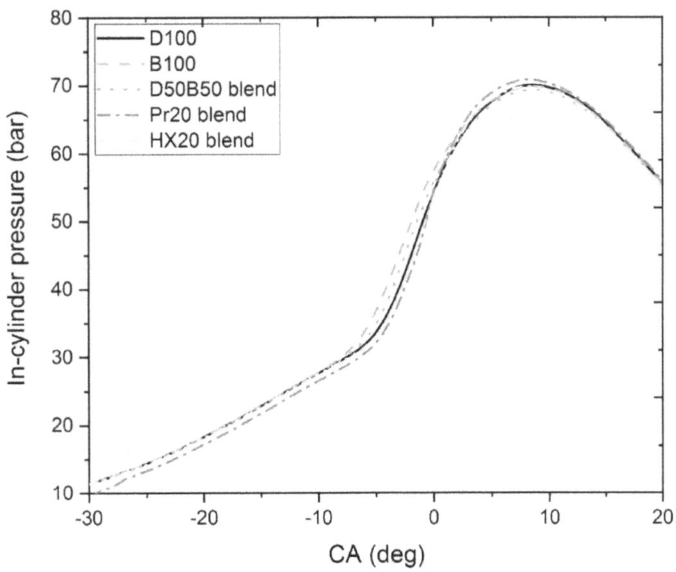

Figure 21. Variation of HRR of test fuels

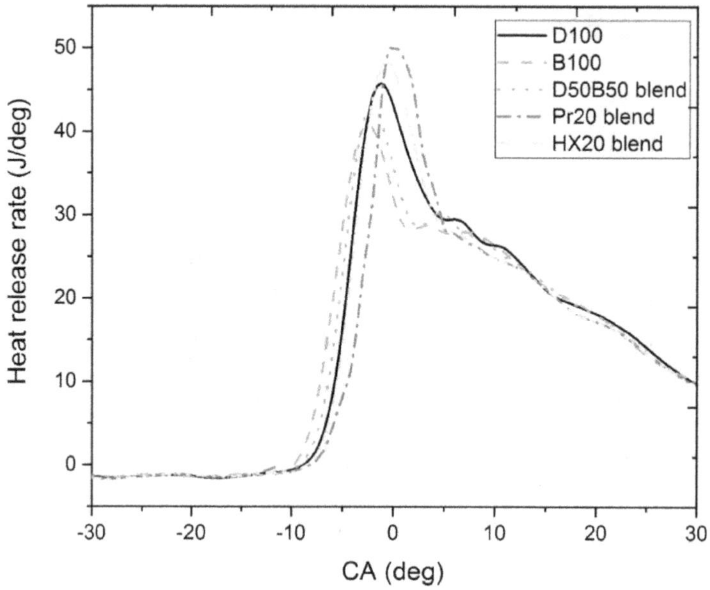

3.2. Using Diestrol (Biodiesel-Diesel-Ethanol)

Several biodiesel blends, including biodiesel-diesel ethanol combinations, diestrol fuels, and diestrol-micro-emulsion fuels, were examined for the simultaneous enhancement of fuel qualities and engine characteristics in the third subdivision of the chapter. In the first part of this phase, the components of diestrol fuel were altered from 0 to 100% in stages of 10% increments to prepare 36 alternative combinations and were then exposed to the stability investigation at various temperatures of 10 to 40°C. Based on miscibility and clarity, three diestrol fuels were chosen for investigation without engine modification: B30D60E10 (30% biodiesel, 60% diesel, and 10% ethanol), B40D50E10 (40% biodiesel, 50% diesel, and 10% ethanol), and B50D30E20 (50% biodiesel, 30% diesel, and 20% ethanol).

The variation of BSEC is presented in Figure 22. The lower BSEC for B30D60E10 is 12.3 MJ/kW h, which is 20.4% less efficient than biodiesel (B100) and just marginally more efficient than diesel at 100% load. All diestrol fuels and biodiesel have a greater BSEC than diesel under all load circumstances. This is because of the reduction in the energy content of diesel and biodiesel.

Figure 23 illustrates the variation in BTE of test fuels. B30D50E20 diestrol fuel demonstrated the lowest brake thermal efficiency among all diestrol fuels. This has happened because diesel fuel has a lower calorific value and a larger latent heat of evaporation (Mathur and Sharma, 1997; Lei Zhu et al., 2010; Raghvendra Gautam and Saket Kumar, 2020; Sadhik Basha et al., 2020).

Figure 22. Variation of BSEC with BMEP for diestrol

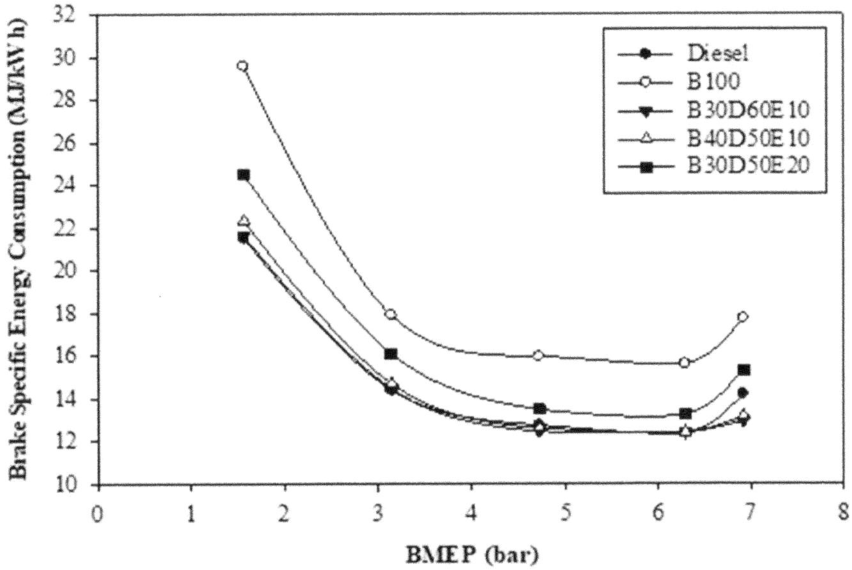

Figure 23. Brake thermal efficiency of diestrol

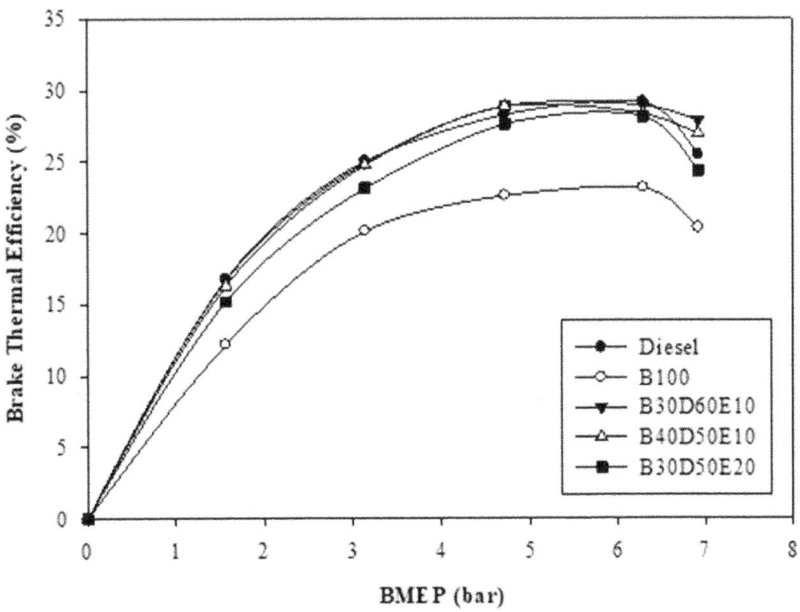

Figure 24 depicts the impact of diestrol fuel on NO emissions under various loading scenarios. For the entire load situation, biodiesel and diestrol fuels had lower NO emissions than diesel fuel. Maximum NO emissions from B30D60E10 diestrol fuel at full load are 508 ppm, whereas they are 485 ppm and 578 ppm for biodiesel and diesel, respectively. The inferior heating value and the higher latent heat of vaporisation of ethanol indicated a lower burning temperature, which resulted in lesser NO emissions. According to the findings of prior studies, biodiesel had lower NO emissions than diesel fuel (Chang and Van Gerpan, 2005).

Figure 24. Nitric oxide of diestrol fuels

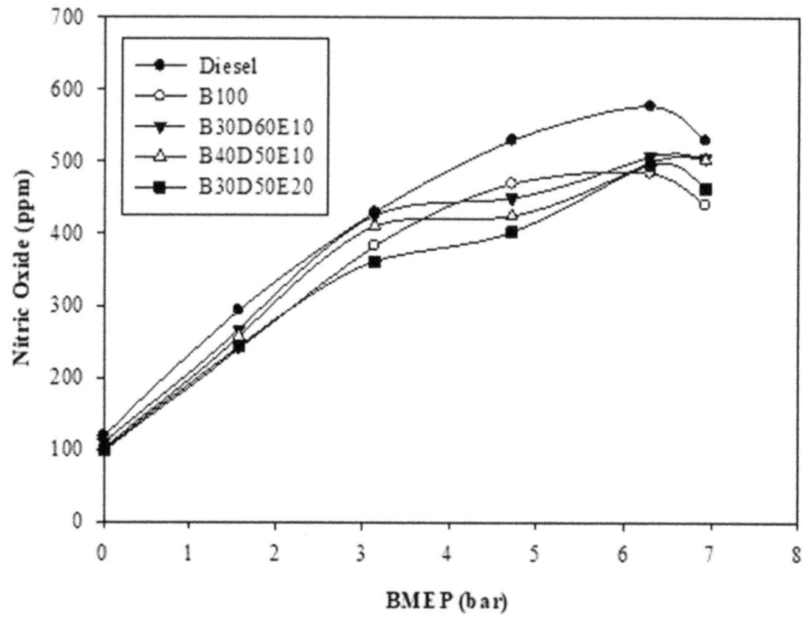

The fluctuation in smoke emissions at different engine loads is depicted in Figure 25. At increased engine load, test fuels emit more smoke. This may be explained by adding additional fuel to the engine, which would produce more smoke when it was operating at full capacity (Cenk Sayin et al., 2009; Fayad, 2019). All diestrol fuels, including biodiesel, emit less smoke than diesel. Diestrol fuel B30D60E10 has a maximum smoke emission rate of 48.9%, which is 3.3% higher and 3.9% lower than diesel and biodiesel, respectively.

The percentage of fuel consumed during the beginning of combustion phase determines the pressure of the cylinder gas. As demonstrated in Figure 26, biodiesel and all diesel fuel had lower cylinder gas pressures than diesel fuel. The B30D60E10 diestrol fuel had a maximum pressure of 73.2 bar at 368.5 oCA, which is extremely close to diesel's value.

Figure 25. Smoke emission of diestrol fuels

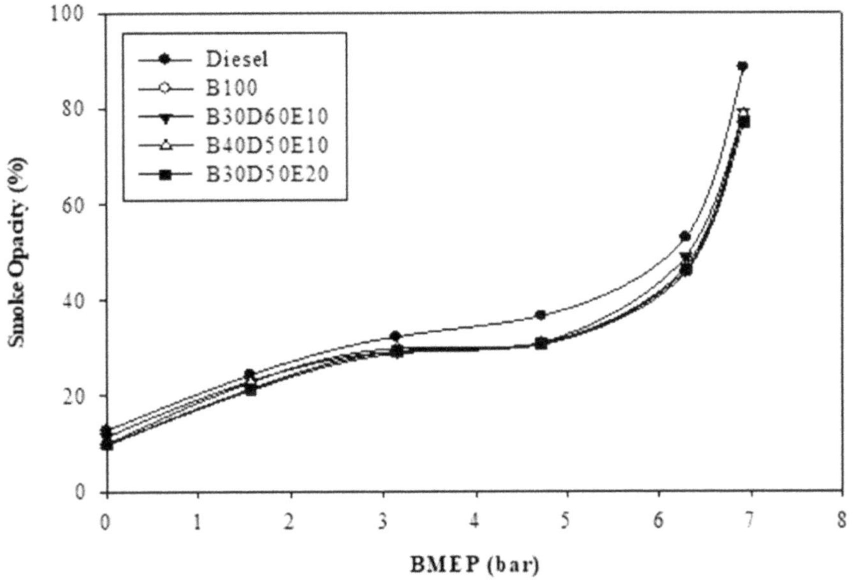

Figure 26. Cylinder gas pressure at 100% load

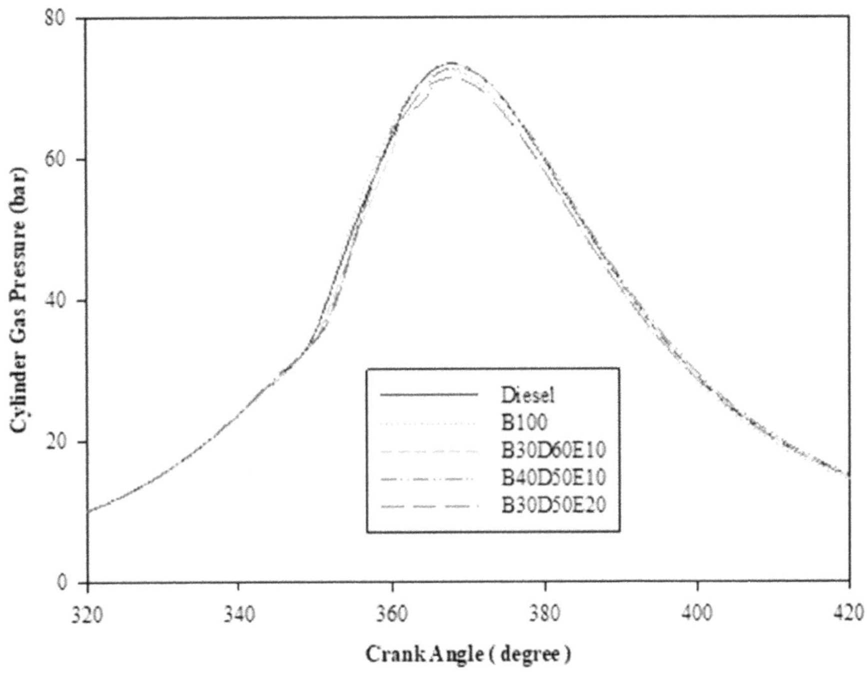

Figure 27 shows the impact of the heat release rate under 100% load conditions. All test fuels demonstrated rapid premixing or an initial combustion phase, followed by carefully regulated mixed combustion (Fayad, 2019). Diestrol fuel B30D60E10 showed maximum heat release rate which is comparable to diesel fuel. This is a result of the rise in diestrol fuel's ignition delay period, which lengthens the time it takes to create an increasing amount of fuel-air mixture and increases the pace at which heat is released under heavy loads.

Figure 27. HRR of diestrol fuels

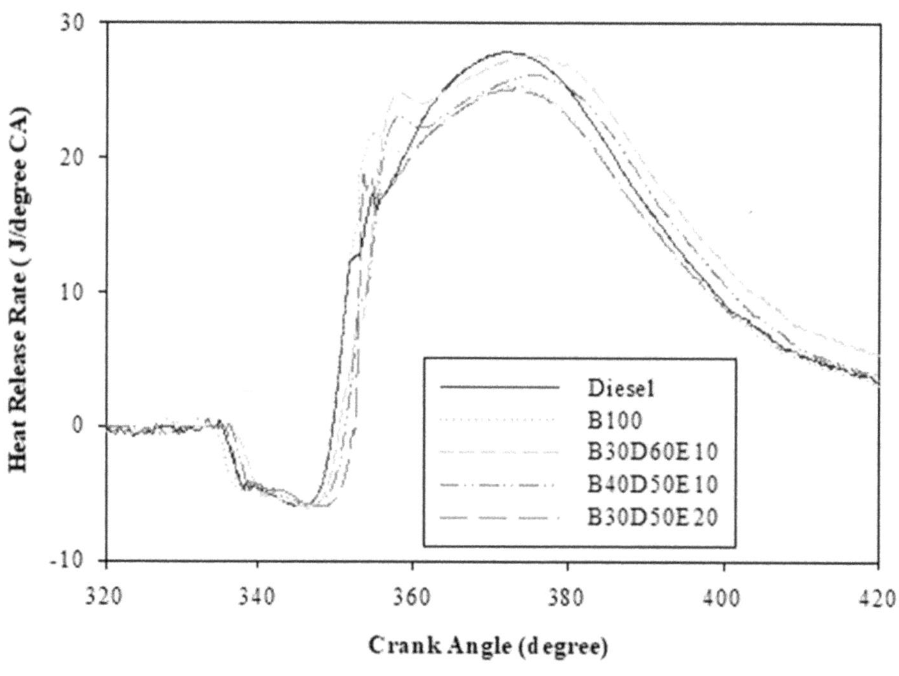

4. ENGINE STUDY USING DIESTROL-MICRO EMULSION

This section outlines the findings of an experimental study on diestrol-water micro-emulsion fuels, specifically B60D20E20M and B70D10E20M, used in DI diesel engine. The outcomes are contrasted with those of diesel and biodiesel under typical operating circumstances. The outcomes obtained using fuels that have been micro-emulsified with diestrol are shown below.

The change in BSEC for the test fuels is shown in Figure 28. Biodiesel was found to have a BSEC, 25.9% higher than diesel, while microemulsion fuels showed a slight improvement in BSEC compared to biodiesel fuel. At 100 percent, the BSEC for microemulsion fuel (B60D20E20M) had the lowest value of 12.25 MJ/kW h. It shows appropriate spraying and atomization that resulted in improved combustion. In addition, adding water improves atomization, thereby improving the fuel blending process (Cheng et al., 2008; Kadota and Yamasaki, 2002; Fayad, 2019).

Figure 28. BSEC with BMEP of micro-emulsion fuels

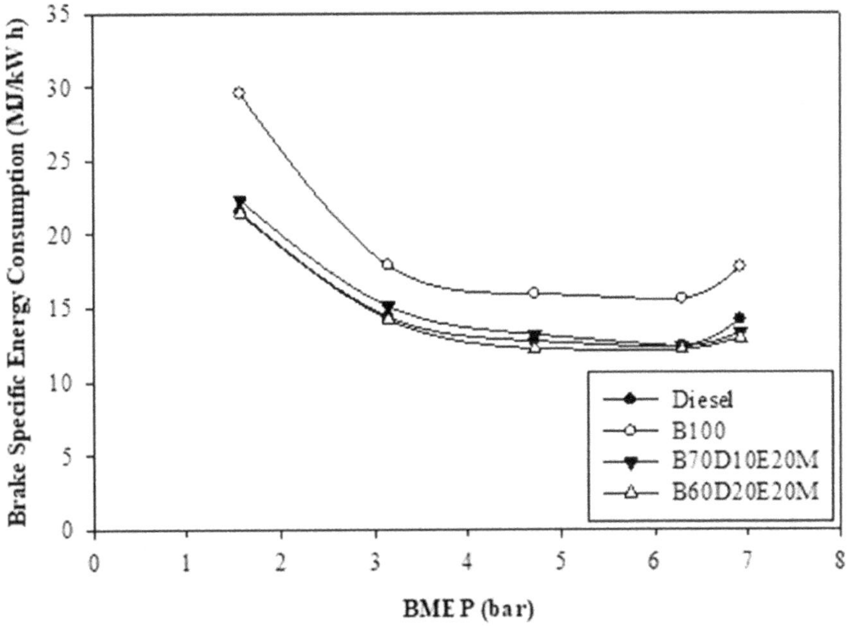

A comparison of the BTE of the various test fuels is given in Figure 29. Up to 75 percent load, biodiesel and micro-emulsion fuels are less efficient than diesel fuel. With greater engine loads, biodiesel and B60D20E20M fuel exhibit significantly enhanced thermal efficiency. B70D10E20M and B60D20E20M had a thermal efficiency of 29.3% and 30.8% respectively, which is almost identical to diesel (29.12%) with 100% load.

Figure 30 displays the variations in NO emissions. In comparison to biodiesel, micro-emulsion fuels emit less NO. When fully loaded, diesel emits 578 ppm of NO and biodiesel emits 485 ppm. This is mostly accounted for by biodiesel's reduced iodine value (57.3). (McCormick et al., 2005). At maximum load, it was found that the emissions of B70D10E20M and B60D20E20M were respectively 21.6% and 18.8% and 6.5% and 3.2% lower than those of diesel and biodiesel. This could be explained by the fact that ethanol and water vaporization's higher latent heat lowers cylinder temperature, which in turn lowers NO emission (Sadhik Basha, 2018).

The smoke emissions are shown in Figure 31. Biodiesel emits 7.2% less smoke than diesel at full load. The oxidation of hydrocarbons is caused by the oxygen component of biodiesel. Figure 31 also demonstrates that utilising B60D20E20M micro-emulsion fuel resulted in smoke emissions that were 10.5% lower than those produced by diesel and 3.3% lower than those produced by biodiesel at 100% load. This is mostly caused by microemulsion fuels' greater oxygen concentration. Moreover, the combination of ethanol and water lengthens the ignition delay and improves combustion (Qi et al., 2010; Sadhik Basha, 2018; Fayad, 2019).

Figure 29. BTE of micro-emulsion fuels

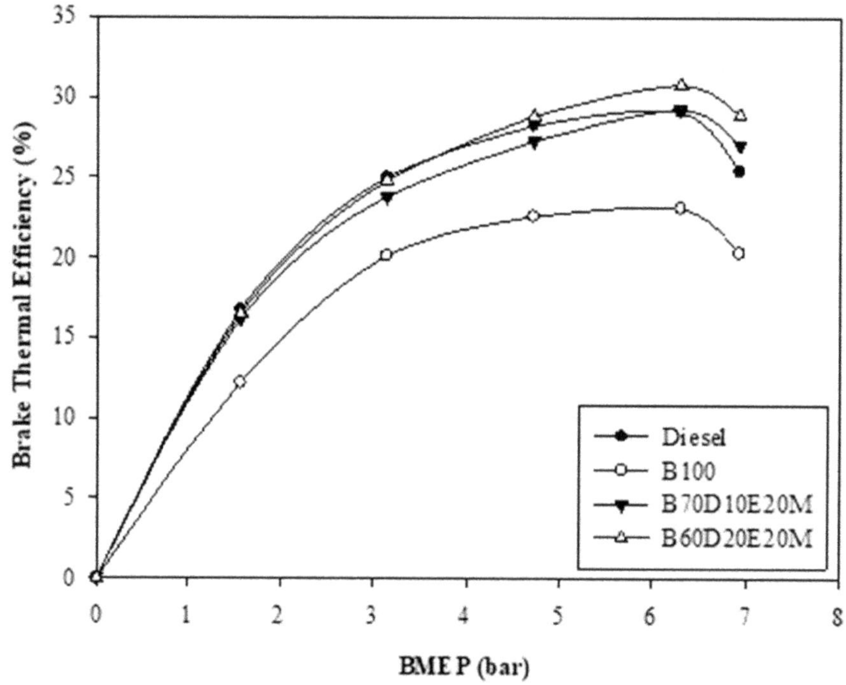

Figure 30. Nitric oxide emission

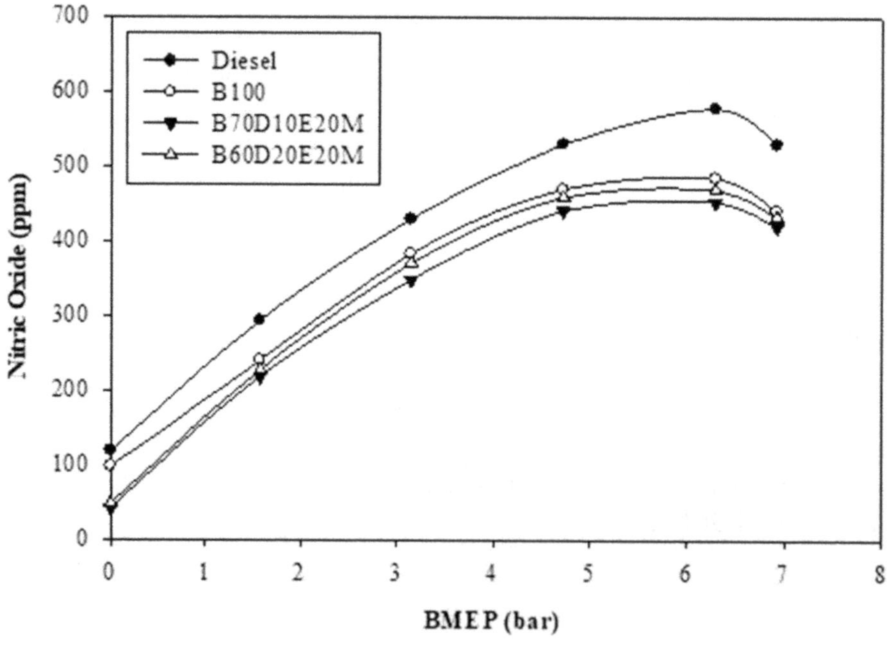

Figure 31. Smoke emission of micro-emulsion fuels

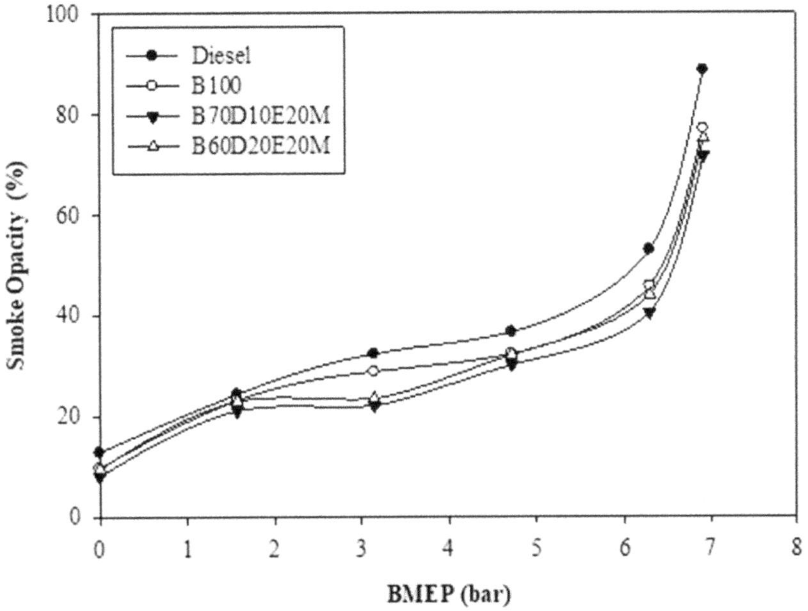

A comparison of the cylinder gas pressure under 100% load is shown in Figure 32. The cylinder gas pressure is higher for micro-emulsion fuels, with the B70D10E20M micro-emulsion having a higher value of 73.7 bar at 371.8 °CA. The lengthy ignition delay of all micro-emulsions, which begin burning closer to the TDC, may help to explain this. The maximum cylinder gas pressure ever measured was 70.5 bar for biodiesel; this pressure happens earlier in the crank angle than it does for microemulsified fuels. The main reason for this is because biodiesel burns more quickly than diesel and micro-emulsified fuels.

Figure 33 displays the HRR at full load. According to research, B70D10E20M releases heat at a rate of 25.16 J/ °CA while diesel releases heat at a rate of 27.5 J/ °CA. The figure shows that for all micro-emulsion fuels, the crank angle that corresponded to the maximum degree of heat release was lower. Because ethanol and water are added, the cetane number is decreased and the ignition delay time is prolonged, causing micro-emulsion fuels to have a higher HRR peak. As a result of the longer delay time and quicker rate of heat release, there is more fuel available for combustion.

5. EFFECT OF VARYING COMPRESSION RATIO AND THERMAL BARRIER-COATED PISTON ON A BIODIESEL-FUELLED DIESEL ENGINE

The standard or non-coated engine (SE) and TBC coated or adiabatic engine (AE) were subjected to experimental investigations employing variable loads ranging from 0% to 100% with increments of 25% and various compression ratios of 17:1, 18:1, and 21:1 based on waste cooking oil derived from sunflower oil (WCOME). At a steady 1500 rpm, the engine characteristics of the various test fuels were studied. In this study, the outcomes of adiabatic or TBC-coated engines with varied compression ratios employing WCOME were compared to the non-coated or conventional engine. In the current investigation, the engine charcateristics of TBC coated engine (AE) and standard engine (SE) utilising diesel at constant compression ratio 17 were also noted and compared with WCOME. Figure 34 shows the photographic perspective of the VCR engine.

Figure 32. Cylinder gas pressure of test fuels

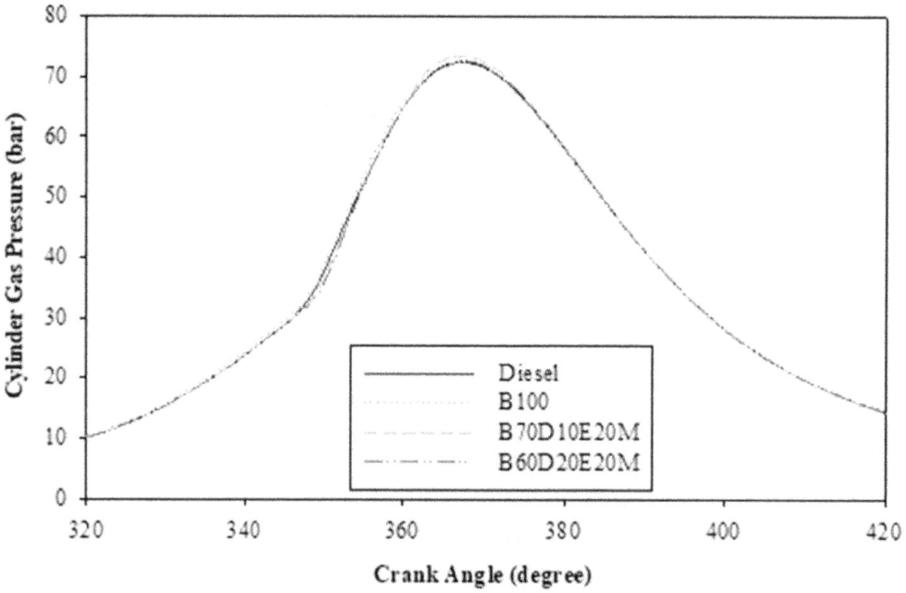

Figure 35 shows the BTE of the conventional engine (SE), thermal barrier-coated engine (AE), and adiabatic engine (AE) utilising biodiesel under various compression ratios and load circumstances. In comparison to CR17-SE-Diesel, CR17-AE-Diesel had a maximum BTE of 28.12%, mostly because of the thermally coated piston ring's better combustion. Despite adjusting CR, standard, and adiabatic engine settings, MOME or biodiesel had a poorer braking thermal efficiency than diesel because of their lower heating values than diesel. By changing the compression ratio between 17:1 and 21:1, it was predicted that biodiesel BTE will rise. With CR21-AE-Biodesel, a maximum TEB of 26.98% was recorded under 100% load. This can be explained by improved combustion as a result of the piston crown's thermal barrier coating, which ensures high temperatures and a high compression ratio, which results in greater atomization when compared to other engine settings (Fayad, 2019).

Nitric oxide (NO) of test fuel are shown in Figure 36. As seen in the image, when compression and loading conditions increase, so do biodiesel and diesel NO emissions. Due to peak pressure, an increase in compression ratio from 17 to 21, a powder-coated engine, and faster, better combustion, as well as greater cylinder gas temperature. Because biodiesel has lower cetane values at different compression ratios than diesel, it emits more NO than diesel. The maximum NO emission of CR21-AE-B100 is more than 137 ppm greater than that of CR17-AE-Diesel due to the higher gas temperature at a higher compression ratio and the higher oxygen content of biodiesel.

Figure 33. HRR of micro-emulsion fuels

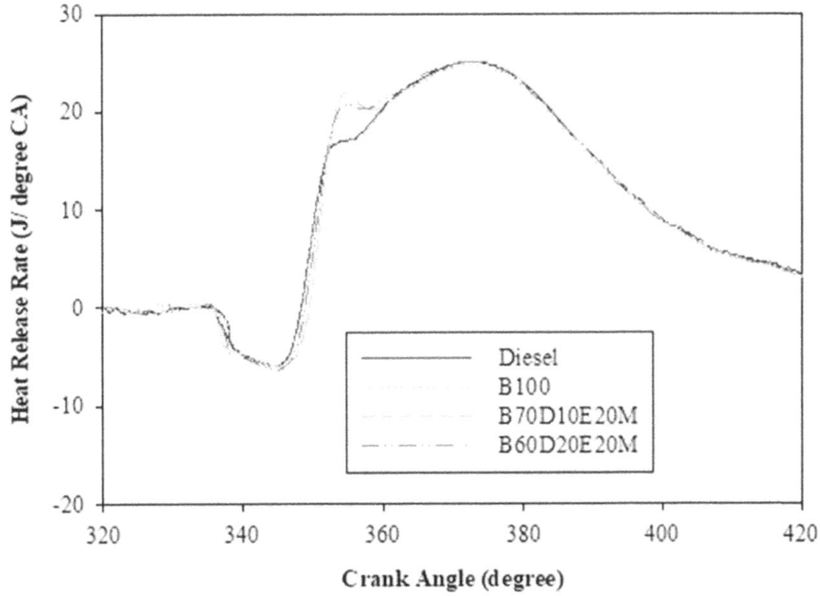

Figure 34. Photographic view of test engine setup

Figure 35. BTE of test fuels with various compression ratios and loads

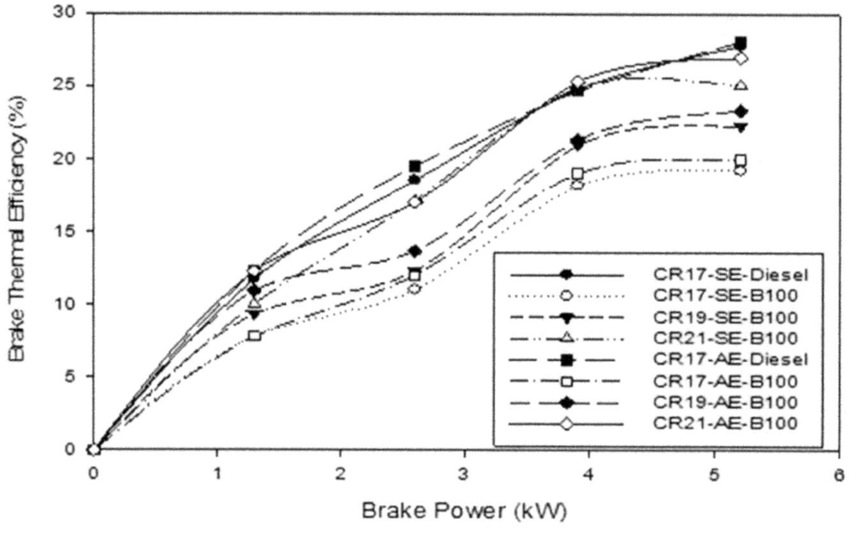

Figure 36. NO emission of test fuels

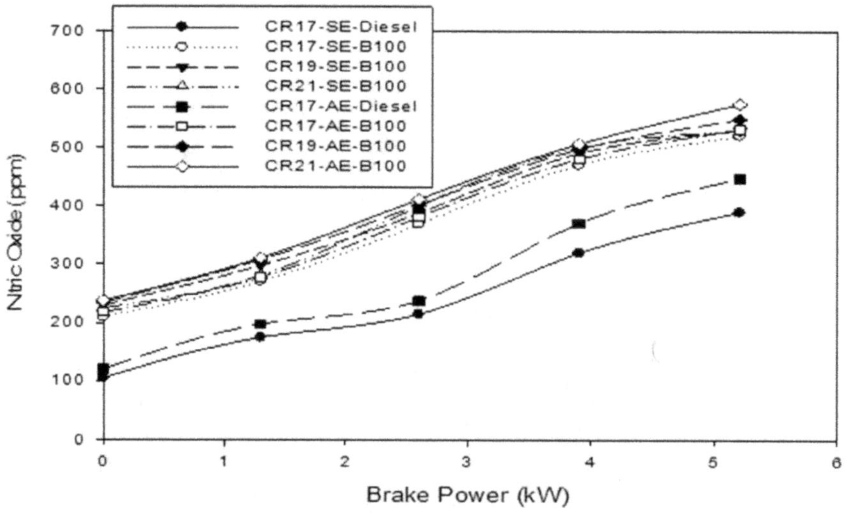

Figure 37 compares the smoke opacity under various compression ratios and load conditions. As the compression ratio rises, the smoke opacity of biodiesel decreases. The CR21-AE-B100 achieves the lowest smoke opacity of 39.1%, which is 26.4% and 31.7% lower than the CR17-AE-Diesel and CR17-SE-Diesel, respectively. This is brought on by both a higher gas temperature and the oxygen concentration of biodiesel. Also, the heat-coated engine displayed a higher gas temperature during the expansion race, which promotes improved combustion by oxidising soot particles. The oxygen content of biodiesel and a greater gas temperature are to blame for this. Also, during the expansion race, the heat-coated engine displayed a higher gas temperature, indicating greater combustion through the oxidation of soot particles.

Figure 37. Smoke opacity of test fuels

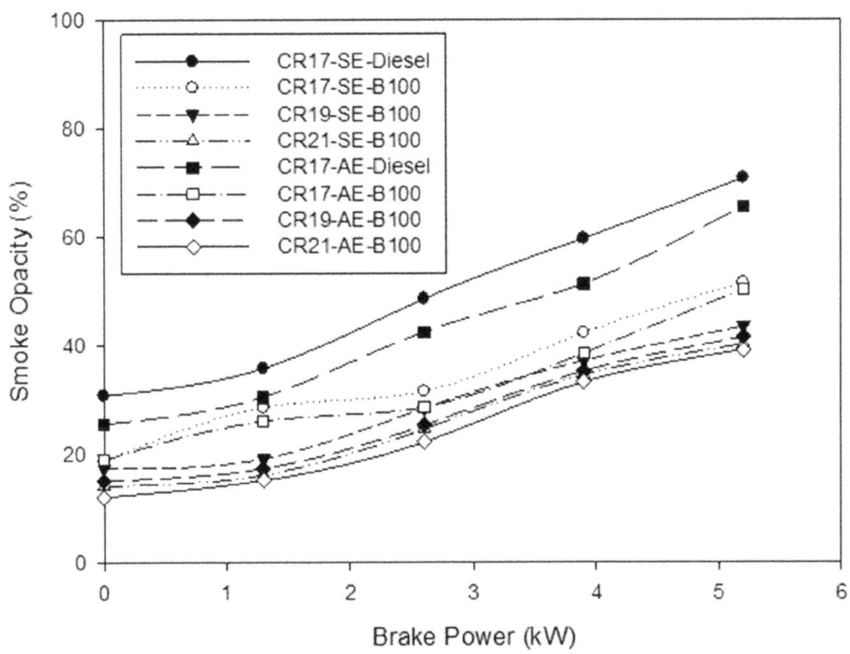

Figure 38 illustrates how CO_2 emissions from test fuels vary depending on the different load circumstances and compression ratios. During 100% charging conditions, maximum CO_2 emissions of 4.3% by volume and 4.5% by volume were recorded for CR17-SE-Diesel and CR17-AE-B100, respectively. The CO_2 emission of test fuels increases with increasing load and compression ratios because of higher combustion chamber temperatures and the oxygen content of biodiesel (Gvidonas Labeckas et al., 2019). The thermal-coated engine is found to emit more CO_2 from the test fuels than the normal engine. An adiabatic engine's greater temperature improved combustion and raised the temperature of the exhaust gases. The maximum CO_2 emissions for the CR21-AE-B100 were 5.3% Vol, which was 1% Vol and 0.2% Vol more than for the CR17-SE-Diesel and CR21-SE-B100. This can be explained by the combination of a thermally coated engine or an adiabatic engine with B100, which led to greater combustion, and the higher compression ratio of 21.

Figure 39 shows cylinder pressure of test fuels for various compression ratios in both regular and adiabatic engines at 100% load. Because of the increasing temperature of the gas from better combustion, the cylinder pressure likewise rises as the compression rate does (Liang et al., 2019). Moreover, the adiabatic engine's cylinder pressure was a little bit greater than the average engine pressure for all test fuels. Due to greater heating values compared to biodiesel, maximum pressures of 68.4 bar and 64.16 bar were recorded for CR17-AE-Diesel and CR17-SE-Diesel, respectively (Gvidonas Labeckas et al., 2019). The maximum cylinder pressure of 62.8 bar was attained for CR21-AE-B100 among biodiesel with varied compression ratios and engine settings, which is 5.9% higher than CR21-SE-B100. This is due to the higher cylinder pressure that was produced by the adiabatic piston's higher gas temperature.

Figure 38. CO_2 emission of test fuels

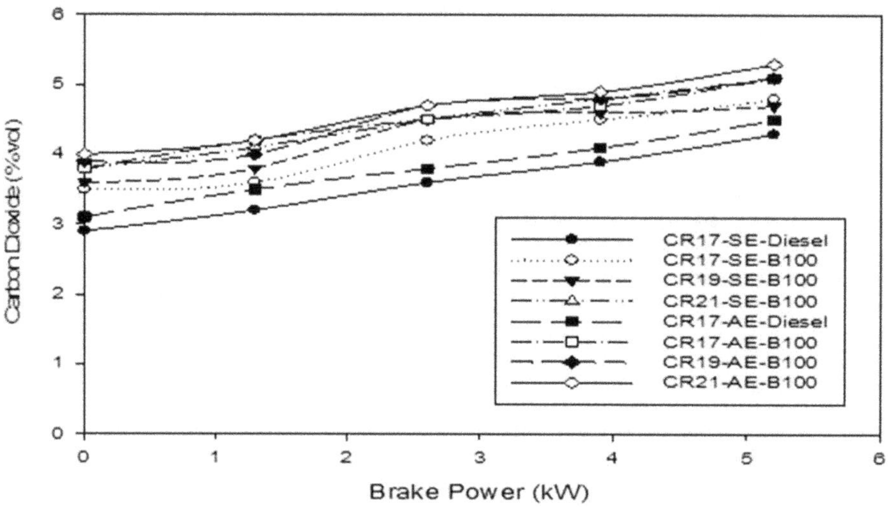

Figure 39. Gas pressure of test fuels for various compression ratios

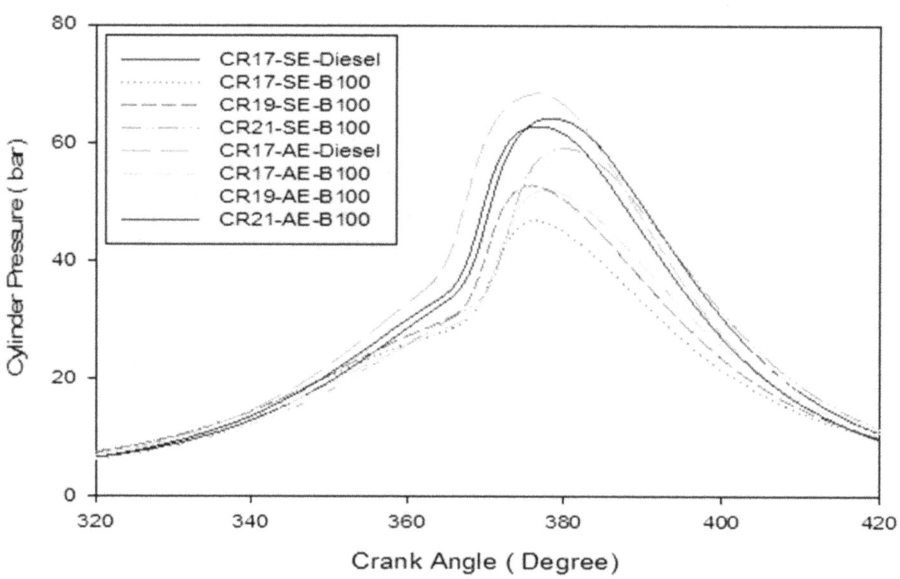

Figure 40 shows the change in the HRR of test fuels for both standard and coated engines at various compression ratios under 100% load. For both normal and adiabatic engines, the rate at which the test fuels release heat rises as the compression ratio does. However independent of SE or AE conditions, biodiesel showed a lower heat release rate at all compression ratio ranges due to its lower heating value as compared to diesel (Liang et al., 2019). The CR21-AE-B100 produced the most heat at 79.7 J/ °CA, which was 5.3% and 3.1% less than the CR17-AE-Diesel and CR17-SE-Diesel, respectively. This is primarily caused by the biodiesel's oxygen concentration, as well as a higher compression ratio and an engine with a thermal barrier coating that displayed a higher biodiesel heat release rate (Gvidonas Labeckas et al., 2019). CR21-SE-B100 showed a significantly lower heat release rate (79.5 J/ °CA) when compared to compared to the CR21-AE-B100,

Figure 40. HRR of test fuels

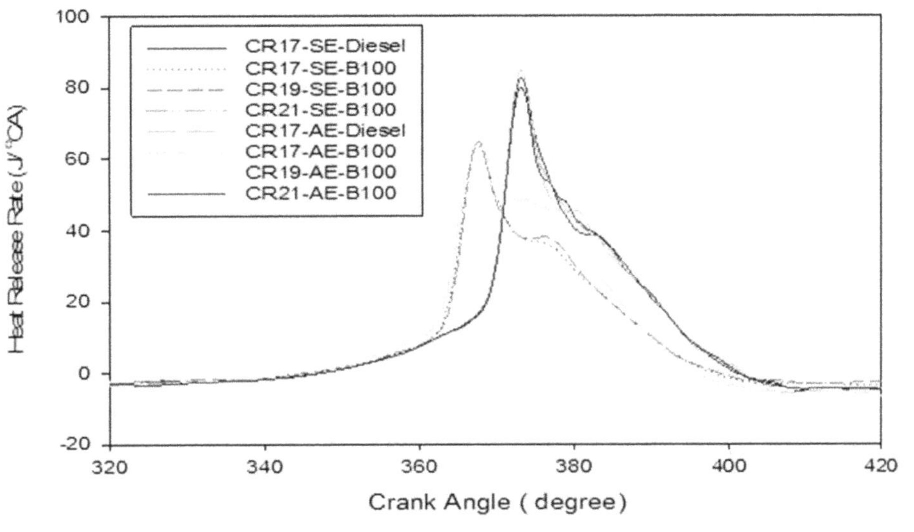

6. CONCLUSION

Based on an experimental survey, the following findings may be drawn.

6.1. Effect of Varying Injection Pressure and Injection Timing on a Diesel Engine Fuelled With Biodiesel

- Biodiesel had a lower emissions pattern than diesel fuel, regardless of operating conditions.
- Biodiesel showed a reduction in CO, NO, and smoke of 0.19% by vol, 38 ppm, and 17.9% compared to diesel under standard operating conditions.
- At 280 bar and 25.5 degrees bTDC, the biodiesel's maximum pressure and heat release rate, respectively, were 2.1% and 1.1% greater than those of the diesel under normal operating circumstances.

6.2 Effect of Fuel Borne Catalyst (FBC) on Diesel Engine Fuelled With Biodiesel

- At optimal operating circumstances, FBC-added biodiesel had a marginally higher NO emission and CO_2 emission than biodiesel without FBC, but it had a marginally lower CO, UHC, and smoke emission.

- The fuel-based catalyst or metal-based additive ($FeCl_3$) is an effective and efficient additive to waste cooking palm oil-based biodiesel.

6.3. Influence of Oxygenate Additives on a Diesel Engine Fuelled With Biodiesel

6.3.1. Using Ternary Blends by the Addition of 1-Propanol and 1-Hexanol

- Of the test samples, pure biodiesel produces the lowest NOx emissions which could be validated against the physical properties of biodiesel. NOx emissions were slightly higher in a binary mixture than in pure biodiesel because of the low net calorific value of the mixture.
- When ternary blends are used instead of binary blends, the smoke opacity value is significantly lower. This is because the alcohols' inherent oxygen molecules effectively provide enough oxygen for full combustion.
- The propanol mixture's maximum HRR is 13.3 percent higher than hexanol's. It increased by 21.5% and 30.5%, respectively, as compared to pure diesel and pure biodiesel.

6.3.2. Using Diestrol (Biodiesel-Diesel-Ethanol)

- B30D60E10 showed marginally reduced NO and smoke emissions compared to diesel.
- The best alternative fuel for DI diesel engines without engine modification was Diestrol fuel, which contains 30% biodiesel, 60% diesel, and 10% ethanol (B30D60E10). This is because it has appropriate temperature stability, has better fuel qualities, and has improved engine characteristics.

6.4. Engine Study Using Diestrol-Microemulsion

- Diestrol-microemulsion fuels produced less NO and smoke than regular diesel fuels. Yet, compared to diesel, all emission metrics for these fuels decreased.
- In comparison to the other test fuels, B60D20E20M produced a maximum pressure of 73.7 bar and a heat release rate of 25.2 J/ °CA at full engine load.
- Without altering the engine, diestrol-micro-emulsified fuels might be regarded as appropriate alternatives to diesel.

6.5. Effect of Varying Compression Ratio and Thermal Barrier Coated Piston on Biodiesel Fuelled Diesel Engine

- The CR17-AE-Diesel displayed a maximum thermal efficiency of 28.12% over the CR17-SE-Diesel. The heat-coated or adiabatic engine reported a substantial improvement in brake thermal efficiency when fuelled with biodiesel.
- CR21-AE-B100 displayed the lowest smoke opacity of 39.1%, 26.4%, and 31.7% below the CR17-AE-Diesel and CR17-SE-Diesel. CR21-AE-B100 exhibited the lowest exhaust gas temperature of 200°C among all test conditions
- When operated at a higher compression ratio of 21 with 100% biodiesel or WCOME, thermal barrier-coated or adiabatic engines showed improved performance, emission, and combustion

characteristics when compared to a standard or non-coated engine at the same compression ratio as well as when compared with conventional diesel.

The use of WCOME as a suitable advanced fuel for engine emission control was demonstrated through experimental research on a diesel engine that was run on waste cooking palm oil-based biodiesel, biodiesel with fuel-borne catalyst added, biodiesel with oxygenate additive added, as well as biodiesel with engine modification by varying injection pressure and injection timing. Also, experimental research on a thermal barrier-coated VCR engine running on biodiesel made from waste cooking sunflower oil demonstrated improved engine performance with with minor engine modifications, paving the path for intriguing alternatives to conventional fuel. So, a potential methodology to adapt WCOME as an advanced fuel in the near future to reduce diesel engine emissions is the combination of fuel and modest engine change.

REFERENCES

Ashok, B., Nanthagopal, K., Saravanan, B., Azad, K., Patel, D., Sudarshan, B., & Ramasamy, R. A. (2019b). Study on isobutanol and Calophyllum inophyllum biodiesel as a partial replacement in CI engine applications. *Fuel*, *235*, 984–994. doi:10.1016/j.fuel.2018.08.087

Ashok, B., Nanthagopal, K., Saravanan, B., Somasundaram, P., Jegadheesan, C., Chaturvedi, B., & Patni, G. (2018). A novel study on the effect lemon peel oil as a fuel in CRDI engine at various injection strategies. *Energy Conversion and Management*, *172*, 517–528. doi:10.1016/j.enconman.2018.07.037

Ashok, B., Raj, R. T. K., Nanthagopal, K., Krishnan, R., & Subbarao, R. (2017). Lemon peel oil–A novel renewable alternative energy source for diesel engine. *Energy Conversion and Management*, *139*, 110–121. doi:10.1016/j.enconman.2017.02.049

Ban-Weiss, G. A., Chen, J. Y., Buchholz, B. A., & Dibble, R. W. (2007). A Numerical Investigation into the Anomalous Slight NOx Increase when Burning Biodiesel: A New (Old) Theory. *Fuel Processing Technology*, *88*(7), 659–667. doi:10.1016/j.fuproc.2007.01.007

Bari, S., Yu, C. W., & Lim, T. H. (2004). The effect of fuel injection timing with waste cooking oil as a fuel in a direct injection diesel engine. *Proc.IMechE, Part D. Journal of Automobile Engineering*, *218*(1), 93–104. doi:10.1243/095440704322829209

Basha, Al Balushi, Al Shibli, Al Fazari, Al Rumhi, Al Mansouri, & Al Balushi. (2020). An emission control strategy in a low capacity single cylinder compression ignition engine powered with DEE blended fuels. *Materials Science for Energy Technologies*, *3*, 770-779.

Cavalcante, F. T. T., Neto, F. S., de Aguiar Falcão, I. R., da Silva Souza, J. E., de Moura, L. S. Junior, da Silva Sousa, P., Rocha, T. G., de Sousa, I. G., de Lima Gomes, P. H., & de Souza, M. C. M. (2021). Opportunities for improving biodiesel production via lipase catalysis. *Fuel*, *288*, 119577. doi:10.1016/j.fuel.2020.119577

Chang, D. Y., & Van Gerpan, J. H. (1997). *Fuel properties and engine performance for biodiesel prepared from modified feedstocks.* SAE Paper No. 971684.

Damodharan, D., Sathiyagnanam, A. P., Kumar, B. R., & Ganesh, K. C. (2018). Cleaner emissions from a DI diesel engine fueled with waste plastic oil derived from municipal solid waste under the influence of n-pentanol addition, cold EGR, and injection timing. *Environmental Science and Pollution Research International*, *25*(14), 13611–13625. doi:10.100711356-018-1558-5 PMID:29497945

Depoures, M. V., Dillikannan, D., & Kaliyaperumal, G. (2019). Alcohol Contribution over Conventional Fuel. In *Introduction to Diesel Emissions*. IntechOpen.

Fayad, M. A. (2019). Effect of fuel injection strategy on combustion performance and NO_x/smoke trade-off under a range of operating conditions for a heavy-duty DI diesel engine. *SN Appl. Sci.*, *1*(9), 1088. doi:10.100742452-019-1083-2

Gautam, R., & Kumar, S. (2020). Performance and Combustion Analysis Of Diesel and Tallow Biodiesel in CI Engine. *Energy Reports*, *6*(6), 2785–2793. doi:10.1016/j.egyr.2020.09.039

Ghadikolaei, M. A., Cheung, C. S., & Yung, K. F. (2018). Study of combustion, performance and emissions of diesel engine fueled with diesel/biodiesel/alcohol blends having the same oxygen concentration. *Energy*, *157*, 258–269. doi:10.1016/j.energy.2018.05.164

Gürü, M., Koca, A., Can, Ö., Çınar, C., & Şahin, F. (2010). Biodiesel production from waste chicken fat based sources and evaluation with Mg based additive in a diesel engine. *Renewable Energy*, *35*(3), 637–643. doi:10.1016/j.renene.2009.08.011

Hoang, A. T. (2021). Prediction of the density and viscosity of biodiesel and the influence of biodiesel properties on a diesel engine fuel supply system. *Journal of Marine Engineering and Technology*, *20*(5), 299–311. doi:10.1080/20464177.2018.1532734

Huang, Z., Lu, H., Jiang, D., Zeng, K., Liu, B., Zhang, J., & Wang, X. (2004). Combustion behaviors of a compression-ignition engine fueled with diesel/methanol blends under various fuel delivery advance angles. *Bioresource Technology*, *95*(3), 331–341. doi:10.1016/j.biortech.2004.02.018 PMID:15288277

Jindal, S., Nandwana, B. P., Rathore, N. S., & Vashistha, V. (2010). Experimental investigation of the effect of compression ratio and injection pressure in a direct injection diesel engine running on Jatropha methyl ester. *Applied Thermal Engineering*, *30*(5), 442–448. doi:10.1016/j.applthermaleng.2009.10.004

John, B. (1988). *Heywood Internal Combustion Engine Fundamentals*. Mc-Graw-Hill.

Keskin, A., Gürü, M., & Altiparmak, D. (2007). Biodiesel production from tall oil synthesized Mn and Ni-based additives: Effects of the additives on fuel consumption and emissions. *Fuel*, *86*(7-8), 139–143. doi:10.1016/j.fuel.2006.10.021

Kumar, B. R., Muthukkumar, T., Krishnamoorthy, V., & Saravanan, S. (2016). A comparative evaluation and optimization of performance and emission characteristics of a DI diesel engine fueled with n-propanol/diesel', n-butanol/diesel and n-pentanol/diesel blends using response surface methodology. *RSC Advances*, *6*(66), 61869–61890. doi:10.1039/C6RA11643D

Labeckas, G., Slavinskas, S., & Kanapkiene, I. (2019). Study of the Effects of Biofuel-Oxygen of Various Origins on a CRDI Diesel Engine Combustion and Emissions. *Energies*, 12, 1241. *Int*, *24*(4), 389–403.

Lapuerta, M., Rodríguez-Fernández, J., & Font de Mora, E. (2009). Correlation for the estimation of the cetane number of biodiesel fuels and implications on the iodine number. *Energy Policy*, *37*(11), 4337–4344. doi:10.1016/j.enpol.2009.05.049

Liang, X., Zheng, Z., Zhang, H., Wang, Y., & Yu, H. (2019). A Review of Early Injection Strategy in Premixed Combustion Engines. *Applied Sciences (Basel, Switzerland)*, *9*(18), 3737. doi:10.3390/app9183737

Mathur, M. L., & Sharma, R. P. (1997). *Internal combustion engines*. Dhanpat Rai & Sons.

McCormick, R. L., Hayes, R. R., Black, S., Ireland, J., McDaniel, T., & Tennant, C. J. (2005) *Regulated emissions from biodiesel tested in heavy duty engines meeting 2004 emission standards*. SAE Paper No. 2005-01-2200.

Mittelbach & Remschmidt. (2004). *Biodiesel—the Comprehensive Handbook* (2nd ed.). Boersedruck,Ges.m.b.H.

Nanthagopal, K., Ashok, B., Saravanan, B., Patel, D., Sudarshan, B., & Ramasamy, R. A. (2018). An assessment on the effects of 1-pentanol and 1-butanol as additives with Calophyllum Inophyllum biodiesel. *Energy Conversion and Management*, *158*, 70–80. doi:10.1016/j.enconman.2017.12.048

Narayana Reddy, J., & Ramesh, A. (2006). Parametric studies for improving the performance of a Jatropha oil-fuelled compression ignition engine. *Renewable Energy*, *31*(12), 1994–2016. doi:10.1016/j.renene.2005.10.006

Qi, D. H., Chen, H., Matthews, R. D., & Bian, Y. Z. H. (2010). Combustion and emission characteristics of ethanol–biodiesel–water micro–emulsions used in a direct injection compression ignition engine. *Fuel*, *89*(5), 958–964. doi:10.1016/j.fuel.2009.06.029

Rajasekaran, S., Damodharan, D., Gopal, K., Kumar, B. R., & De Poures, M. V. (2020). Collective influence of 1-decanol addition, injection pressure and EGR on diesel engine characteristics fueled with diesel/LDPE oil blends. *Fuel*, *277*, 118166. doi:10.1016/j.fuel.2020.118166

Rajesh, A., Gopal, K., Kumar, B. R., Sathiyagnanam, A. P., & Damodharan, D. (2020). Effect of anisole addition to waste cooking oil methyl ester on combustion', emission and performance characteristics of a DI diesel engine without any modifications. *Fuel*, *278*, 118315. doi:10.1016/j.fuel.2020.118315

Rakopoulos, D. C., Rakopoulos, C. D., Mavropoulos, G. C., & Kosmadakis, G. M. (2018). Investigating the EGR rate and temperature impact on diesel engine combustion and emissions under various injection timings and loads by comprehensive two-zone modeling. *Energy*, *157*, 990–1014. doi:10.1016/j.energy.2018.05.178

Sadhik Basha, J. (2018). Impact of Carbon Nanotubes and Di-Ethyl Ether as additives with biodiesel emulsion fuels in a diesel engine – An experimental investigation. *Journal of the Energy Institute*, *91*(2), 289–303. doi:10.1016/j.joei.2016.11.006

Sadhik Basha, J. (2022). Applications of Nano-Additives in Internal Combustion Engines: A Critical Review. *Journal of Thermal Analysis and Calorimetry*, *147*(17), 9383–9403. doi:10.100710973-022-11199-6

Sayin, C., Ilhan, M., Canakci, M., & Gumus, M. (2009). Effect of injection timing on the exhaust emissions of a diesel engine using diesel-methanol blends. *Renewable Energy*, *34*(5), 1261–1269. doi:10.1016/j.renene.2008.10.010

Schmidt, K., & Van Garpen, J. H. (1996). *The Effect of Biodiesel Fuel Composition on Diesel Combustion and Emission.* SAE Paper No. 961086.

Selvakumar Ramalingam & Mahalakshmi. (2020). Influence Of *Moringa Oleifera* Biodiesel–Diesel–Hexanol And Biodiesel–Diesel–Ethanol Blends On Compression Ignition Engine Performance, Combustion And Emission Characteristics. *RSC Advances*, (10), 4274–4285. PMID:35495259

Shan, X., & Que, L. Jr. (2006). High-valent nonheme iron-oxo species in biomimetic oxidations. *Journal of Inorganic Biochemistry*, *100*(4), 421–433. doi:10.1016/j.jinorgbio.2006.01.014 PMID:16530841

Simpson, T., & Depcik, C. (2022). Multiple Fuel Injection Strategies for Compression Ignition Engines. *Energies*, *15*(14), 5214. doi:10.3390/en15145214

Sukumar Puhan, R. (2009). Effect of injection pressure on performance, emission and combustion characteristics of high linolenic linseed oil methyl ester in a DI engine. *Renewable Energy*, *34*(5), 1227–1233. doi:10.1016/j.renene.2008.10.001

Yang, H. H., Lee, W. J., Mi, H. H., Wong, C. H., & Chen, C. B. (1998). *PAH emissions influenced by Mn based additive and turbo charging from a heavy-duty diesel engine*. Environ.

Zhu, L. (2010). Emission characteristics of a diesel engine operating on biodiesel and biodiesel blended with ethanol and methanol. *The Science of the Total Environment*, *408*(4), 914–921. doi:10.1016/j.scitotenv.2009.10.078 PMID:19926114

Chapter 8
The Use of Alternative Fuels and Emission Control Methods for Compression Ignition Engines

T. Shaafi
Institute of Agricultural Engineering, Saveetha School of Engineering, India & Saveetha Institute of Medical and Technical Sciences, India

J. Sadhik Basha
Department of Process Engineering, National University of Science & Technology, Oman

Abidoye Luqman Kolawole
Osun State University, Nigeria

V. Siva Shankar
Institute of Agricultural Engineering, Saveetha School of Engineering, India & Saveetha Institute of Medical and Technical Sciences, India

G. Velmurugan
Institute of Agricultural Engineering, Saveetha School of Engineering, India & Saveetha Institute of Medical and Technical Sciences, India

ABSTRACT

Diesel engines are widely used and have a lot of benefits, but they also contribute significantly to worldwide environmental pollution problems. Many regulations have recently been passed all over the world to minimize the emissions from diesel engines to protect the environment and people's health. Efforts are being made to reduce fossil fuel consumption and maximize the utilization of environmentally friendly alternative fuels and reduce the emissions from compression ignition engines. In this context, the standard emissions for diesel engines, pollutant emissions from diesel engines, the necessity for alternative fuels, various types of alternative fuels employed, and emission control techniques are all discussed in this chapter.

DOI: 10.4018/978-1-6684-7303-0.ch008

INTRODUCTION

Energy is necessary for economic progress, which is beneficial for emerging nations. The primary sources of energy are those that can be found or are stored in nature. Coal, oil, natural gas and biomass are the common primary energy sources. Coal, oil and solid biomass collectively provide more than 80% of India's energy needs (International Energy Agency [IEA] 2021). The relationship between rising energy demand and economic growth is not always a simple linear one. Energy is indispensable for living and vital for development of a country and also it's directly contributes in reducing the poverty, increasing the productivity and improving the quality of life. The energy availability and its per capita consumption are two of the most crucial indicators of a nation's citizenry's living conditions. India has been working gradually over the years to increase per capita energy usage with an emphasis on ending poverty in the nation (Society of Indian Automobile Manufacturers [SIAM], 2019). The rate of economic growth, global industrialization and motorization are the factors to decide the energy consumption pattern in India. Indian economy is agriculture based and modern agriculture system mostly depends on the machineries operated with internal combustion engines. The Indian economy is predicted to expand at a rate of above 6% annually, necessitating an increase in energy demand to 166 MT by 2019 and 622 MT by 2047 (Dwivedi, Sharma, & Kumar,2014). Due to the tremendous worldwide threat posed by rising emissions from the burning of fossil fuels and the constantly increasing rate of crude oil imports, the growth of the Indian economy suffers significantly. India has highly used road transportation infrastructure in the world. Transport plays a significant role in the overall development of a nation's economy. The transport sector is also the second largest consumer of energy next to industry. The most cost effective and preferred mode of transport, for both freight and passengers are road transport in India. Approximately 87% of all passenger and 60% of all freight traffic in the nation is carried by road (Ministry of Road Transport and Highways [MRTH], 2021). Road transportation is advantageous for a number of reasons including its simplicity of accessibility, capacity to accommodate to individual requirements and cost savings. Therefore, road sector is vital for economic development and social integration of the country. In modern life style the diesel engines have become an indispensable part because of their role in transportation and mechanized agriculture sector. Depending upon the usage, the diesel engines are available in various sizes and speeds. Due to its high compression ratio, fuel efficient operation and low maintenance costs, diesel engines have greater fuel economy, thermal efficiencies which has boosted the adoption of diesel vehicles. The largest drivers of diesel demand growth in India are commercial vehicles including trucks and public transport and their consumption of fuel is linked to the overall economy. About 98 percentage of which is in the form high speed diesel (HSD) and gasoline used in the form of commercial energy consumption. India is established via the use of fossil fuels. Many nations now import all of their oil needs as a result of the depletion of fossil fuels. India is heavily reliant on crude oil imports.

India would surpass China and Japan to become the third largest oil importer in the world by 2025 according to International Energy Agency. In 2022, India's oil demand is expected to rise 8.2% to 5.15 million barrels per day (PTI / Mar 17). In 2022- 2023 fuel consumption is expected to increase from 203.2 million tonnes to 214.5 million tonnes (Petroleum Planning and Analysis Cell [PPAC], 2022).

By 2029-30, diesel consumption in India is predicted to increase to 163 MT, with diesel and gasoline accounting for 58% of India's oil demand by 2045 (Indian Oil and Gas Industry Analysis Presentation [IBEF] 2022).

Figure 1. Total primary energy demand in India
Source: IEA (2021)

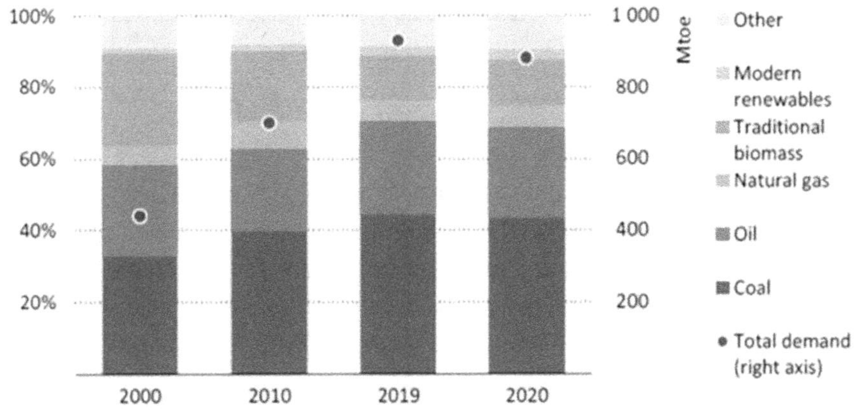

Figure 2. India's fuel consumption in the road transportation sector
Source: SIAM (2019)

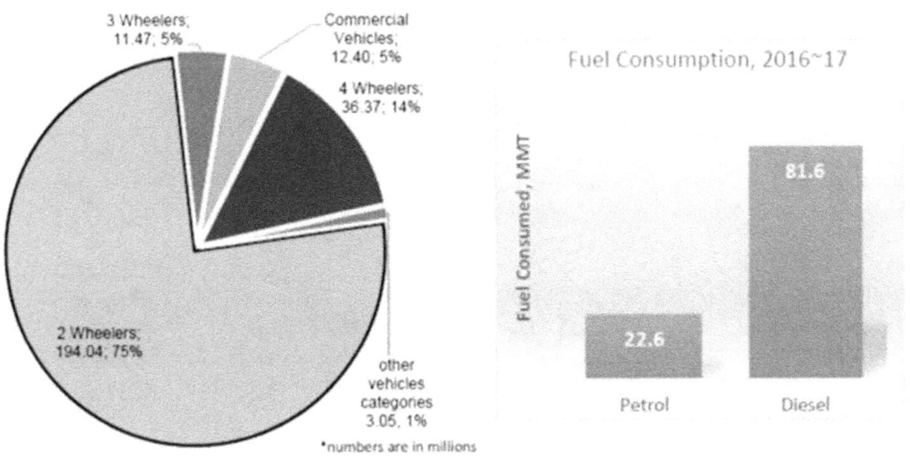

However, transport sector contributes a significant and increasing amount of air pollution (Misra, Shaafi, Gayathri, Elayaperumal & Velraj 2017). Diesel pollutants, particularly diesel particulate matter, can have a major negative impact on health. The oxides of nitrogen (NO_X) emission come out from the diesel engine is the critical task for the diesel engine manufacturers. Moreover, they need to upgrade the new emission standards in their new design diesel engine. Table 1 illustrates the new emission standards for the diesel engine (Borkhade et al., 2022).

Table 1. Emission norms (g/km) of diesel engine

Emission Norms	CO	HC	NOx	PM
BS-III	0.64	0.50	0.56	0.050
BS-IV	0.50	0.25	0.30	0.025
BS-VI	0.50	0.06	0.17	0.005

The reserves of mineral oil, such as gasoline and diesel, are being depleted due to the ongoing use of fossil fuels (Shaafi Tajudeen & Velraj, 2014). The dwindling sources of conventional fossil fuels, their ever increasing demand, reduce the emission and prices have prompted the engine manufacturers and researcher's first step in lowering the diesel emission is to switch to biodiesel or renewable diesel mixes instead of regular diesel fuel (Basha et al., 2021). This chapter focused on the availability of various alternative fuels used and methods to control the emissions of a compression ignition engines.

EMISSIONS FROM DIESEL ENGINES

The compression ignition engine is a self-igniting engine.

Figure 3. The components of inlet air and exhaust gas of a diesel engine

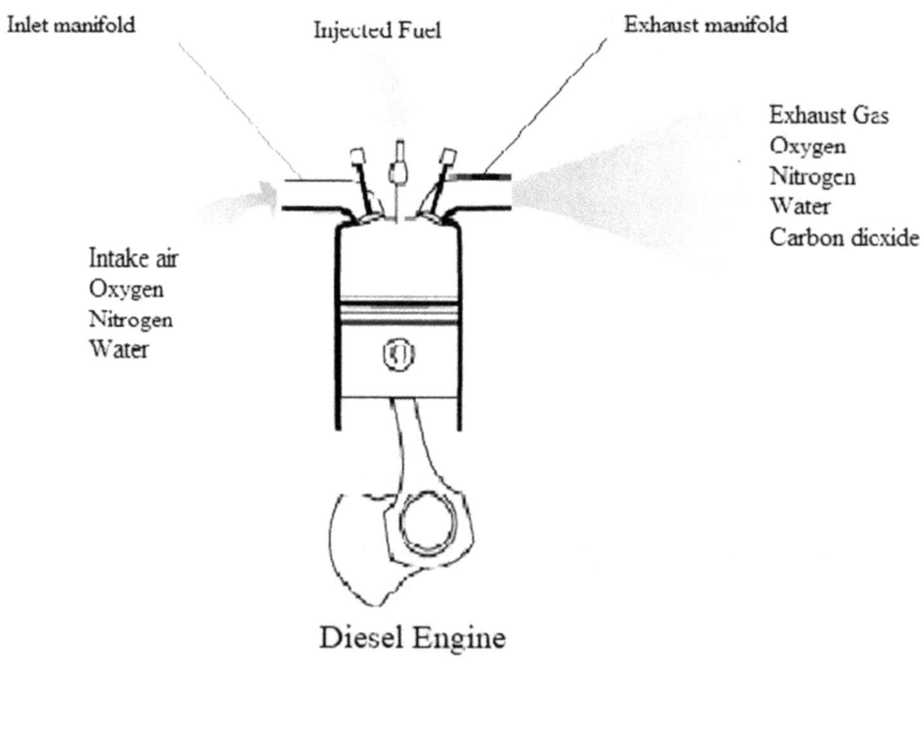

Diesel exhaust pollutants may be dangerous to the environment or to human health. (Basha et al., 2020). Pollutant emissions from diesel exhaust gas are less than 1%. NO_X has consistently accounted for the greatest proportion of diesel pollution emissions accounting for even more than 50%. PM emissions are the most commonly emitted pollutant after the NO_X emissions. The quantity of CO and HC emitted is low due to the use of lean combustion in diesel engine. However, depending upon the fuel characteristics and quality, emission levels may contain a trace of SO_2. Nowadays, most of the oil distributors and customers use the ultra sulphur diesel for compression ignition engines to prevent the undesirable effects of SO_2. The following section describe about the four principal pollutant emissions from diesel engines are: CO, HC, PM and NO_X.

Carbon Monoxide (CO)

Carbon monoxide is a gas that is neither colourless nor odorless. It is created during the combustion process, which actually occurs whenever the oxidation process is incomplete. CO is heavily influenced by the air/fuel combination. As engines are starting and accelerating, rich mixtures are required since the most CO is created at these times. The entire amount of carbon could be oxidized to CO_2, instead being created as CO concentration due to the absence of oxygen and the amount of reactant in the rich mixture. CO emission from the lean running compression ignition engines are typically low.

Hydrocarbons (HC)

Diesel engines typically emit only small traces of hydrocarbons. Due to the low temperature near the cylinder wall, unburned fuels are present in hydrocarbon emission (Machado Corrêa & Arbilla, 2008). Hydrocarbon emissions are more common at low loads in diesel engines. Lean air fuel mixture is the main contributor to HC emissions from light loads. Diesel engine design, engine modification, and fuel type all have an impact on the amount of hydrocarbons generated by the engine. Both the human health and environment are endangered by the hydrocarbons. Moreover, hydrocarbon along with other pollution sources, greatly contribute to the formation of ground level ozone.

Particulate Matter (PM)

The combustion process the produces PM emissions in the exhaust gas. They might have evolved as a result of the accumulation of tiny particles of partially burned fuel, lubricating oil and cylinder lube oil ash content or sulphates and water. The majority of particulate matter is formed by the incomplete combustion of HC found in lubricating oil and fuel. The diameter of diesel particulate matter ranges from 15 to 40 nm. More than 90 percent of the total of PM have sizes under 1 micron. The following variables affect how PM emissions are formed: 1. the process of combustion and expansion 2. Fuel quality 3. temperature of combustion and 4. Soot, organic substances fraction, and inorganic fraction are the three primary components of diesel particulate emissions. Soot, which resembles black smoke, makes more than fifty percent of all PM emissions.

Nitrogen Oxides (NO_X)

Since nitrogen in the air does not react with oxygen during combustion, oxides of nitrogen are present in diesel engine exhaust and are emitted into the environment by the engine. The majority of this will comprise Nitrogen Oxide (NO) with small amount of Nitrogen Dioxide (NO_2). The NO_X production parameters are the combustion's temperature and oxygen concentration. The highest flame temperature in the combustion chamber causes more NO_x to be produced. The environment and public health are seriously endangered by nitrogen oxide emissions. NO_X emissions play a role in nutrient enrichment, ozone generation, acidification, and smog development (Reşitoğlu et al., 2014).

NEED OF ALTERNATIVE FUELS

Diesel is required heavy duty engines and transportation. Because of their high thermal efficiency, diesel engines have a good reputation. In the near future, diesel engines may overtake other sources of air pollution. An air pollutant is any substance, whether chemical, physical, or biological, that modifies the atmosphere's inherent qualities. Carbon monoxide, ozone, nitrogen dioxide, particulate matter and Sulphur dioxide are pollutants of great public health concern. Air pollution has become a major concern for every country. Combustion by fossil fuel releases huge amount of greenhouse gases and carbon dioxide is the most significant of all these gases. Because the usage of conventional fossil fuels can contribute to climate change which could result in ecological disasters in some location. Emission control has now become the primary driving factor behind the development of diesel engines. Efforts are being made to reduce the fossil fuel consumption and maximize the utilization of environment friendly alternative fuels and reduce the emissions from compression ignition engines. The impact of global warming due to the air pollution imposed the world wide emission regulations and the researchers and engine manufacturers to shown their interest towards the design of engines or fuel modification and emission control techniques (Shaafi etal.,2015). Due to their inherent advantages in terms of economics, high quality energy, compatibility with the environment, and renewability, alternative fuels are gaining attention throughout the globe. It is essential to replace diesel with alternative fuels like biodiesel, CNG and bioethanol in order to eliminate dangerous emissions and significant import deals. Alternative fuels can either completely replace diesel or be blended with diesel (Muvva, Shaafi & Arunkumar, 2020) (Velmurugan, & Shaafi, 2019).

Benefits of Alternative Fuels

Alternative fuels provide the following benefits: 1. Energy security 2. Social empowerment 3. Job creation. 4. Significant foreign exchange savings

Selection Criteria for Alternative Fuels

The following selection criteria to be considered for choosing the alternative fuels such as: 1. Cheap and available in everywhere, 2. Produce less emission, 3. High calorific value, 4. Less modification in the existing compression ignition engine, 5. Increase the engine life, 6. Easy to handling and store.

Types of Alternative Fuels Used for Compression Ignition Engines

The following alternative fuels used in compression ignition engines are the substitute to the Diesel: 1. Biodiesel, 2. Butanol, 3. Producer gas, 4. Dimethyl ether, and 5. Hydrogen.

Biodiesel

The essential materials used to produce biodiesel are short chain alcohols, animal fats and vegetable oils. Biodiesel is a term used to describe a liquid manufactured from animal or vegetable oils and alcohol. The primary oils used in the production of biodiesel around the world are palm, rapeseed, sunflower, soybean oil and animal fats. Animal fats are another intriguing possibility with abundant livestock resources (Dinesh Babu & Shaafi, 2020). Although they must undergo preparatory treatment due to their solid nature. Microalgae appear to be a critical option for future biodiesel generation due to its high oil output. Through a transesterification reaction, biodiesel is produced and this chemical reaction produces a variety of esters of the fatty acids that make up the oil from an ester of vegetable oil or animal fat. By purifying a mixture of fatty acid methyl esters, biodiesel is produced. The transesterification reaction is accelerated using a catalyst. Depending upon the catalyst used, the transesterification process might be basic, acidic or enzymatic. Biodiesel can be blended with diesel or alone and it is used as a fuel for compression ignition engines. (Shaafi & Velraj, 2015:Sadhik Basha & Anand, 2014). Nontoxic and biodegradable are the predominant one for biodiesel to get the more attraction as a fuel in a diesel engines. Without any changes of biodiesel releases considerably low aromatic compounds, chemical matters and sulfates to the atmosphere. Nearly 75% of the price of the biodiesel are due to the availability of viable feed stock. It is very essential to pick out a not expensive feedstock for enhancing the economics of biodiesel production. (Sani et al., 2012). The first generation feedstock of biodiesel sources are the edible oil such as sunflower, rice bran, palm oil, rapeseed, soybean, peanut and coconut and the second generation feedstock are the nonedible oils, animal fat, waste oils and greases.

Plantation of feedstock's for biodiesel may additionally require deforestation, reduction of cultivatable land and damage to the soil resources. The preparation cost of biodiesel will depends upon the vegetable oil cost. Vegetable oils, animal fats and short chain alcohols are the foremost raw substances for biodiesel production. Palm, sunflower, soybean, rapeseed, caster, jatropha, pongamia vegetable oils are used for the production of biodiesel in worldwide.

Palm Oil Biodiesel

In comparison to several other vegetable oils, palm oil may be a more desirable and useful alternative as a source of biodiesel. Despite the fact that palm oil is a widely used vegetable oil for everything from food to biofuels (Tan, Lee, Mohamed, & Bhatia, 2009). The experimental findings used to assess the impact of the anticorrosion ingredient in biodiesel derived from palm oil on the operation, emissions, and wear characteristics of diesel engines (Kalam, & Masjuki, 2002). The advantages of palm oil biodiesel is to releasing significantly less carbon dioxide than conventional petro-diesel. Moreover, palm oil biodiesel is also nontoxic and biodegradable. Furthermore, palm oil has several advantages over other types of oil crops because of its large oil content, and high production capacity (Dey, Reang, Das, & Deb, 2021). A higher cetane number is definitely advantageous, especially in terms of engine efficiency and clean emissions. Kinematic viscosity, iodine value, saponification number, and greater heating value are a few additional crucial characteristics of Palm oil biodiesel (Zahan, & Kano, 2018).

Sunflower Oil Biodiesel

Based on its extensive cultivation, sunflower oil is one of the options for large-scale biodiesel manufacturing. The factors influencing the yield and properties of the biodiesel made from the sunflower vegetable oils were studied by (Thirumarimurugan et al., 2012). In order to choose the best method for preparing biodiesel through the transesterification of sunflower oil, four different enhancing techniques were investigated: mechanical stirring, flat plate ultrasonic irradiation, flat plate ultrasonic irradiation with mechanical stirring and probe ultrasonic irradiation by (Yin et al., 2012). (Akhihiero et al., 2019) were investigated the flash point, density, and kinematic viscosity characteristics of the various blends by blending sunflower biodiesel with gasoline and diesel at various ratios. The feedstock is the most key component required for the production of biodiesel. Due of the restricted supply of feedstock, the use of hybrid feedstock has been studied by (Saydut et al., 2016). Sunflower biodiesel has a very low sulphur content, which gives it a significant advantage because it prevents the production of sulfur-containing compounds when it is burned in diesel engines.

Soybean Biodiesel

The sustainability objectives for soy biodiesel are to provide a locally derived fuel, maintain and improve the quality of the environment and produce an economically viable fuel, and contribute to energy security (Milazzo et al., 2013). The tribological performance of soybean biodiesel was evaluated using HFRR by (Xiao et al., 2019). Thermal analysis is a useful method for determining features like thermal and oxidative stabilities and crystallisation temperatures (Zhao et al., 2012). The results and prospects for the future of soy biodiesel synthesis as determined by about 30 life-cycle assessments (LCAs) on a global scale. In all global contexts, soy biodiesel is not the most sustainable product. Life-cycle energy is influenced by the agro- and processing methods utilised, as well as by particular meteorological circumstances (Milazzo et al., 2013). A stable water-in-soybean biodiesel emulsion fuel is created, and its performance in terms of emissions in a single cylinder diesel engine is examined by (Vellaiyan & Partheeban, 2018).

Rapeseed Biodiesel

One of the most widely produced and commercialised edible oil-based biodiesel fuels is rapeseed oil-based biodiesel (Konur 2021). Rapeseed biofuel manufacturing has a relatively low energy efficiency (Van Duren et al., 2015) (Firrisa et al., 2014). Under various storage circumstances, the rate of physical and chemical degradation of used frying oil and rapeseed-based biodiesel was investigated and there were no appreciable distinctions between frying oil that had been used previously and fresh rapeseed oil produced into undistilled biodiesel (Mittelbach, & Gangl, 2001).

Castor Oil Biodiesel

Due to the cost and consumption for edible oils, non-edible oils may be chosen in India for the production of biodiesel (Shrirame et al., 2011). Due to its low cloud and pour points, biodiesel generated from castor oil is a good alternative during cold weather. (Keera et al., 2018). By neutralising the crude oils, the quality of biodiesel made from jatropha and castor oils was greatly enhanced (Okullo et al., 2012). Castor oil was converted into biodiesel (ethyl ester) via the transesterification process; its physical characteristics were then assessed using ASTM standards, and the effects of its various mixes (B5 and B20) were investigated by (Hajlari et al., 2019). Several techniques, including transesterification, mixing with diesel, and blending with low viscosity biofuel called pine oil were used to improve the combustion properties of neat castor oil (Prakash et al., 2019).

Jatropha Biodiesel

Jatropha biodiesel emits substantially fewer emissions, making it more environmentally friendly (Prabhu et al., 2021). Jatropha-based biodiesel may face competition from third-generation biofuels, particularly algal biodiesel. India is an agro-economic country, therefore if jatropha is grown more scientifically, it might end up being more cost-effective than algae. (Datta, & Mandal, 2014). The different biodiesel samples (mahua biodiesel, jatropha biodiesel and mineral diesel) were held under certain circumstances for 8 fortnights (120 days) for the storage stability study and the various parameters density, kinematic viscosity, flash point, and acid number were then determined (Acharya et al., 2019). The oxidation stability of biodiesel by mixing it with petro-diesel, because the oxidation stability of biodiesel from J.curcas oil is relatively low (Jain & Sharma 2011).

Pongamia Biodiesel

Pongamia is a forest tree with non-edible oil and a high potential for biodiesel production (Gaurav & Sharma 2014). The fundamental problem with the development of Pongamia biodiesel as a diesel substitute is its cold flow qualities. Poor cold flow feature causes gum development and fuel particle crystallisation, although it can be improved by winterization, blending, and the inclusion of cold flow improvers (Dwivedi & Sharma 2015). Weak thermal and oxidative stability causes a decline in quality and prevents it from being stored for an extended period of time. The poor thermal stability of the biodiesel also has an impact on the functioning of high temperature engines. The oxidation and thermal stability of Pongamia biodiesel are enhanced by pyrogallol (Dwivedi & Sharma 2016). Research on Pongamia will be conducted to standardise agricultural techniques, develop affordable and effective mechanical oil extraction devices, and identify cultivars with high yields and high oil content that are economically viable and suitable for India's many agro-climates. (Dwivedi et al., 2011).

Producer Gas

Compression ignition engine requires the engine modification when the producer gas used as a fuel (Raman, & Ram, 2013) (Aung, 2008). Producer gas from the biomass is a feasible substitute to conventional fuels. Wood, charcoal, rice husks and coconut shell are the viable dry materials for the production of producer gas. The gasifying of dry carbonaceous materials may be used to produce the producer gas. In this process with the aid of heat the solid biomass cracked down. Suction gas is another name for producer gas. The way an IC engine sucked air in to the gas generator is referred to as suction. The producer gas mostly comprises of nitrogen in significant amounts, hydrogen and carbon monoxide and it has a low heat value mainly due to its high level of nitrogen.

Hydrogen

Hydrogen stands out as a very good potential substitute for the transportation and electricity generation. Hydrogen is a component of water, which covers 70% of the planet's surface and this may be an endless source of the element. In nature, hydrogen is found as the diatomic molecule H_2 which is odourless, colourless, tasteless and flammable gaseous substance and which produces enormous amount of heat and water when it is combusted with oxygen at high pressure and temperatures. It is the most common element in the universe and the light weight chemical element. Compared to other types of fuels, hydrogen is said to have high combustion efficiency, good combustion characteristics and its having high energy content, making it the cleanest fuel for internal combustion engines. The simplest and most adaptable way to use hydrogen in a diesel engine is through dual-fuel technology (Gopal, Rao, Gopalakrishnan, & Murthy,1982). It has been observed that employing hydrogen as fuel decreases the level of diesel engine pollution emissions. Storage of hydrogen as a gas or a liquid requires high pressure tanks 350 to 700 bar and cryogenic temperatures because the boiling point of hydrogen at one atmospheric pressure is -252.8 degree centigrade respectively.

EMISSION CONTROL METHODS

The three main categories for reducing emissions from the compression ignition engines are covered in this section: 1. Engine design modification, 2. Technologies related to the reformulation of fuel, 3. Exhaust gas after treatment.

Engine Design Technologies for Emission Reduction

Engine modifications involve altering the combustion process to describe the generation of pollutants. Even though the changes that reduce the NO_X and increase in engine-out PM emissions and vice versa. NO_X tradeoff is referred to as reduces the NO_X emission during combustion, but at the same time hinders full oxidation of soot, increasing particle emission. This NO_X tradeoff poses a significant obstacle to reduce the emission measures.

Fuel Injection Strategies

The sophisticated fuel injection systems are extremely advantageous for diesel engines. The key factors in controlling NO_x and PM are injection pressure, injection rate, and injection time. This system must able to do the following: 1) maintain fuel injection pressure from 1500 to 2000 fold atmospheric pressure 2) flexible injection rate shaping and 3) while meeting strict durability, cost etc., Better air-fuel mixing is achieved by increasing injection pressure, which diminishes fuel particles the size hence improves fuel penetration into the cylinders. By the use of improved nozzle and piston bowls, fuel systems may improve fuel mixing and combustion process. The fuel flow is changed during the injection time with rate shaping of injection effectively controlling the injection process to achieve low emission. A small amount of fuel which is injected early and very late in the expansion stroke in pilot and port injection respectively. This raises the quantity of hydrocarbons in the engine exhaust, resulting in more effective emission control systems. Electronic fuel injection control provides for accurate, varied fuel timing and metering and this can also be used for delayed injection in after treatment systems with active regeneration.

Common Rail Injection System

One of the most effective injection technologies is the common rail system. Since they are flexible and fulfil the stringent pollution control standards, contemporary diesel engines use common rail direct injection fuel systems.

Figure 4. The common-rail fuel injection system

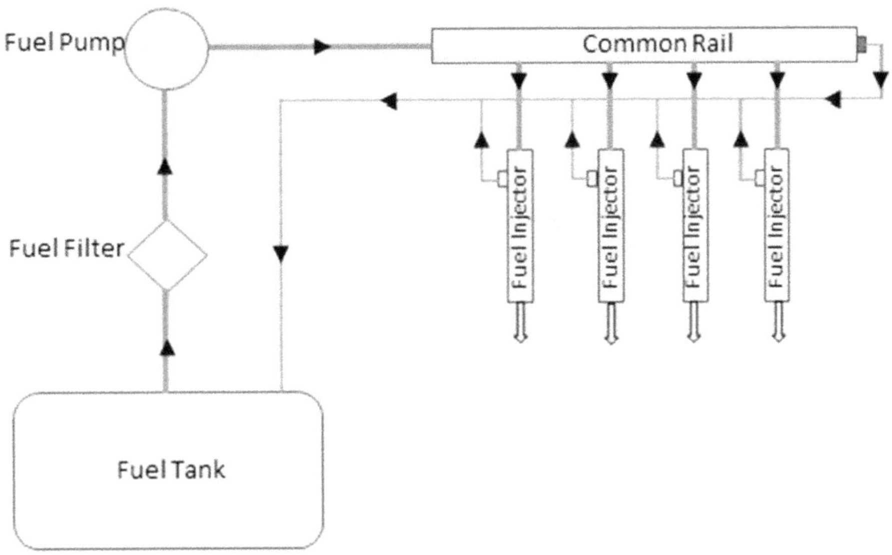

The rail functions as a diesel accumulator, allows the engine to maintain relatively constant pressure at all fueling rates. The fuel is fed to the injectors using high pressure pipes from the rail. Timing valves control the timing and amount of diesel delivered to cylinder injectors. The control units regulate the injection timing, volume, and injection pattern shape.

Exhaust Gas Recirculation (EGR)

The NO_X emission reductions are controlled with the aid of Exhaust Gas Recirculation (EGR) for most of the light and heavy duty engine applications. Carbon dioxide, nitrogen and other elements are the principal constituents of exhaust gases, and the mixture has a higher specific heat than outside air. Exhaust gas recirculation is the process of directing exhaust gases back toward the engine after the combustion process. Before the catalytic converter and after the exhaust manifold, the EGR valve is fixed. A portion of the engine exhaust are redirected back to the inlet manifold by opening this valve, where they are combined with fresh air. The quantity of exhaust gas flow back in to the engine can be fixed as 30% of the total intake. Further lower the temperature of combustion, an EGR cooler may be used before the mixture of exhaust gas with fresh air.

Figure 5. The exhaust gas recirculation

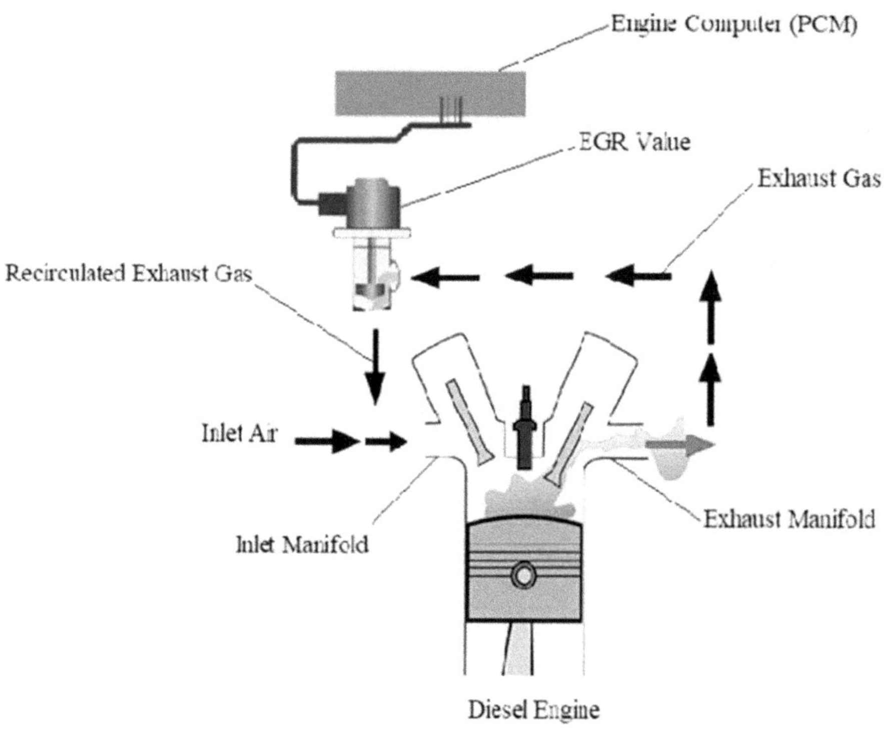

EGR cooler and sensor are used in order to exert even greater control over the exhaust gases. The major challenge in this is regulating the mass of the exhaust gases and recirculating them as needed. The air entering the combustion chamber is replaced by the recirculated exhaust gas, which also contains water vapour and carbon dioxide from the engine. The exhaust gases are now inert gases since they have previously been burned, and lower the oxygen concentration in the air, which slows down combustion, lower maximum temperatures and reduces the generation of NO_X. Benefits of EGR include decreased fuel usage, temperature and pressure management with in the combustion chamber and decrease in hazardous and toxic emissions.

Catalytic Converter

The catalytic converters are used to reduce the pollutant emissions from gaseous such as carbon monoxide (CO) nitrogen oxides (NO_x) and hydrocarbons (HCs). Catalytic converters are looks like a chambers and through which the exhaust gases from the engine pass through and it is encourages the oxidation of the emission contained in the exhaust. The carbon monoxide and hydrocarbons can be oxidized in to carbon dioxide and water respectively. The base ceramic material used for catalytic converter is the aluminium oxide. Platinum, palladium and rhodium are the mostly used for catalytic materials.

Figure 6. The catalytic converter

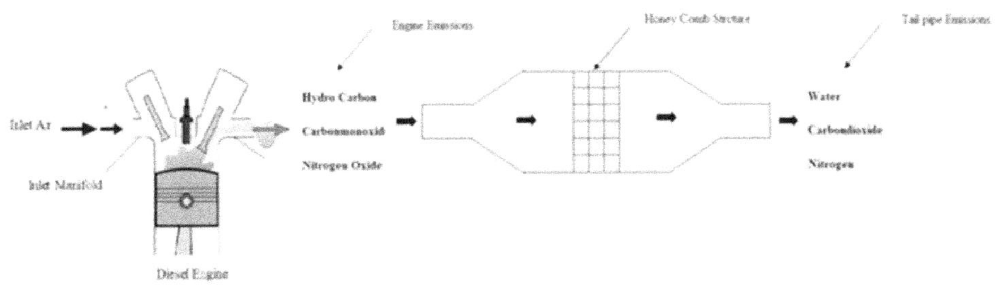

When the catalytic converters are efficient and it is to be operated at hot condition and at the same time it is not efficient when they are at very cold condition. The temperature at which the catalytic converter becomes 50% efficient is described as the light-off temperature.

Reformulation of Fuel

Due to the effects of global warming, it is essential that researchers concentrate their attention on either the design of engines or fuel related solutions. This could include alternative oxygenated or biodegradable fuels that can minimize particulate emissions. It is possible to use biodiesel at a variety of concentrations when combined with neat diesel. The most popular blends are B5 and B20, which stand for 5% and 20% of diesel respectively. Compared to B100, B20 would have good emission reduction. Pure biodiesel (B100) is a rarely utilized transportation fuel.

Exhaust After Treatment

A method or device for lowering detrimental emissions from internal combustion engines is referred as an aftertreatment system.

Diesel Oxidation Catalyst (DOC)

Nowadays, DOC's are still an essential element for the diesel engine. Recent light –duty, heavy-duty and passenger car mounted diesel engines all come equipped with DOC's. It is estimated that 60 to 90 percentage of HC and CO are reduced using of DOC. Diesel oxidation catalyst primary task is to oxidize hydrocarbon and carbon monoxide emissions. Carbon monoxide, hydrocarbons, organic component of diesel particles and unrestricted emissions like aldehydes or PAHs can all be oxidized to innocuous by-products when they are passed over an oxidation catalyst. The capacity of the diesel oxidation catalyst to boost the oxidation of exhaust gas constituents by oxygen, which is abundant in diesel engine exhaust. The temperature has a significant impact on DOC effectiveness. Light-off temperature is the temperature at which the catalyst starts to react and it changes with flow rate, catalyst composition, and exhaust composition. Effectiveness of the diesel oxidation catalyst is dependent on HC and CO oxidation, which is above “light off” temperature for the catalytic function.

Figure 7. The diesel oxidation catalyst (DOC)

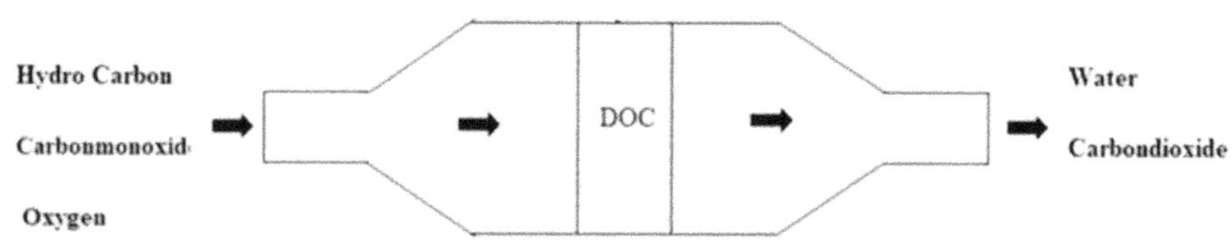

Diesel oxidation catalysts are constructed using ceramic or metal monolith honeycomb structures. Aluminum oxide, cerium oxide and zirconium oxide are among the active catalytic noble metals used to make the wash coat or oxide mixture. The wash coat’s main purpose is to greatly expand the area of the noble metal and diminish the rates of catalytic sintering, which happen at high temperatures and causes a depletion of catalyst activity. The most widely used oxidation catalysts comprising Pt and Pd are available on the global market but at the same time set back of these catalysts are carry reactions of SO_2 to SO_3 and it is subsequently produces forms of sulfates and sulphuric acid when it reacts with water. This issue could not be fully resolved, even though 500 ppm or ultra low sulfer diesel (USLD) is utilized. So far, no technology exists to stop and get rid of these forms of sulfates and sulphuric acid. Moreover, these forms of sulfates and sulphuric acid are destructive the after treatment emission control systems and also produces environmental and health issues. By using the alternative fuels like biodiesel, methyl alcohol etc., can improve the conversion efficiency of DOC. Diesel oxidation catalyst may also be employed in conjunction with SCR catalysts.

Diesel Particulate Filter (DPF)

Due to the combustion process in a diesel engine releases the harmful pollutants, particulate emissions. Nowadays, the particulate emissions reductions from diesel engine is a great challenge. The performance of vehicles that run on diesel fuel depends on diesel particulate filters. DPF is used to eliminate the particle pollutants from the exhaust gas using physical filter. It is made up of either silicon carbide or cordierite honeycomb structure with the channels blocked at an alternate ends. Soot particles are carried in to the pore walls via diffusion as they travel through the filter walls. To enable the exhaust gases to pass through the filter walls without much hindrance the filter walls have more porosity.

Figure 8. The schematic of diesel particulate filter

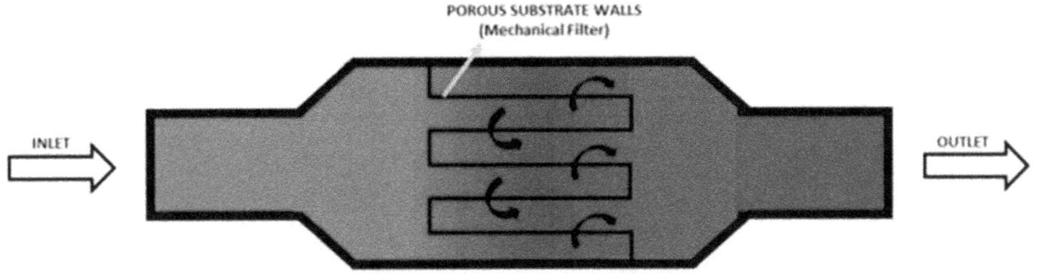

DPF filter successfully reduce the particulate matter in the exhaust flow by more than 95%. As particulate matter builds up in the filters, backpressure forms, which has a number of detrimental effects like higher consumption fuel, filter stress and failure of engine. To prevent these negative effects, the DPF must be refreshed by burning trapped particle matter. Burns these excess soot deposited in the diesel particulate filter is called regeneration and this regeneration can be categorized as passive regeneration or active regeneration. The diesel particulate filter lifespan may varied depends upon the application.

Passive Regeneration

The exhaust gas is allowed to pass over the diesel oxidation catalyst first, then it passes through the particulate filter, which traps the soot particles. Using the nitrogen dioxide as an oxidizer, the deposited materials are burned is called passive regeneration. Small quantity of NO_2 will boost the continual oxidation of the deposited carbon particles at temperatures between 200 and 450 degree centigrade.

Active Regeneration

The serious complications for active regeneration is i) higher regeneration temperature and ii) the amount of energy utilized for heat supply. Diesel particulate filter failure is carried by the temperatures over the filter boiling point. Active regeneration occurs when the engine is not generating sufficient heat. Due to the complex additions required for energy heating increases the manufacturing cost of the system. This is the reasons the active regeneration are not preferred. Whenever the soot level crosses a certain limit, the fuel injects in to the exhaust system, which oxidises the fuel and produces heat and this subsequently utilized to transfer soot to carbon dioxide.

Selective Catalytic Reduction (SCR)

Selective catalytic reduction is the dynamic emission control system and it reduces the emissions of nitrogen oxides (NO_X) to zero level. One of the most affordable and fuel efficient solutions for effectively eliminating emissions from diesel engine is SCR technology. SCR technology is used by all heavy duty diesel truck engines to meet the most recent emissions standards. SCR has the capability to reduce NO_X emissions by up to 90% while also reducing HC and CO emissions by 50% to 90% and PM emissions by 30% to 50%. This systems can also be used in collaborations with a diesel particulate filter to reduce PM emissions.

Figure 9. The selective catalytic reduction (SCR)

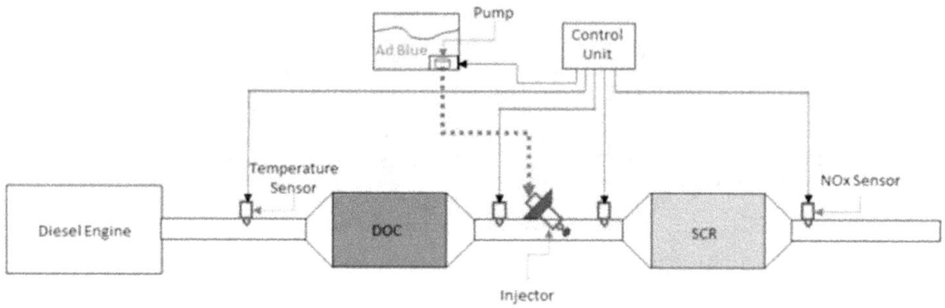

The effectiveness of SCR system at reducing the NO_X emissions enables the engine manufacturers to optimize fuel efficiency while balancing the engine performance. The diesel engine exhaust gas is first admitted in to the SCR system then the diesel exhaust fluid comprises of aqueous urea is sprayed on to a catalyst. It is termed selective because it suppresses NO_X levels inside a catalyst system by utilizing ammonia as a reductant. In the exhaust stream, the DPF is rapidly broken down to create oxidizing ammonia. The nitrogen oxides in the exhaust are turned in to nitrogen, water and trace amounts if carbon dioxide (CO_2) by a chemical reaction that is initiated by the DPF and occurs on a specific catalyst. The SCR allows the nitrogen oxide (NO_X) reduction processes to take in an oxidizing environment.

Particle Oxidation Catalyst (POC)

Emissions from non-road diesel engines pose a major risk to both users and the environment (Feng et al., 2022). Particulate oxidation catalysts is a new PM emission control method, offer a greater particulate control efficiency than DOC but a lower particulate control efficiency than diesel particulate filters. Particle oxidation catalyst are devices that catch and hold the carbonaceous PM particles prolonged enough for catalytic oxidation. The soot accumulated in the DPF raises exhaust back pressure, resulting in increased fuel consumption. As a result, the filter requires frequent regeneration and ash disposal by external methods to prevent clogging of the exhaust pipe. The regeneration process is done with the help of catalysis. Instead of the above, enhancing particle accumulation through an oxidation catalyst is another way for eliminating soot particles. Particle oxidation catalysts are not intended to remove as much soot as DPFs. Instead they aim to achieve the statuary emission limits by eliminating a portion of the small solid particles while drastically lowering the volatile percentage of the particles. The particles that have been collected must be removed from the device by oxidizing them to gaseous products, a process called as regeneration.

Feng et al. (2022) conducted the study is to determine how using an integrated system (DOC + DPF/POC + SCR + ASC) affects emissions from non-road diesel engines during steady-state and transient operating situations. For a light-duty application, Bielaczyc et al. (2012) conducted a research was done on fuels with three different levels of sulphur, ranging from 6 to 340 parts per million. Feng et al. (2014) conducted the studies using a particulate oxidation catalyst (POC) to assess the impacts on the nitrogen dioxide (NO_2) and particulate matter (PM) emissions from diesel engines. The average increase in the NO_2/nitrogen oxides (NO_x) ratio was 4.5 times across all trial modes with the POC. The effects of a particle oxidation catalyst (POC) on the emission characteristics of hydrocarbons (HC), carbon monoxide (CO), nitrogen oxides (NO_x), and particulate matter (PM) were investigated by Guan, Li, Liao, and Huang (2016) and the findings demonstrate that fuel injection techniques have less of an impact on CO and HC emissions. CO and HC emissions are greatly decreased by employing the POC.

Figure 10. The particle oxidation catalyst (POC)

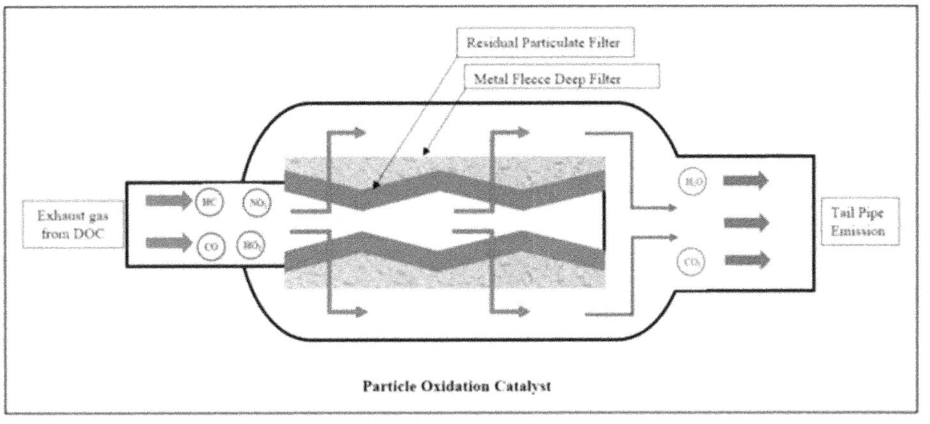

NO_X Adsorbers Catalyst

To reduce the NO_X emissions produced by diesel engines, solutions such as recirculation of exhaust gases (EGR) and selective catalytic reduction (SCR) have been deployed. However, the EGR is ineffective and SCR requires a constant supply of reductant to the exhaust. A NO_X adsorber is intended to minimize nitrogen oxides generated in the engine exhaust of a low burn engine. The high quantities of O_2 in the exhaust flow, lean burn diesel engines pose a unique challenge to pollution control system designers, NO_X adsorbers were invented by utilizing an acid-base (washcoat chemistry). It requires strong NO_X on the catalyst under lean exhaust circumstances and releasing it during rich operation and or high temperatures. NO_X adsorber systems are divided based on their NO_X release strategy: 1) active NO_X adsorbers, 2) paasive NO_X adsorbers.

Active NO_X Adsorbers

In this method the accumulated NO_X is released on a regular basis. This is similar to that seen over three way catalysts (TWC) in stoichiometric gasoline engines and the emitted NO_X is catalytically transformed to nitrogen. Under lean exhaust circumstances, when oxygen can be found in the exhaust gas, the three-way catalysts often do not convert NO_X. The three-way catalyst's applicability to lean burn engines has been increased by switching between both the rich release and lean storage phases. Active NO_X adsorbers finds minimal benefit in heavy duty truck engines due to their diminishing NO_X reduction capacity at increasing exhaust temperature.

Passive NO_X Adsorbers

Without rich regeneration, adsorb NO_X during vehicles cold start and liberate this when exhaust temperature goes up and it is to be transferred via a downstream pipeline. Because of this, passive NO_X adsorbers can be employed to enhance the systems low temperature performance when they are used conjunction with urea SCR after treatment. 70 to 90 percentage of the NO_X reduction is achieved through this.

CONCLUSION

The information in this chapter covers the necessity of alternative fuels, the different kinds of alternative fuels used and emission control techniques. Diesel pollutants, particularly diesel particulate matter, can have a major negative impact on health and environment. Hence the impact of global warming due to the air pollution imposed the world wide emission regulations and the researchers and the engine manufacturers to shown their interest towards the design of engines or fuel modification and emission control techniques to reduce the emissions.

REFERENCES

Acharya, N., Nanda, P., Panda, S., & Acharya, S. (2019). A comparative study of stability characteristics of mahua and jatropha biodiesel and their blends. *Journal of King Saud University-Engineering Sciences*, *31*(2), 184–190. doi:10.1016/j.jksues.2017.09.003

Akhihiero, E. T., Omorewa, Y. G., & Ebhodaghe, S. O. (2019). Effect of blending ratio on the properties of sunflower biodiesel. *J. Mater. Environ. Sci*, *10*(10), 987–994.

Aung, N. Z. (2008). Modification of diesel engine to producer gas engine. *Jurnal Teknologi Energi*, *1*(6).

Basha, J. S., Al Balushi, M., Al Shibli, K., Al Fazari, M., Al Rumhi, S., Al Mansouri, S., & Al Balushi, R. (2020). An emission control strategy in a low capacity single cylinder compression ignition engine powered with dee blended fuels. *Materials Science for Energy Technologies*, *3*, 770–779. doi:10.1016/j.mset.2020.09.004

Basha, J. S., Jafary, T., Vasudevan, R., Bahadur, J. K., Ajmi, M. A., Neyadi, A. A., Soudagar, M. E., Mujtaba, M. A., Hussain, A., Ahmed, W., Shahapurkar, K., Rahman, S. M., & Fattah, I. M. (2021). Potential of utilization of renewable energy technologies in Gulf Countries. *Sustainability*, *13*(18), 10261. doi:10.3390u131810261

Bielaczyc, P., Keskinen, J., Dzida, J., Sala, R., Ronkko, T., Kinnunen, T., Matilainen, P., Karjalainen, P., & Happonen, M. J. (2012). Performance of particle oxidation catalyst and particle formation studies with sulphur containing fuels. *SAE International Journal of Fuels and Lubricants*, *5*(2), 611–619. doi:10.4271/2012-01-0366

Borkhade, R., Bhat, K. S., & Mahesha, G. T. (2022). Implementation of sustainable reforms in the Indian automotive industry: From Vehicle Emissions Perspective. *Cogent Engineering*, *9*(1), 2014024. Advance online publication. doi:10.1080/23311916.2021.2014024

Datta, A., & Mandal, B. K. (2014). Use of Jatropha biodiesel as a future sustainable fuel. *Energy Technology & Policy, 1*(1), 8–14. doi:10.1080/23317000.2014.930723

Dey, S., Reang, N. M., Das, P. K., & Deb, M. (2021). A comprehensive study on prospects of economy, environment, and efficiency of palm oil biodiesel as a renewable fuel. *Journal of Cleaner Production, 286*, 124981. doi:10.1016/j.jclepro.2020.124981

Dinesh Babu, G., & Shaafi, T. (2020). Use of animal fat methyl ester blend as a fuel in a CI engine. *IOP Conference Series. Materials Science and Engineering, 992*(1), 012008. doi:10.1088/1757-899X/992/1/012008

Dwivedi, G., Jain, S., & Sharma, M. P. (2011). Pongamia as a source of biodiesel in India. *Smart Grid and Renewable Energy, 2*(3), 184–189. doi:10.4236gre.2011.23022

Dwivedi, G., & Sharma, M. P. (2015). Investigation and improvement in cold flow properties of Pongamia biodiesel. *Waste and Biomass Valorization, 6*(1), 73–79. doi:10.100712649-014-9330-7

Dwivedi, G., & Sharma, M. P. (2016). Experimental investigation on thermal stability of Pongamia Biodiesel by thermogravimetric analysis. *Egyptian Journal of Petroleum, 25*(1), 33–38. doi:10.1016/j.ejpe.2015.06.008

Dwivedi, G., Sharma, M. P., & Kumar, M. (2014). Status and Policy of Biodiesel Development in India. *International Journal of Renewable Energy Research, 4*(2), 246–254.

Feng, R., Hu, X., Li, G., Sun, Z., & Deng, B. (2022). A comparative investigation between particle oxidation catalyst (POC) and diesel particulate filter (DPF) coupling aftertreatment system on emission reduction of a non-road diesel engine. *Ecotoxicology and Environmental Safety, 238*, 238. doi:10.1016/j.ecoenv.2022.113576 PMID:35512474

Feng, R., Sun, Z., Hu, X., Li, G., & Deng, B. (2022). Role of particle oxidation catalyst on emission reduction of a non-road diesel engine: A multi case study. *Chemical Engineering Science, 260*, 260. doi:10.1016/j.ces.2022.117914

Feng, X., Ge, Y., Ma, C., Tan, J., Yu, L., Li, J., & Wang, X. (2014). Experimental study on the nitrogen dioxide and particulate matter emissions from diesel engine retrofitted with particulate oxidation catalyst. *The Science of the Total Environment, 472*, 56–62. doi:10.1016/j.scitotenv.2013.11.041 PMID:24291555

Firrisa, M. T., van Duren, I., & Voinov, A. (2014). Energy efficiency for rapeseed biodiesel production in different farming systems. *Energy Efficiency, 7*(1), 79–95. doi:10.100712053-013-9201-2

Gaurav, D., & Sharma, M. P. (2014). Prospects of biodiesel from pongamia in India. *Renewable & Sustainable Energy Reviews, 32*, 114–122. doi:10.1016/j.rser.2014.01.009

Gopal, G., Rao, P. S., Gopalakrishnan, K. V., & Murthy, B. S. (1982). Use of hydrogen in dual-fuel engines. *International Journal of Hydrogen Energy, 7*(3), 267–272. doi:10.1016/0360-3199(82)90090-8

Hajlari, S. A., Najafi, B., & Ardabili, S. F. (2019). Castor oil, a source for biodiesel production and its impact on the diesel engine performance. *Renewable Energy Focus, 28*, 1–10. doi:10.1016/j.ref.2018.09.006

IEA. (n.d.). *Energy in India Today – india energy outlook 2021 – analysis*. IEA. Retrieved December 1, 2022, from https://www.iea.org/reports/india-energy-outlook-2021/energy-in-india-today

India Brand Equity Foundation. (n.d.). *Indian oil and Gas Industry Analysis Presentation: IBEF*. Retrieved November 30, 2022, from https://www.ibef.org/industry/indian-oil-and-gas-industry-analysis-presentation

Jain, S., & Sharma, M. P. (2011). Oxidation stability of blends of Jatropha biodiesel with diesel. *Fuel, 90*(10), 3014–3020. doi:10.1016/j.fuel.2011.05.003

Kalam, M. A., & Masjuki, H. H. (2002). Biodiesel from palmoil—An analysis of its properties and potential. *Biomass and Bioenergy, 23*(6), 471–479. doi:10.1016/S0961-9534(02)00085-5

Keera, S. T., El Sabagh, S. M., & Taman, A. R. (2018). Castor oil biodiesel production and optimization. *Egyptian Journal of Petroleum, 27*(4), 979-984.

Konur, O. (2021). Rapeseed oil-based biodiesel fuels: A review of the research. *Biodiesel Fuels Based on Edible and Nonedible Feedstocks, Wastes, and Algae*, 497-514.

Liu, Z., Ge, Y., Tan, J., He, C., Shah, A. N., Ding, Y., Yu, L., & Zhao, W. (2012). Impacts of continuously regenerating trap and particle oxidation catalyst on the NO2 and particulate matter emissions emitted from diesel engine. *Journal of Environmental Sciences (China), 24*(4), 624–631. doi:10.1016/S1001-0742(11)60810-3 PMID:22894096

Machado Corrêa, S., & Arbilla, G. (2008). Carbonyl emissions in diesel and biodiesel exhaust. *Atmospheric Environment, 42*(4), 769–775. doi:10.1016/j.atmosenv.2007.09.073

Milazzo, M. F., Spina, F., Cavallaro, S., & Bart, J. C. J. (2013). Sustainable soy biodiesel. *Renewable & Sustainable Energy Reviews, 27*, 806–852. doi:10.1016/j.rser.2013.07.031

Milazzo, M. F., Spina, F., Primerano, P., & Bart, J. C. J. (2013). Soy biodiesel pathways: Global prospects. *Renewable & Sustainable Energy Reviews, 26*, 579–624. doi:10.1016/j.rser.2013.05.056

Misra, A., Tajudeen, S., Venkataramani, G., Ayyasamy, E., & Ramalingam, V. (2017). Role of biodiesel with nanoadditives in port owned trucks and other vehicles for emission reduction. *Thermal Science, 21*(1 Part B), 605–614. doi:10.2298/TSCI160613295M

Mittelbach, M., & Gangl, S. (2001). Long storage stability of biodiesel made from rapeseed and used frying oil. *Journal of the American Oil Chemists' Society, 78*(6), 573–577. doi:10.100711746-001-0306-z

Okullo, A. A., Temu, A. K., Ogwok, P., & Ntalikwa, J. W. (2012). Physico-chemical properties of biodiesel from jatropha and castor oils. *International Journal of Renewable Energy Research, 2*(1), 47-52.

Prabhu, A., Venkata Ramanan, M., & Jayaprabakar, J. (2021). Production, properties and engine characteristics of Jatropha biodiesel–a review. *International Journal of Ambient Energy, 42*(15), 1810–1814. doi:10.1080/01430750.2018.1557548

Prakash, T., Geo, V. E., Martin, L. J., & Nagalingam, B. (2019). Evaluation of pine oil blending to improve the combustion of high viscous (castor oil) biofuel compared to castor oil biodiesel in a CI engine. *Heat and Mass Transfer, 55*(5), 1491–1501. doi:10.100700231-018-2519-8

PTI. (n.d.). *India's oil demand likely to jump 8% in 2022 - Times of India*. The Times of India. Retrieved December 1, 2022, from https://timesofindia.indiatimes.com/business/india-business/indias-oil-demand-likely-to-jump-8-in-2022/articleshow/90293976.cms

Qi, D. H., Chen, H., Geng, L. M., & Bian, Y. Z. (2011). Effect of diethyl ether and ethanol additives on the combustion and emission characteristics of biodiesel-diesel blended fuel engine. *Renewable Energy*, *36*(4), 1252–1258. doi:10.1016/j.renene.2010.09.021

Raman, P., & Ram, N. K. (2013). Performance analysis of an internal combustion engine operated on producer gas, in comparison with the performance of the natural gas and diesel engines. *Energy*, *63*, 317–333. doi:10.1016/j.energy.2013.10.033

Reşitoğlu, İ. A., Altinişik, K., & Keskin, A. (2014). The pollutant emissions from diesel-engine vehicles and exhaust aftertreatment systems. *Clean Technologies and Environmental Policy*, *17*(1), 15–27. doi:10.100710098-014-0793-9

Road Transport Year Book. (n.d.). Retrieved November 30, 2022, from https://morth.gov.in/sites/default/files/RTYB-2017-18-2018-19.pdf

Sadhik Basha, J., & Anand, R. B. (2014). Performance, emission and combustion characteristics of a diesel engine using carbon nanotubes blended jatropha methyl ester emulsions. *Alexandria Engineering Journal*, *53*(2), 259–273. doi:10.1016/j.aej.2014.04.001

Sani, Y. M., Daud, W. M. A. W., & Aziz, A. R. A. (2012, December 3). *Biodiesel feedstock and production technologies: Successes, challenges and prospects*. IntechOpen.

Saydut, A., Erdogan, S., Kafadar, A. B., Kaya, C., Aydin, F., & Hamamci, C. (2016). Process optimization for production of biodiesel from hazelnut oil, sunflower oil and their hybrid feedstock. *Fuel*, *183*, 512–517. doi:10.1016/j.fuel.2016.06.114

Shaafi, T., Sairam, K., Gopinath, A., Kumaresan, G., & Velraj, R. (2015). Effect of dispersion of various nanoadditives on the performance and emission characteristics of a CI engine fuelled with diesel, biodiesel and blends—A review. *Renewable & Sustainable Energy Reviews*, *49*, 563–573. doi:10.1016/j.rser.2015.04.086

Shaafi, T., & Velraj, R. (2015). Influence of alumina nanoparticles, ethanol and isopropanol blend as additive with diesel–soybean biodiesel blend fuel: Combustion, engine performance and emissions. *Renewable Energy*, *80*, 655–663. doi:10.1016/j.renene.2015.02.042

Shrirame, H. Y., Panwar, N. L., & Bamniya, B. R. (2011). Bio diesel from castor oil-a green energy option. *Low Carbon Economy*, *2*(1), 1–6. doi:10.4236/lce.2011.21001

Society of Indian Automobile Manufacturers. (n.d.). Retrieved December 1, 2022, from https://www.siam.in/uploads/filemanager/159SIAMWhitePaperonAlternativeFuelsforvehicles.pdf

Tajudeen, S., & Velraj, R. (2014). Nanoadditives: Enhancement of combustion and performance characteristics of a CI diesel engine. *International Journal of Applied Environmental Sciences*, *9*(4), 1727–1741.

Tan, K. T., Lee, K. T., Mohamed, A. R., & Bhatia, S. (2009). Palm oil: Addressing issues and towards sustainable development. *Renewable & Sustainable Energy Reviews, 13*(2), 420–427. doi:10.1016/j.rser.2007.10.001

Teja Muvva, R., Shaafi, T., & Arunkumar, M. (2020). Experimental investigation by utilizing nano alumina with waste cooking oil biodiesel fuel in CI engine. *Materials Today: Proceedings*. Advance online publication. doi:10.1016/j.matpr.2020.10.185

Thirumarimurugan, M., Sivakumar, V. M., Xavier, A. M., Prabhakaran, D., & Kannadasan, T. (2012). Preparation of biodiesel from sunflower oil by transesterification. *International Journal of Bioscience, Biochemistry, Bioinformatics, 2*(6), 441–444. doi:10.7763/IJBBB.2012.V2.151

Van Duren, I., Voinov, A., Arodudu, O., & Firrisa, M. T. (2015). Where to produce rapeseed biodiesel and why? Mapping European rapeseed energy efficiency. *Renewable Energy, 74*, 49–59. doi:10.1016/j.renene.2014.07.016

Vellaiyan, S., & Partheeban, C. A. (2018). Emission analysis of diesel engine fueled with soybean biodiesel and its water blends. *Energy Sources. Part A, Recovery, Utilization, and Environmental Effects, 40*(16), 1956–1965. doi:10.1080/15567036.2018.1489911

Velmurugan, K., & Shaafi, T. (2019). Performance and emission characteristics of a twin cylinder di diesel engine running on diesel-waste cooking methyl ester blend. *International Journal of Mechanical and Production Engineering Research and Development, 9*(6), 183–190.

Xiao, H., Zou, H., Liu, S., & Li, C. (2019). An investigation of the friction and wear behavior of soybean biodiesel. *Tribology International, 131*, 377–385. doi:10.1016/j.triboint.2018.10.037

Yin, X., Ma, H., You, Q., Wang, Z., & Chang, J. (2012). Comparison of four different enhancing methods for preparing biodiesel through transesterification of sunflower oil. *Applied Energy, 91*(1), 320–325. doi:10.1016/j.apenergy.2011.09.016

Zahan, K. A., & Kano, M. (2018). Biodiesel production from palm oil, its by-products, and mill effluent: A review. *Energies, 11*(8), 2132. doi:10.3390/en11082132

Zhao, H., Cao, Y., Orndorff, W., Cheng, Y. H., & Pan, W. P. (2012). Thermal behaviors of soy biodiesel. *Journal of Thermal Analysis and Calorimetry, 109*(3), 1145–1150. doi:10.100710973-012-2551-8

Chapter 9
Production of Biogas From Biowaste and Its Potential as a Future Alternative Energy for Power Generation

Mohamed Musthafa M.
SASTRA University (Deemed), India

Ashok Kumar T.
SASTRA University (Deemed), India

ABSTRACT

Due to reducing petroleum reserves' scarcity, the engineer has to consider alternative fuel sources for IC engines. Vegetable oils of the non-edible category can be used as a substitute fuel to diesel fuel and use gaseous fuel as a partial supplement to petrol fuel. Biogas technology offers a desirable direction to utilize certain categories of biomass for meeting partial energy needs. This study is to describe the production of biogas from cow dung in the fabricated floating drum type digester and was stored in the tube. Stored biogas was used as supplement fuel to run the single cylinder 305 cc variable speed engine under no load. The performance and emission characteristics of engines running on biogas and petrol blend were compared to petrol fuel operation. Results revealed that the performance of the engine was improved with the increases in the amount of gas substitution. Brake-specific fuel consumption was found to improve. However, emissions of HC and CO were increased with the increases in gas substitution, and reduced NOx emission was observed.

INTRODUCTION

Fossil fuels are non-renewable and require finite resources, which cause environmental pollution and high cost. Cheap and renewable resources are in great need. An efficient and feasible option is to depend on alternative fuel. Alternative fuels are obtained from waste other than petroleum sources. These can be

DOI: 10.4018/978-1-6684-7303-0.ch009

produced domestically, reducing our dependence on fossil fuels. They produce less pollution than petrol or diesel. Alternative fuels can be used by slight modifications in the operating engine as an only fuel or as a supplementary fuel in dual fuel mode. Gaseous fuels are a good alternative in SI engines because they are easily miscible with air. Biogas is a good choice fuel and is obtained from cow dung, waste from poultry, plant, industrial and domestic is an easily available source of energy. Modern advancement in technology and biogas storage has to be made to use the fuel for IC engines. Hybrid engines that run biogas as supplementary fuel have a separate fuel storage system. The process is widely used in biogas production from animal waste using a digestion tank and anaerobic digestion. Anaerobic digestion is a microbial isolate from the digesters including Bacillus licheniformis, Escherichia coli and Clostridium, which are responsible for the breakdown of the complex substance. They observed the highest total gas yield (15.60 cm^3) in the digesters which were left at ambient temperature and in which gas was collected over limewater. Research on biogas as a supplementary fuel which was used either in SI engines or CI engines is less exploded. To accomplish the research gaps this study has been performed to produce biogas from a simple mixture of cow dung and water in 3:1 proportion through anaerobic digestion. Bacteria present in cow dung help to break down complex substances resulting in a yield of biogas. It was stored temporarily and then tested on a single-cylinder petrol engine fuelled by biogas as a supplementary fuel along with petrol.

As conventional energy is to be depleted in near future, the development of renewable energy is very important for the variety of the energy supply. The global energy demand is produced from fossil resources. The depletion of fossil fuels and environmental pollution is the main focus of research into renewable energy mainly from waste organic resources. Alternate energy to fossil fuels is a challenge to improve security. Most countries are accepting the new promise for renewable energy supply and providing a possibility for sustainable evolution. Production of biogas from biomass is soured to reduce environmental pollution and provide extra revenue to the farmers (Amon et al., 2007). Biogas is produced from manure, plants, food wastes, offal, etc. through anaerobic fermentation under an anaerobic condition (absence of oxygen). Biogas from various sources like sewage, organic waste and landfills have a mixture of various quantities like CH_4, CO_2 and N_2. Biogas from sewage contains CH_4 54 - 64%, CO_2 34- 44% and N_2 is less than 1%. Biogas from organic waste, contains CH_4 62- 72%, CO_2 is 32- 42% and N_2 is less than<1% while in landfills CH_4 is 44- 55%, CO_2 is 31- 42% and N_2 is 3- 12%. Apart from these gases it also contains H_2S and other sulphur siloxanes, aromatic and halogenated compounds (Amon et al., 2007; De aere, 2000; Rasi et al., 2007). The average calorific value of biogas is 6,0 kWh/m^3 and depends on the methane percentage in the gas. From the safety point of view, the biogas ignition temperature is 700 °C and the stoichiometric air requirement for complete combustion is 5-7 m^3 per m^3 of biogas (Vindis et al., 2008; Lungkhimba et al. 2010). While biogas produces a temperature, pressure, and Solid/water ratio, pH value should be properly followed and maintained.

Biogas production from anaerobic digestion (AD) provides energy efficiency and reduces greenhouse gas emissions and sustainability development over other forms of energy. The by-product digestate during the process is a good fertilizer for crop cultivation. (Rotmans et al., 1992). Few experimental studies (Eze et al., 2010; Awogbemi et al., 2013; Omid et al., 2011) were carried out on biogas as a partial substitution fuel in SI engines and their results make a promising brake power, brake thermal efficiency reduced fuel consumption and improved emissions.Even biogas is good alternative, technological advancement is needed to produce gas efficiently,financial assistance from government is also required to develop biogas production. Addition heat source is required to produce biogas continuously in cold weather condition. This study aims to produce biogas from cow dung on a small scale and stored it in a tube. Stored biogas can be used as a partial supplement fuel in a single-cylinder 350cc capacity variable speed petrol engine at no load conditions. The performance and emission characteristic of the engine run by petrol and biogas as a partial substitution with petrol was compared.

MATERIALS AND METHODS

Biogas Production

Biogas is produced from cow dung via anaerobic digestion in a floating drum digester. It consists of an outer and inner drum. The Centre of the outer drum has a long shaft through which the inner drum is inserted. Drums are made of GI sheets. A conventional diagram of the fabricated digester is shown in Figure 1. The outer drum is provided with inlet and outlet pipes for cow dung feeding and manure let out. The inner drum is provided with a gas outlet.

25kg of cow dung was taken into the outer drum and mixed with a one-third volume of water then make into the slurry. The slurry occupied two third the volume of the outer drum and was closed with an inner drum. The setup is exposed to sunlight throughout the day. After 2 weeks period, biogas started yielding from cow dung. The biogas was stored in a bigger size tube through the hose. One-time feed produced biogas five times of tube volume. Manure that comes out from the outlet in the bottom and can be used as a fertilizer. The photographic view of the digester with biogas storage in a tube is shown in figure 2.

Figure 1. Schematic diagram of fabricated floating drum type digester

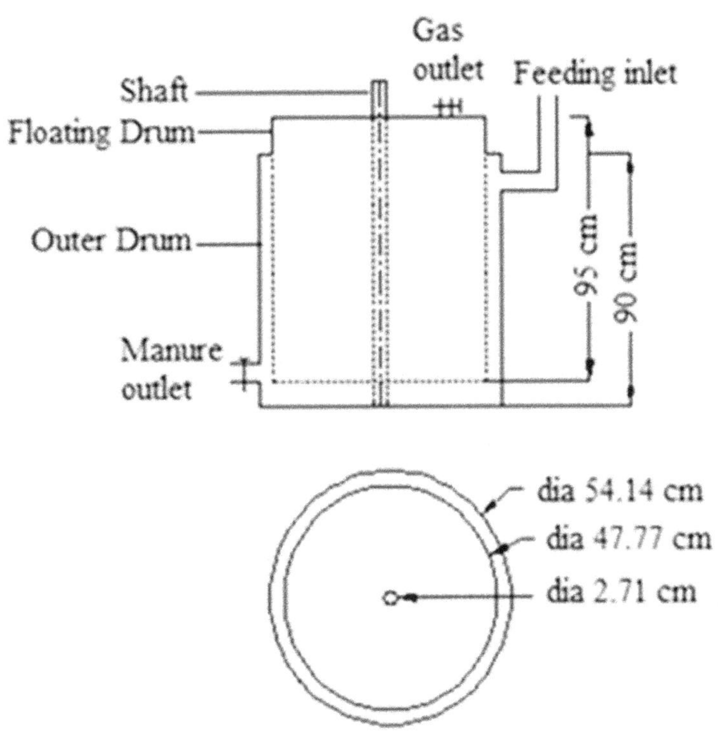

Figure 2. Photographic view of fabricated digester with storage

Figure 3. Schematic diagram of engine set up

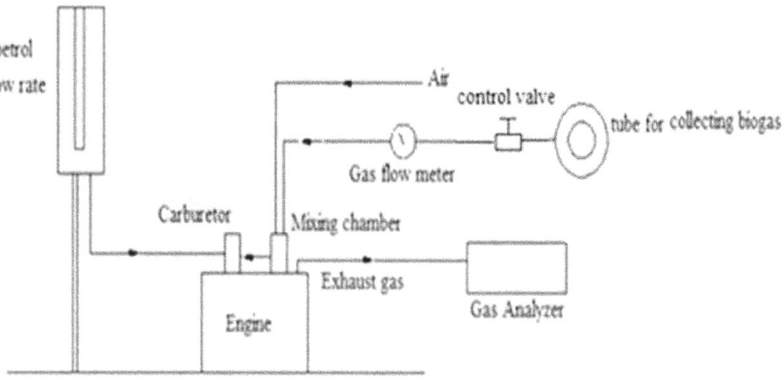

Experimental Setup

Briggs and Stratton Engine 1450 Series (SAE Baja Engine) Horizontal OHV Engine was used for testing biogas as a partial substitution fuel. The detailed specification of the engine is given the Table.2. The schematic diagram and photographic view of the test setup are shown in Figures 2 and 3. Tests were performed on the engine at variable speeds from 2000 to 3200 rpm under no load conditions. The engine was run with petrol at a different speed. For the same speed, the engine was run with partial substitution of biogas sent via carburettor inlet from the tube through the hose with the control valve to adjust the flow rate. Consumption of petrol and gas was measured in each speed condition. Petrol consumption was measured using a cc graduated tank and stopwatch. Gas consumption was measured using a wet-type gas flow meter. Delta 1600 gas analyzer was used to measure CO, CO2, HC and NOx levels in the exhaust in both operating conditions.

Figure 4. Photographic view of test engine

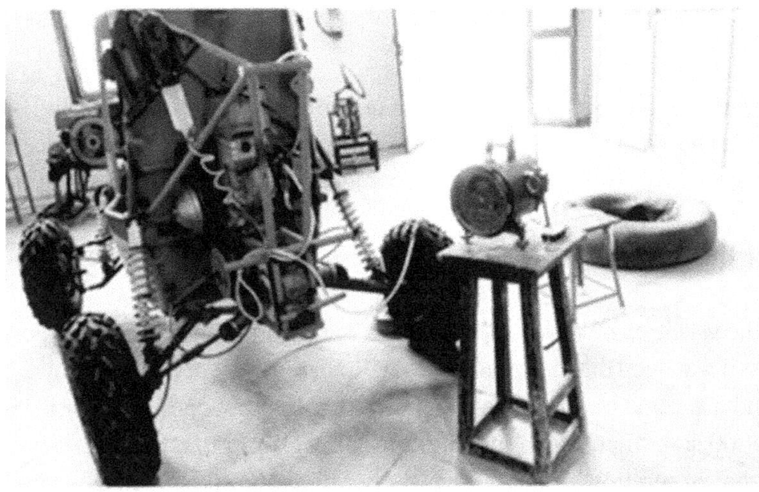

Figure 5. Engine speed vs. brake power

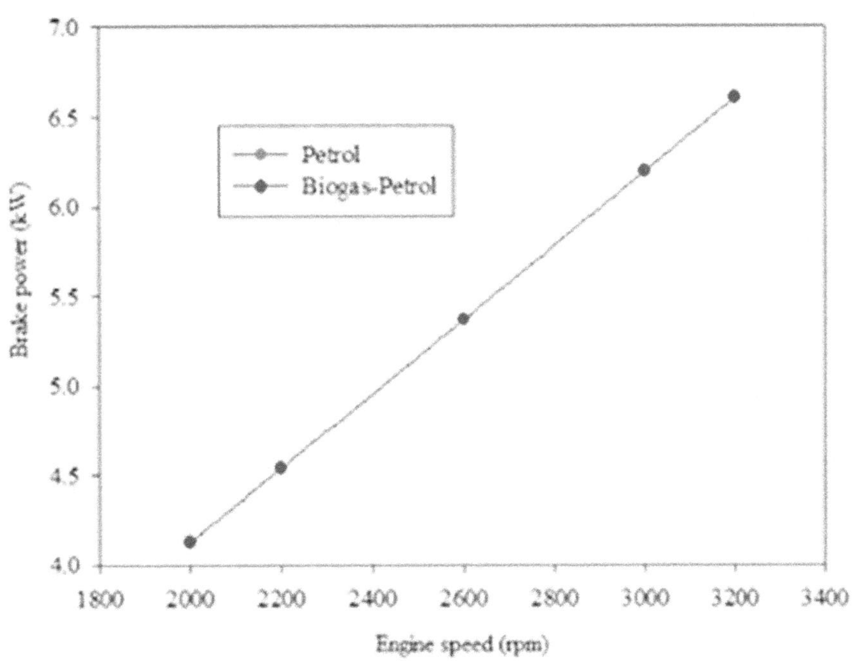

RESULTS AND DISCUSSION

The experiment was performed in a single cylinder variable speed engine with two fuel modes at no load conditions. The performance and emission results were compared in a graphical representation and are shown in the following sections.

Brake Power

The variations of engine speed versus brake power when the engine runs on petrol and biogas-petrol blend at no load condition are shown in Figure 5. Brake power increases with an increase in engine speed for the test fuels. But no variation was observed in both test fuels due to constant torque at no load conditions.

Brake-Specific Fuel Consumption

Figure 6 represents the variation of brake-specific fuel consumption (bsfc) against speed for the test fuels at no load conditions. It was seen that bsfc decreases with increasing speed in the range of 2000-3000 rpm for both test fuels and a reverse trend of slightly increasing fuel beyond 3000 rpm. The power required at the speed above 3000 rpm imposed an increase in fuel consumption due to the decrease in quality of fuel atomization in the carburettor in case of petrol operation and the increase in substitution of biogas decreased oxygen required for complete combustion in dual fuel mode (Eze et al., 2010). Petrol consumption in dual fuel mode is lower than that of an engine operating with 100% petrol. It was noted that approximately 30% of petrol consumption is replaced by biogas with an insignificant variation in brake thermal efficiency. The increase in petrol for dual fuel operation is credited to the low energy content, low combustion temperature, and low flame propagation speed of biogas. Higher petrol consumption in dual fuel mode is required to deliver equal brake power as petrol fuel operation.

Hydrocarbon Emission

Figure 7 depicts the variation of hydrocarbon emission versus engine speed. HC emission was measured at no load with different speeds of the engine for the test fuels. HC emission decreased with an increase in speed in both fuels. It was observed that higher HC emissions in biogas–petrol blend fuels compared to petrol fuel. As the speed increases, the biogas flow rate also increases which results in a rich mixture or less concentration of fresh air. In addition, biogas also has lower flame velocity and higher methane concentration these reasons could result in the incomplete combustion process in the combustion chamber and make a higher unburnt hydrocarbon in the exhaust gas (Omid et al., 2011; Huang et al., 1998; Oubaud et al., 2005). The admission of biogas in the suction of the engine causes a rich mixture in the cylinder and decreases the quantity of oxygen in the mixture also biogas lowers the burning velocity. These brought about insufficient combustion in dual-fuel operations hence HC increased.

Figure 6. Engine speed vs. brake specific fuel consumption

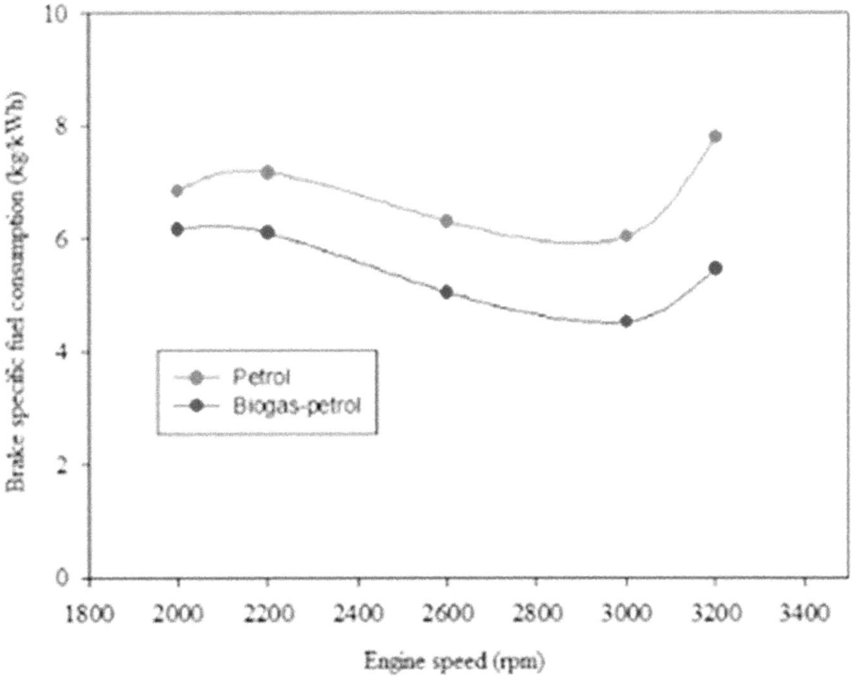

Figure 7. Engine speed vs. hydrocarbon emission

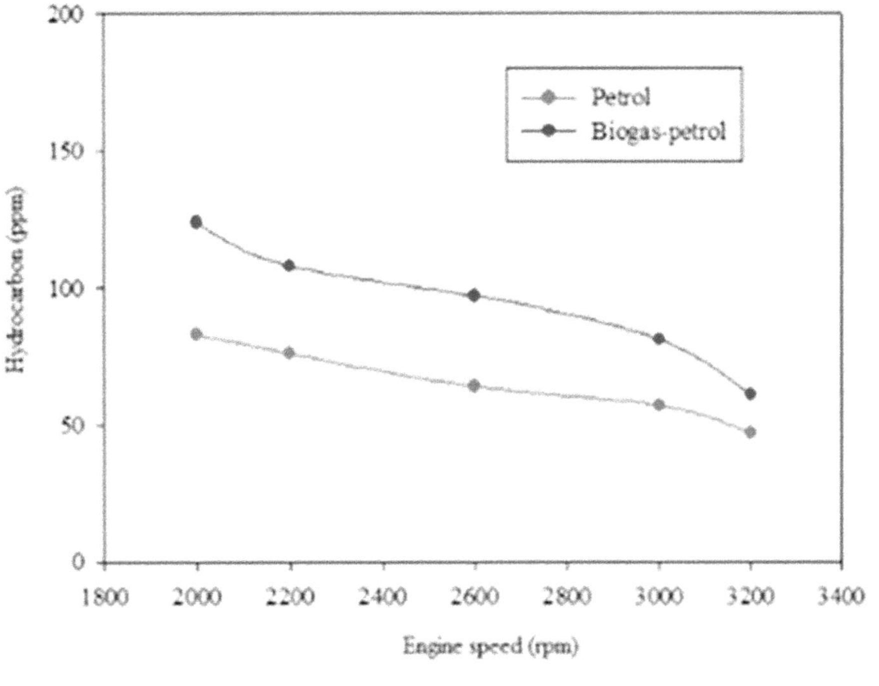

Carbon Monoxide Emissions

Carbon monoxide emission in the exhaust gas is due to a deficiency of oxygen in the fuel-air mixture in the combustion chamber which leads to incomplete combustion. Figure 8 shows CO emissions for petrol and dual fuel mode. It was noted that CO emissions for the biogas-petrol fuel mode were considerably higher than those of the petrol mode under different engine speeds at no-load conditions. When the biogas is mixed with air before the carburettor, the percentage of oxygen in the air is decreased and fuels have mix ability challenges, which results in incomplete combustion and CO emissions. However, the CO2 content in the biogas may influence the level of carbon oxide emissions (Chandra et al., 2011; Lacour et al., 2012; Porpatham et al., 2008). The increase in CO_2 emissions from petrol fuel operations can be attributed to the increase in combustion temperature that resulted in the oxidation of greater amounts of CO into $CO_{2.}$

NOx Emission

NOx formation by the Zeldovich mechanism which governs factors is local temperature, the concentration of oxygen and retention time (Heywood, 1988). Figure 9 illustrates NOx variation with engine load. Due to the increase in equivalence ratio, NOx emissions increase with engine speed for both fuel modes. Biogas supply to the engine along with air reduces O_2 concentration for combustion, the existence of high specific heat CO_2 in biogas to decrease the combustion temperature and decrements in the duration of combustion results in lower NOx emissions. (Nirendra et al., 2013; Nathan et al., 2010). The higher HC emission is apparent that the reduction in the hydrocarbon oxidation processes.

Figure 8. Engine speed vs. carbon monoxide emissions

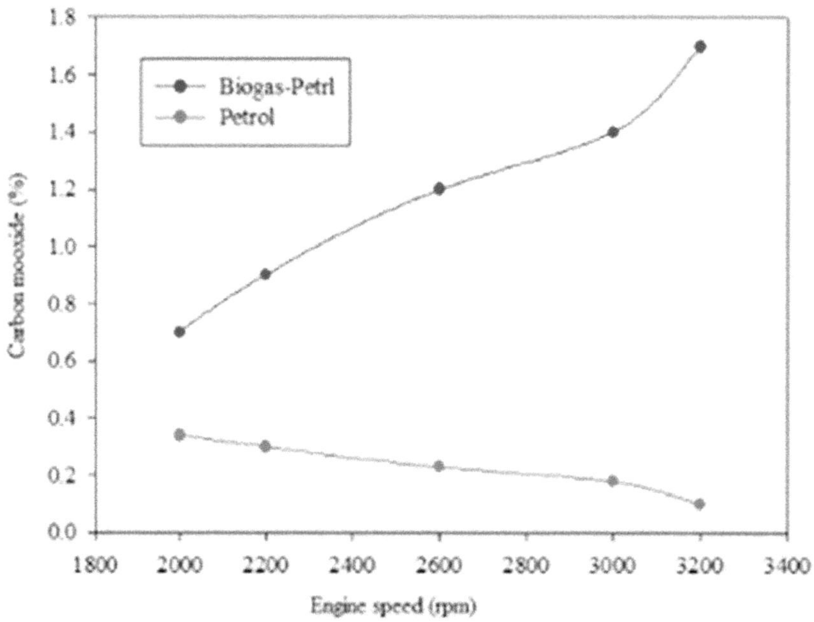

Figure 9. Engine speed vs. NOx emissions

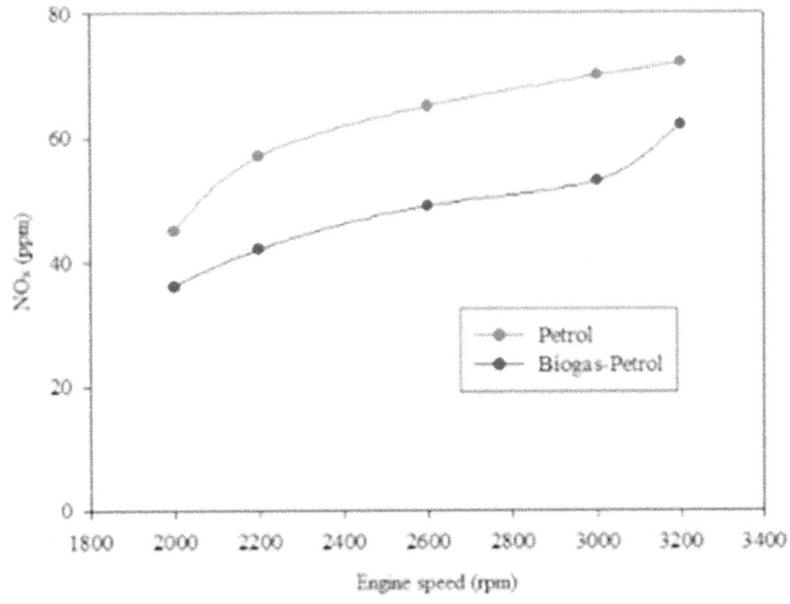

CONCLUSION

In this study, a single-cylinder 350cc capacity variable speed SI engine has been tested by using biogas as a supplement fuel with petrol without any significant modification. The biogas used in the experiments was made from cow dung by using a digester. The conclusions of this study are as follows.

- A percentage reduction in specific fuel consumption was observed as 11-32% in the speed range 2000-3000rpm for biogas- petrol fuel than the petrol.
- HC and CO emissions were found to increase in biogas- petrol fuel compared to petrol fuel operation
- NOx emissions were decreased to a maximum of 25% at 2000 rpm and a minimum of 16% at 3000 rpm in dual fuel mode than petrol.

Based on the performance and emission characteristics, biogas may be utilized as an efficient partial supplement fuel in SI engines. A minor alteration is required for low-density biogas fuel as supplementary fuel in the dual-fuel engine. Biogas as a future fuel for IC engines is possible when the cost of fuel is relatively low-cost and readily available. Through methane enrichment, alteration of ignition parameters, increasing the compression ratio and the addition of hydrogen, better performance and improved emission can be obtained.

REFERENCES

Amon, T., Amon, B., Kryvoruchko, V., Zollitsch, W., Mayer, K., & Gruber, L. (2007). Biogas production from maize and dairy cattle manure Influence of biomass composition on the methane yield. *Agriculture, Ecosystems & Environment, 118*(1-4), 173–182. doi:10.1016/j.agee.2006.05.007

Awogbemi, O., & Adeyemo, S. B. (2013). Investigation of the influence of some parameters on biogas yield. *Elixir International Journal, 54A*, 12825–12828.

Chandra, R., Vijay, V. K., Subbarao, P. M. K., & Khura, T. K. (2011). Performance evaluation of a constant speed IC on CNG, methane enriched biogas and biogas. *Applied Energy, 88*(11), 3969–3977. doi:10.1016/j.apenergy.2011.04.032

De aere, L. (2000). Anaerobic digestion of solid waste: state-of-theart. *Wat. Sci. Technol., 41*(3), 283-290.

Heywood, J. B. (1988). *Internal combustion engine fundamentals. McGraw-Hill series in mechanical engineering*. McGraw-Hill Inc.

Huang, J., & Crookes, R. (1998). Assessment of simulated biogas as a fuel for the spark ignition engine. *Fuel, 77*(15), 1793–1801. doi:10.1016/S0016-2361(98)00114-8

Islam, R., Banerjee, G., & Ali, F. (2014). Design and Benefit Analysis of Biogas Plant for Rural Development in Bangladesh. *International Journal of Engineering and Advanced Technology, 3*(3).

Lacour, S., Chinese, T., Alkandee, D., Perilhon, C., & Descombes, G. (2012). Energy and environmental balance of biogas for dual-fuel mobile applications. *Renewable & Sustainable Energy Reviews, 16*(3), 1745–1753. doi:10.1016/j.rser.2011.11.029

Lungkhimba, H. M., Karki, A. B., & Shrestha, J. N. (2010). Biogas Production from Anaerobic Digestion of Biodegradable Household Wastes. *Nepal Journal of Science and Technology, 11*, 167–172. doi:10.3126/njst.v11i0.4140

Nathan, S. S., Mallikarjuna, J. M., & Ramesh, A. (2010). An experimental study of the biogas–diesel HCCI mode of engine operation. *Energy Conversion and Management, 51*(7), 1347–1353. doi:10.1016/j.enconman.2009.09.008

Nirendra, N. M., Robert, R. R., & Sebastian, V. (2013). Combustion and emissions characteristics of a dual fuel engine operated on alternative gaseous fuels. *Fuel, 109*, 669–678. doi:10.1016/j.fuel.2013.03.007

Omid, R., Nima, M., & Mohsen, A. (2011). Literature review and road map for using biogas in internal combustion engines. *Third International Conference on Applied Energy*.

Porpatham, E., Ramesh, A., & Nagalingam, B. (2008). Investigation on the effect of concentration of methane in biogas when used as a fuel for a spark ignition engine. *Fuel, 87*(8-9), 1651–1659. doi:10.1016/j.fuel.2007.08.014

Rasi, S., Veijanen, A., & Rintala, J. (2007). Trace compounds of biogas from different biogas production plants. *Energy, 32*(8), 1375–1380. doi:10.1016/j.energy.2006.10.018

Rotmans, J., Den Elzen, M.G.J., Krol, M.S., Swart, R.J., & Van Der Woerd, H.J. (1992). Stabilizing atmospheric concentrations – Towards international methane control. *Ambio, 21*(6), 404–413.

Roubaud, A., Rothlisberge, R., & Favrat, D. (2005). Improving performances of a lean burn cogeneration biogas engine equipped with precombustion chambers. *Applied Thermodynamics*, *5*, 169–175.

Sheanut Oil blends in Compression Ignition Engine. (2010). *Global Journal of Emerging Research in Engineering*, *10*(1), 84–92.

Sita Rama Raju, A. V. (2001). *Experimental investigations on the performance of a lean burn spark ignited gas engine* [PhD thesis]. IIT Madras, Chennai, India.

Vindis, P., Mursec, B., Rozman, C., Janzekovic, M., & Cus, F. (2008). Biogas production with the use of mini digester. *Journal Of Achievements in Materials and Manufacturing Engineering*, *28*(1), 99–102.

Chapter 10
Type 2 Intelligent Controller for Grid-Tied Solar Electric Vehicle Charging Stations:
Novel Controller for PV FED Grid-Tied EV Chargers

Thomas Thangam
International Maritime College, Oman

Abdul Hameed Hameed Kalifullah
International Maritime College, Oman

ABSTRACT

Electric vehicles (EVs) are becoming a popular alternative to gas-powered cars. These cars need "full" batteries to run. Solar-powered chargers are an exciting alternative to grid-based EV charging. These chargers give electric vehicles pollution-free electricity, which benefits the environment. Solar PV is the most popular renewable energy source. This chapter establishes a solar EV charging station, which charges EVs. Bi-directional batteries store photovoltaic (PV) energy for use during power outages. PV overproduction is transferred to the grid for later use. Cascaded interval type 2 fuzzy logic controller (CIT2FLC) boosts voltage using KY converter to track maximum photovoltaic power. To accomplish grid synchronization, a DC voltage is delivered to a grid-connected 1 phase VSI and optimized using a PI controller. During peak hours, EV gets power from the grid via 1 phase VSI, with the KY converter in buck mode. The suggested work ensures uninterrupted charging. The complete structure was tested using MATLAB Simulink and yielded 94.7% efficiency and 3.9% THD.

DOI: 10.4018/978-1-6684-7303-0.ch010

1. INTRODUCTION

The advancement of current millennium in all sectors has abundantly raised the need for energy sources, especially in transportation. Deterioration of air quality, ongoing depletion of fossil fuels, climate crisis and the authoritarian state of global warming has induced the wide acceptance of transportation system that uses renewable energy. The progressive use of renewable energy sources (RES) in manufacturing vehicles offers a sustainable solution to these problems. The development of battery technology and demand for environmentally friendly transport has attracted electric vehicles as a potential form of transportation (Shariff et al., 2019). As a result, manufacturers have prioritized EVs that consume clean, RES instead of fossil fuel consuming vehicles (Harini et al., 2022; Xia et al., 2018). Battery Energy Storage (BES) is an energy storage system that forms a part of current and future vehicle technologies. Although the growth of EVs reduces the use of fossil fuels, due to their nonlinear features and unequal distribution, they strain the utilities and power grid operators as EVs produce harmonic currents and cause problems with power quality (Chen et al., 2020; B. Singh et al., 2019).

Particularly in rural areas, PV technology is a significant renewable energy source. In PV systems, inverters play a dominant role in energy transmission. A PV system requires an inverter to convert a DC power source into usable AC electricity. Due to nonlinearity and output volatility, design of a photovoltaic based inverter is complicated as the interaction between the inverters and the DC grid cause instability in output voltage reference. The power conversion system of a solar photovoltaic system is provided with a DC-DC boost converter for optimizing output voltage regulation. Considering estimates, a rising number of EV's leads to immense charging demands that significantly would exceed the capacity of currently available charging infrastructure (Hannan et al., 2019; Pokharel et al., 2019; Zhou et al., 2019). A DC link is created by connecting a bidirectional DC-DC converter between solar PV outputs and grid interfaced converter output, followed by Maximum Power Point Trackers (MPPT). Finally, the battery is linked between DC link and. Centralized MPPT systems are more appealing for larger PV plants as it has higher efficiency and is cost effective (Liu et al., 2019; A. Singh et al., 2020).

The primary goal of Grid Integrated PV Battery System is to provide grid with steady electricity and sustainable energy storage in the battery. System reliability, load and power balancing are achieved by charging and discharging battery. Typically, there is significant potential difference between Photovoltaic and DC bus voltage in grid connected or independent PV systems. As the DC voltage needs to be high enough to ensure over modulation in bridge inverters, a transformer is used in a standard stage to raise the voltage to the grid level. Nevertheless, this is challenging in RES without a separate voltage enhancing method, frequently performed by a boost converter (Dutta et al., 2017; Karthikeyan et al., 2019; Saxena et al., 2017). Before the bidirectional converter, the boost converter had a decoupling capacitor connected to lower the power of electrical gadgets (Xu et al., 2019). In the PV system, a standard multi-port converter is utilized. Several criteria must be observed for designing an efficient converter. The topology of multi-port converters is examined using CUK and SEPIC. Nevertheless, dependability is compromised as standard parts serve as single points of failure for the converters as a whole (Chandrasekar et al., 2020). A converter has an output port, a bidirectional port for backup battery and unidirectional ports for PV module's input and output. The converter produces a highly regulated output voltage by wiring RES in series with the battery and utilizing a coupled inductor (Al-Soeidat et al., 2019; Xia et al., 2018; Yamaguchi et al., 2018).The typical Boost DC-DC converter is the most used non-isolated step up DC-DC converter. The converter's complexity is reduced by the presence of a single power switch.

Nevertheless, the circuit's intrinsic components limit the voltage gain. The overall controller's equivalent circuit model is created by merging the buck converter, battery and PV models (Venkatramanan et al., 2019; Zhang et al., 2016). Due to its buck boost power, the Cuk converter is used in independent, and grid tied operations with time varying and uncertain load needs. The main characteristics of the Cuk converter are that it produces an identical voltage waveform on inductor and continuous current at source and load port. Due to these advantages, modified Cuk converters with high step-up potential are preferred over traditional Cuk or boost converters. These modified Cuk converters use switched inductors and capacitors. Comparing SEPICs with a single regulated switch and more contemporary converters, an improved SEPIC converter offers a more considerable voltage gain (Chandrasekar et al., 2019; Maroti et al., 2019). In the proposed work, for increasing the voltage acquired from PV, KY converter has been employed.

For a converter to function effectively controller approaches are required which results in making the converter voltage stable (Manikandan et al., 2015). Comparing the Fuzzy Logic control to the conventional technique, accuracy and performance are improved. Additionally, utilizing P&O and fuzzy, a dual MPPT method is introduced, providing high operating point adaptability (Rezk et al., 2019). The components provide a quick current control reaction and a progress MPPT algorithm for enhanced peak tracking. It ensures control even with significant variations in irradiance. The MPPT is implemented and the DC bus is passed through a diode rectifier and boost converter (Raiker et al., 2021; Rezkallah et al., 2018). For attaining maximum power, the proposed work utilizes CIT2FLC for extracting maximum power from PV system. Similarly, the on-grid side PI controller is utilized to control.

In the proposed KY converter is employed for effective boosting of PV voltage along with CIT2FLC control approach. The excess energy from PV is supplied to grid via converter coupled. The PI controller is effectively utilized on the grid side for VSI control. Hence framework focuses on uninterrupted energy supply for both EV battery and grid.

Five sections make up the remainder of the article. The designed solar PV system's operational technique and description of suggested control strategies are all explained in Section 2. The findings and discussions are presented in Section 3. The suggested system's conclusion is described in Section 4 of the paper.

2. PROPOSED SYSTEM

The control method for PV grid connected systems, which enables feeding of a battery EV at the same time and location as the energy PV production, is the primary focus of this research as shown in Figure 1.

The proposed framework aims to extract maximum power from PV system and regulates power transfer through electrical vehicle battery. Since, it is a grid connected system, the excess energy from PV source is tranferred to the grid. In such instance, the voltage is enhanced with the utilization of DC-DC KY converter. Although PV has beneficial features, its low efficiency at peak hours is an flaw for effective EV charging. Hence, CIT2FLC which is an soft computing MPPT approach to extract maximum power is introduced along with KY converter. The maximized power is supplied to for grid supply. Thewith the assistance of PI controller provide enhanced supply to the grid. Thus, ceaseless power transfer is performed without any lag. The proposed EV charging setup is simulated using MATLAB and the corresponding responses is obtained with reduced THD of 3.91% along with 1:8 voltage gain.

Figure 1. Block diagram of proposed system

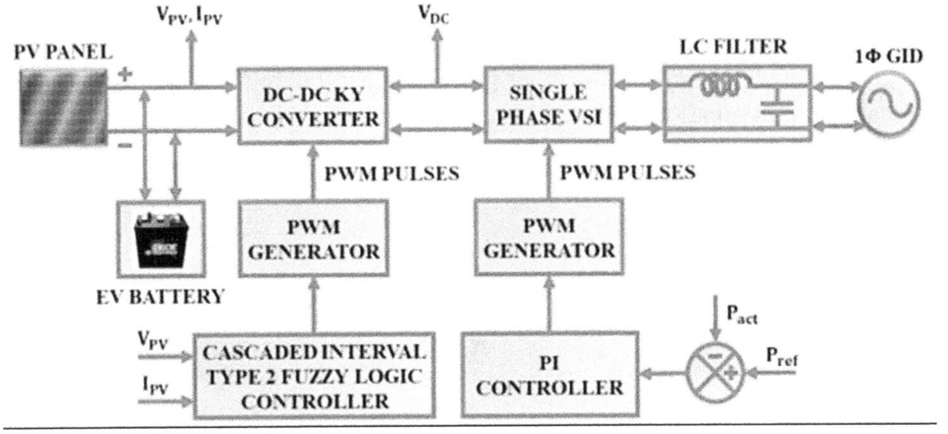

2.1. Modelling of PV

The solar energy is directly converted into electricity by a PV system. Any photovoltaic device comprised a number of basic cells that is mentioned to be an array. For supplying load, modules are connected to form arrays because a single module rarely produces enough power for commercial use. The connection between modules in an array is same as the connection between cells in a module. Additionally, modules are linked in parallel or series to boost current or voltage, respectively. A PV device's terminal voltage and current has the potential to directly power small loads including DC motors and lighting systems. To convert energy from PV device for more advanced applications, electronic converters are essential. In grid connected systems, these converters are utilised to control power flow by controlling current and voltage at load. A spectral distribution of solar radiation affects the efficiency of a PV device. A PV device evaluations are typically performed in accordance to a standard spectral distribution. The Photovoltaic cell equivalent circuit model is implemented in Figure 2.

Figure 2. Equivalent circuit of single diode PV cell

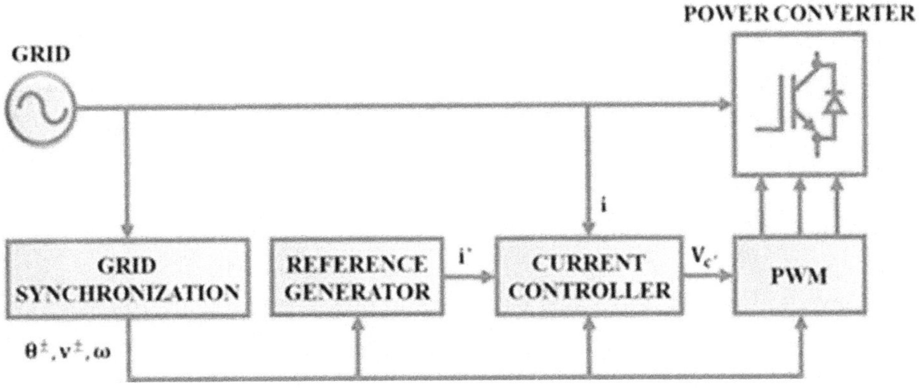

An anode coupled to the positive terminal of current source serves as the representation of an ideal PV cell, which consists a source of current parallel to PN junction. The current source designated I_{ph} is employed to model the photoelectric effect. A two pathways are taken by the leakage current, one goes through PN junction denoted as I_D and the other passes via parallel resistor named R_{sh} and is signified as I_{sh}. An internal power dissipation of the cell is controlled by a series resistor R_s. The formulation for an output current I_{PV} is expressed as,

$$I_{PV} = I_{ph} - I_D - I_{sh} \tag{1}$$

Substituting I_D for diode expression and I_{sh} for $\frac{V_D}{R_{sh}}$, equation (1) is changed by,

$$I_{PV} = I_{ph} - I_D \left\{ \exp\left[\frac{\left(V_{PV} + I_{PV} \times R_s\right)}{\left(\frac{\eta k_b T}{q}\right)} \right] - 1 \right\} - \frac{V_{PV} + I_{PV} \times R_s}{R_{sh}} \tag{2}$$

Where, diode reverse saturation current is denoted by I_O, a charge of an electron is q, Boltzmann constant is k_b, T considers the kelvin scale of cell temperature and η indicates the ideal factor. The behaviour of the panel is then approximately represented as,

$$I_{PV} = N_p I_{ph} - N_p I_O \left\{ \exp\left[\frac{\left(V_{PV} + I_{PV} \times R_s \frac{N_s}{N_p}\right)}{N_s * \left(\frac{\eta k_b T}{q}\right)} \right] - 1 \right\} - \frac{V_{PV} + I_{PV} \times R_s \frac{N_s}{N_p}}{\frac{N_s}{N_p} R_{sh}} \tag{3}$$

When an irradiation is homogeneous across the panels, solar panels produce their maximum power. A partial shadowing is a phenomena that happens when there are close physical objects, generates partial shade on the PV panels. An output power reduction and reduction in panel lifespan are the main effects of partial shade. The MPPT technique have difficulty recognizing the distinction among local peak and maximum global power peak because under these operating conditions the PV curve includes several local peaks, which results in large power generation. Hence, a maximum power generated by the PV system is regulated employing a Cascaded Interval Type 2 Fuzzy Logic Controller (CIT2FLC).

2.2. DC-DC KY Converter

A switch mode converter from DC to DC with a behaviour of KY Boost Converter as shown in Figure 3, has a higher output voltage that of input voltage. Whenever the input necessitates a step up to a larger value, it is more appropriate for DC load applications. With a voltage gain of 1+D, a KY Boost Converter enhances the voltage. Wherein, D represents the converter's duty cycle. An input circuit's inductor resists abrupt changes in input current, which is KY Boost Converter's primary operating mechanism. Inductors store magnetic energy when switch Q1 is ON and switch Q2 is OFF and subsequently releases it when Q1 is turned off and Q2 is turned on by the switch. An output circuit's capacitor is apparently large enough for the output stage's RC circuit to have a high time constant.

Operating Modes of the KY Boost Converter

There are two modes of operation for KY Boost Converter.

1. **Continuous Conduction Mode (CCM):** Here, inductor current never reaches zero; rather, it partly discharges before the switching cycle begins.
2. **Discontinuous Conduction Mode:** Consequently, the inductor's current is zero, meaning that it is fully drained at the termination of switching cycle.

Cir'cuit Analysis of KY Boost Converter at CCM

Consider throughout an entire study, which a voltage fluctuation through the capacitor and a current swing via inductor are both relatively small, allowing them to change linearly. The analysis has been conducted easier as a consequence and the conclusions we obtain from is quite accurate when compared to real statistics.

Figure 3. Equivalent circuit of KY converter

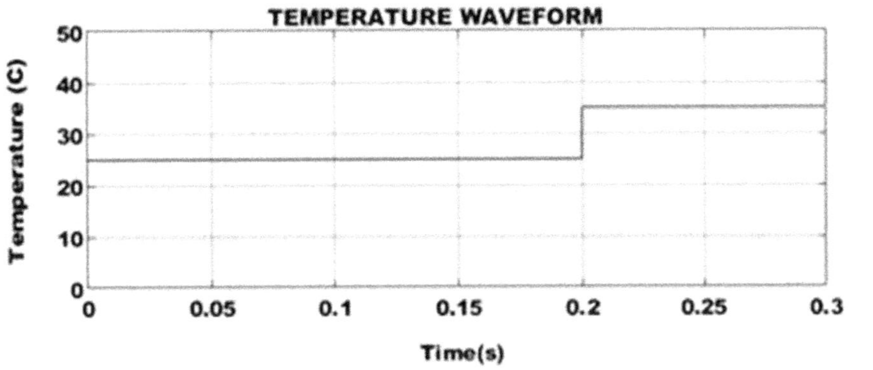

Case 1: Switch Q1 is turned ON while Q2 is turned OFF.

As n side of the diode is at higher voltage than the p side, which is shorted to ground by the switch, a diode would be open circuited when the switch is turned ON. Thus, a novel representation of KY Boost Converter is as presented in Figure 4 (a). This state causes the inductor to charge and results in an increase in inductor current. Consider that the inductor current is i_L, off before the switch swings. This results in input voltage' consistency. According to D*T_s, where D is a duty cycle and T_s is the switching period of time, switch Q1 is ON and switch Q2 is OFF for t_{on} seconds. The disparity between the currents flowing via L and R is equivalent to the current passing through C. Furthermore, C_b is released in this manner. When the capacitors are turned on, the voltage across inductor and currents across them are given as,

$$V_{LO} = V_g + V_{Co} - V_o \quad (5)$$

$$i_{co} = i_L - i_o \quad (6)$$

$$i_{cb} = i_{in} \quad (7)$$

Figure 4. Converter on period

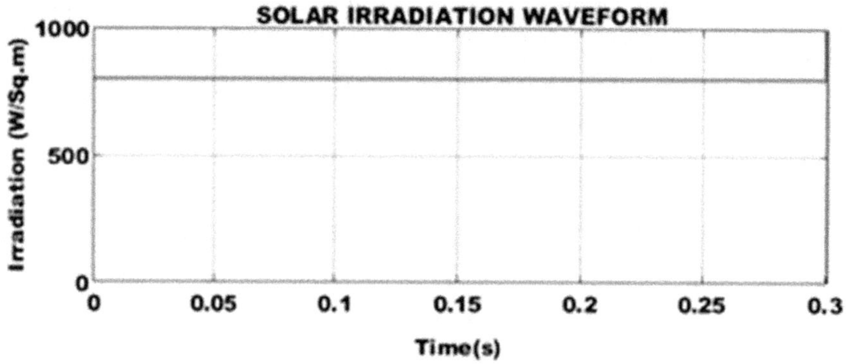

Case 2: When switch Q2 is ON and Q1 is OFF.

Switch Q1 short circuit the diode when it is off, allowing the KY Boost Converter circuit to be redrawn as illustrated in Figure. 4 (b). Consequently, the capacitor C_b and RC combination discharge an inductor. Assuming the inductor current is i_L, off immediately before the switch closes. A negative symbol indicates that an inductor is discharging as depicted in figure 4 (b). Consider the switch is on for a duration of t_{off} seconds, which is equal to (1-D)*T_s. During on phase, the voltage across inductor and currents flowing through the capacitors are expressed as,

$$V_{Lo} = V_g - V_o \quad (8)$$

$$i_{co} = i_L - i_o \quad (9)$$

$$i_{cb} = i_{in} - i_L \quad (10)$$

As long as there are no sudden changes in inductor's current during steady state. There ought to be an equal difference between the current at the end of switch on phase and switch to off state. Additionally, the currents at the beginning of the switch off mode and the end of the turn phase has to be equivalent.

$$V_{out}/V_g = (1+D) \quad (11)$$

Figure 5. Converter on t_{off} period

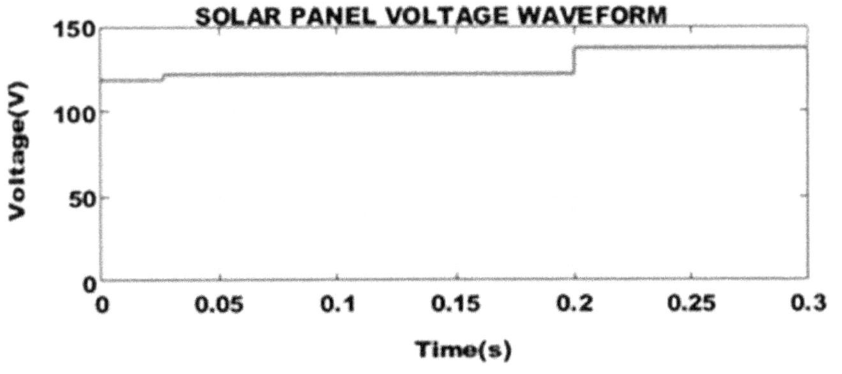

As 0< D<1, V_{out}>V_g. Employing the law of energy conservation and presuming there are no losses in the circuit.

$$V_{out}*I_{out} = V_g*I_{in} \tag{12}$$

Which, indicates, I_{out}/I_{in}=(1–*D*), therefore, I_{out}<I_{in}.

3.3. Cascaded Interval Type 2 Converter

Type 2 fuzzy logic system (FLS) has a capacity of handling rule uncertainties. Type 2 FLS comprises inference, fuzzification and output processing. An objective of the suggested proposal is to provide a fundamental idea essential for comprehending methods and algorithms that are discussed. The fuzzy set output of type 1 is produced by an output processor, which also comprises defuzzifier and type reducer. In this proposed work cascaded Type 2 FLC employed as shown in Figure 6, for effective controlling of converter.

Figure 6. Cascaded interval type 2 FLC

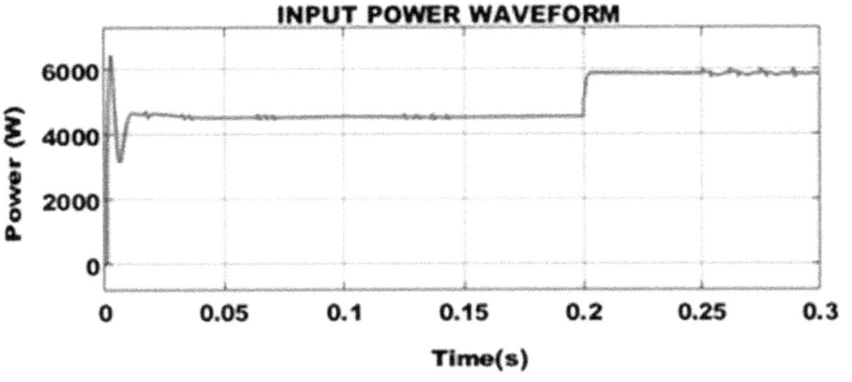

The cascaded Type 2 membership function (MF) has been formulated via blurring type 1 MF to left and right. In this scenario, the MF (u) takes on several values for a specific value x', which are not equally weighted are depicted in Figure 7 for a single Type 2 FLC.

Figure 7. Type 2 membership function as a blurred membership function of Type 1

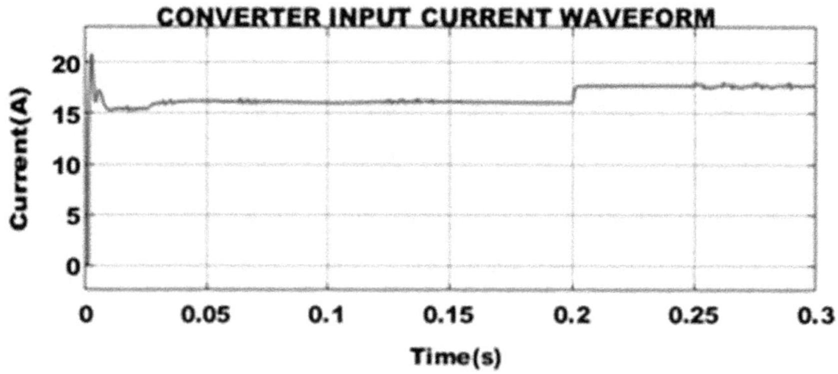

Type 2 MF is formed as a result, which features fuzzy set of type 2 in three dimensions for all *x*∈*X*. A membership function defines type 2 fuzzy set $\widetilde{A}$ as,

$$\tilde{A} = \left\{\left((x,u), \mu_{\tilde{A}}(x,u)\right) \mid \forall x \in X, \forall u \in J_x \subseteq [0,1]\right\} \tag{13}$$

The primary MF of x is indicated in the given formulation as $J_x \subseteq [0,1]$ and a secondary set, which has been termed as type 1 fuzzy is signified as $\mu_{\tilde{A}}(x,u)$. For each primary membership, the secondary membership defines a potential outcome for primary membership.

Figure 8. Interval Type 2 membership function

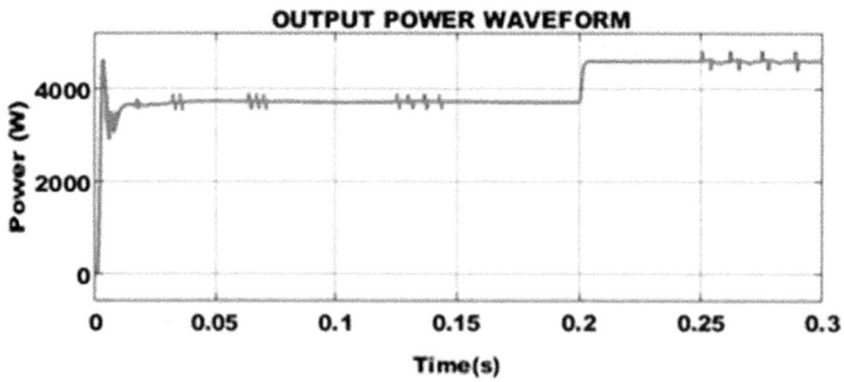

The upper MF $\mu_{\tilde{A}}$ and lower MF $\mu_{\tilde{A}}$ illustrated in Figure 8, are employed to explain a uniform shading for FOU (Footprint of Uncertainty), which represents the complete interval fuzzy set of type 2. Since they rely on type 1 fuzzy sets, which are known as existing, type 1 FLSs are incapable of managing uncertainty rule directly. If-THEN rules that have type 2 FLS as their consequent or antecedent sets define a type 2 FLS. When it is impossible to estimate membership values exactly, type 2 FLSs are employed. An output processor, fuzzy inference engine, rule base and fuzzifier are all components of a type 2 FLS that are well explicated in the following sections.

Fuzzifier

A fuzzifier converts numeric vector $X = \{x_1, \ldots, (x_p)^T \in X_1 \times X_2 \times \ldots \times X_p \equiv X$ into an interval type 2 fuzzy set $\tilde{A}_x$ in X. A single point with nonzero membership constitutes a singleton in fuzzy set as input, hence we employ singleton fuzzifiers of type 2. A fuzzy singleton of type 2 has been depicted as $\tilde{A}_x$, if $\mu\tilde{A}_x$ (x) = 1/1 for x=x^1 and $\mu\tilde{A}_x$ (x) = 1/0 for all other x$\neq x'$.

Rules

Type 2 FLS has a same rule structure as type 1 FLS, but fuzzy sets of type 2 are employed for representing an antecedents and consequents. Assuming the M rules, type 2 FLS's *l* th rule is expressed as,

$$\boldsymbol{R}^1\text{: IF } x_1 \text{ is } \tilde{F}_p^l, \text{THEN y is } \tilde{G}^l\, l = 1, \ldots, \text{M} \tag{14}$$

Inference

A FLS type 2 inference engine combines principles and delivers mapping to generate type 2 fuzzy sets from type 2 fuzzy input sets. Computing compositions of type 2 relations with extended sup stars, join (unions) and meet (intersections) are essential. Considering $\tilde{F}_1^l \times \ldots \times \tilde{F}_p^l \rightarrow \tilde{G}^l = \tilde{A}^l$ equation (15) is written as,

$$R^l : \tilde{F}_1^l \times \ldots \times \tilde{F}_p^l \rightarrow \tilde{G}^l = \tilde{A}^l \rightarrow \tilde{G}^l \quad l = 1, \ldots, M \tag{15}$$

The membership function $\mu_{R^l}(\mathbf{x}, y) = \mu_{R^l}(x_1 \ldots, x_p, y)$ defines R^l, where $\mu_{R^l}(\mathbf{x}, y) = \mu_{\tilde{A}^l} \rightarrow \tilde{G}^l(\mathbf{x}, y)$ is specified as,

$$\mu_{R^l}(x, y) = \mu_{\tilde{A}^l} \rightarrow \tilde{G}^l(x, y) = \mu_{\tilde{F}_1^l}(x_1)\Pi\ldots\Pi\mu_{\tilde{F}_p^l}(x_p)\Pi\mu_{\tilde{G}_p^l}(y) = \left[\Pi_{i=1}^p \mu_{\tilde{F}_i^l}(x_i)\Pi\mu_{\tilde{G}_p^l}(y)\right] \tag{16}$$

Type 2 fuzzy set $\tilde{A}_x$, whose MF is typically given as *p* dimensional input to R^l becomes,

$$\mu_{\tilde{A}_X}(X)=\mu_{\tilde{X}_1}(x_1)\Pi\ldots\Pi\mu_{\tilde{X}_p}(x_p)=\Pi_{i=1}^{p}\mu_{\tilde{x}i}(x_i) \tag{17}$$

Every R^l characterises fuzzy set of type 2 $\tilde{B}^l = \tilde{A}_x \circ R^l$ such,

$$\mu_{\tilde{B}}(y)=\mu_{\tilde{A}_x \circ R^l}=\sqcup_{x\in\mathbf{X}}\left[\mu_{\tilde{A}_x}(\mathbf{x})\sqcap\mu_{\tilde{R}^l}(\mathbf{x},y)\right] \quad y\in Y\ l=1,\ldots,M \tag{18}$$

A relationship between type 2 fuzzy set at an output of an inference engine and type 2 fuzzy set, which activates specific feature in engine is an input or output relation depicted in Figure 7. An Intersection under product t norm and an interval type 2 fuzzy set is employed in FLS, resulting in interval type 1 set as an output of input and an antecedent operation, which are included in firing set $\Pi_{i=1}^{p}\mu_{\tilde{x}i}(x_i)=F^l(\mathbf{X}')$ as,

$$F^l(\mathbf{X}')=\left[\underline{f}^l(\mathbf{X}'),\bar{f}^l(\mathbf{X}')\right]\equiv\left[\underline{f}^l,\bar{f}^l\right] \tag{19}$$

Where

$$\underline{f}^l(\mathbf{X}')=\underline{\mu}_{\tilde{F}_p^l}(x_1')*\ldots*\underline{\mu}_{\tilde{F}_p^l}(x_p') \tag{20}$$

$$\bar{f}^l(\mathbf{X}')=\bar{\mu}_{\tilde{F}_1^l}(x_1')*\ldots*\bar{\mu}_{\tilde{F}_p^l}(x_p') \tag{21}$$

Here, a product operation is signified as *.

Type Reducer

After being converted to a numeric output by the defuzzifier, type 1 fuzzy set is an output of type reducer. A centre of sets (cos) type reduction, commonly known Y_{cos} has been employed in the case of FLS, because the given fuzzy set is also considered as an interval set as specified as,

$$Y_{cos}(\mathbf{X})=[y_l,y_r]=\int_{y^1\subseteq[y_l^1,y_r^1]}\cdots\int_{y^M\in[y_l^M,y_r^M]}\int_{f^1\in[\underline{f}^1,\bar{f}^1]}\cdots\times\int_{fM\in[\underline{f}^M,\bar{f}^M]}\frac{1}{\Sigma_{i=1}^{M}f^iy^i/\Sigma_{i=1}^{M}f^i} \tag{22}$$

Consequently, a centroid of type 2 interval set $\tilde{G}^i$ has been determined via contrasting two end points of an interval set Y_l and Y_r as,

$$C_{\tilde{G}^i}=\int_{\theta_N\in J_{yN}}\frac{1}{\Sigma_{i=1}^{N}y_i\theta_i/\Sigma_{i=1}^{N}\theta_i}=[y_l^i,y_r^i] \tag{23}$$

It has to be evaluated Equation (23) and its two end points, Y_l and Y_r, before computing $Y_{cos}(x)$. From Eq. (24), if the values of f_i and y_i associated with y_i are represented as f_l^i and y_l^i, correspondingly, as well as the values of f_i and y_i related with y_r are expressed as f_r^i and y_r^i, respectively, we obtain,

$$y_l = \frac{\sum_{i=1}^{M} f_l^i y_l^i}{\sum_{i=1}^{M} f_l^i} \tag{24}$$

$$y_r = \frac{\sum_{i=1}^{M} f_r^i y_r^i}{\sum_{i=1}^{M} f_r^i} \tag{25}$$

Type 2 fuzzy systems output interval, which is utilized for evaluating whether testing or training data are included in output is defined by Y_l and Y_r values.

Defuzzifier

The interval singleton type 2 FLS's defuzzified output is an interval set Y_{cos}, which we receive from the type 2 reducer and defuzzifier using the average of Y_l and Y_r is,

$$y(\mathbf{x}) = \frac{y_l + y_r}{2} \tag{26}$$

Thus, for non-pulsating output current generation, output voltage ripples minimization and the output capacitors current stress are all involved in the employment of DC-DC KY converter, which has the capability of converting energy to the grid from PV are discussed in the subsequent section.

2.4. Single Phase VSI

A type of inverter circuit known as voltage source inverter or VSI, converts DC input voltage to its output AC equivalent. A simple inverter for single phases provide a square type AC output from a DC input. They operate at substantially lower frequencies and possess simple on off switching devices.

A schematic representation of single-phase bridge VSI is depicted in Figure 9. An output ac waveform has been generated by using four switches. Since the load current i_o and output voltage v_o are in phase, four switches are adequate for a resistive load. Furthermore, in cases, where i_o and v_o are out of phase, diodes coupled in anti-parallel to the switch enables the current flow even when the main switch is off. These diodes are termed as feedback diodes; thus, the dc source absorbs the energy back.

A single-phase inverter with full bridge architecture that consist of four switching devices is illustrated in Figure 9. Considering an equivalent input voltage, a full bridge inverter generates an output power, which is twice that of half bridge inverter. In this part, three PWM switching techniques that enhance the inverter's performance are examined. In order to ensure the devices, which are clamped to either positive or negative dc rail, a negative voltage sequence has to be added to modulation signals. The bottom devices in figure 8 are designated as S_{12} and S_{22}, while the top devices are S_{11} and S_{21}. For Single Phase VSI, the voltage equations are as follows,

Figure 9. Circuit of single phase VSI

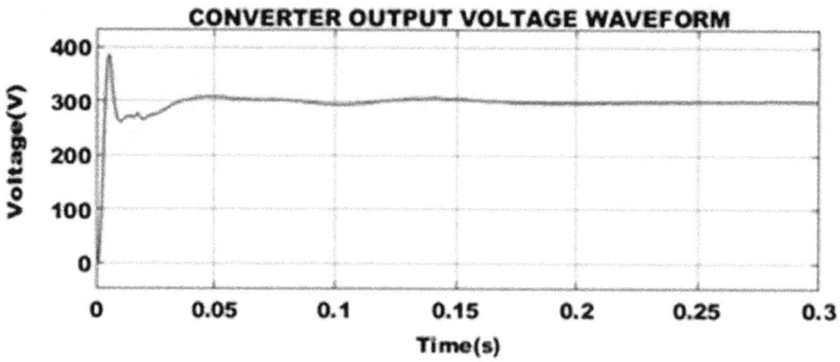

$$\frac{V_d}{2}\left(S_{11}-S_{12}\right)=V_{an}+V_{no}=V_{ao} \tag{27}$$

$$\frac{V_d}{2}\left(S_{21}-S_{22}\right)=V_{bn}+V_{no}=V_{bo} \tag{28}$$

$$V_{ab} + V_{an} = V_{bn} \tag{29}$$

Here, the phase's *A* and *B*'s output voltages to an arbitrary point *n* are represented by voltages V_a and V_b and between point *n* and middle of the DC source, there is a neutral voltage denoted by V_o respectively.

2.5. PI Controller

The PI controller, which compensates for current inaccuracy, is the most frequently used control approach. The value of error is determined by a PI controller as the disparity between desired grid injected current and measured inverter output current. The integral constant K_i and proportional constant K_p are two distinct constants that are employed in the technique. By multiplying the error signal by K_p gain, a proportional term of the controller has been generated. Consequently, the proportional term's impact does not entirely eliminate the inaccuracy and some steady state error still remains. An integral output term of PI controller is obtained through multiplying and integrating the error by K_i constant. In addition to possessing a clearly defined harmonic content, PI current control typically provides a lower current ripple, an excellent steady state response and consistent frequency of switching.

The PI controller's fundamental block diagram is illustrated in Figure 10. By using a sensor or transducer, the error values are computed and compared to the desired set point. The various types of detection include voltage, current, temperature, movement and angle. By changing the value of K_i to zero, the PI controller is manually adjusted. Furthermore, instability in the system causes an enhancement in the overshoot. By ensuring the system to endure overshoot while minimizing settling time and maintaining stability, the value of K_i has to be modified in a specified amount. The PI controller is derived as follows:

Figure 10. Block diagram of PI controller system

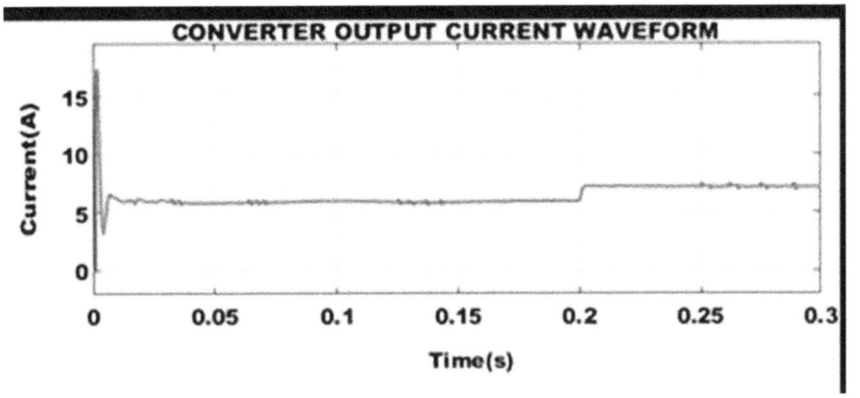

$$u(t) = K_p(t) + K_t \int_0^t e(\tau)\, d\tau \tag{26}$$

Hence, by implementing PI controller, the grid synchronization has been attained, which is described in the following sequence.

2.6. Grid Synchronization

Grid synchronisation, which enables the grid and the synchronised power converter to operate in unison, is a critical issue in connecting power converters to the grid. The control system of a grid connected converter employs data produced by the grid synchronisation method at various levels. For the grid variables to be converted from the natural frame of reference to synchronous reference frame, it is essential to comprehend a phase angle of grid voltage. A regulation of grid connected electronic power converters has a significant impact on grid synchronisation.

The block diagram of grid synchronization has been presented in Figure 11. Methodologies employed in power systems for harmonic detection and grid synchronisation are quite similar. A frequency and time domain are two classifications of the suggested approaches. The frequency domain detection algorithms typically rely on a discrete Fourier analysis implementation. In addition for an internal oscillator to monitor the signal component of interest, the time domain detection techniques depend on some sort of adaptive loop.

For synchronising the grid and monitoring multiple grid conditions, it is essential to measure the utility voltage's phase, amplitude and frequency precisely. Furthermore, the primary objective of synchronisation is to minimise the frequency, voltage and phase angle variances between correlating phases of generator output and grid supply in addition to monitoring, access, enable and automatically execute the control signal to minimise voltage and frequency irregularities.

Figure 11. Diagram of grid synchronization

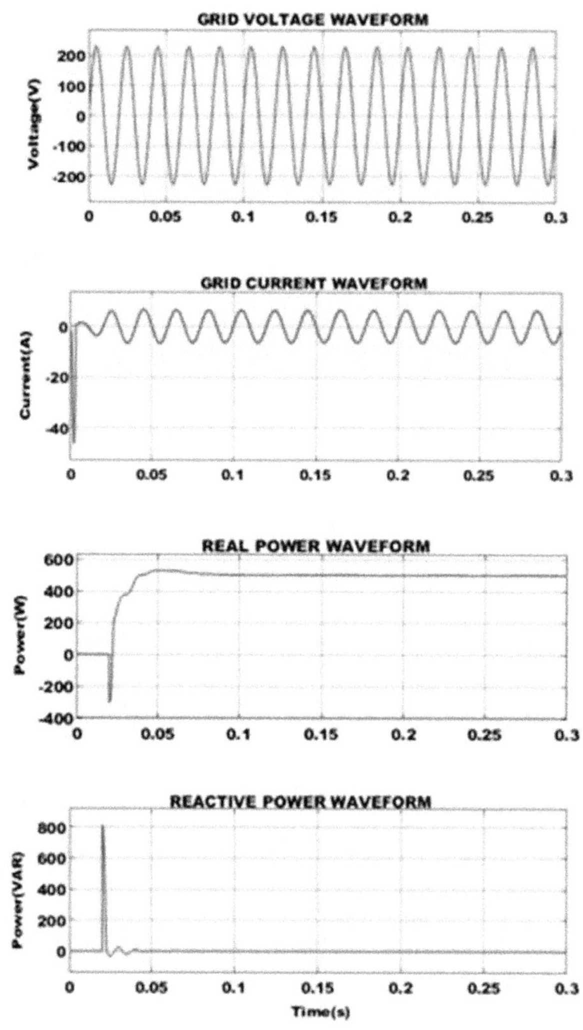

3. RESULTS AND DISCUSSION

The proposed work is simulated using MATLAB simulink and results corresponding to the suggested work are decipted in figures shown below. Table 1 determines the detailed specification of proposed project.

The temperature waveform of Solar PV system is depcited in Figure 12. It is noticed that, temperature varies with respect to time. Initially, a constant of 25°C is maintained constant till 0.2s. After 0.2s the temperature increases by 35°C and maintained constant throughtout the system. Generally, when temperature is high, it results in improved power.

The power absorbed in the form of PV is illustrated as waveform in Figure 13. It is observed from figure that, a constant irradiance of 750W is maintained till 0.3s and further gets raises in accordance with temperature.

Table 1. Specification of parameters

Parameters	Specifications
Solar PV System	
Peak Power	2000W
Series Connected PV Cells	36
No. Of Panels	100W, 20Panels
KY Converter	
L_0	5mH
R_0	100Ω
C_b, C_0	870μF
Switching Frequency	10KHZ

Figure 12. PV temperature waveform

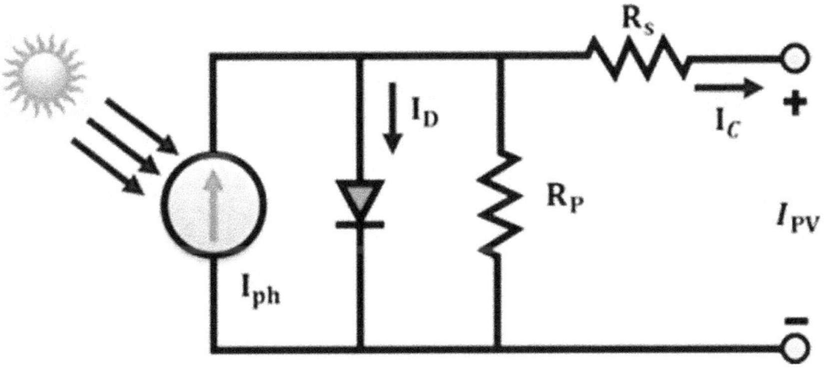

Initially, with minor fluctuation the power form PV system is maintained at a constant of 120V till 0.2s. As depicted in Figure 14, after 0.2s with increase in temperature the solar panel voltage increase and attains 140V, which is further maintained constant.

As like Solar PV voltage, the input power waveform is illustrated in Figure 15. From figure it is observed that, with initial rise a stable power of 4500W is maintained. After 0.2s with respect to temperature power gets increased to 5900W with slight fluctuations.

The converter's waveform for input current with harmonics has been illustrated in Figure 16. The current initially reaches a peak value of 20*A* and starts fluctuation till 0.2s. Due to increase in temperature a constant current of 17*A* current is maintained constant with fluctuations after 0.25s

The converter output power with the assistance of KY converter is shown in Figure 17. An input power reaches a peak value of 4600W and fall down to 3500W, which is maintained with minor distortions until 0.2s. At 0.2s, the power increase to 4500W and maintained constant with further slight distortions.

The output voltage obtain with the aid of KY converter has been depicted in Figure 18. Initially, a peak voltage of 590V is obtained and falls down to 300V at 0.05s with is maintained with minor fluctuations. After 0.2s a stable voltage of 300V is obtained which is maintained constant throughout the system.

Figure 13. Solar irradiation waveform

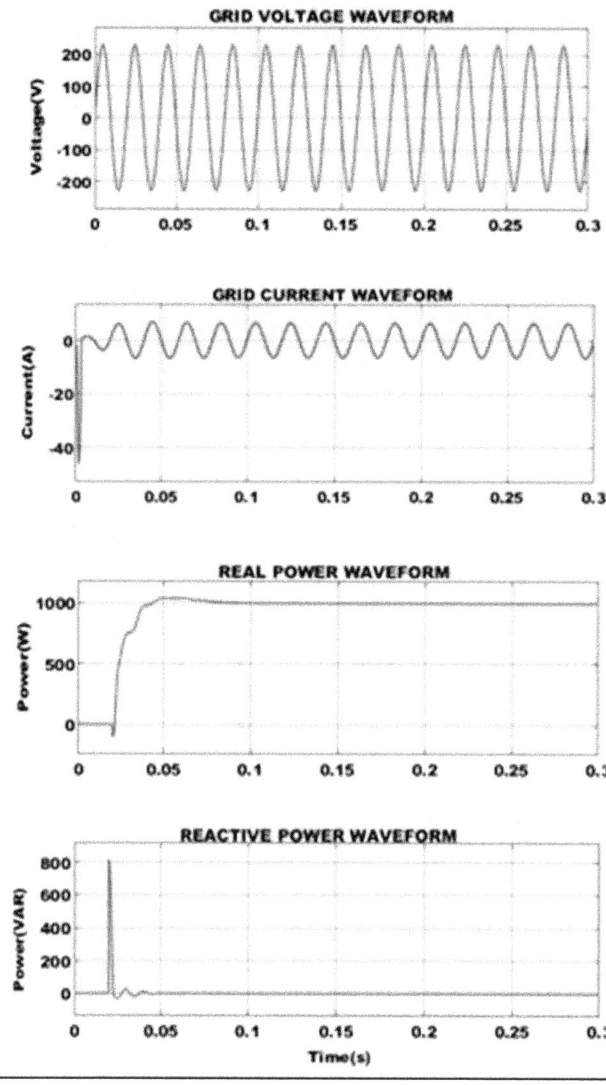

The output waveform with the support of KY converter along with control approach is demonstrated in Figure 19. With peak raise of 17*A* at the beginning, a constant current of 6*A* with fluctuations is attained till 0.2s. Due to increase in temperature and with respect to converter a constant current of 7*A* is maintained stable with slight distortions after 0.25s.

Figure 19 shows the waveform description for grid voltage, current, real and reactive power with a 500W reference. It is evident from the aforementioned Figure that the system maintains a consistent grid voltage of 230V. While the 5*A* matching grid current is stabilised after 0.02s. The system's actual power drops to -300W at 0.02s and rises to a 500w peak at 0.05s before remaining constant. Accordingly, the reactive power peaks at 800W and then rapidly decreases, maintaining a minimal power that is steady throughout the system.

Figure 14. PV voltage waveform

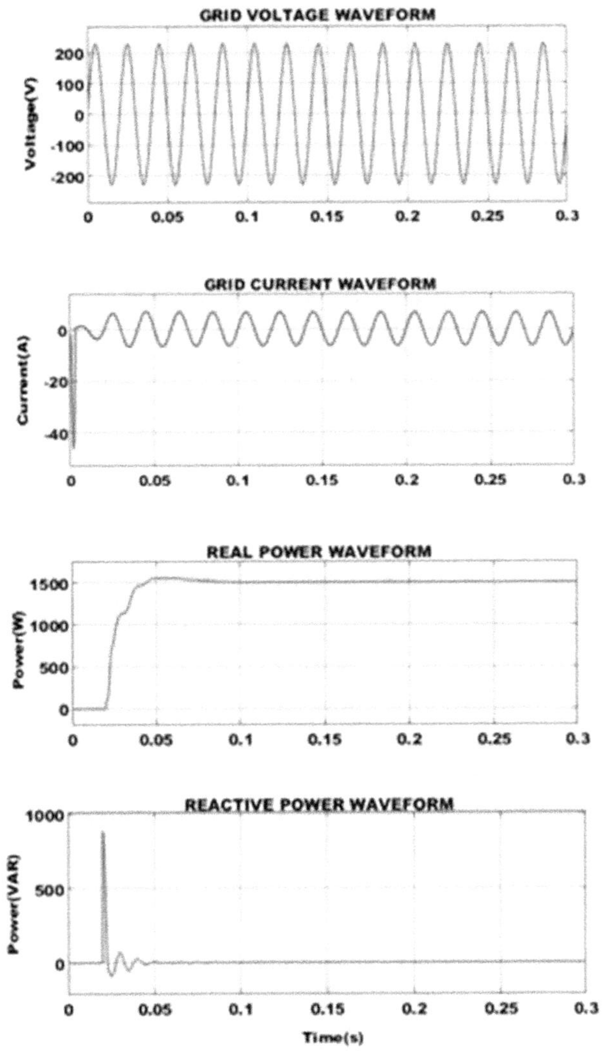

The grid voltage, current, actual power and reactive power are all shown as waveforms in Figure 20. With the assistance of PI controller, the generated PWM pulses are sent from VSI to grid. A steady input voltage of 230V is produced by the 1000W reference power that is supplied to the grid. After 0.02 seconds, a constant current of 10A is maintained with just minor fluctuations. It is evident from the third and fourth waveforms that actual power rises and decreases slightly before remaining constant at 1000W for 0.03s while reactive power is kept to a minimum.

Voltage, current, real power and reactive power metrics for the grid are depicted in Figure 21. Initial grid voltage of 230V is maintained stable throughout the system. In correspondence, with initial drop in current a constant of 5A is attained with 1500W grid power. Accordingly, the real power reaches the value same as that of reference power that is 1500W. In contrast to real power a minimum reactive power is generated after 0.1s, which is further maintained constant.

Figure 15. Solar power waveform

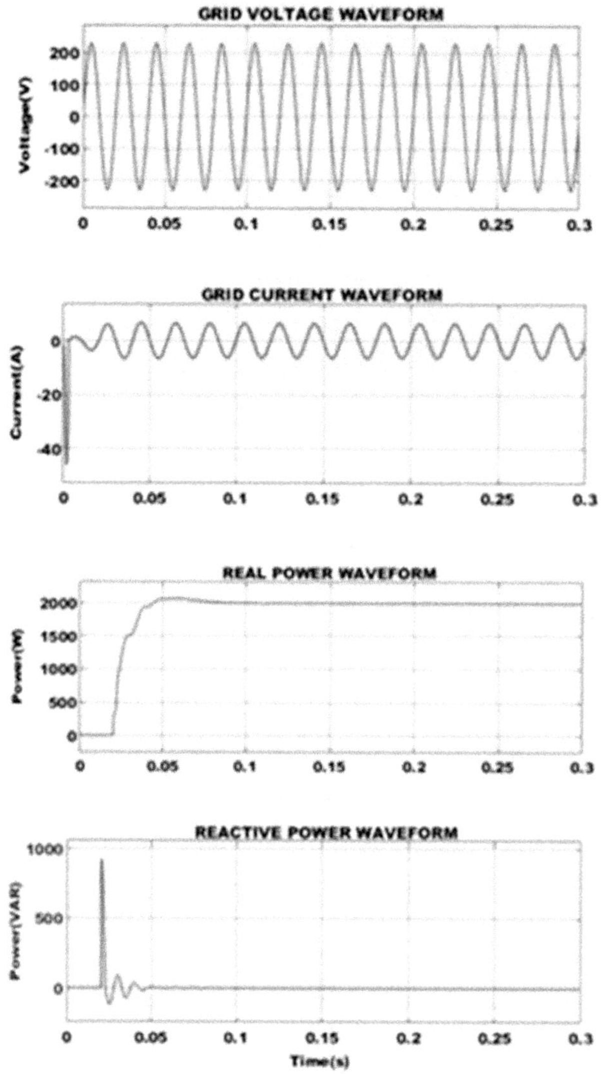

In the following Figure 22, grid parameters are represented in terms of voltage, current, real power and reactive power. Throughout the entire system, a voltage of 230V is maintained as the initial grid voltage. In correlation, with a drop in current at the beginning, a constant current of 5A is achieved with 2000W of grid power at a drop in current. Thus, the real power of the system reaches a value that is equal to the reference power of 2000W and is therefore the same as real power. In divergence to real power a least possible reactive power is produced later at 0.1s, which remains constant.

The THD Waveform for proposed work is illustrated in Figure 24. THD is obtained with respect to frequency. The THD obtained is 3.91% which shows reduced harmonics and satisfies the IEEE standards. Table 2 displays the comparative analysis of converter in terms of efficiency, voltage gain and THD.

Figure 16. Converter input current waveform

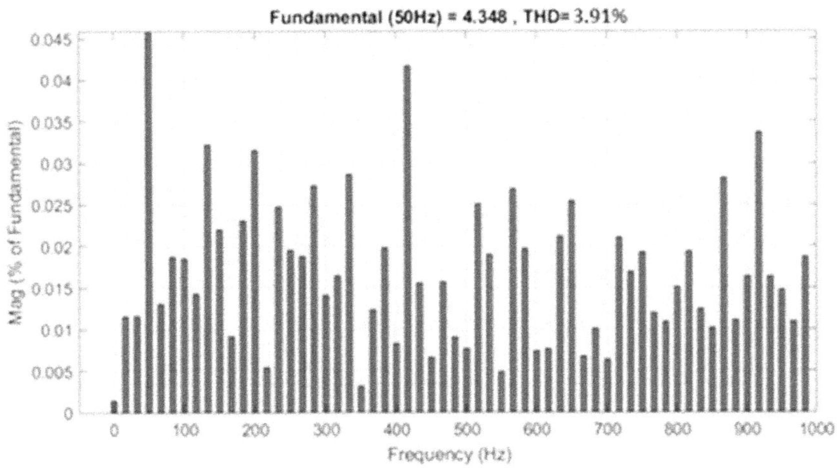

Figure 17. Output power waveform

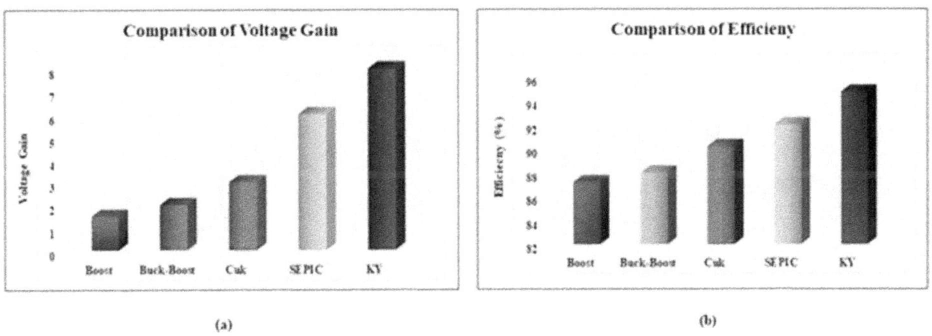

Figure 18. Converter output voltage

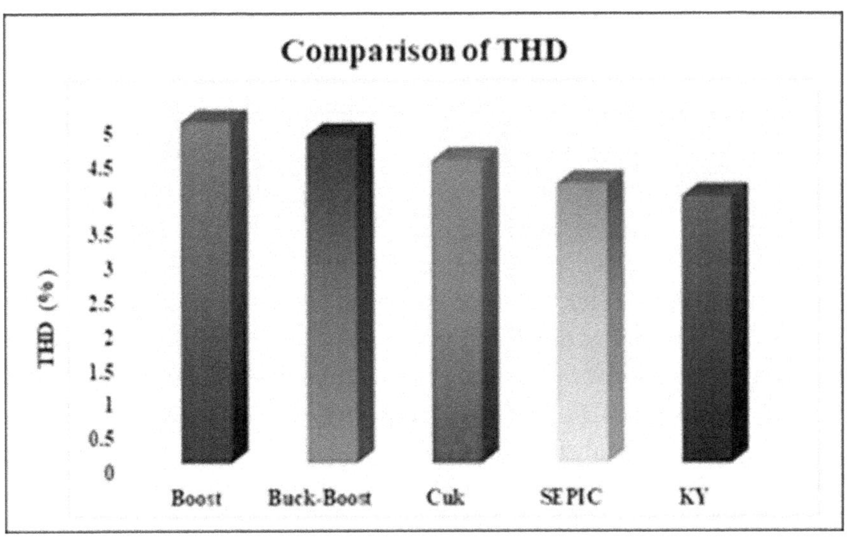

Figure 19. Converter output current waveform

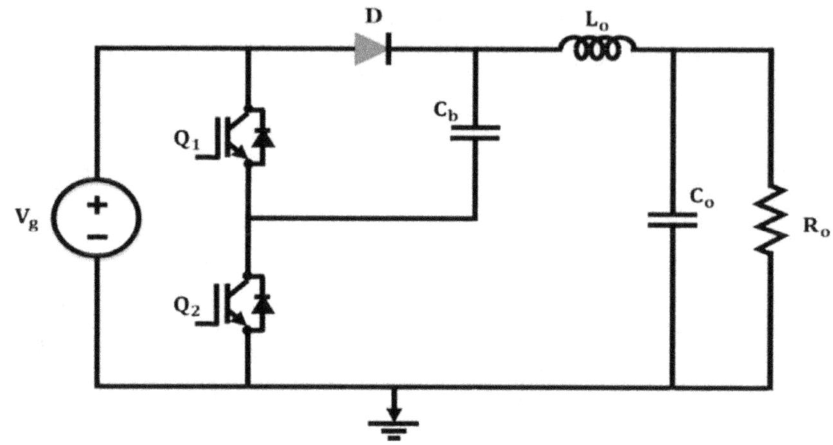

Figure 20. Grid parameter with reference 500W

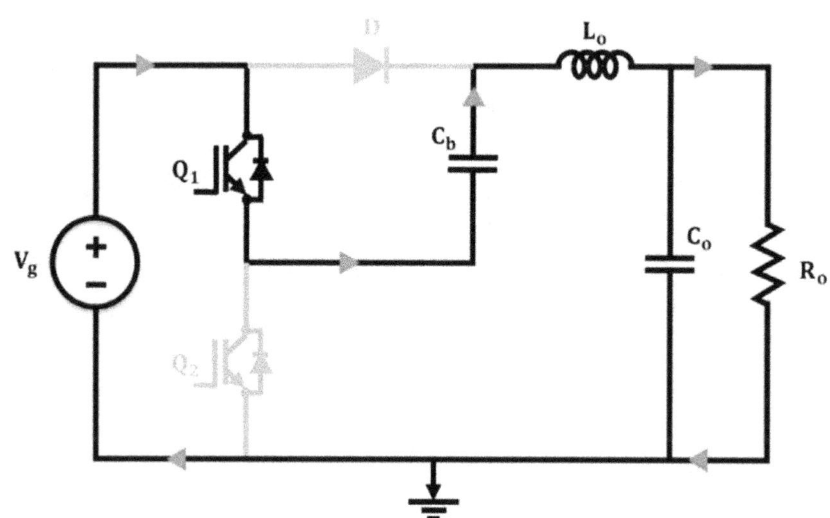

For converters including Boost, Buck Boost, Cuk, SEPIC and KY converters, a comparison analysis is carried out in Figure 25with regards to voltage gain and efficiency. From comparison, it has bee analyzed that the suggested KY converter shows enhanced results with 1:8 voltage gain and efficiency of 94.7%. In contrst voltage gain of Boost, Buck Boost, Cuk and SEPIC converter is observed to be, 1:1.5, 1:2, 1:3 and 1:6. Similarly, there efficiency is noticed to be 87.2%, 88%, 90.2% and 92% respectively.

The comparion graph showing THD analysis betweeen converters including Boost, Buck Boost, Cuk, SEPIC and proposed KY converter is illustrated in Figure 25. It is observed that, a minimum THD of 3.91% is obtained by the KY converter where other converter attains THD value of 5% by Boost, 4.8% by Buck Boost, 4.43% by Cuk and 4.1% by SEPIC converter.

From overall comparsion it is analyzed that the proposed work posses high stability of voltage, improved power compensation accompained with reduced switching loss.

Figure 21. Grid parameter with reference 1000W

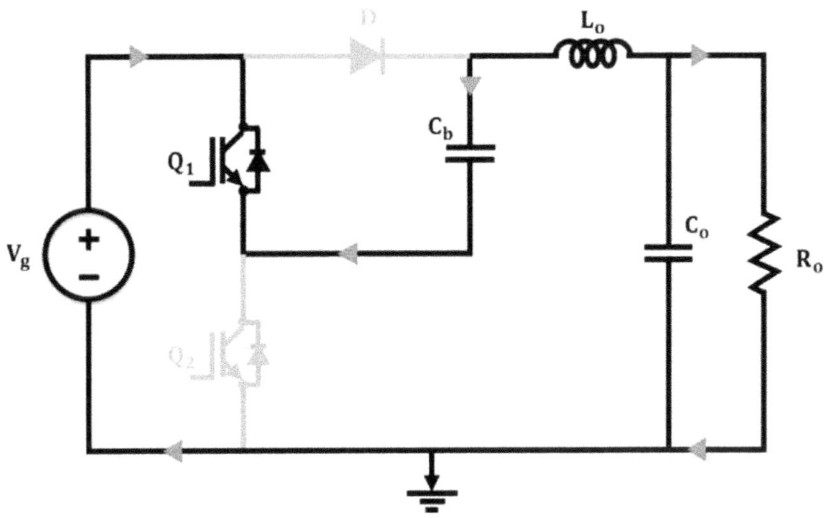

Figure 22. Grid parameter with reference 1500W

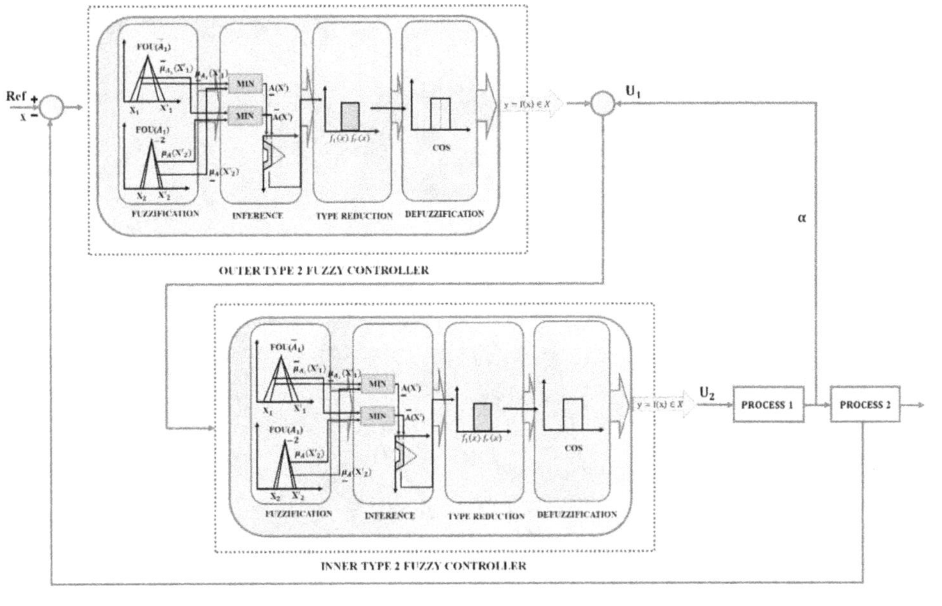

Table 2. Comparative analysis of converter

Converters	Efficiency	Voltage Gain	THD
Boost	87.2	1:1.5	5%
Buck-Boost	88	1:2	4.8%
Cuk	90.2	1:3	4.43%
SEPIC	92	1:6	4.1%
KY	94.7	1:8	3.91%

Figure 23. Grid parametes with reference 2000W

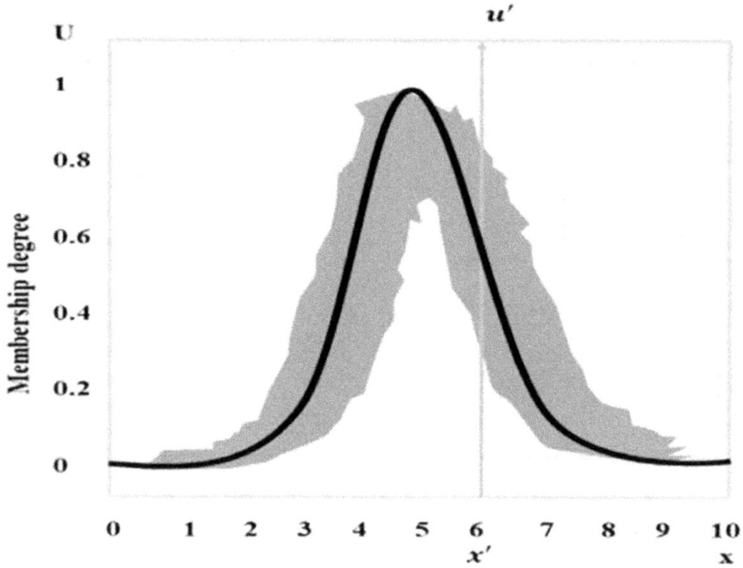

Figure 24. THD waveform

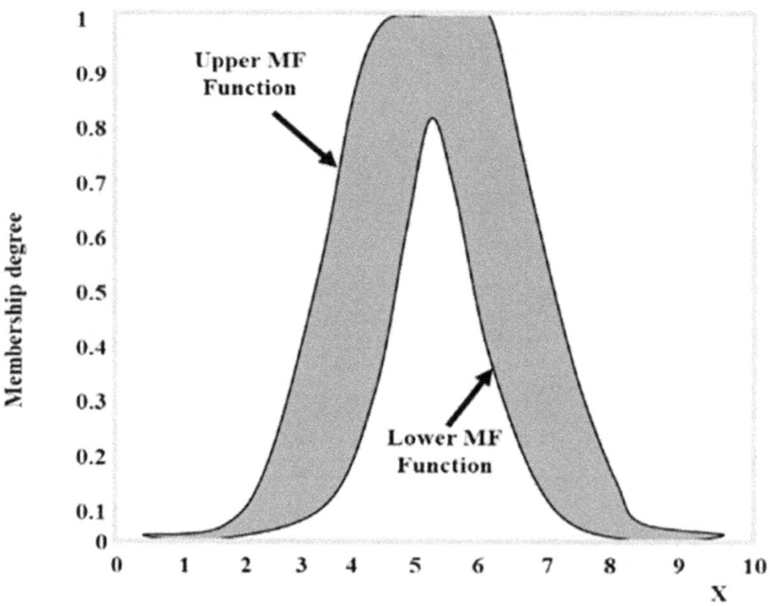

Figure 25. (a) voltage gain, (b) efficiency (%)

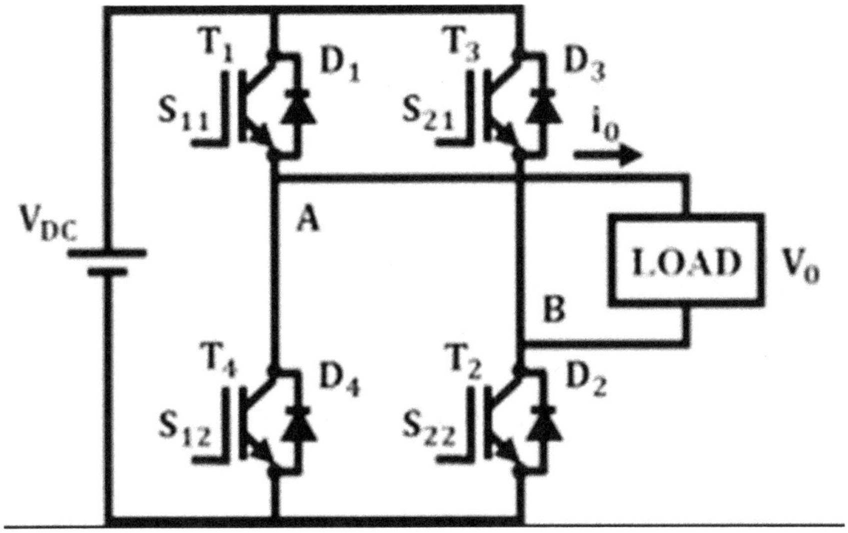

Figure 26. THD comparison

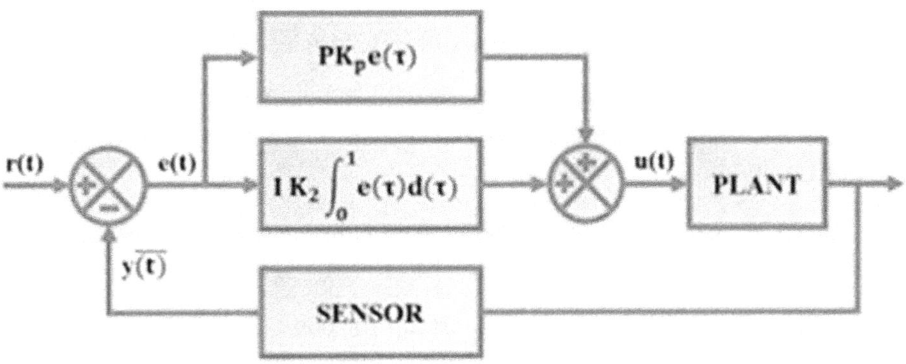

4. CONCLUSION

Utilization of renewable sources replaced fuel based vehicles to EV, which emits no emission towards environment. In order to make EVs' total carbon impact zero, solar electricity is used at many charging stations. In this study, the advantages of combining solar energy generation with EV charging infrastructure is examined. The initial voltage from PV system is fed to the battery and excess charge transferred to grid through KY converter and 1φVSI. The converter voltage with the aid of KY converter along with CIT2FLC approach generates enhanced voltage for grid. The 1φVSI using PI control charges the grid effectively. At instance, when there is no enough supply from PV, the energy stored in grid starts discharging and EV gets charged. Depending on availability, the EV receives uninterrupted power supply from grid or PV panel, resulting in better energy conservation. The overall framework is simulated using MATLAB shows improved converter efficiency of 94.7%, with 1:8 voltage gain. Also effective grid synchronization is achieved with reduced THD of 3.91%.

REFERENCES

Al-Soeidat, M. (2019). *A reconfigurable three-port DC–DC converter for integrated PV-battery system.* https://ieeexplore.ieee.org/abstract/document/8839067/

Chandrasekar, B. (2019). *Non-isolated unidirectional three-port Cuk-Cuk converter for fuel cell/solar PV systems.* https://www.koreascience.or.kr/article/JAKO201926838276891.page

Chandrasekar, B. (2020). *Non-isolated high-gain triple port DC–DC buck-boost converter with positive output voltage for photovoltaic applications.* https://ieeexplore.ieee.org/abstract/document/9120014/

Chen, X., Zhang, T., & Ye, W. (2020). *Blockchain-based electric vehicle incentive system for renewable energy consumption.* https://ieeexplore.ieee.org/abstract/document/9097307/

Dutta, S., & Debnath, D. (2017). *A grid-connected single-phase transformerless inverter controlling two solar PV arrays operating under different atmospheric conditions.* https://ieeexplore.ieee.org/abstract/document/7938358/

Hannan, M., Ghani, Z., & Hoque, M. (2019). *Fuzzy logic inverter controller in photovoltaic applications: Issues and recommendations.* https://ieeexplore.ieee.org/abstract/document/8642881/

Harini, S., & Chellammal, N. (2022). *A Novel High Gain Dual Input Single Output Z-Quasi Resonant (ZQR) DC/DC Converter for Off-Board EV Charging.* https://ieeexplore.ieee.org/abstract/document/9847236/

Karthikeyan, V., Sundaramoorthy, K., Kumar, G. G., & Babaei, E. (2019). Regenerative switched-inductor/capacitor type DC–DC converter with large voltage gain for PV applications. *Wiley Online Library, 13*(1), 68–77. doi:10.1049/iet-pel.2019.0408

Liu, J., Wang, K., Zheng, Z., & Li, C. (2019). A dual-active-clamp quasi-resonant isolated boost converter for PV integration to medium-voltage DC grids. doi:10.1109/JESTPE.2019.2924145

Manikandan, R., &Saha, N. (2015). A control algorithm for nonlinear offshore structural dynamical systems. *Proceedings of the Royal Society of London A: Mathematical, Physical and Engineering Sciences, The Royal Society, 471*(2184).

Maroti, P., & Padmanaban, S. (2019). *A new structure of high voltage gain SEPIC converter for renewable energy applications.* https://ieeexplore.ieee.org/abstract/document/8750785/

Pokharel, M., & Hildebrandt, N. (2019). *A fast-dynamic unipolar switching control scheme for single-phase inverters in DC microgrids.* https://ieeexplore.ieee.org/abstract/document/8322262/

Raiker, G. (2021). *Current control of boost converter for PV interface with momentum-based perturb and observe MPPT.* https://ieeexplore.ieee.org/abstract/document/9435036/

Rezk, H., Aly, M., & Al-Dhaifallah, M. (2019). *Design and hardware implementation of new adaptive fuzzy logic-based MPPT control method for photovoltaic applications.* https://ieeexplore.ieee.org/abstract/document/8786102/

Rezkallah, M., & Hamadi, A. (2018). *Design and implementation of active power control with improved P&O method for wind-PV-battery-based standalone generation system.* https://ieeexplore.ieee.org/abstract/document/8119845/

Saxena, N., & Hussain, I. (2017). *Implementation of a grid-integrated PV-battery system for residential and electrical vehicle applications.* https://ieeexplore.ieee.org/abstract/document/8023763/

Shariff, S., Alam, M., & Ahmad, F. (2019). *System design and realization of a solar-powered electric vehicle charging station.* https://ieeexplore.ieee.org/abstract/document/8822448/

Singh, A., & Mishra, A. (2020). *An integrated converter with reduced components for electric vehicles utilizing solar and grid power sources.* https://ieeexplore.ieee.org/abstract/document/9104749/

Singh, B., & Kushwaha, R. (2019). A PFC based EV battery charger using a bridgeless isolated SEPIC converter. doi:10.1109/TIA.2019.2951510

Venkatramanan, D. (2019). *Dynamic modeling and analysis of buck converter based solar PV charge controller for improved MPPT performance.* https://ieeexplore.ieee.org/abstract/document/8815948/

Xia, Y., Wei, W., & Yu, M. (2018). *Stability analysis of PV generators with consideration of P&O-based power control.* doi:10.1109/TIE.2018.2864695

Xu, S., & Chang, L. (2019). *Single-phase voltage source inverter with voltage boosting and power decoupling capabilities.* https://ieeexplore.ieee.org/abstract/document/8805155/

Yamaguchi, D. (2018). *A new PV converter for a high-leg delta transformer using cooperative control of boost converters and inverters.* https://ieeexplore.ieee.org/abstract/document/8249823/

Zhang, Y., Shi, J., Zhou, L., & Li, J. (2016). *Wide input-voltage range boost three-level DC–DC converter with quasi-Z source for fuel cell vehicles.* https://ieeexplore.ieee.org/abstract/document/7736069/

Zhou, S., Qiu, Y., Zou, F., He, D., Yu, P., & Du, J. (2019). *Dynamic EV charging pricing methodology for facilitating renewable energy with consideration of highway traffic flow.* https://ieeexplore.ieee.org/abstract/document/8928490/

Chapter 11
Spent Coffee Ground Briquettes: A Critical Review

Gokul Raghavendra Srinivasan
Steamax Envirocare Private Limited, India

Aditya Mahajan
Steamax Envirocare Private Limited, India

Rajiv Seth
Steamax Envirocare Private Limited, India

Rakesh Mahajan
Steamax Envirocare Private Limited, India

ABSTRACT

Coffee is regarded as highly consumed beverage throughout the world and has established a key spot in the world economy as an important commodity for trading. In general, it is produced by brewing its roasted and ground beans, which release aromatic coffee as well as produce an equivalent amount of spent coffee grounds (SCG) as waste. In recent times, they are valorized into SCG briquettes owing to their high calorific value, which gained wide attention for supplying energy renewably, especially to the rising energy demand; and also have been identified as an effective measure to reduce their pollution. With this in mind, this chapter focuses on reviewing the availability and chemistry involved in these SCG wastes, pre-treatments, and preparations required for their briquetting, compacting techniques followed, and fuel characteristics of their briquettes, from various available works of literature. Eventually, all the reported data were in accordance with their permissible standards and suggested these SCG wastes as a highly renewable solid biofuel.

DOI: 10.4018/978-1-6684-7303-0.ch011

INTRODUCTION

To begin with, Coffea are genus of flowering shrubs from Rubiaceae family, indigenous to tropical African and Asian lands; and are primarily cultivated as commodity crops for their seeds, that hold key significance in beverage industries (Brunerová et al., 2019). Though, these plants occur in 70 different species, only *Coffea Arabica* and *Coffea canephora* (or) *Coffea robusta* are cultivated purposely, as they contribute up to 75% and 25% of global coffee production, respectively (Belitz et al., 2009); and are cultivated in large scale by over 80 countries (Murthy & Naidu, 2012). In fact, these coffee seeds have been processed into wide varieties of beverages, especially coffee for more than 1000 years, and are seen as the second largest traded commodity in the world (Peshev et al., 2018). As a result, these seeds are seen as an essential commodity having greater importance in international trade and relationships, economics, and even politics for many developing countries (Potip & Wongwuttanasatian, 2018). Eventually, the cultivation and trading process offers a wide variety of job opportunities, and benefits millions of people during different stages of its production that includes cultivation of plants, harvesting of cherries, processing of beans, packaging and storage, and shipping and handling (Mussatto et al., 2006).

In general, coffee is brewed from ground roasted coffee beans by adding hot water or steam, which extracts their soluble compounds and aromas; leaving behind a large quantity of residual wastes, commonly termed as spent coffee grounds (SCG) (Fehse et al., 2021). The amount of this waste generated depends on their rate of consumption and source, which vary from domestic households to coffee shops or cafeterias, sometimes even large-scale coffee beverage industries (Blinová & Sirotiak, 2019); and on average, constitutes up to 60% of the processed beans (Acevedo et al., 2013; Fithratullah, 2022). Besides SCG residues, other coffee based wastes include their pulps removed from the beans, their silver skin produced during roasting or hulling, and coffee wastewater obtained during handling of their beans (Seco et al., 2020; Nabais et al., 2008). Discouragingly, these wastes have no market value, or scope as value-added products having any direct application (Kourmentza et al., 2018); and are either disposed in landfills, or discarded in open environment or burnt off in incinerators (Lee et al., 2021; Fermoso & Mašek, 2018). However, this simply results in environmental pollution because of their high pollutant potential, harbourage for insects, and odour (Rivera et al., 2020). Despite this, these SCG wastes have promising calorific content with good thermal efficiency and permissible emission range, which makes them ideal for processing into fuels (Allesina et al., 2017); and in general, are valorised into biofuels by undergoing different chemical and biotechnological processes (Karmee, 2018). In specific, these SCG residues exhibit superior performance in their solid form, especially in form of densified briquettes or pellets, and showcases fuel characteristics similar to existing solid fossil fuels, along with high renewability and low ash content (Fehse et al., 2021) at minimal cost (Rajaseenivasan et al., 2016). Besides, these SCG residues are also used as composts and manures for plants, as animal feed, and as raw material for synthesing numerous organic compounds (Colantoni et al., 2021; Atabani et al., 2019; Kang et al., 2017).

About densified SCGs, these wastes are compacted into briquettes or pellets under application of high pressure that agglomerates their particles and are held together by the binders (Moreno-Ariasa et al., 2021). Here, these binders are organic substances that hold back the SCG particles or fibres together through cohesion or adhesion, and can be any natural gum or starch; and theoretically, SCG briquettes can be developed using only their high lignin content as natural binders (Brunerová et al., 2019). However, in reality, it is highly impossible to develop binder less or 100% raw SCG briquettes, especially at lower compaction pressures citing their high lipid content and their coarse particle morphology (Bejenari et al., 2021; Seco et al., 2020). Yet, this can be managed by introducing a calculated amount of these binders and moisture content to the SCG wastes during briquetting, which allows for compactions at low pressure and temperatures (Espuelas et al., 2020). Besides, adding a portion of other biomasses also helps in enhancing the durability and handability of these briquettes; and accordingly, Kang et al. (2017) added appropriate proportion of spent coffee ground, coal fines, sawdust to produce eco-friendly briquettes for domestic applications capable of producing very low toxic emissions (Kang et al., 2017; Roy & Corscadden, 2012; Chen et al., 2017). Worth mentioning, increasing the moisture content in the briquetting dough helps increase the compaction efficiency by simultaneously softening the biomass and activating the binder, as a result producing highly durable briquettes (Huang et al., 2017); but are reduced post-compaction by drying the compacted products (Brunerová et al., 2019). In addition, briquettes developed from SCG-biochar were also found promising for energy generation, citing their superior fuel and combustion characteristics (Lee et al., 2021).

Accounting for their potential, these SCG-based briquettes are expected to support and contribute a significant part to the present energy demand, rising steadily due to modernization and exponential human population growth (Chia et al., 2018). Unfortunately, this demand has simultaneously started depleting the existing fossil fuel resources and harming the planet's atmosphere by causing events such as climate change and global warming (Tan et al., 2019). Hence, relying on these SCG wastes can be seen as a fruitful option, as they are highly renewable as well as calorific, and easily available. Available in abundance, these wastes give rise to second-generation solid biofuels, which this present book chapter will focus on by summarising their availability, chemistry involved in these wastes, pre-processing techniques followed to prepare them and various briquetting methods used for their compaction, along with their fuel and combustions characteristics.

AVAILABILITY OF SCG WASTES

Cultivation and Consumption of Coffee

As mentioned earlier, Coffea has been used in numerous cuisine applications, besides being traded as a commodity crop globally (Fithratullah, 2022). Traditionally, Brazil, Vietnam, Colombia, and Indonesia are considered the global pioneers in producing coffee, collectively contributing up to 50% of world's coffee production (ICO, 2020; Brunerová et al., 2019). Looking at their consumption rate, about 9.08 million tons of coffee beans were consumed during year 2015/2016, and about 9.44 million tons, were estimated as on November 2017 (Potip & Wongwuttanasatian, 2018; ICO, 2018). Even more, International Coffee Organization (ICO) data suggested that about 10.3 billion tonne of coffee were harvested during 2020, and about 166,628 thousand 60 kg bags of coffee beans were consumed globally during 2020/2021 (ICO, 2021). This data also estimated that roughly 5.03 billion tons of coffee was imported to the European Union countries through marine transportation (ICO, 2019). In terms of consumption post-brewing, almost 400 billion cups of coffee are consumed annually and are expected to increase in coming years (Grigg, 2002). Table 1 summarises the annual worldwide coffee production during the year 2014-2017, as recorded by ICO (Brunerová et al., 2019). Moreover, these data display a steady rise in their production and consumption; and can be explained by the newly found interest and taste for these beverages among the younger generation, thus increasing their demand considerably (Fithratullah, 2022).

Table 1. Annual global coffee production rate (2014-2017), along with their cultivation area and yield (FAO, 2019; ICO, 2010; Brunerová et al., 2019)

Harvest Year	Total Area Used for Harvesting (X 10^6 Ha)	Average Yield (Tonnes/Ha)	Amount of Green Beans Harvested (X 10^6 Tonnes)	Amount of Coffee Beans Produced (X 10^6 60 kg Bags)
2014	10.52	8.37	8.80	0.154
2015	10.95	8.10	8.87	0.150
2016	10.95	8.59	9.32	0.153
2017	10.84	8.50	9.21	0.159

Source: (FAO, 2019; ICO, 2010; Brunerová et al., 2019)

Coffee to SCG Equivalence

It is fairly evident that brewing coffee beans produces aromatic coffee extract, and leaves behind a large quantity of residue coffee dust as SCG wastes. In fact, literature reported different volumes of these SCG wastes, post-brewing; with their concentration typically exceeding 50%, depending on the extraction process employed. Explaining this, about 550-650 grams of SCG waste are produced while brewing 1 kg of raw coffee beans, for a moisture content added between 55 and 88% (D.B) (Gómez-de la Cruz et al., 2015; Santos et al., 2017; Srinivasan et al., 2022); whereas instant coffee preparation yields only 45-50% of SCG wastes (Murthy & Naidu, 2012). Meanwhile, 1 kg of ground coffee and green coffee can produce up to 900 grams and 650 grams of SCG wastes; whereas, 1 kg of soluble or instant coffee can produce wet SCG wastes upto 2 kilograms (Luz et al., 2018; Blinová & Sirotiak, 2019).

Generation of SCG Wastes

As such, almost 50% of coffee feedstock is rejected as SCG wastes; and availability of these wastes can be accounted for based on this conversion value (Tsai et al., 2012). Recollecting from earlier facts, these SCGs are mostly produced in coffee stalls and shops, apart from coffee-based industries (Mata et al., 2018); and are rich in sugar molecules polymerized into cellulose and hemicellulose structures (Fu and Chena, 2020). In fact, these wastes are generated in both urban and rural parts of the country citing the boom in coffee business; and are always readily available, mostly in their wet form (Potip & Wongwut-tanasatian, 2018). On average, about 6-8 million tonnes of SCG wastes are generated annually, and was reported as 9.3 million tonnes during the year 2017 (Srinivasan et al., 2022; Zabaniotou & Kamaterou, 2019; Echeverria & Nuti, 2017). About 50% of these residual wastes are disposed directly into landfills, while the remaining portion are usually discarded in open garbage bins and drainages, and even in dump yards (Tsai et al., 2012). However, with appropriate waste management techniques, around 4.40-4.65 million tonnes of SCG wastes, normally disposed of in landfills can be valorised into suitable biofuel (Lee et al., 2021). Research suggests about 2.3 PJ of energy; equivalent to 1.16% of bioenergy potential in Poland was contributed by these SCG wastes alone, thus signifying these SCG wastes as valuable feedstock (Ciesielczuk et al., 2015).

CHEMICAL CHARACTERISTICS OF SCG WASTES

Besides assessing their availability, it is also equally important to understand the chemical characteristics of these SCG wastes to ensure their renewability. To start with, SCG wastes are primarily made up of cellulose and hemicellulose compounds (Wei et al., 2019), followed by significant amount of lignin, amino and fatty acids, caffeine, flavonoids and Millard products, polysaccharides and polyphenols and tannins (Srinivasan et al., 2022; Atabani et al., 2019; Pujol et al., 2013). In specific, the lignocellulosic composition of these wastes report 11.6%-13.2% of cellulose, 37.2%-41.0% of hemicellulose, and 22.2-34.94% of lignin (Srinivasan et al., 2022; Lee et al., 2017); and varies along with coffee species and coffee preparation technique (Blinová & Sirotiak, 2019). Here, high concentration of lignin content favours their torrefaction during the conversion of these wastes into biochar; besides acting as natural binder during briquetting (Lee et al., 2017). Figure 1 depicts the distribution of lignocellulosic compounds in these SCG wastes (Lee et al., 2021).

Following this, proximate analysis on coffee beans, before brewing reported its water content as 2.1%, volatile matter as 76.75%, ash content as 3.88% and fixed carbon as 17.27%; while, proximate composition of dried SCG wastes was measured as follows: water content-11.69%, volatile matter-70.03%, ash content-2.06% and fixed carbon-16.22% (Kang et al., 2017). However, moisture content in these SCG wastes was reported up to 55-60%, and are always dried to 8-10%, before briquetting (Kang et al., 2017; Potip & Wongwuttanasatian, 2018). Meanwhile, higher heating value (HHV) of these SCG wastes ranged between 18.8 and 26.9 MJ/kg, thus making these wastes a successful substitute for existing fossil coal (Murthy & Naidu, 2012; Srinivasan et al., 2022).

About their elemental composition, SCG wastes are identified as one of the potential sources of high carbon content biomass, by reporting the carbon content as 54% for both Coffee beans and dried spent coffee grounds (Tsai et al., 2012). In addition, elemental composition of coffee beans, pre-brewing and dried SCG wastes were measured as hydrogen- 6.94 and 7.19%,nitrogen-1.62 and 1.45%, sulphur-0.06 and 0.05%, and oxygen- 34.25% and 36.2%, respectively (Kang et al., 2017; Srinivasan et al., 2022). Besides, EDX analysis on these SCG wastes also showed mixed presence and traces of numerous macronutrients like aluminium, calcium, magnesium, phosphorous, potassium, silicon, sodium, and sulphur in them; and are summarised in Table 2 along with their composition (Colantoni et al., 2021; Bejenari et al., 2021). In case of SCG biochar, concentration of carbon increased from 56.25% to 78.82%, while oxygen reduced from 29.75% to 10.59% (Lee et al., 2021).

Looking into their particle size and morphological properties, almost 90% of SCG waste particles are sized between 320 and 420 μm, with irregular shapes and intra-particle gaps, thus making these wastes highly porous. On average, these pores range between 20 and 30 μm, with extreme irregularity in the shapes; however, similar morphological characteristics were exhibited among the SCG wastes of other known coffee species (Bejenari et al., 2021). Eventually, these results were supported by the SEM images of SCG particles, which comprised flakes with shallow pores all over, giving a sponge-like appearance. Again, similar morphological characteristics were noted for the bio-chars of SCG wastes, strongly concluding that thermal degradation induced chemical alteration, without affecting their physical appearances including their particle size (Lee et al., 2021).

Figure 1. Distribution of lignocellulosic compounds in raw SCG wastes
Source: Lee et al. (2021)

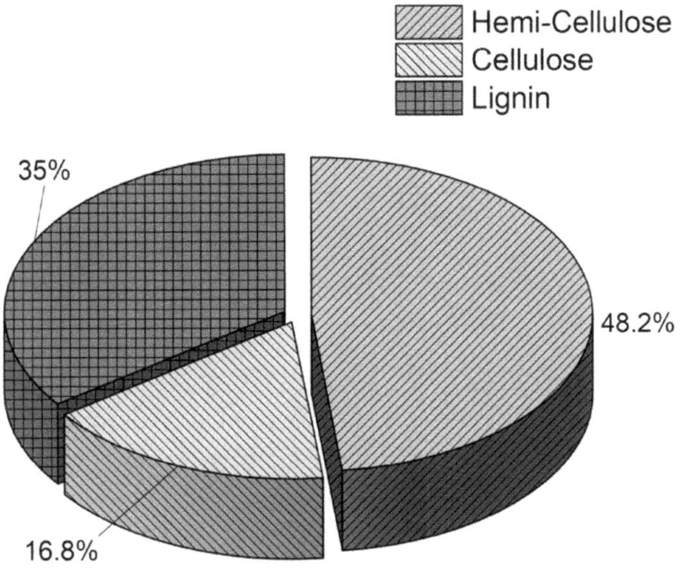

Table 2. Micronutrients and mineral composition of SCG wastes

Micronutrient	Availability	Micronutrient	Availability
Metallic Micronutrients			
Lithium	0.53 ± 0.5 mg/kg	Nickel	1.23 ± 0.5 mg/kg
Sodium	37.15 ± 1.2 mg/kg	Copper	9.82 ± 3.5 mg/kg
Magnesium	1.98 ± 0.018 g/kg	Zinc	3.57 ± 1.04 mg/kg
Aluminium	21.44 ± 3.6 mg/kg	Strontium	10.51 ± 2.01 mg/kg
Potassium	1.69 ± 0.007 g/kg	Barium	3.17 ± 0.9 mg/kg
Calcium	0.25 ± 0.004 g/kg	**Non-metallic Micronutrients**	
Manganese	10.52 ± 2.6 mg/kg	Boron	10.6 ± 4.5 mg/kg
Iron	9.87 ± 1.5 mg/kg	Gallium	0.27 ± 0.3 mg/kg

Source: (Colantoni et al., 2021)

Next up, Thermo-gravimetric studies on raw SCG wastes generated two exothermic peaks signifying the burning of their volatile matter and fixed carbon content, with highest peak noted during the stable combustion of their fixed carbon content (Seco et al., 2020). Also, this study supported the results from proximate analysis on SCG wastes claiming the distribution of low fixed carbon and high volatile matter content in them. On the contrary, SCG with binders reported slightly reduced volatile matter and increased fixed carbon content (Espuelas et al., 2020). Moreover, this study briefed about the stability of their lignocellulosic compounds, with hemicellulose exhibiting poor stability followed by cellulose,

Table 3. Chemical functional groups detected in SCG and optimised SCG-Biochar

Designated and Noted Wavenumber Range (in cm^{-1})	Reported Bond Activities of Functional Groups
Range: 3600-3400; **Raw SCG:** 3331; **SCG char:** 3404	Strong stretching of O-H bonds in Alcohols and Phenols
Range: 3000-2850; **Raw SCG:** 2929-2852; **SCG char:** 2989	Stretching of C-H bonds in Alkanes
Range:2140-2100; **Raw SCG:** not noted; **SCG char:** 2143	stretching of C≡C bonds in Alkynes
Range: 2000-1660; **Raw SCG:** 1679-1775; **SCG char:** 1947	Bending of C-H bonds in Aromatic Compounds
Range:1610-1550; **Raw SCG:** not noted; **SCG char:** 1554	Medium stretching of C=O bonds in Carboxylic Acids
Range:1555-1485; **Raw SCG:** 1492; **SCG char:** not noted	Strong stretching of N-O bonds in Aromatic Nitro Compounds
Range:1470-1350; **Raw SCG:** 1410; **SCG char:** 1359	Bending of C-H bonds in alkanes
Range:1340-1250; **Raw SCG:** not noted; **SCG char:** 1302	Strong stretching of C-N bonds in aromatic amines
Range:1350-1000; **Raw SCG:** 1209-1290; **SCG char:** 1151	Strong stretching of C-F bonds in alkyl fluoro compounds
Range:1320-1000; **Raw SCG:** 1084; **SCG char:** 1077	Strong stretching of C-O bonds in primary alcohols and phenols
Range: 995-850; **Raw SCG:** 925; **SCG char:** 985	Stretching of P-O-C bonds in aromatic phosphate compounds
Range: 865-810; **Raw SCG:** 859; **SCG char:** 854	Medium Bending of C-H bonds in aromatic compounds, towards out of plane
Range:850-750; **Raw SCG:** 774; **SCG char:** 735	Strong stretching of C-Cl bonds in alkyl chloro compounds
Range:680-500; **Raw SCG:** not noted; **SCG char:** 573	Strong stretching of C-Br bonds in alkyl bromo compounds

Source: Lee et al. (2021)

and lignin showcasing highest stability (Fermoso & Mašek, 2018). And, significant loss of mass in the SCG wastes was noted during the volatilization and burning phase of their combustion. Here, other phases include burning of chars and burnout, and dewatering phase before the volatilization and burning phase (Seco et al., 2020).

Moving on, FT-IR characterization on these wastes generated few distinct peaks corresponding to various bond activities of O-H bond, aliphatic C-H bonds, C=O (esters) bonds, C=C (aromatic) bonds, and C-O bonds. Table 3 reports the various bond activities corresponding to different functional groups identified in SCG and SCG biochar wastes, along with their wavelength range and numbers. Besides, sharp peaks corresponding to polysaccharides like galactose, arabinose, glucose, mannose and other polysaccharides are noted in between the range of 950-700 cm^{-1}; and intensity of these peaks varies with coffee species analyzed. Adding to this, peak noticed at 1637 cm^{-1} signifies the stretching of C=C bond of cyclic hydrocarbons, C=O bond, and C=N bond that corresponds to caffeine; is a marker for confirming the presence of residual caffeine content in the SCG wastes depending on its intensity (Zarrinbakhsh et al., 2016; Paradkar & Irudayaraj, 2002).

Accounting their unique chemical characteristics, these SCG wastes are used for producing biodiesel (due to their high bio-oil content, upto 16.12%) (Goh et al., 2020), biohydrogen or ethanol (Kondamudi et al., 2008), and synthesize adsorbents and sorbent for metal removal (Colantoni et al., 2021), develop anti-bacterial devices, and increase conductivity of lithium-ions batteries (Luna-Lama et al., 2019). Here, Figure 2 portrays the Proximate (PC), and elemental composition, using EDX and Ultimate analysis (UC) of Raw SCG wastes and SCG biochar produced under optimised carbonization. Besides, Table 4 tabulates the additional thermal and physicochemical properties of SCG and optimised SCG-Biochar (Lee et al., 2021).

Figure 2. Proximate (PC), and elemental composition, using EDX and Ultimate analysis (UC) of Raw SCG wastes and SCG bio-char produced under optimised carbonization
Source: Lee et al. (2021)

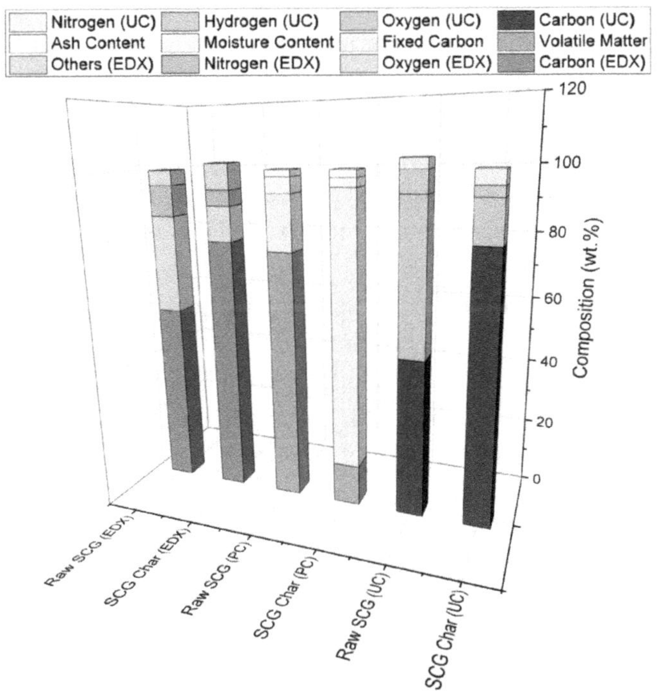

PRE-TREATMENT AND PROCESSING OF SCG WASTES

Any biomass feedstock needs to under certain processing or pre-treatments or both, prior to briquetting to develop high-quality briquettes. Even, SCG also needs certain pre-treatment before being compacted into briquettes, owing to their irregular morphology and residual oil content that hinders proper binding between their particles (Bejenari et al., 2021). Any pre-treatment of SCG begins with roasting of its coffee beans, prior to brewing at 200-230 ^{0}C for 10-15 minutes until their colour changes from bluish-green to dark down (Colantoni et al., 2021). Post-brewing, the residual wet SCG wastes are dried at 105 ^{0}C, in hot air oven for 24 hours to reduce their overall moisture content below 8-10% (Seco et al., 2020).

Table 4. Physicochemical characteristics of SCG and optimised SCG-biochar

Physicochemical Characteristics	SCG	Optimised SCG-Biochar
Biochar yield (%)	-	29.94
HHV (MJ/kg)	22.59	31.41
Energy yield (%)	-	41.6
Volatile matter/fixed carbon	4.55	0.15
H/C atomic ratio	1.65	0.44
O/C atomic ratio	0.68	0.12
Fuel ratio	0.22	6.75

Source: Lee et al. (2021)

Meanwhile, large-scale drying is carried out at slightly higher temperatures (~ 200 ^{0}C), accounting for the large quantity of SCG wastes to be dehydrated (Setiawan et al., 2016). Following this, the dehydrated SCG wastes are treated with solvents like hexane or toluene to remove their lipid fraction (Setter et al., 2020; Tongcumpou et al., 2019), which later on can be used in the production of SCG oil biodiesel (Bui et al., 2022). In general, these wastes are refluxed with these solvents for 4-6 hours, at temperatures maintained slightly above the corresponding boiling point of the solvent used; and are treated in a Soxhlet extractor for lab scale studies. However, solvent-based pre-treatments are least considered for large-scale processing due to the requirement of heavy operating costs and highly organised operational techniques, including solvent recovery and storage.

Yet, this can be overcome by converting these SCG wastes into high calorific bio-chars through pyrolysis (Lee et al., 2021), which later on can be compacted into briquettes adding suitable additives. Most commonly practised pyrolysis techniques includes low and high-temperature pyrolysis, fast and slow pyrolysis, and catalytic pyrolysis (Bok et al., 2012; Romeiro et al., 2012; Kan et al., 2014; Kelkar et al., 2015); are decided based on the availability of resources, and volume of SCG wastes needs to be processed. To begin with, Setiawan et al. (2016) carbonated these SCG wastes via low-temperature pyrolysis, by heat treating them at 250 ^{0}C inside a hot air oven in the absence of oxygen for 120 minutes (Setiawan et al., 2016); whilst, Cho et al. (2015) carried out the pyrolysis of SCG again, in the absence of oxygen, using CO_2 instead of air/N_2 for promoting the pyrolysis yield (Cho et al., 2015). Likewise, Lee et al. (2021) carbonized these SCG wastes into bio-char through pyrolysis, for the following parameters: Pyrolysis temp- 500.00; Pyrolysis time- 30.00 mins; Heating rate- 6.00 ^{0}C/min; Nitrogen flow rate- 250.00 mL/min; and recorded their biochar yield and high energy yield as 29.94% and 41.60%, respectively. Eventually, these bio-chars exhibited superior elemental composition (carbon- 80.35%, oxygen- 12.43%, H/C ratio- 0.44, O/C-0.12 and fuel ratio- 6.75), and high thermal efficiencies with low activation energies (63.24–122.93 kJ/mol)(Lee et al., 2021).

With both SCG wastes and SCG bio-chars already existing in their pulverised form, they do not require any further milling or crushing processes (Brunerová et al., 2019); and can be directly compacted into briquettes using different additives, techniques and process parameters, as discussed in the following section.

BRIQUETTING OF SCG WASTES

Briquetting of SCG wastes deals with densifying and agglomerating their particles into solids, which in turn enhances their handability and fuel quality. Generally, these briquettes are developed by compacting the SCG wastes along with binder and additives (if any), under optimum compaction pressure and temperature, and moisture content; and are carried out with help of various tools that includes, manual compaction tool, hydraulic and pneumatic press, briquetting machines, and screw press. Most importantly, the quality of these briquettes depends entirely on compaction pressure, binder dosage and particle size; and even on compaction technique followed (Lisowski et al., 2019; Roy & Corscadden, 2012). Following this, numerous researchers developed these SCG briquettes, and works related to their briquetting are tabulated in Table 5, which includes details about the briquette type, additives and binder added, compaction technique and conditions, and key highlights in their work.

FUEL CHARACTERISTICS OF SCG BRIQUETTES

Post drying, the compacted briquettes are evaluated for their fuel characteristics, to ensure their compatibility and similarity to existing fossil coals; and in general, are assessed as per international testing standards. Though these results vary with different testing standards, the suitability of these briquettes is decided, if found in between their permissible range. Likewise, SCG briquettes are also assessed as per international standards, especially ASTM standards, to ensure their fuel characteristics meet the essential requirements of a commercial biofuel.

To begin with, moisture content of the dried compacted SCG briquettes ranged in between 10-15%, depending on the particle size, water absorptivity, brewing technique and drying method (Srinivasan et al., 2022; Kang et al., 2017). Though SCG wastes are dried after brewing, an adequate amount of water is added again during compaction for activating the binders (Seco et al., 2020). Even though, wet SCG wastes and briquettes report their moisture content between 20 and 60%, they required less drying time (on average 24 hours) owing to their morphological characteristics; however, binders used for binding these SCG particles required more time drying (Ciesielczuk et al., 2015). Again, this porous morphology of the SCG particles resulted in the low absorption rates of their briquettes. Moreover, briquettes with high moisture content report high density, perform poorly during combustion, and consume more time and energy for drying; besides losing their durability (Seco et al., 2020).

Moving on, SCG briquettes are widely recognised for their very low ash content; and on average, report about 1-3%, depending on the particle size and rate of combustion (Srinivasan et al., 2022; Allesina et al., 2017; Somnuk et al., 2017). In general, ash content of these SCG wastes is around 1.5% and tends to increase with addition of binders or additives (Brunerová et al., 2019). Supporting this, addition of xanthum gum as binder increased the ash content of the SCG briquettes from 0.66% to 0.97% (Seco et al., 2020); and, increased from 0.81% to 0.97%, upon increasing the binder concentration from 5% to 10%. Meanwhile, SCG briquettes noted reduction in their ash content from 0.66% to 0.57% and 0.52%, when added 5% and 10% of Guar gum as binder (Espuelas et al., 2020). In short, reduced ash content for these SCG briquettes can be explained by their low metal concentrations (Colantoni et al., 2021); and is predominantly contributed by their binders or additives, instead.

Table 5. Tabulated summary of briquette type, additive and binder added, compaction technique and parameters, and key highlights of different studies reported in the literature related to SCG briquettes

Briquette Type	Feedstock	Additives	Binder	Briquetting Technique	Briquetting Conditions	Remarks	References
Compound / Composite	25% of SCG	75% of Larch sawdust	Lignin (natural binder)	High-pressure hydraulic briquetting press	Compaction pressure: 80 -100 MPa	• Mass ratios (Larch sawdust + SCG): (100% + 0%), (50% + 50%), (75% + 25%) • Briquette dimensions: diameter- 50.51 ± 0.20 mm; height-53.94 ± 4.19 mm; mass: 113.76 ± 14.39 g	(Brunerová et al., 2019)
Compound / Composite	50% of SCG	50% of Spruce shavings	Lignin (natural binder)	High-pressure hydraulic briquetting press	Compaction pressure: 80 -100 MPa	• Mass ratios (Larch sawdust + SCG): (100% + 0%), (50% + 50%), (75% + 25%) • Briquette dimensions: diameter- 50.51 ± 0.20 mm; height-53.94 ± 4.19 mm; mass: 113.76 ± 14.39 g	(Brunerová et al., 2019)
Homogeneous	SCG	-	Lignin (natural binder)	High-pressure hydraulic briquetting press	Compaction pressure: 80 -100 MPa	• Briquette dimensions: diameter- 50.51 ± 0.20 mm; height-53.94 ± 4.19 mm; mass: 113.76 ± 14.39 g • Failed mechanically owing to their weak binding forces amongst their particles	(Brunerová et al., 2019)
Homogeneous	90% of SCG	-	10% of crude glycerol	Hydraulic Briquettizer	Compaction pressure: 10 MPa; Moisture content: 1.5%	• Mass ratios (SCG: crude glycerol)-100:0 (G0), 95:5 (G5) and 90:10 (G10) • Briquette dimensions: diameter- 53 mm; length-52 mm; mass: 100 g • SCG-glycerol briquettes recorded their highest temperature and combustion rate of 533.4°C and 0.20 g/s, respectively	(Potip & Wongwuttanasatian, 2018)
Homogeneous	90% of SCG	-	10% of Xanthan gum	Laboratory press	compaction pressure: 12 MPa; compaction temperature: 30-35 °C (room temperature); moisture content: 15%	• Moisture content: 10, 15, 20, 25, and 30% • Compaction pressure: 8 MPa, 10 MPa, and 12 MPa • Briquette dimensions: diameter- 65 mm; cylindrical mold • Sample size: 100 g of SCG+ glycerol slurry mix • Drying of briquettes: dried at room temperature for 24 h	(Seco et al., 2020)
Homogeneous	SCG	5% of water	-	Hydraulic stamp press (Zeulenroda Presstechnik PYXE 250 F and PYE 100SS)	Compaction pressure: 140 MPa; compaction temperature: 60oC; compression rate: 10 mm/s; Briquetting time: 3 seconds	• Briquettes developed without any binder • Moisture content: 5-10%; • Compaction pressure: 120-160 MPa; • Briquette dimensions: diameter- 50 mm; height-20 mm; mass: 40 g • Optimum moisture content: 5%	(Fehse et al., 2021)
Homogeneous	SCG	-	5% of cellulose	Hydraulic stamp press (Zeulenroda Presstechnik PYXE 250 F and PYE 100SS)	Compaction pressure: 140 MPa; compaction temperature: 60 oC; compression rate: 10 mm/s; Briquetting time: 3 seconds	• Optimum moisture content:5% • Compaction pressure: 120-160 MPa • Briquette dimensions: diameter- 50 mm; height-20 mm; mass: 40 g	(Fehse et al., 2021)

continues on following page

Table 5. Continued

Briquette Type	Feedstock	Additives	Binder	Briquetting Technique	Briquetting Conditions	Remarks	References
Homogeneous	SCG (Extracted using percolation extractor- Fischer Extraction plant)	-	-	Hydraulic stamp press (Zeulenroda Presstechnik PYXE 250 F and PYE 100SS)	Compaction pressure: 140 MPa; compaction temperature: 60 oC; compression rate: 10 mm/s; Briquetting time: 3 seconds	• Optimum moisture content:5% • 5 Litres of toluene added into the percolation extractor, and refluxed with SCG for 3 hours • Moisture content: 5-10% • Optimum moisture content:5% • Compaction pressure: 120-160 MPa • Briquette dimensions: diameter- 50 mm; height-20 mm; mass: 40 g	(Fehse et al., 2021)
Homogeneous	SCG biochar (pyrolyzed SCG)	-	10% of Tapioca starch	Hydraulic stamp press (Zeulenroda Presstechnik PYXE 250 F and PYE 100SS)	Compaction pressure: 140 MPa; compaction temperature: 60 oC; compression rate: 10 mm/s; Briquetting time: 3 seconds	• Carbonization of SCG at 850 oC for 1 hour (rise in temperature by 0.83 oC/min upto 320 oC, and 2.85 oC/min upto 850 oC) • Moisture content: 5-10% • Optimum moisture content:5% • Compaction pressure: 120-160 MPa • Briquette dimensions: diameter- 50 mm; height-20 mm; mass: 40 g	(Fehse et al., 2021)
Homogeneous	SCG biochar (pyrolyzed SCG)	-	40% of Tapioca starch	Manual compaction (Hand held hammer)	-	• Carbonization at 500 °C • Starch:Water- 2:1 • Bio-char:tapioca starch- (40% w/v) @ mass ratio of 4:1 • Briquette dimensions: diameter- 35 mm; height-45 mm • Drying of briquettes: dried in hot air oven at 105 °C for 24 hours	(Tongcumpou et al., 2019)
Homogeneous	SCG	-	5% of Xanthan gum	Laboratory press	Compaction pressure: 12 MPa; Compaction temperature: 30-35 °C (room temperature); Optimum moisture content: 30%	• Moisture content: 15%, 20%, 25% and 30% • Compaction pressure 8 MPa, 10 MPa and 12 MPa • Binder dosage: 5 and 10% • Briquette dimensions: diameter- 65 mm; height- 12 mm; mass: 100 g	(Espuelas et al., 2020)
Homogeneous	SCG	-	5% of Guar gum	Laboratory press	Compaction pressure: 12 MPa; Compaction temperature: 30-35 °C (room temperature); Optimum moisture content: 30%	• Moisture content: 15%, 20%, 25% and 30% • Compaction pressure 8 MPa, 10 MPa and 12 MPa • Binder dosage: 5 and 10% • Briquette dimensions: diameter- 65 mm; height- 12 mm; mass: 100 g	(Espuelas et al., 2020)

continues on following page

Table 5. Continued

Briquette Type	Feedstock	Additives	Binder	Briquetting Technique	Briquetting Conditions	Remarks	References
Homogeneous	SCG	10% of Sodium chloride (NaCl)	Lignin (natural binder)	Molding machine (HEBI; XL140118, China)	Compaction pressure: 0.1 MPa; Compaction temperature: 30-35 °C (room temperature)	• Briquette dimensions: diameter- 40 mm; height- 5 mm; mass: 5 ± 0.1 g • Drying time of SCG + NaCl briquettes reduced by 18%, compared to raw SCG briquettes	(Fu & Chen, 2020)
Homogeneous	SCG	10% of Sodium sulphate (Na2SO4)	Lignin (natural binder)	Molding machine (HEBI; XL140118, China)	Compaction pressure: 0.1 MPa; Compaction temperature: 30-35 °C (room temperature)	• Briquette dimensions: diameter- 40 mm; height- 5 mm; mass: 5 ± 0.1 g • Drying time of SCG + Na2SO4 briquettes reduced by 38%, compared to raw SCG briquettes	(Fu & Chen, 2020)
Homogeneous	SCG	10% of Lignite dust	Lignin (natural binder)	Molding machine (HEBI; XL140118, China)	Compaction pressure: 0.1 MPa; Compaction temperature: 30-35 °C (room temperature)	• Briquette dimensions: diameter- 40 mm; height- 5 mm; mass: 5 ± 0.1 g • Drying time of SCG + lignite briquettes increased by 9%, compared to raw SCG briquettes • Here, lignite had negative effect on drying kinetics of SCG briquette	(Fu & Chen, 2020)
Homogeneous	SCG	-	Lignin (natural binder)	Molding machine (HEBI; XL140118, China)	Compaction pressure: 0.1 MPa; Compaction temperature: 30-35 °C (room temperature)	• Briquette dimensions: diameter- 40 mm; height- 5 mm; mass: 5 ± 0.1 g • Microwave drying was proposed as an efficient technique for drying the SCG briquettes • During microwave drying, two distinct falling rates were reported, with moisture diffusivities reported as 7.89 x 10-9 to 1.75 x 10-8 m2/s during first falling rate period, and 1.09 x 10-9 to 1.81 x 10-8 m2/s during second falling rate period; and average activation energy of both falling rate period as 24.43 W/g • Besides, microwave drying provided sterilization effect, which prolonged the storage of these briquettes for a very long time without any contamination	(Fu & Chen, 2020)

Next up, fixed carbon content and volatile matter of these SCG briquettes were found in the range of 18-20% and 75-77%, respectively (Srinivasan et al., 2022; Espuelas et al., 2020), and were primarily responsible for their stable combustion. Studies reported that addition of organic binders increased the fixed carbon content and reduced the volatile matter of these briquettes slightly, which, however, favoured their steady burning and heat release rate. Illustrating this, Espuelas et al. (2020) noted rise in fixed carbon content from 18.00% to 19.01% and 19.05%, and reduction in volatile matter to 76.35% and 75.85%, while adding 5% and 10% of xanthan gum during compaction. Likewise, similar concentration of guar gum resulted in increased fixed carbon content to 19.07% and 19.20% and reduced volatile matter to 76.43% and 76.33%, respectively (Espuelas et al., 2020).

Regarding their calorific content, SCG wastes recorded their heating value between 18.8–26.9 MJ/kg (Srinivasan et al., 2022; Sermyagina et al., 2021), while SCG briquettes maintained their heating value above 22 MJ/kg (Espuelas et al., 2020; Kang et al., 2017). Amongst known biomasses, SCG wastes and their briquettes exhibit fuel characteristics equivalent to lignite (Fu & Chen, 2020); and produce very limited emission levels during emission. In spite of this, slight reductions in calorific content were noted upon adding binders; yet, their impact is least regarded accounting their contribution to developing high-quality briquettes with good durability and strength (Espuelas et al., 2020). For instance, addition of xanthum gum reduced the calorific value of these SCG briquettes from 25.399 MJ/kg to 23.503 MJ/kg (Seco et al., 2020). Supporting this, Figure 3 portrays the reduction in calorific value of these SCG briquettes, with addition of xanthum (XG) and guar (GG) gum (Espuelas et al., 2020). Besides, addition of SCG increased the overall calorific content of the compound briquettes, as their counterparts like sugarcane bagasse, rice husk, saw dust reported inferior CV (Pallavi et al., 2013; Colantoni et al., 2021).

Figure 3. Variation in calorific values of raw SCG briquettes, Xanthan and Guar gum-binded briquettes
Source: Espuelas et al. (2020)

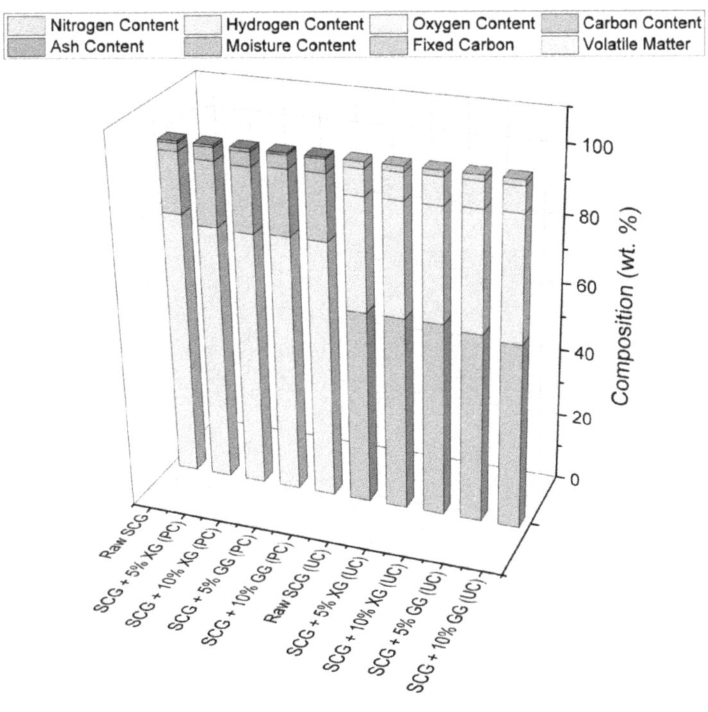

Following this, density of SCG briquettes was measured between 669 and 1078 kg/m^3; and was influenced by moisture content, particle size, compaction pressure, and type of binder (Srinivasan et al., 2022; Seco et al., 2020; Espuelas et al., 2020). In specific, pure SCG briquettes with 10% moisture content reported their density between 669 and 735 kg/m^3, while SCG briquettes with 10% organic binders reported between 672 and 819 kg/m^3 (Seco et al., 2020; Espuelas et al., 2020); and compound briquettes developed from SCG wastes with sawdust and wood shavings reported in between 899.44-1077.49 kg/m^3 (Brunerová et al., 2019). Plainly evident, high moisture content and compaction pressure yielded high-density briquettes, and briquettes with low moisture content or poor compaction crumbled during testing and handling.

Besides this, these SCG briquettes reported their durability up to 98% (Brunerová et al., 2019); and again, were dependent on moisture content, particle size, binder concentration, and compaction pressure. Here, both water and binder content helps establish van der Waals' forces between the SCG particles under compaction; thus explaining the impossibility of developing raw SCG briquettes with good durability and strength (Rajaseenivasan et al., 2016). Infact, briquettes developed by adding 10% of moisture content failed even before the testing; while, bonded briquettes held back for a majority portion of the durability test (Seco et al., 2020; Srinivasan et al., 2022). Supporting this, Espuelas et al. (2020) revealed that SCG briquettes compacted using 5% of xanthan and 30% of moisture, at 12 MPa withstood for 300 seconds with a minimal loss of mass of 3.9% (Espuelas et al., 2020). Likewise, another study recommended using 10% tapioca starch as binder, and compaction pressure of 140 MPa for developing briquettes with high durability, beyond which resulted in reduced durability (Fehse et al., 2021).

Figure 4. Proximate and elemental composition of raw SCG briquettes, Xanthan and Guar gum-binded briquettes
Source: Espuelas et al. (2020)

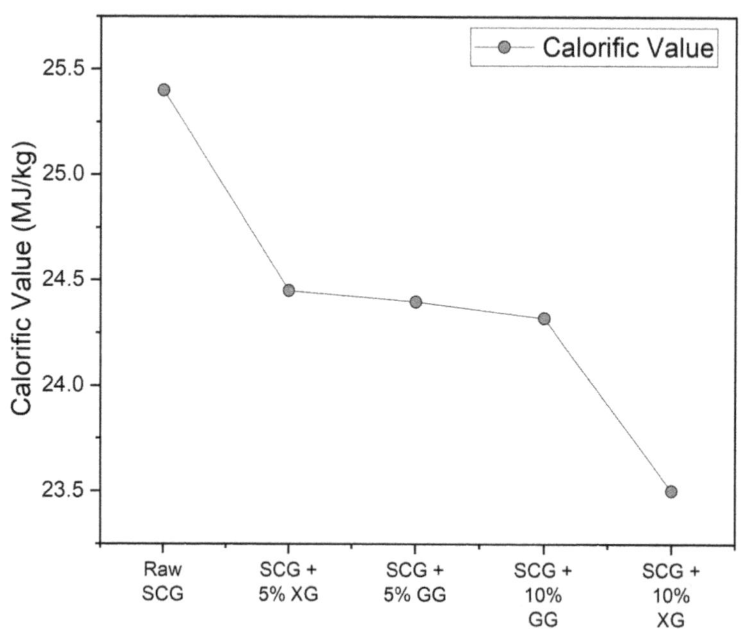

About their elemental composition, pure SCG briquettes reported their carbon, hydrogen, oxygen, and nitrogen as 57.29%, 7.52%, 33.18%, and 2.01%, respectively. Studies reported that carbonized SCG biochar exhibited increased carbon content; while SCG wastes blended with binders reported briquettes with reduced nitrogen content. Supporting this, Espuelas et al. (2020) concluded that average reduction in nitrogen content for xanthan gum and guar gum was estimated to be 13.93% and 16.42%, respectively. Figure 4 illustrates the proximate and elemental composition of briquettes developed from raw SCG, SCG with xanthan gum, and SCG with guar gum (Espuelas et al., 2020).

COMBUSTION CHARACTERISTICS OF SCG BRIQUETTES

Standardly, combustion of any briquettes takes place in three different phases, namely the pre-heating phase, degassing phase and charcoal phase, as defined by Smit and Meincken, 2012; and similar phases were noted for SCG-based briquettes, and even pellets (Smit and Meincken, 2012; Potip & Wongwuttanasatian, 2018). Recollecting from earlier facts, it is impossible to produce 100% raw SCG briquettes without any additives and binders; yet, it is important to understand the behaviour of their combustion, especially accounting their commercial and large-scale applications. However, this can be accomplished by studying the combustion characteristics of these briquettes developed from raw SCG with minimal amount of additive or binder. Responding to this, Potip & Wongwuttanasatian (2018) studied the combustion behaviour of the briquettes, produced by compacting raw SCG wastes with 10% of crude glycerol, for varied inlet air supply (Potip & Wongwuttanasatian, 2018). Here, this study varied the inlet air supply for combustion of these briquettes at 550%, 700%, 850% and 1000% theoretical air; and chose 850% theoretical air as the most ideal inlet air supply. Preliminary results suggested that these SCG briquettes required inlet air supply higher than 500% theoretical air for effective combustion, and were influenced by both SCG and glycerol molecules (Limousy et al., 2015). This was because briquettes reporting density between 400 and 700 kg/m^3 use only 150%–250% theoretical air; whereas, density of these briquettes ranged between 870-880 kg/m3, thus seeking high inlet air supply (Mitchell et al., 2016). Besides, the rate and temperature of the combustion increased proportionally with concentration of crude glycerol. And, phases of SCG-glycerol briquettes' combustion conveyed that the preheating phase lasted for about 28–35 minutes, while, degassing phase lasted for about 15-25 minutes, with highest rate of combustion noted during the preheating phase, after 9-18 minutes.

Looking into the stoichiometric combustion equation for 100g SCG (equation 1) and 1 mole of crude glycerol (equation 2) at 100% theoretical air, 200 g of each sample would require an air flow rate of 2.259 and 1.684 m3/h, respectively, and 32 minutes for their complete combustion (Potip & Wongwuttanasatian, 2018). Figure 5 depicts the calculated air flow rate required for SCG-glycerol briquettes for their varying mixing ratios and inlet air supply.

$$(4.551C + 3.295H_2 + 1.088O_2 + 0.142N_2) + 5.111(O_2 + 3.76N_2) + Ash \rightarrow 4.551CO_2 + 3.295H_2O + 19.359N_2 + Ash \tag{1}$$

$$C_3H_8O_3 + 3.5\,(O_2 + 3.76N_2) \rightarrow 3CO_2 + 4H_2O + 13.16N_2 \tag{2}$$

Furthermore, heating value of these SCG briquettes was measured as 21.14-21.55 MJ/kg, and their combustion rate and temperatures (Figure 6a and 6b), for varying inlet air supply, were reported between 0.09 and 0.18 g/s, and 463 and 533°C, respectively. Similar results were noted upon studying the combustion characteristics of briquettes developed from SCG wastes mixed with 10% of xanthan gum; which reported superior rate of combustion along with minimal amount of ash content (Seco et al., 2020). Moreover, addition of binders, especially organic gums reduced the calorific content of the SCG briquettes; nevertheless, favoured for their steady combustion (Espuelas et al., 2020). Moving on, study on combustion performance of SCG-anthracite briquettes, blended in 60:40 ratio suggested the deformation temperature, softening temperature, hemispherical temperature and flow temperature of the raw SCG wastes as 1252.85 ^{0}C (1526 K), 1352.85 ^{0}C (1626 K), 1413.85 ^{0}C (1687 K) and 1517.85 ^{0}C (1791 K), respectively, and were improved by introducing anthracite along with it

In addition, burning characteristics (Figure 7a and 7b) depicted that raw SCG briquettes exhibited higher rate of burning than anthracite at all temperatures, owing to their high volatile content.

Figure 5. Required air flow rate, as computed with respect to theoretical air supply, for SCG + glycerol briquettes
Source: Potip and Wongwuttanasatian (2018)

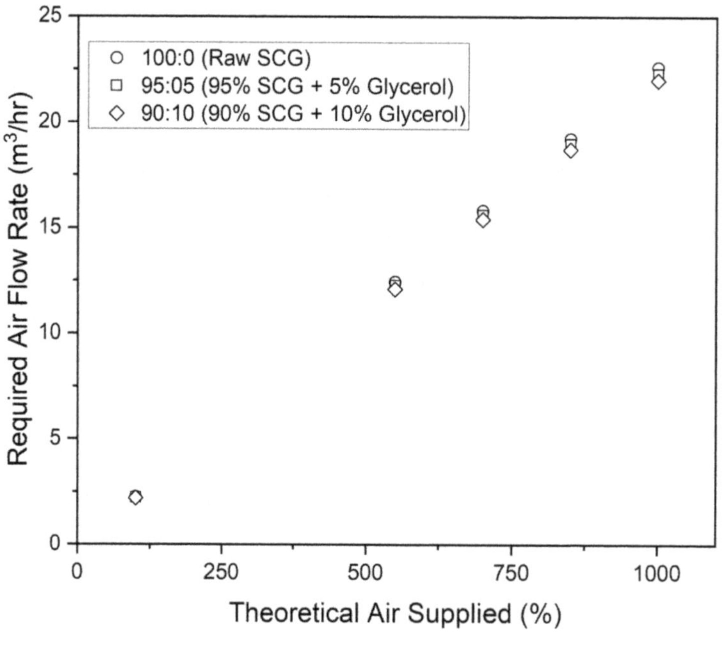

Figure 6. Average combustion rate of SCG and SCG-Glycerol briquettes
Source: Potip and Wongwuttanasatian (2018)

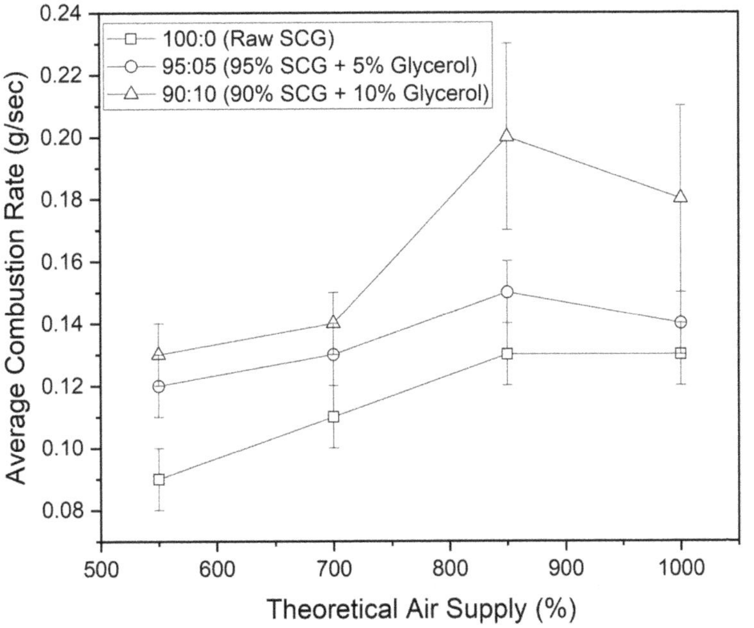

Figure 7. Average combustion temperature of SCG and SCG-Glycerol briquettes
Source: Potip and Wongwuttanasatian (2018)

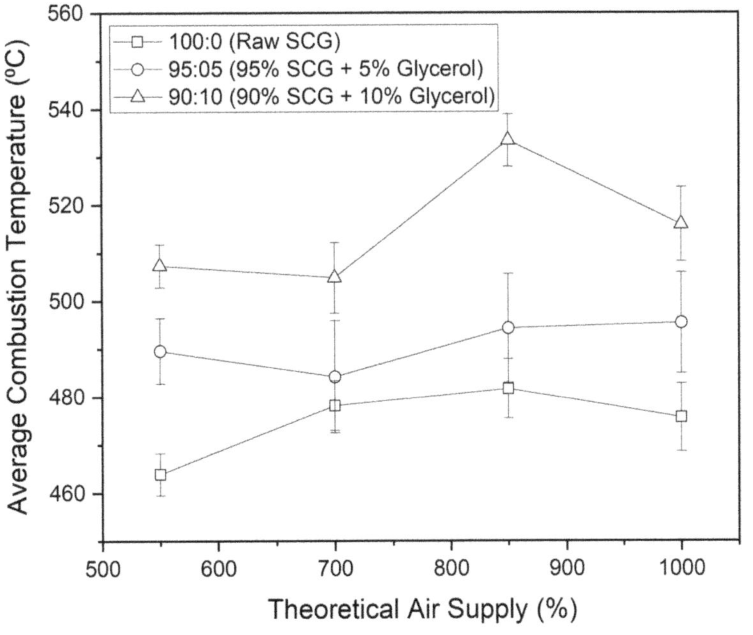

Figure 8. Average and maximum burning rates of raw SCG and SCG-Anthracite briquettes
Source: Wei et al. (2019)

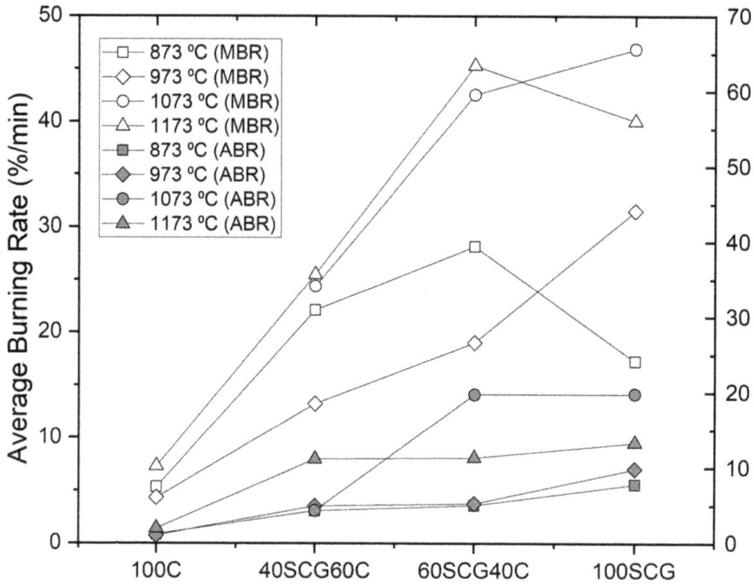

Figure 9. Burnout time of raw SCG and SCG-Anthracite briquettes
Source: Wei et al. (2019)

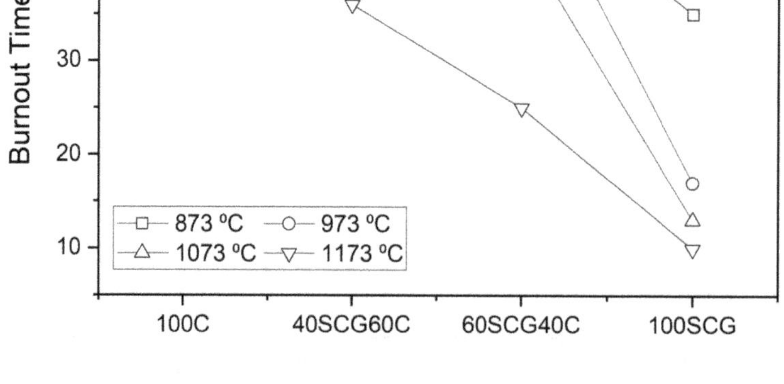

Table 6. Emission levels in flue gas emission of SCG briquettes and pellets, ranging within the permissible standards

Emission	SCG + 10% Glycerol*	SCG Char With Bottom Ash and Kaolinite Clay Briquettes	100% SCG Char Pellets	100% SCG Char Pellets	50% SCG + 50% Pine Dust Pellets	50% SCG + 50% Pine Dust Pellets
		(Kim et al., 2022)	(Limousy et al., 2013)			
CsO Emission	1262 mg/m³	8-48 ppm	1785 ppm	3069 ppm	353 ppm	606 ppm
CO_2 Emission	19%	929-2180 ppm	-	-	-	-
NO_X Emission	38 mg/m³	3.7-0.7 ppm	178 ppm	79 ppm	8 ppm	22 ppm
O_2 Emission	21%	-	-	-	-	-
HC Emission	270 mg/m³	-	1071 mg/m³	1472 mg/m³	310 mg/m³	426 mg/m³
SO_2 Emission	0	7-1.4 ppm	0	0	0	0
*(Potip and Wongwuttanasatian, 2018)						

Source: Potip and Wongwuttanasatian (2018), Kim et al. (2022), Limousy et al. (2013)

Looking into their emission characteristics, SCG briquettes reported better emission (CO_2, O_2, CO and NOx) levels than compared to commercial gas and wooden briquettes (Kang et al., 2017); and were well within the MCP, ELVs and REL-STEL (NIOSH) Standards (Potip & Wongwuttanasatian, 2018). Table 6 consolidates the concentrations of different emissions from the SCG briquettes and even pellets, measured within their permissible range as per international standards (Potip & Wongwuttanasatian, 2018; Kim et al., 2022; Limousy et al., 2013). Here, high concentration of NOx and CO2 emissions were contributed by the fuel-bound carbon and nitrogen molecules; whereas, zero sulphur emissions signified the absence of sulphur content in the SCG molecules (Potip & Wongwuttanasatian, 2018). Meanwhile, all blends of SCG-anthracite briquettes reported rise in NOx emission between 2.80 and 20.98 PPM, and reduction in SOx emissions, with increase in concentration of SCG wastes. Again, reduced ash content was noticed in raw SCG briquettes, and increased with rising anthracite concentrations. Figure 8 demonstrates the composition of different metal oxides characterised from the residual ash content of anthracite and SCG (Wei et al., 2019).

CONCLUSION

Thus, a comprehensive review comprising of summarised list of data relevant to the availability and chemistry, pre-treatments and productions, fuel and burning characteristics of SCG wastes and their briquettes have been compiled successfully. Following are the key conclusions deduced from this literature study:

1. SCG wastes are abundantly available biomass, having good potential as raw feedstock for valorising into biofuels, with least concerns over food vs. fuel conflict
2. With high carbon and hydrogen content, these wastes reported high calorific content, almost similar to that of existing fossil coal

Figure 10. Residual composition of ash (in %) from pure anthracite and raw SCG briquettes
Source: Wei et al. (2019)

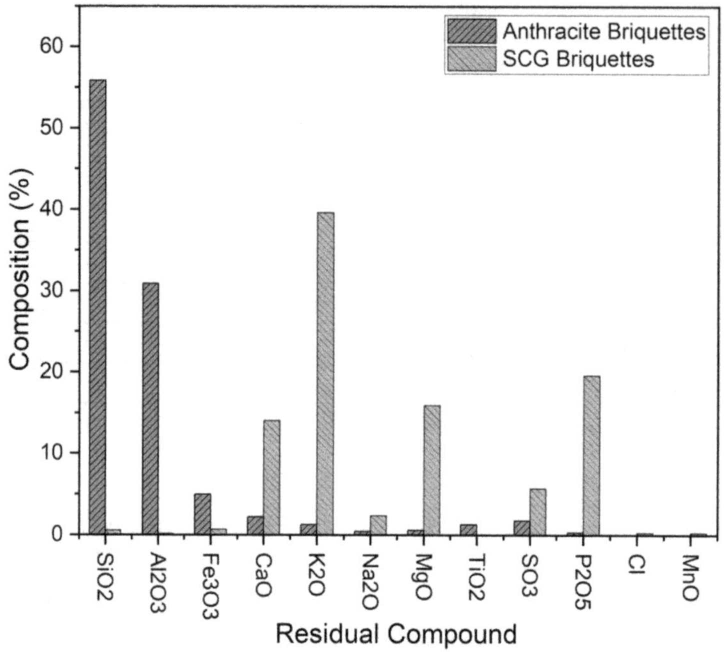

3. Owing to their oil content and irregular morphology, it is always highly impossible to develop 100% raw SCG briquettes with their fuel characteristics meeting the standards; inspite of their high lignin content.
4. Citing this, these SCG wastes were either carbonised or mixed with additive for compaction into briquettes, which required both binder and water content to establish higher level of bonding between their particles.
5. Besides, increase in binder dosage and reduction in particle size had significant effect on fuel properties of the SCG briquettes; and resulted in increased moisture and ash content, fixed carbon content, density and compressive strength.
6. About their combustion chararctertsitcs, high volatility and fixed carbon content favoured stable combustion of these briquettes; and reported superior rate of combustion along with minimal amount of ash content.
7. Summing up, it was fairly evident that these briquettes can be regarded as successful replacement or supplement for fossil coals, and is expected to cut short the fuel demand and also impact to environment.

Besides, this chapter also shares knowledge related to the entire life cycle of these SCG briquettes, helps individual in identifying the possible opportunities in terms of employment, capital investment and trading, and educates the readers about choosing these SCG briquettes for right application. As always, introducing renewable biofuels for meeting the energy demand helps in preserving the environment safely and develop sustainably.

REFERENCES

Acevedo, F., Rubilar, M., Scheuermann, E., Cancino, B., Uquiche, E., Garcés, M., . . . Shene, C. (2013, March). Bioactive compounds of spent coffee grounds, a coffee industrial residue. In *Proceedings of the III Symposium on* (*Vol. 212*). Academic Press.

Allesina, G., Pedrazzi, S., Allegretti, F., & Tartarini, P. (2017). Spent coffee grounds as heat source for coffee roasting plants: Experimental validation and case study. *Applied Thermal Engineering*, *126*, 730–736. doi:10.1016/j.applthermaleng.2017.07.202

Atabani, A. E., Ala'a, H., Kumar, G., Saratale, G. D., Aslam, M., Khan, H. A., ... Mahmoud, E. (2019). Valorization of spent coffee grounds into biofuels and value-added products: Pathway towards integrated bio-refinery. *Fuel*, *254*, 115640. doi:10.1016/j.fuel.2019.115640

Bejenari, V., Marcu, A., Ipate, A. M., Rusu, D., Tudorachi, N., Anghel, I., Şofran, I.-E., & Lisa, G. (2021). Physicochemical characterization and energy recovery of spent coffee grounds. *Journal of Materials Research and Technology*, *15*, 4437–4451. doi:10.1016/j.jmrt.2021.10.064

Belitz, H. D., Grosch, W., & Schieberle, P. (2009). Coffee, tea, cocoa. *Food Chemistry*, 938–970.

Blinová, L., & Sirotiak, M. (2019). Utilization of spent coffee grounds for removal of hazardous substances from water: A Review. *Vedecké Práce Materiálovotechnologickej Fakulty Slovenskej Technickej Univerzity v Bratislave so Sídlom v Trnave, 27*(44), 145-152.

Bok, J. P., Choi, H. S., Choi, Y. S., Park, H. C., & Kim, S. J. (2012). Fast pyrolysis of coffee grounds: Characteristics of product yields and biocrude oil quality. *Energy*, *47*(1), 17–24. doi:10.1016/j.energy.2012.06.003

Brunerová, A., Roubík, H., Brožek, M., Haryanto, A., Hasanudin, U., Iryani, D. A., & Herák, D. (2019). Valorization of bio-briquette fuel by using spent coffee ground as an external additive. *Energies*, *13*(1), 54. doi:10.3390/en13010054

Bui, H. N., Do, H. Q., Duong, H. T. G., Perng, Y. S., Dam, V. N., Nguyen, V. T., & Bui, H. M. (2022). Taguchi optimization and life cycle assessment of biodiesel production from spent ground coffee. *Environment, Development and Sustainability*, *24*(11), 12900–12916. doi:10.100710668-021-01974-0

Chen, W. H., Peng, J., & Bi, X. T. (2015). A state-of-the-art review of biomass torrefaction, densification and applications. *Renewable & Sustainable Energy Reviews*, *44*, 847–866. doi:10.1016/j.rser.2014.12.039

Chia, S. R., Ong, H. C., Chew, K. W., Show, P. L., Phang, S. M., Ling, T. C., Nagarajan, D., Lee, D.-J., & Chang, J. S. (2018). Sustainable approaches for algae utilisation in bioenergy production. *Renewable Energy*, *129*, 838–852. doi:10.1016/j.renene.2017.04.001

Cho, D. W., Cho, S. H., Song, H., & Kwon, E. E. (2015). Carbon dioxide assisted sustainability enhancement of pyrolysis of waste biomass: A case study with spent coffee ground. *Bioresource Technology*, *189*, 1–6. doi:10.1016/j.biortech.2015.04.002 PMID:25864025

Ciesielczuk, T., Karwaczyńska, U., & Sporek, M. (2015). The possibility of disposing of spent coffee ground with energy recycling. *Journal of Ecological Engineering*, *16*(4), 133–138. doi:10.12911/22998993/59361

Colantoni, A., Paris, E., Bianchini, L., Ferri, S., Marcantonio, V., Carnevale, M., Palma, A., Civitarese, V., & Gallucci, F. (2021). Spent coffee ground characterization, pelletization test and emissions assessment in the combustion process. *Scientific Reports*, *11*(1), 1–14. doi:10.103841598-021-84772-y PMID:33664428

Echeverria, M. C., & Nuti, M. (2017). Valorisation of the residues of coffee agro-industry: Perspectives and limitations. *The Open Waste Management Journal*, *10*(1), 13–22. doi:10.2174/1876400201710010013

Espuelas, S., Marcelino, S., Echeverría, A. M., Del Castillo, J. M., & Seco, A. (2020). Low energy spent coffee grounds briquetting with organic binders for biomass fuel manufacturing. *Fuel*, *278*, 118310. doi:10.1016/j.fuel.2020.118310

FAO. Food and Agriculture Organization of the United Nations. (2019). Available online: http://www.fao.org/ faostat/en/?#data

Fehse, F., Kummich, J., & Schröder, H. W. (2021). Influence of pre-treatment and variation of briquetting parameters on the mechanical refinement of spent coffee grounds. *Biomass and Bioenergy*, *152*, 106201. doi:10.1016/j.biombioe.2021.106201

Fermoso, J., & Mašek, O. (2018). Thermochemical decomposition of coffee ground residues by TG-MS: A kinetic study. *Journal of Analytical and Applied Pyrolysis*, *130*, 358–367. doi:10.1016/j.jaap.2017.12.007

Fithratullah, R. (2022). The Study of Making Biomass Briquettes from Spent Coffee Ground. *Journal of Environmental Engineering and Waste Management*, *7*(1), 34–53.

Fu, B. A., & Chen, M. Q. (2020). Microwave drying performance of spent coffee grounds briquette coupled with mineral additives. *Drying Technology*, *38*(15), 2094–2101. doi:10.1080/07373937.2019.1692862

Goh, B. H. H., Ong, H. C., Chong, C. T., Chen, W. H., Leong, K. Y., Tan, S. X., & Lee, X. J. (2020). Ultrasonic assisted oil extraction and biodiesel synthesis of Spent Coffee Ground. *Fuel*, *261*, 116121. doi:10.1016/j.fuel.2019.116121

Gómez-de la Cruz, F. J., Cruz-Peragón, F., Casanova-Pelaez, P. J., & Palomar-Carnicero, J. M. (2015). A vital stage in the large-scale production of biofuels from spent coffee grounds: The drying kinetics. *Fuel Processing Technology*, *130*, 188–196. doi:10.1016/j.fuproc.2014.10.012

Grigg, D. (2002). The worlds of tea and coffee: Patterns of consumption. *GeoJournal*, *57*(4), 283–294. doi:10.1023/B:GEJO.0000007249.91153.c3

Huang, Y., Finell, M., Larsson, S., Wang, X., Zhang, J., Wei, R., & Liu, L. (2017). Biofuel pellets made at low moisture content–Influence of water in the binding mechanism of densified biomass. *Biomass and Bioenergy*, *98*, 8–14. doi:10.1016/j.biombioe.2017.01.002

ICO. (2010). Available online: http://www.ico.org/historical/1990%20onwards/PDF/1a-total-production.pdf

ICO. (2018). *Coffee market recovers slightly from December slump*. http:// www.ico.org

ICO. (2019). *Coffee Imports by Selected Importing Countries, 2019*. http://www.ico.org/historical/1990 onwards/PDF/2b-imports.pdf

ICO. (2020). *Crop Year Production by Country, 2020*. http://www.ico.org/prices/po-production.pdf

ICO. (2021). *Coffee Market Report January 2021*. https://www.ico.org/documents/cy2020-21/cmr-0121-e.pdf

Kan, T., Strezov, V., & Evans, T. (2014). Catalytic pyrolysis of coffee grounds using NiCu-impregnated catalysts. *Energy & Fuels*, *28*(1), 228–235. doi:10.1021/ef401511u

Kang, S. B., Oh, H. Y., Kim, J. J., & Choi, K. S. (2017). Characteristics of spent coffee ground as a fuel and combustion test in a small boiler (6.5 kW). *Renewable Energy*, *113*, 1208–1214. doi:10.1016/j.renene.2017.06.092

Karmee, S. K. (2018). A spent coffee grounds based biorefinery for the production of biofuels, biopolymers, antioxidants and biocomposites. *Waste Management (New York, N.Y.)*, *72*, 240–254. doi:10.1016/j.wasman.2017.10.042 PMID:29150258

Kelkar, S., Saffron, C. M., Chai, L., Bovee, J., Stuecken, T. R., Garedew, M., Li, Z., & Kriegel, R. M. (2015). Pyrolysis of spent coffee grounds using a screw-conveyor reactor. *Fuel Processing Technology*, *137*, 170–178. doi:10.1016/j.fuproc.2015.04.006

Kim, Y., Park, T., & Hong, D. (2022). Heating and emission characteristics of briquettes developed from spent coffee grounds. *Environmental Engineering Research*, *27*(4), 210063. doi:10.4491/eer.2021.063

Kondamudi, N., Mohapatra, S. K., & Misra, M. (2008). Spent coffee grounds as a versatile source of green energy. *Journal of Agricultural and Food Chemistry*, *56*(24), 11757–11760. doi:10.1021/jf802487s PMID:19053356

Kourmentza, C., Economou, C. N., Tsafrakidou, P., & Kornaros, M. (2018). Spent coffee grounds make much more than waste: Exploring recent advances and future exploitation strategies for the valorization of an emerging food waste stream. *Journal of Cleaner Production*, *172*, 980–992. doi:10.1016/j.jclepro.2017.10.088

Lee, X. J., Lee, L. Y., Gan, S., Thangalazhy-Gopakumar, S., & Ng, H. K. (2017). Biochar potential evaluation of palm oil wastes through slow pyrolysis: Thermochemical characterization and pyrolytic kinetic studies. *Bioresource Technology*, *236*, 155–163. doi:10.1016/j.biortech.2017.03.105 PMID:28399419

Lee, X. J., Ong, H. C., Gao, W., Ok, Y. S., Chen, W. H., Goh, B. H. H., & Chong, C. T. (2021). Solid biofuel production from spent coffee ground wastes: Process optimisation, characterisation and kinetic studies. *Fuel*, *292*, 120309. doi:10.1016/j.fuel.2021.120309

Limousy, L., Jeguirim, M., Dutournié, P., Kraiem, N., Lajili, M., & Said, R. (2013). Gaseous products and particulate matter emissions of biomass residential boiler fired with spent coffee grounds pellets. *Fuel, 107*, 323–329. doi:10.1016/j.fuel.2012.10.019

Lisowski, A., Olendzki, D., Świętochowski, A., Dąbrowska, M., Mieszkalski, L., Ostrowska-Ligęza, E., Stasiak, M., Klonowski, J., & Piątek, M. (2019). Spent coffee grounds compaction process: Its effects on the strength properties of biofuel pellets. *Renewable Energy, 142*, 173–183. doi:10.1016/j.renene.2019.04.114

Luna-Lama, F., Rodríguez-Padrón, D., Puente-Santiago, A. R., Muñoz-Batista, M. J., Caballero, A., Balu, A. M., Romero, A. A., & Luque, R. (2019). Non-porous carbonaceous materials derived from coffee waste grounds as highly sustainable anodes for lithium-ion batteries. *Journal of Cleaner Production, 207*, 411–417. doi:10.1016/j.jclepro.2018.10.024

Luz, F. C., Volpe, M., Fiori, L., Manni, A., Cordiner, S., Mulone, V., & Rocco, V. (2018). Spent coffee enhanced biomethane potential via an integrated hydrothermal carbonization-anaerobic digestion process. *Bioresource Technology, 256*, 102–109. doi:10.1016/j.biortech.2018.02.021 PMID:29433044

Mata, T. M., Martins, A. A., & Caetano, N. S. (2018). Bio-refinery approach for spent coffee grounds valorization. *Bioresource Technology, 247*, 1077–1084. doi:10.1016/j.biortech.2017.09.106 PMID:28969966

Mitchell, E. J. S., Lea-Langton, A. R., Jones, J. M., Williams, A., Layden, P., & Johnson, R. (2016). The impact of fuel properties on the emissions from the combustion of biomass and other solid fuels in a fixed bed domestic stove. *Fuel Processing Technology, 142*, 115–123. doi:10.1016/j.fuproc.2015.09.031

Moreno-Ariasa, C. A., Chamarraví-Guerraa, Ó., Navarro-Árqueza, E. F., & Jiménez-Jiméneza, M. F. (2021). International overview of the use of spent coffee ground as solid fuel. A review of worldwide studies. *Gestion y Ambiente, 24*(2), 95261.

Murthy, P. S., & Naidu, M. M. (2012). Sustainable management of coffee industry by-products and value addition—A review. *Resources, Conservation and Recycling, 66*, 45–58. doi:10.1016/j.resconrec.2012.06.005

Mussatto, S. I., Dragone, G., & Roberto, I. C. (2006). Brewers' spent grain: Generation, characteristics and potential applications. *Journal of Cereal Science, 43*(1), 1–14. doi:10.1016/j.jcs.2005.06.001

Nabais, J. M. V., Nunes, P., Carrott, P. J., Carrott, M. M. L. R., García, A. M., & Díaz-Díez, M. A. (2008). Production of activated carbons from coffee endocarp by CO2 and steam activation. *Fuel Processing Technology, 89*(3), 262–268. doi:10.1016/j.fuproc.2007.11.030

Pallavi, H. V., Srikantaswamy, S., Kiran, B. M., Vyshnavi, D. R., & Ashwin, C. A. (2013). Briquetting agricultural waste as an energy source. *Journal of Environmental Science. Computer Science and Engineering & Technology, 2*(1), 160–172.

Paradkar, M. M., & Irudayaraj, J. (2002). A rapid FTIR spectroscopic method for estimation of caffeine in soft drinks and total methylxanthines in tea and coffee. *Journal of Food Science, 67*(7), 2507–2511. doi:10.1111/j.1365-2621.2002.tb08767.x

Peshev, D., Mitev, D., Peeva, L., & Peev, G. (2018). Valorization of spent coffee grounds–A new approach. *Separation and Purification Technology*, *192*, 271–277. doi:10.1016/j.seppur.2017.10.021

Potip, S., & Wongwuttanasatian, T. (2018). Combustion characteristics of spent coffee ground mixed with crude glycerol briquette fuel. *Combustion Science and Technology*, *190*(11), 2030–2043. doi:10.1080/00102202.2018.1482888

Pujol, D., Liu, C., Gominho, J., Olivella, M. À., Fiol, N., Villaescusa, I., & Pereira, H. (2013). The chemical composition of exhausted coffee waste. *Industrial Crops and Products*, *50*, 423–429. doi:10.1016/j.indcrop.2013.07.056

Rajaseenivasan, T., Srinivasan, V., Qadir, G. S. M., & Srithar, K. (2016). An investigation on the performance of sawdust briquette blending with neem powder. *Alexandria Engineering Journal*, *55*(3), 2833–2838. doi:10.1016/j.aej.2016.07.009

Rivera, X. C. S., Gallego-Schmid, A., Najdanovic-Visak, V., & Azapagic, A. (2020). Life cycle environmental sustainability of valorisation routes for spent coffee grounds: From waste to resources. *Resources, Conservation and Recycling*, *157*, 104751. doi:10.1016/j.resconrec.2020.104751

Romeiro, G. A., Salgado, E. C., Silva, R. V. S., Figueiredo, M. K., Pinto, P. A., & Damasceno, R. N. (2012). A study of pyrolysis oil from soluble coffee ground using low temperature conversion (LTC) process. *Journal of Analytical and Applied Pyrolysis*, *93*, 47–51. doi:10.1016/j.jaap.2011.09.006

Roy, M. M., & Corscadden, K. W. (2012). An experimental study of combustion and emissions of biomass briquettes in a domestic wood stove. *Applied Energy*, *99*, 206–212. doi:10.1016/j.apenergy.2012.05.003

Santos, C., Fonseca, J., Aires, A., Coutinho, J., & Trindade, H. (2017). Effect of different rates of spent coffee grounds (SCG) on composting process, gaseous emissions and quality of end-product. *Waste Management (New York, N.Y.)*, *59*, 37–47. doi:10.1016/j.wasman.2016.10.020 PMID:28340969

Seco, A., Espuelas, S., Marcelino, S., Echeverría, A. M., & Prieto, E. (2020). Characterization of biomass briquettes from spent coffee grounds and xanthan gum using low pressure and temperature. *BioEnergy Research*, *13*(1), 369–377. doi:10.100712155-019-10069-8

Sermyagina, E., Martinez, C. L. M., Nikku, M., & Vakkilainen, E. (2021). Spent coffee grounds and tea leaf residues: Characterization, evaluation of thermal reactivity and recovery of high-value compounds. *Biomass and Bioenergy*, *150*, 106141. doi:10.1016/j.biombioe.2021.106141

Setiawan, B., Himawanto, D. A., Budiana, E. P., & Widodo, P. J. (2016). Analisa thermogravimetry pada pirolisis limbah pertanian. *ETHOS: Jurnal Penelitian dan Pengabdian kepada Masyarakat*, 49-56.

Setter, C., Silva, F. T. M., Assis, M. R., Ataíde, C. H., Trugilho, P. F., & Oliveira, T. J. P. (2020). Slow pyrolysis of coffee husk briquettes: Characterization of the solid and liquid fractions. *Fuel*, *261*, 116420. doi:10.1016/j.fuel.2019.116420

Smit, H. C., & Meincken, M. (2012). Time/temperature combustion profiles of various wood-based biofuels. *Biomass and Bioenergy*, *39*, 317–323. doi:10.1016/j.biombioe.2012.01.021

Somnuk, K., Eawlex, P., & Prateepchaikul, G. (2017). Optimization of coffee oil extraction from spent coffee grounds using four solvents and prototype-scale extraction using circulation process. *Agriculture and Natural Resources (Bangkok)*, *51*(3), 181–189. doi:10.1016/j.anres.2017.01.003

Srinivasan, G. R., Mahajan, A., Seth, R., & Mahajan, R. (2022).A Critical Review on Briquettes Developed from Spent Coffee Ground Wastes. Preprints, 2022120138

Tan, S. X., Lim, S., Ong, H. C., & Pang, Y. L. (2019). State of the art review on development of ultrasound-assisted catalytic transesterification process for biodiesel production. *Fuel*, *235*, 886–907. doi:10.1016/j.fuel.2018.08.021

Tongcumpou, C., Usapein, P., & Tuntiwiwattanapun, N. (2019). Complete utilization of wet spent coffee grounds waste as a novel feedstock for antioxidant, biodiesel, and bio-char production. *Industrial Crops and Products*, *138*, 111484. doi:10.1016/j.indcrop.2019.111484

Tsai, W. T., Liu, S. C., & Hsieh, C. H. (2012). Preparation and fuel properties of biochars from the pyrolysis of exhausted coffee residue. *Journal of Analytical and Applied Pyrolysis*, *93*, 63–67. doi:10.1016/j.jaap.2011.09.010

Wei, Y., Chen, M., Li, Q., Niu, S., & Li, Y. (2019). Isothermal combustion characteristics of anthracite and spent coffee grounds briquettes. *Journal of Thermal Analysis and Calorimetry*, *136*(3), 1447–1456. doi:10.100710973-018-7790-x

Zabaniotou, A., & Kamaterou, P. (2019). Food waste valorization advocating Circular Bioeconomy-A critical review of potentialities and perspectives of spent coffee grounds biorefinery. *Journal of Cleaner Production*, *211*, 1553–1566. doi:10.1016/j.jclepro.2018.11.230

Zarrinbakhsh, N., Wang, T., Rodriguez-Uribe, A., Misra, M., & Mohanty, A. K. (2016). Characterization of wastes and coproducts from the coffee industry for composite material production. *BioResources*, *11*(3), 7637–7765. doi:10.15376/biores.11.3.7637-7653

KEY TERMS AND DEFINITIONS

Biodiesel: A fatty acid ester molecule derived from plant and animal biomass, with fuel properties similar to petro diesel

Briquettizer: A mechanical press used for compacting waste biomass into briquettes under the application of huge force/pressure, and temperature.

Durability: Ability of a material to withstand external force or damage, and signifies the strength of the material.

Lignin: A complex organic cross-linked polymer that provides rigidity and structural support to plants, and is found in their cell walls.

Micronutrients: Essential chemical compounds consumed by living organisms for their metabolic activities.

Pellet: A small cylindrical shaped particles with uniform size, developed by compacting biomass under pressure.

Press: A mechanical device used for extracting oil from them by means of external force and pressure.

Chapter 12
Bio-Briquettes From Processed Cashew Nut Shell Waste: A Review

Gokul Raghavendra Srinivasan
Steamax Envirocare Private Limited, India

Aditya Mahajan
Steamax Envirocare Private Limited, India

Rajiv Seth
Steamax Envirocare Private Limited, India

Rakesh Mahajan
Steamax Envirocare Private Limited, India

ABSTRACT

Cashew nut shells (CNS) are the primary waste produced during the processing of cashew nuts and need constant attention to handle or valorize these wastes effectively. As a result, these CNS wastes are processed into solid briquettes citing their significant calorific content, thus making them a promising renewable biofuel for combustion-based applications. In most cases, these wastes are pre-treated either through de-oiling or carbonizing prior to compaction, thus removing the harmful hydrocarbons present in them in form of CNS liquids. Presently, this chapter focuses on summarising various data related to these CNS wastes and their briquettes in terms of their availability, chemical characteristics, pre-treatment, processing technique, and fuel and combustion properties as reported in various works of literature. Moreover, all the reported results and data included in this present study were in accordance with the international testing standards and ranged in between their permissible range.

DOI: 10.4018/978-1-6684-7303-0.ch012

INTRODUCTION

To begin with, Cashew (*Anacardium occidentale*) is a pseudo-fruit bearing tropical evergreen tree, indigenous to North-eastern Brazil and South-eastern Venezuela that was later spread across the globe during late sixteenth century with an intention of conserving soil. This tree is mainly cultivated for its nuts (cashew seed) which grows at the base of its peduncle or pseudocarp (cashew apple); and are consumed as direct snacks, as an ingredient in cuisines, and are even used as the raw material for producing cashew oil and butter (Mgaya et al., 2019). Apart from this, these nuts (either in roasted or dried form) are used for making pastries, confectionaries, ice creams and chocolates (Tuates Jr et al., 2020); and contribute substantially in international and foreign exchanges market in many countries (Rabany et al., 2015; Maia et al., 2000). Presently, many countries are cultivating these exotic nuts, with United States and European Union countries being established well in this market (Mubofu & Mgaya, 2018). Infact, India is also a leading contributor in this well established market, and is known to be the largest producer, processor, exporter and second largest consumer of cashew in the world (Raghavendra Prasada, 2014; Sanger et al., 2011).

In general, these nuts are deshelled from their pseudocarp, and are then roasted or dried before consumption; hence, producing a large volume of cashew nut shell (CNS) wastes (Sawadogo et al., 2018; Sanger et al., 2011). These wastes are usually discarded into the environment; and causes acid pollution to the soil, besides possessing threats in form of fire risks citing their high organic content (Godjo et al., 2015). As a primary waste from cashew nut processing industries, they are often disposed quickly at very low prices (Singh et al., 2006; Huko et al., 2015); however, these shells hold certain calorific content, which makes them suitable for energy applications. Infact, these CNS wastes have been identified as most versatile organic renewable material available in abundance, and are observed as a raw material for numerous renewable energy resources that can replace fossil-based petroleum, value-added chemicals, and polymers (Kimutai & Kimutai, 2019; Mgaya et al., 2019; Pandiyan et al., 2020).

Though, these wastes do not pose any food competition threats, direct use or burning them produces toxic emissions and releases highly corrosive chemicals, which have adverse effect on the boilers, combustors and heating equipments. Hence, these chemicals in form of cashew nut shell liquid (CNSL) must be removed from these CNS, thus leaving behind residual shell cakes that can be processed into biofuels (Sawadogo et al., 2018). However, these CNS cakes are low density residues, and occupy large areas during their transportation and storage which increases their handling costs (Ngusale et al., 2014; Chen et al., 2015). As a result, the CNS cakes are compacted into briquettes that introduce their superior energy characteristics like low moisture content, and high density and calorific content (Stolarski et al., 2013). Apart from the de-oiled or pressed cakes, CNS wastes are also carbonized into bio-chars using pyrolysis or carbonization process, which later on are compacted into charcoal briquettes. These thermochemical treatments vaporise the harmful and volatile hydrocarbons responsible for toxic emissions from these shells, thus leaving behind significant amount of bio-char and mild traces of CNSL, depending upon the process adopted (Ifa, 2019; Sanger et al., 2011). Hence, briquetting of these CNS wastes helps in developing cheap source for clean energy compatible for both domestic and industrial applications, and is seen as an effective solution for disposing these wastes in a sustainable manner (Lubwama and Yiga, 2017; Mousa et al., 2019; Tuates Jr et al., 2020).

In general, these briquettes are priced lower than coals, and are highly renewable than compared to non-renewable fuel resources citing the absence of sulphur content in them, thus contributing very minimal to environmental pollution. Besides, these briquettes report higher calorific values than their raw feedstocks and even other solid fuels, thereby exhibiting better and uniform combustion, and high boiling efficiencies with low ash content. Interestingly, these solid biofuels have good mechanical properties including low MC, high density and durability that make them resistant to impacts and microbial activities allowing them to get transported, handled and stored easily (Kathuria & Grover, 2012; Sharma et al., 2015; Tuates Jr et al., 2020; Chungcharoen & Srisang, 2020).

With improvised energy conversion technologies like briquetting, around 36.6 tg/yr of carbon dioxide (CO_2) and 76,000 ha of woodlands can be conserved by 2030, thus saving about 8.1 million tons of woody biomass and even reducing the dependence on them (Huko et al., 2015). Infact, this practise will tend to reduce the pressure on low income countries to rely on firewood or charcoal harvested from local forests; which eventually will mitigate the partial destruction of their forest cover, thereby conserving both environment and climate change from any adverse effects (Mitchel et al., 2007). Besides, demand for energy is also rising steadily citing the modernisation amidst human population and revolutionary industrialisation; thus consuming more fossil fuel every day, only to release increased CO_2 emission, resulting in global warming and climate change (Srinivasan et al., 2020; Srinivasan & Jambulingam, 2018). Hence, identifying an ideal source of energy to meet this energy demand renewably, to avoid any further damage to the environment is highly appreciated.

Summing up, it is fairly evident that these CNS wastes are seen as high potential feedstock for making briquettes, which in turn are regarded as suitable alternative or additive for the existing solid fossil fuels. Taking this into account, this present chapter will focus on briefing enriched information about the availability of these wastes, the anatomy of these shells and nuts, the chemistry involved in these wastes, pre-treatment techniques (de-oiling and carbonizing) followed, briquetting methods practised, and the fuel and combustion characteristics of these CNS briquettes in a consolidated manner.

AVAILABILITY OF CNS WASTES

As mentioned earlier, this cashew tree (*Anacardium occidentale*) is indigenous to South America, and has been cultivated predominantly in northeast Brazil, East-west Africa, South East Asia and southern Indonesia after being introduced by Portuguese during 1500s (Dendena & Corsi, 2014). Though edible, these trees have been mainly planted for reforestation purposes since 1970s citing their adaptability to harsh terrain and climatic conditions (Tinlot, 2010). Besides reforestation, these trees were also planted for conserving soil. About their continental distribution, these trees are primarily grown in continents of Africa, South America and Asia; with India, Vietnam and western African countries being the leading producers for past 5 years (Mgaya et al., 2019). Table 1 lists the countries prominently known for cultivating these cashew nut trees.

Looking into their annual production rate (in Figure 1), the overall production of these nuts (in shells) grew from 3.99 million tonnes to 4.18 million tonnes between 2010 and 2020 (Zafeer and Bhat, 2023); however, there were significant fluctuations in the production rate ranging between 3.18 and 4.41 million tonnes (Shahbandeh, 2022). Previously, the overall annual production rate of these nuts was estimated to be only 3.21 million tonnes during 2005 (Mubofu & Mgaya, 2018). Besides, the global production rate of these exotic nuts (in kernel) was estimated to be 0.79 million tonnes during 2017-18, which grew by 32%

than compared to the production rate during 2007-08. Figure 2 illustrates the country wise distribution of production rate (in kernel) for the cultivation season 2017-18 (Mgaya et al., 2019). As far as India is concerned, overall productivity area was estimated to be in between 6,66,000-8,54,000 hectares, with an annual production rate ranging between 0.47-0.62 million tonnes/year, and with highest productivity contributed by the states of Tamil Nadu, Maharashtra, Kerala, and Karnataka (Sanger et al., 2011; Singh et al., 2006). In recent times, the net production rate increased from 0.70 million tonnes to 0.77 million tonnes during the cultivation season 2019-20 and 2021-22.

Table 1. List of countries cultivating cashew nuts (continent wise)

Continent	Countries
Asia	India, Indonesia, Malaysia, Philippines, Sri-Lanka, Thailand (Southern Thailand),Vietnam (Mgaya et al., 2019)
Africa	Benin, Côte d'Ivoire, Ghana, Guinea-Bissau, Kenya (Kwale, Kilifi, Lamu, Mombasa, Tana River, and Taita Taveta counties), Madagascar, Mozambique, Nigeria, Senegal, Tanzania (Kimutai, 2021; Mgaya et al., 2019; Mgaya et al., 2019)
South America	Brazil, Columbia, Costa Rica, El Salvador, Honduras (Mgaya et al., 2019)

Source: Mubofu and Mgaya (2018), Mgaya et al. (2019), and Sawekwiharee et al. (2015)

Figure 1. Annual production rate of cashew nuts (in million tonnes) between 2010 and 2020
Source: Shahbandeh (2022)

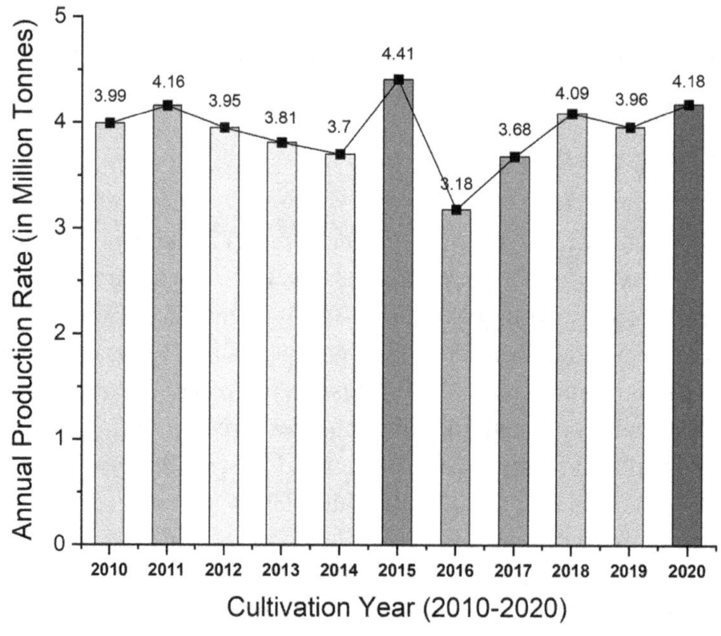

Figure 2. Country wise distribution of production rate of cashew nuts (in kernels) during 2017-2018
Source: Mgaya et al. (2019)

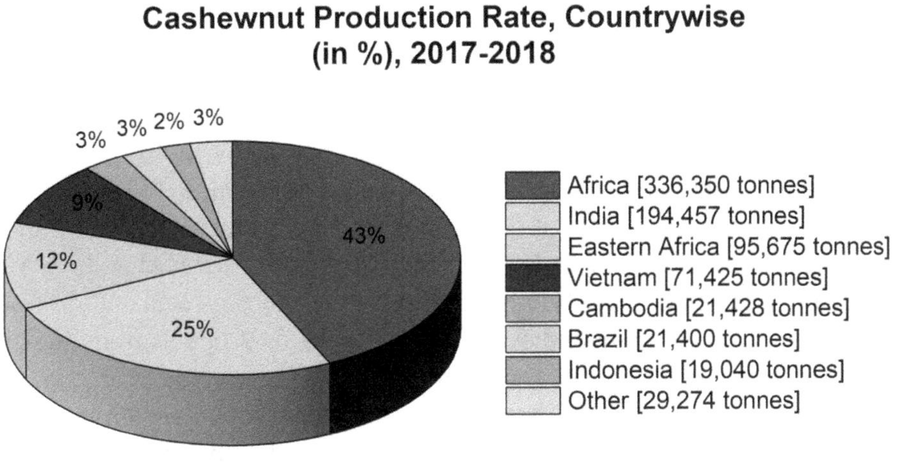

With increasing rate of production and consumption, the amount of waste generated from these nuts will also increase equally; and on average, each raw cashew nut fruit will generate wastes upto 70% of its weight, in form of shell wastes. Inevitably, this explains the surplus availability of these wastes globally, and can be found in abundance in countries primarily associated with cashew cultivation and processing. To illustrate this, districts of Kollam and Thiruvananthapuram in Kerala, India alone generates almost 75,000–80,000 tonnes of these wastes, while Konkan region at India's western coast generates as much as 20,000 tonnes of these shell wastes during de-shelling of cashew kernels (Singh et al., 2006; Sanger et al., 2011). Meanwhile, nearly 2.26-3.09 Million tonnes of shell wastes would have been generated during the cultivation year between 2010 and 2020, as per the estimated annual production rate (in shells).

Fairly evident, these data ensures sufficient availability of these shell wastes, with promising potential for supply chain in briquette production; however, geographical distribution of these trees makes their availability uneven. Yet, these challenges can be overcome by strategically planned shipping and handling.

CHEMICAL CHARACTERISTICS OF CNS WASTE

Looking into their anatomy, these cashew fruits comprise of an apple and a nut enclosed inside a shell and skin, commonly known as cashew testa; and falls off from the tree when they ripen (Sawadogo et al., 2018; Singh et al., 2006). Furthermore, these nuts consists of kidney shaped kernel constituting upto 35-45% of its raw weight; and are enveloped inside its shell that contributes to the remaining weight, and comprises of 15–35% cashew nut shell liquid, a dark reddish brown viscous liquid deposited in their honeycomb structure (Mubofu & Mgaya, 2018). In general, processing of these nuts involves roasting them, which releases the kernels from their hard shells; and results in generating large volume of these shells, CNS liquid, cashew testa and even press cakes depending upon the process undertaken and scale of the processing units (Agyemang et al., 2016).

To begin with, cashew shells have always been seen as a potential fuel for domestic and small to medium scale energy applications owing to their high fixed carbon and calorific content. About their proximate composition, waste CNS reported its volatile matter (VM), fixed carbon (FC), moisture content (MC), ash content (AC), and calorific value (CV) as 80.8%, 16.7%, 7.8%, 2.5%, and 21.58 MJ/kg, respectively; and its elemental composition as follows- carbon: 59.06%, hydrogen: 6-7%, nitrogen: 0.43% and oxygen: 29-33.52% (Sanger et al., 2011; Himbane et al., 2018). Accounting this, they are widely used in boilers replacing fossil fuel, thus ensuring high renewability and sustainability. However, direct combustion of these wastes simultaneously produces carcinogenic smoke and releases phenolic lipids, which normally corrodes the boiler components (Sawadogo et al., 2018). This can be explained by the presence of cashew nut shell liquid (CNSL), which comprises of unsaturated phenolic compounds namely anacardic acid, cardanol, cardol and mild traces of 2-methyl cardol (Srinivasan et al., 2022; Mubofu & Mgaya, 2018; Quirino et al., 2014; Gandhi et al., 2012; Rodrigues et al., 2011; Subbarao et al., 2011; Tyman, 1977).

Eventually, this can be overcome by removing these harmful shell liquids from these shells using suitable extraction techniques, which would leave behind residual press cakes. Like CNS, even de-oiled press cakes have also been identified as ideal raw material for fuelling boilers and furnaces, besides being used as primary ingredient for briquetting. This was supported by their proximate composition, which reported their VM, FC, MC, AC, and CV as follows: 77.81%, 19.14%, 9.72%, 3.05% and 20.78 MJ/kg (Sawadogo et al., 2018). Inevitably, there was a slight yet significant reduction in the CV upon de-oiling the waste shells, which raised concerns over this energy loss; however, this can be addressed by carbonizing these shells (raw or de-oiled) using pyrolysis or carbonization into bio-chars. On average, pyrolysis of these waste CNS yields about 41% of biochar, 39.3% of volatile liquids smokes, and rest as gas (Ifa et al., 2020); whereas, carbonization of these shells produce 21.1-23.8% of oil content, 18.3-21.04% of char content, and 1.27-3.34% of AC depending upon the method followed (direct or indirect carbonization) (Sanger et al., 2011). Interestingly, these CNS bio-char exhibited superior fuel characteristics and promising combustion behaviour. Supporting this, these biochar reported its proximate and elemental composition as follows: VM: 8.93-29.65%, FC: 64.55-78.02%, AC: 5.8-13.05% and CV: 27.31-27.73 MJ/kg, respectively; carbon content: 73-78.27%, oxygen content: 13-19%, hydrogen content: 1.24-5%, and nitrogen content: 1-2% (Sanger et al., 2011; Himbane et al., 2018). Lastly, liquids derived from cashew testae have good potential as leather tannin in tanning industries as they contain about 25% and 11% of tannin and non-tannin materials, respectively (Singh et al., 2006).

Fairly evident, these wastes can be valorised into numerous energy products or as raw materials for producing numerous commercial chemical products like varnishes and paints. However, waste cashew shells are of prime interest for this present study as they are primarily involved in briquette production. On the other hand, no detailed focus has been given for cashew testae liquid or shell liquid as they don't play any significant contribution in briquetting process.

PREPARATION AND PROCESSING OF CNS WASTE

In general, these shell wastes needs to undergo certain pre-treatments and pre-processing prior densifying them into briquettes; that includes roasting, de-oiling, drying, pyrolysing and carbonizing, and milling and sieving (Bellamy et al., 1984; Njenga et al., 2014). It is worth mentioning that many biomasses do not need any carbonizing and milling prior to briquetting; however, cashew shell wastes needs to undergo both of these pre-treatment methods in order to produce high quality CNS briquettes (Sawadogo et al., 2018).

Roasting of Cashew Nuts

As mentioned earlier, edible cashew kernels are enveloped inside their nut shell, and are cracked open by roasting them using heat, that make these shells turn brittle, and less corrosive, thus allowing them to be handled easily. Traditionally, these nuts are roasted directly on an open pan under continuous heating and stirring (to avoid scorching), followed by throwing them into sand to extinguish the fire, and ensure slow heat loss. On the other hand, modernised large scale cashew industries used hot oil bath for roasting these nuts at 192 °C for about 90 seconds; however, roasting temperature and time duration were dependent on the size of the nuts roasted (Mubofu & Mgaya, 2018).

De-Oiling of CNS Waste

De-oiling of these CNS ensures the removal of CNSL from them and is regarded as an essential step in the briquetting process as it helps in isolating the harmful chemicals from these wastes. In general, this de-oiling can be carried out in numerous methods; however, the most commonly preferred methods include mechanical extraction, thermal extraction, solvent extraction and pyrolysis method (Mgaya et al., 2019). Table 2 consolidates the detailed description of the most commonly practised de-oiling techniques (Mubofu and Mgaya, 2018).

Drying of CNS Waste

In case of mechanical or manual extraction, these nuts do not undergo any heat treatment, and are cracked open using external force. As a result, they carry some residual MC along with them, and needs to be removed before processing further. In relevance to that, these waste shells are dried under direct sun light or in open dry air to reduce their MC; and have been proven to the most economical way for pre-processing these wastes (FAO, 1985). On average, these shells are dried under sun between 1 and 4 days; and are very effective in bringing down their MC below 10% with minimal energy and cost (Ifa et al., 2020; Kimutai, 2021). Likewise, even de-oiled and solvent treated CNS also report upto 13% of MC, and needs to be sun dried for 3 to 5 days in order to reduce it below 7% (Tuates Jr et al., 2020). In fact, even briquettes are also dried in similar manner; and supporting this, Sawekwiharee et al. (2015) reported the average weight of the sun dried CNS briquettes, with dimension of 5×4.5 cm as 42.49 g (Sawekwiharee et al., 2015). Moreover, any foreign particles like stones and metal pieces must be removed from these shells prior drying them (Kimutai & Kimutai, 2019).

Table 2. Detailed description of the most commonly practised de-oiling techniques

De-Oiling Technique	Approach	Comment
Solvent extraction (Suitable when CNSL is rich in anacardic acid) (Mubofu and Mgaya, 2018)	Static Solvent* Extraction (Tyman et al., 1989), Soxhlet Extraction (Senthil kumar et al., 2009; Idah et al., 2014); Ultrasonic Extraction, Two-Step Extraction (Yuliana et al., 2012), Subcritical Water (SCW) Extraction (Yuliana et al., 2012), and Supercritical Carbon Dioxide (Patel et al., 2006; Smith et al., 2003)	Dissolves all aromatic lipids and leaves behind non-lipid residues
Thermal extraction (Suitable if end product needed is cardanol-rich CNSL, as anacardic acid is decarboxylated into cardanol at higher temperatures)	Drum Roasting, Hot Oil Roasting and Open Pan Roasting (Mubofu and Mgaya, 2018)	Imparts embrittlement to the shells by forcing out the oil from their cells
Mechanical extraction (Suitable when CNSL is rich in anacardic acid) (Rodrigues et al., 2011; Himabindu et al., 2015)	Screw Press Method (Tapered Compression Screw)	Exert high pressure on shells, which helps in release of CNSL from them
*Acetone, Carbon tetrachloride, Cyclohexane, Diethyl ether, Ethanol, Ethyl acetate, Hexane, Methanol, Petroleum ether, Polar solvents Toluene, Xylene (Gandhi et al., 2013)		

Source: Mubofu and Mgaya (2018)

Pyrolysis and Carbonizing of CNS Waste

In most cases, these waste shells cannot be directly compacted into briquettes, due to their minimal lignin content, and residues of CNSL which are carcinogenic during its combustion. In order to overcome these challenges, these waste shells are carbonized into biochar in an inert atmosphere through pyrolysis, which increases the concentration of carbon compounds having high activation energy (Sanger et al., 2011). Following this, waste CNS are thermally degraded in absence of air or oxygen, and produce solid bio-char, liquid smoke, and tar depending upon the reaction parameters of pyrolysis. This can be explained by the breaking down of the aromatic hydrocarbons (anacardic acid, cardol, and cardanol) into by-products with relatively smaller molecules (Basu, 2018). In general, pyrolysis of these waste CNS are carried out in between 250 and 400 °C (Huko et al., 2015; Kimutai & Kimutai, 2019; Kimutai, 2021; Gimba & Turoti, 2008); however, the most promising yields are reported at an optimum temperature of 350 ^{0}C producing about 41% of biochar, 39.3% of liquid smoke and 19.7% of gas upon pyrolysing CNS (Ifa et al., 2020; Sawadogo et al., 2018). In most cases, pyrolysis of CNS at lab scale using muffle furnace last long upto 60 minutes; but prolongs beyond that during commercial processing (Ifa, 2019). Other works include burning of CNS (1 kg) in muffle furnace at 600 ^{0}C for 60 minutes; burning in open air for 75 minutes, followed by closed combustion in absence of oxygen for another 30 minutes (Sawekwiharee et al., 2015), burning in developed prototype kiln (capacity of 5 kg) using direct carbonization at 452.2 ^{0}C, and indirect carbonization at 458.8 ^{0}C (Sanger et al., 2011). Besides, these shells can also be carbonized in customised metal box, maintained at 250 °C for 3 hours (Sen et al., 2016; Sanger et al., 2011); and a drum type carbonizer operated at 250-350 °C (Tuates Jr et al., 2020). Eventually, carbonizing these wastes into biochar enhances their overall combustion efficiency and eases the briquetting process (Sanger et al., 2011).

Milling and Sieving of CNS Char

Post carbonizing, these CNS char were milled into fine powder until the particle size reached below 2 mm; and in general, are milled using traditional tools like manual grinders, electrical blenders, and even milling machines (Ifa et al., 2020; Kimutai & Kimutai, 2019). In practise, these pulverised CNS char are differentiated either based on their particle size or on the mesh size of the sieve used. In first case, these CNS char are pulverised into fine powder having particle sizes of 0.5mm, 1.0mm and 2.0mm (Kimutai & Kimutai, 2019); whereas in second case, they are sieved in 70, 140, and 200 mesh sizes (Ifa, 2019). Supporting this, Sawadogo et al. (2018) used char particles of carbonised CNS press cake with grain size of 0.5 mm; and developed briquettes with high durability and superior fuel efficiency (Sawadogo et al., 2018). However, any residual particles bigger than 2.0 mm or that could not pass through any of these sieves were pulverised and milled again (Ifa et al., 2020; Ifa, 2019; Kimutai, 2021; Kimutai & Kimutai, 2019). Apart from this, different set of sieve mesh sizes namely U.S. sieve No. 3/8 (particle size of 9.51 mm), No. 4 (particle size of 4.76 mm), and No. 8 (particle size of 2.38 mm) (Chungcharoen & Srisang, 2020); and particle sizes namely 3mm, 5mm, 7mm and 10mm were followed in differentiating the pulverised CNS char (Zhang et al., 2012). Moreover, Sawekwiharee et al. (2015) estimated the overall time duration of the milling process as 1 hour, as it was fairly enough to produce fine powdered char particles (Sawekwiharee et al., 2015).

BRIQUETTING OF CNS BRIQUETTES

To begin with, CNS is valorised by compacting them into briquettes; Recollecting from earlier facts, raw CNS contain harmful hydrocarbons that are carcinogenic in nature, hence must be processed by removing their shell liquids before densification. Thus, only de-oiled or carbonised CNS wastes are used in the production of briquettes. In general, an ideal briquette requires a potential biomass as feedstock having appropriate particle size and MC, a binder with high starch content and good adhesive property that does not affect the overall renewability, an additive with similar fuel characteristics of feedstock used, and appropriate compaction pressure and temperature, to induce proper binding among these particles. In common practise, these compaction are performed using different techniques that includes direct compacting or ramming, and are carried out using numerous tools and machinery like screw press, hydraulic and pneumatic press, manual compaction tool, briquetting machines. In fact, quality of these briquettes relies on compaction technique, followed by dosage of the binder; and varies with the feedstock used (Chen et al., 2015; Roy & Corscadden, 2012; Kaliyan & Morey, 2009). Following this, numerous researchers have carried out detailed study on compacting these processed CNS wastes into CNS briquettes; and works related to their briquetting have been tabulated in Table 3, comprising details related to briquette type, additive and binder added, compaction technique and conditions, and key highlights in their work.

Table 3. Tabulated summary of briquette type, additive and binder added, compaction technique and parameters, and key highlights of different studies reported in the literature related to CNS briquettes

Briquette Type	Feedstock	Additive	Binder	Briquetting Technique	Briquetting Conditions	Remarks	Reference
Homogeneous	CNS biochar	50% ethanol, and 50% warm water	8-12% of tapioca flour	Briquetting press (Krisbow hydraulic press floor type 10T)	Compaction pressure- 29.4 MPa (300 kgf/cm^2); Dwell time- 5 minutes; Mesh size: 200 mesh	• Briquette shape: Hollow cylinder • Mesh sizes used: 70,140, 200 mesh • Optimum mesh size: 200 mesh • Briquettes dried in the hot air oven at 50 °C for 4–6 hours • binder concentration: 8-12% of slurry (flour + ethanol + warm water) mix	(Ifa et al., 2020; Ifa, 2019)
Composite/ compound	CNS Char	Mango seed shells (MSS)	50% of banana peel paste	Briquetting press	Compaction pressure- 5 MPa; Compaction temperature: 30-35 °C (room temperature)	• CNS wastes sun dried separately to 5% MC • carbonization at 400 °C for 5 minutes in the muffle furnace • Particle sizes: 3mm, 5mm, 7mm and 10mm • 50g mix of CNS char and MSS • Briquette dimensions: diameter - 50 mm, Length-100 mm	(Huko et al., 2015)
Composite/ compound	65 wt.% of CNS Char	25 wt.% of Areca nut shells (ANS)	10 wt.% of cassava flour	Electrically operated Screw type press machine	Compressed screw speed: 70-90 RPM	• Semi-carbonized at 300 °C • CNS: MC-8%, ANS:MC-12% • CNS particle size: U.S. sieve No. 3/8 (particle size of 9.51 mm), No. 4 (particle size of 4.76 mm), and No. 8 (particle size of 2.38 mm). • CNS: ANS: Flour (% by weight)-65:25:10 (Sample A), 45:45:10 (Sample B), 25:65:10 (Sample C), 58:22:20 (Sample D), 40:40:20 (Sample E), and 22:58:20 (Sample F). • Briquette dimensions: diameter - 50 mm, Length-100 mm, weight - 210 g • Average production rate: 245 pieces/h (52 kg/h) • Influence of compression speed on production rate was highly significant, especially at 90 RPM	(Chungcharoen and Srisang, 2020)

continues on following page

Table 3. Continued

Briquette Type	Feedstock	Additive	Binder	Briquetting Technique	Briquetting Conditions	Remarks	Reference
Homogeneous	CNS Char	-	10-30% of Cassava starch	Hydraulic press	Compaction parameters varied for optimization purposes; Particle size: 0.5 mm; Dwell time: 120 sec	• CNS wastes were sun dried under the sunlight for 2–3 days • Carbonization at 250°C for 3 hours • Particle sizes: 0.5 mm, 1.0 mm and 2.0 mm • Compaction pressures: 100 Kg/cm^2, 200 Kg/cm^2 and 300 Kg/cm^2 • Binder dosage: 10%, 20% and 30% • Dwelling time: 0 seconds, 60 seconds and 120 seconds • Drying of briquettes: stored at room temperature for 6 to 7 days	(Kimutai and Kimutai, 2019)
Homogeneous	55% of CNS press cake char	35% of water	10% of Cassava starch	Engine driven conic screw press	Production output: 50 kg/h; compression ratio: 1/45	• CNS press cake was carbonized at 350 °C • Particle size: 0.5 mm • Briquette dimensions: diameter - 50 mm, Length-100 mm • Water content: 25%, 30% and 35% of the total mass (charcoal + starch + water) • Binder concentration: 5%, 10%, 15%, 20% and 25% of the total mass • Drying of briquettes: dried for two weeks at ambient atmosphere (30-31°C), • Final MC: 8% (after 2 weeks) • Cassava starch chosen for its physical-chemical properties, low price, easy availability	(Sawadogo et al., 2018)
Homogeneous	CNS Char	-	Cassava starch	Hydraulic press	Compaction parameters varied for optimization purposes; Optimum particle size: 0.5 mm; Optimum compaction pressure: 29.42 MPa; Optimum dwell time: 120 second (based on highest relaxed density)	• CNS wastes: sun dried for 2–4 days • Carbonization at 250°C for 3 hours • Particle size: 0.5 mm, 1 mm, and 2 mm • Compaction pressure: 9.81 MPa, 19.6 MPa, and 29.42 MPa • Binder dosage: 10%, 20% and 30% • Dwell time: 0 min, 1 minutes, and 2 minutes • Drying of briquettes: stored at room temperature	(Kimutai, 2021)

continues on following page

Table 3. Continued

Briquette Type	Feedstock	Additive	Binder	Briquetting Technique	Briquetting Conditions	Remarks	Reference
Homogeneous	De-oiled CNS; De-oiled and hexane treated CNS; Carbonized de-oiled CNS char (Sample size: 100 g)	-	4-10% of Starch binder [Screw type: 4%; Piston type: 5-10%]	Piston-type briquetting machine; and Screw-type briquetting machine	-	• De-oiled, De-oiled hexane treated, and Carbonized de-oiled char samples: sun dried for 3 to 5 days (MC: 6-8%) • Particle size: reduced to less than 2 mm using commercially available hammer mill. • Binding concentration (piston-type): 0%, 5% and 10% • Binding concentration (screw-type): 0%, 2% and 4% • Briquette dimensions: inner diameter- 16 mm, outer diameter- 50 mm, Length- 50 mm	(Tuates Jr. et al., 2020)
Homogeneous	1 kg of CNS Char	1L of water	200 g of cassava starch	Briquetting press	-	• Method 1: carbonization of CNS at 600°C for 60 minutes, then cooled at 5°C/min to room temperature • Method 2: Burning of CNS in open air, using natural firewood for 90 minutes and sealed to drive out oxygen for 30 minutes • Milled for 60 minutes, resulting in a fine powder • Drying of briquettes: sun dried for 3 days • Briquette dimensions: diameter - 45 mm, Length-50 mm, weight - 42.49 g	(Sawekwiharee et al., 2015)
Homogeneous	CNS Char	-	Clay and Arabic gum	Manual Briquetting (Hand held hammer)	-	• Carbonization at 250°C for 3 hours • Particle size: 1 mm • Briquette dimensions: diameter-53 mm	(Himbane et al., 2018)

FUEL AND BURNING CHARACTERISTICS OF CNS BRIQUETTES

All produced CNS briquettes needs to be evaluated for their fuel characteristics, to ensure their suitability for commercial applications. Generally, these briquettes are assessed following the international testing standards, as par with ASTM standards (MC- ASTM D3173; AC- ASTM D3174; VM- ASTM D3175; FC- by difference; CV- ASTM D5865; Compressive strength (CS): ASTM D 2166-85; density: ASAE S269.4)(Huko et al., 2015). Table 4 summarises and compares the elemental composition of CNS wastes, CNS bio-char and CNS bio-char briquettes (Himbane et al., 2018).

Firstly, MC of CNS briquettes, post drying were estimated to be less than 10% (Ifa et al., 2020); while, MC of fresh CNS briquettes reached upto 72% (Sawadogo et al., 2018; Sawekwiharee et al., 2015). Here, high water content in fresh briquettes was explained by their necessity during briquetting of CNS wastes, in order to avoid formation of cracks and resist compression (Sawadogo et al., 2018). However, CNS briquettes, upon drying reported very low MC than compared to other biomass briquettes (Ifa et al., 2020), and were deeply influence by the method of drying and particle size. Explaining this, many studies suggested drying these briquettes at 105 ^{0}C for 24 hours, either using hot air oven or dryers (Ifa et al., 2020; Ifa, 2019; Sawadogo et al., 2018; Sawekwiharee et al., 2015); yet, certain studies recommended sun drying during summer (@ 34-36 ^{0}C for 7 days) as an effective, simple and low-cost method, especially during large scale production (Chungcharoen & Srisang, 2020). Next up, increase in particle size reduced the MC; and according, Huko et al. (2015) noted reduction in MC from 7.56% to 6.88%, as particle size increased from 3mm to 11mm (Huko et al., 2015). On the other hand, CNS char briquettes reported their MC in between 5-7%; and presence of MC in these char particles were explained by the difficulty in vaporisation of absorbed water, post carbonization, that are trapped in their small pores. Eventually, absorption of these wasters can be rectified by adding binders (Kimutai & Kimutai, 2019; Ifa, 2019; Tuates Jr. et al., 2020).

Table 4. Proximate and elemental composition of CNS wastes, CNS bio-char and CNS bio-char briquettes

Sample	Proximate Composition (Dry Basis)(in %)			
	Fixed Carbon	**Volatile Matter**	**Moisture Content**	**Ash Content**
CNS Wastes	16.7	80.8	7.8	2.5
CNS Char	78.02	8.93	10.6	13.05
CNS Briquettes	62.89	19.13	5.43	17.98
	Elemental Composition (Dry Basis)(in %)			
	Carbon Content	**Oxygen Content**	**Hydrogen Content**	**Nitrogen Content**
CNS Wastes	59.06	33.52	6.99	0.43
CNS Char	78.27	19.35	1.24	1.14
CNS Briquettes	71.76	24.57	2.72	0.95
	Calorific Value (Dry Basis) (MJ/kg)			
CNS Wastes	21.58			
CNS Char	27.31			
CNS Briquettes	26.52			

Source: Himbane et al. (2018)

Moving on, AC of these CNS briquettes defines the amount of burnt resudes left behind after their combustion (Huko et al., 2015; Tamilvanan, 2013; Abdillahi et al., 2017; Ujjinappa & Sreepathi, 2018; Ifa et al., 2020; Sawekwiharee et al., 2015; Glushankova et al., 2018); and in general, ranges between 2-5% citing the presence of numerous minerals, which don't oxidise with oxygen and remain in their solid form even after burning (Olanders & Steenari, 1995; Zhu et al., 2011; Tamilvanan, 2013; Ujjinappa & Krishnamurthy, 2018; Tuates Jr. et al., 2020). Supporting this, only 4.3% of AC were noted upon combusting these CNS briquettes at 750°C at 6 hours; and were recorded for the particle size of 200 mesh, thus claiming that AC reduces with increasing particle size (Sawekwiharee et al., 2015; Huko et al., 2015; Chungcharoen & Srisang, 2020). Moreover, CNS char briquettes reported higher AC than its raw counterpart; but, still remained below 8% as prescribed by fuel standards (SNI 016235-2000) meant for bio-char briquettes (Ifa, 2019). Most importantly, these AC have adverse effect on both environment and CV of these briquettes, and on environment, through harmful emissions (Sawadogo et al., 2018; Akowuah et al., 2012; Onchieku et al., 2012; Onukak et al., 2017).

Following this, volatile and FC content in these CNS briquettes was reported in between 65 to 80%, and 17 to 20% (Sanger et al., 2011; Sawadogo et al., 2018; Chungcharoen & Srisang, 2020); and varied inversely for CNS bio-char briquettes, thus reporting only 25-30% of volatile content, 70-80% of FC content (Ifa, 2019; Sawekwiharee et al., 2015; Sanger et al., 2011; Tamilvanan, 2013; Ujjinappa & Krishnamurthy, 2018). Illustrating this, the FC and volatile content in the raw CNS briquettes were measured as 19.14% and 77.81% (Sawadogo et al., 2018); however, the CNS wastes carbonised at 350 ^{0}C, produced bio-char with FC and volatile content as 71.19-72.62%, and 17.16-19.56% (Ifa, 2019). Here, high volatile content in these raw briquettes tends to reduce the utilization of their calorific content effectively, thus reducing their thermal efficiency and making them slightly uneconomical (Kimutai & Kimutai, 2019; Karunanithy et al., 2012); yet, this can be rectified by carbonizing these wastes prior briquetting, which eliminates their volatile content significantly, thus enhancing their FC content (Sanger et al., 2011; Sotannde et al, 2010; Zapusek et al, 2003). Besides, addition of binders simultaneously increased the FC content, and reduced the volatile content of their resultant briquettes, thus favouring their complete combustion (Kimutai & Kimutai, 2019). Moreover, reduction in the particle size of CNS and their char particles increases their FC content, but again decreases their volatile content (Chungcharoen & Srisang, 2020).

Next up, calorific content of this CNS briquetes estimate their quantity required for producing the desired amount of energy (Sotannde et al, 2010); and generally, is reported in between 18.81-30.5 MJ/kg (Ifa et al., 2020; Ifa, 2019; Kimutai & Kimutai, 2019; Sanger et al., 2011; Chungcharoen & Srisang, 2020). The calorific content of these briquettes are inversely proportional to their volatile content (Suhartini et al., 2011), hence, briquettes developed from carbonised CNS wastes exhibited significant CV owing to their high FC content (Ismaila et al, 2013 ; Sanger et al., 2011). Explaining this, briquettes developed from CNS bio-chars recorded their CV in between 28.3 and 29.49 MJ/kg (Kimutai & Kimutai, 2019; Ifa, 2019). Besides, studies have claimed that addition of binders increased the CV of these briquettes; and relating this, rise in CV from 28.3 MJ/kg to 30.5 MJ/kg upon introducing 30% of cassava starch as binder, with CNS char (Kimutai & Kimutai, 2019). Moreover, no significant variations in CVs were noticed in these briquettes, with respect to change in their particle size (Huko et al., 2015; Ifa, 2019). For instance, CVs of briquettes from CNS bio-char with particle size of 70 and 140 mesh had minimal difference of 0.04 MJ/kg (Sutrisno et al., 2017).

Looking into their physical characteristics, these CNS briquettes report their density in range of 765-910 kg/m^3 (Tuates Jr. et al., 2020; Sawadogo et al., 2018); and is influenced by compaction pressure and particle size, and plays a crucial role in their combustion (Himbane et al., 2018). To begin with, high compaction pressure established strong inter-particle attraction forces amongst the CNS particles; and maximum pressure was achieved using piston type compaction (Oladeji, 2015). To illustrate this, briquettes developed from mechanical pressed and hexane treated de-oiled CNS reported their as 0.87 g/ cm^3, for both cases, upon using piston type briquetting machine for compaction along with 10% binding agent; meanwhile, briquettes compacted using screw type briquetting machines measured only 0.83 g/ cm^3, for both samples (Tuates Jr. et al., 2020). With respect to particle size, density of these briquettes reduced with increasing size of CNS particles; and accordingly, Huko et al. (2015) noted significant drop in density of CNS briquettes from 729.08 kg/m^3 to 492.41 kg/m^3, while particle size increased from 3 mm to 11 mm (Huko et al., 2015). This can be explained by the reduced inter- and intra- particle distances between the CNS particles due to their tight packing (Kimutai, 2021). And, addition of binders increased the density of these briquettes, as they induced adhesion and close packing of the CNS particles, post compaction; and was explained when briquettes with 30% binder recorded higher density than briquettes developed using 10% binder content (Kimutai & Kimutai, 2019).

Compressive strength is regarded as an indicator for briquettes' durability (Wilaipon, 2007), and ranges in between 0.04-29.42 MPa (Sawekwiharee et al., 2015; Kimutai & Kimutai, 2019; Sawadogo et al., 2018; Tuates Jr. et al., 2020), depending on their briquette dimension. In general, CS of these briquettes depends directly on compaction pressure and binder concentration, and inversely with their particle size (Kimutai & Kimutai, 2019). Explaining this, high compaction pressure induced effective compaction of the CNS particles into briquettes; and responding to this, Kimutai & Kimutai (2019) noted highest and lowest CS of 7.84 MPa (80 kg/cm^2) and 2.94 MPa (30 kg/cm^2) for a compaction pressure of 300 kg/cm^2 and 100kg/cm^2, respectively (Kimutai & Kimutai, 2019). Furthermore, de-oiled CNS briquettes, with 10% of binding agent developed using screw type briquetting machine reported high CS of 101.82 kPa opposed to screw type briquetting machine reporting only 43.97 kPa (Tuates Jr. et al., 2020). Besides, reduced particle size of these CNS particles reduced the formation of voids between them, thus increasing their CS significantly; however, larger particles created voids between them, thereby affecting their strength adversely (Huko et al., 2015). And, increased concentration of binders helps in filling these voids, apart from bridging the CNS particles; thus, producing high quality briquettes (Rahman et al., 2003). Infact, numerous studies have suggested that increased use of binders produced CNS briquettes with high strength and durability (Ifa, 2019; Kimutai & Kimutai, 2019; Tuates Jr. et al., 2020). Eventually, these facts were supported upon noticing a rise in CS from 40 to 60 kg/cm^2, with increase in binder concentration from 10% to 30%, in CNS char briquettes compacted at 200 kg/ cm^2 (Kimutai & Kimutai, 2019).

For better understanding, variation in fuel characteristics of the waste CNS char briquettes with respect to their particle size and binder concentration have been illustrated in Figure 3 and Figure 4, respectively. Moreover, comparative study on these CNS wastes along with other wastes biomass like Peanut Shell (PS) and Millet Stalk (MS), focused on their proximate and ultimate composition, and physical characteristics signified the superior fuel qualities of the cashew shell based briquettes. Figure 5 portrays the comparison in the proximate and ultimate composition of CNS, PS and MS, their bio-chars and char briquettes; wherein figure 6 depicts the comparison in the physical characteristics of CNS, PS and MS char briquettes.

About their burning characteristics (from Table 5), CNS briquettes reported their heat utilization efficiency and average burning rate as 18.01% and 11.90 g/min, respectively; with a weight loss rate of 61.9-62.5%, thus exhibiting burning characteristics equivalent to firewood or fuel (Sawekwiharee et al., 2015). On the other hand, water boiling tests on these briquettes, as tabulated in Table 6, showcased combustion performance similar to that of wood charcoal (Sawadogo et al., 2018). Moreover, CNS briquettes with adequate binder concentration ignited faster and combusted rapidly than compared to raw CNS briquettes (Kimutai & Kimutai, 2019). Relating these facts, briquettes developed using CNS bio-char with 35% of MC and 10% of binder dosage exhibited favourable burning characteristics; and following were the parameters deduced from their water boiling test: Boiling duration- 47 mins, specific fuel consumption- 131.09 KJ/L/min, thermal efficiency- 33.9%, firepower- 1.34 kW (Sawadogo et al., 2018; Angulo-Mosquera et al., 2021). Lastly, studies reported that CNS briquettes underwent burned well without any cracks; and produced similar results of wood or char biomass, in terms of their combustion and emission (fumes and soot) (Sawekwiharee et al., 2015).

Figure 3. Variation in fuel characteristics of CNS char briquettes for different particle size
Source: Huko et al. (2015)

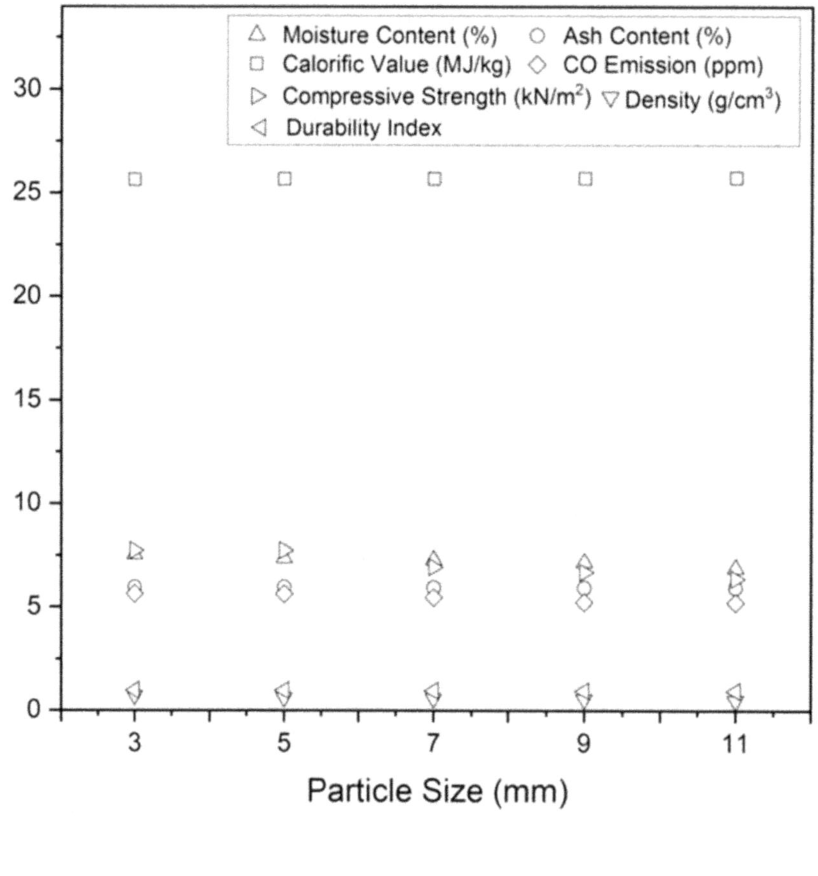

Figure 4. Variation in fuel characteristics of CNS char briquettes for different binder concentration (VM- Volatile Matter, FC-Fixed Carbon, CV- Calorific Value, CS- Compressive Strength, IR- Impact Resisance,30- 30% water content added,35- 35% water content added)
Source: (Sawadogo et al., 2018)

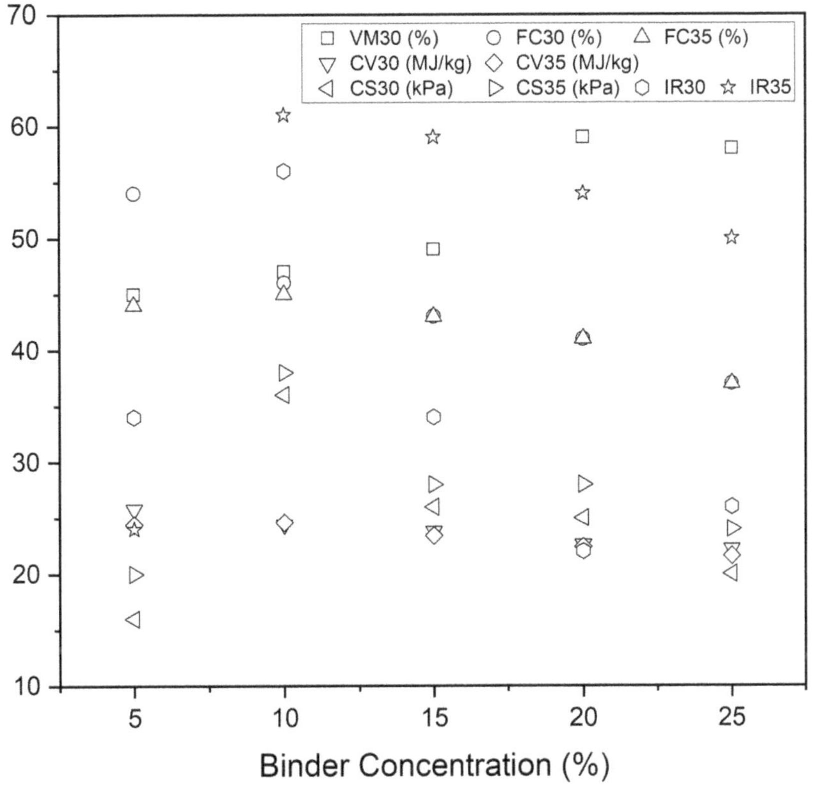

ECONOMICS OF CNS BRIQUETTES

Besides fuel characteristics, economic features also decide the choice of these briquettes, which makes these studies more essential, especially for attracting consumers having high concerns over renewable energy, and less financially supported and remote backgrounds. With this in mind, Ifa et al., 2020 studied the techno-economic aspects of bio-briquettes production from CNS wastes, sourced from Muna Regency, Southeast Sulawesi, Indonesia; for a production capacity of 2,000 tons/year. This study recorded the overall production and selling cost of these CNS briquettes as USD 842,304 and USD 1,052,878, respectively, with a net profit of USD 147,402. Besides, the investment rate, break-even point (BEP) and pay-out time (PoT) of the overall production process were estimated as 23.55%, 50.09% and 3.42 years, respectively. Moreover, replacement of traditional LPG fuel with these CNS briquettes for heating applications in the households of Muna Regency community cut short their annual fuel expense by 37.00% (Ifa et al., 2020). Previously, Singh et al., 2006 have claimed these CNS wastes as cheaply available raw material for briquette production, and estimated their average selling price ranging in between INR 300 and 400

Figure 5. Proximate and elemental composition of different biomasses, and their biochar and biochar briquettes (PSC-peanut shell char, CSC- cashew shell char, MSC-millet stalk char, PSCB-peanut shell char briquettes, CSCB- cashew shell char briquettes, MSCB-millet stalk char briquettes)
Source: (Himbane et al., 2018)

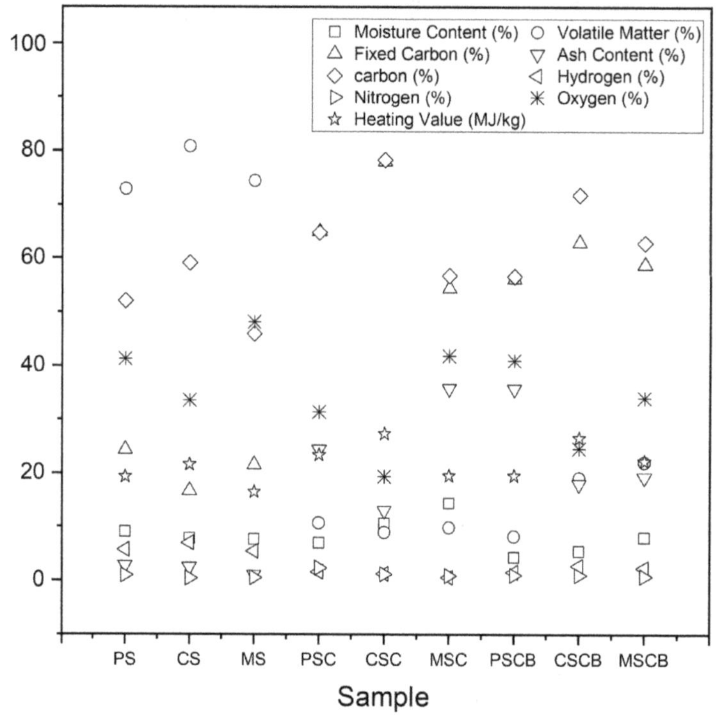

per ton, if purchased directly from the factory premises (Singh et al., 2006). In short, it is fairly evident that the overall cost of these briquettes is primarily contributed by the cost involved during production, followed by their shipping and handling charges, and marginally, by direct expenses of raw materials.

CONCLUSION

Hence, the reported set of data related to availability, chemical characteristics, pre-treatment and processing technique, fuel and combustion properties of the CNS briquettes have been successfully complied into a comprehensive review; and following were the main findings drawn from this study:

1. CNS wastes have been proven to be an ideal choice as the raw material for producing briquettes, available abundantly with zero food vs. fuel conflict issues.
2. Citing their high aromatic hydrocarbons, it is highly recommended to pre-treat these wastes by de-oiling them, followed by carbonising for producing high quality briquettes with minimal emission levels.

Figure 6. Densities and compressive strength of char briquettes developed from different biomasses
Source: Himbane et al. (2018)

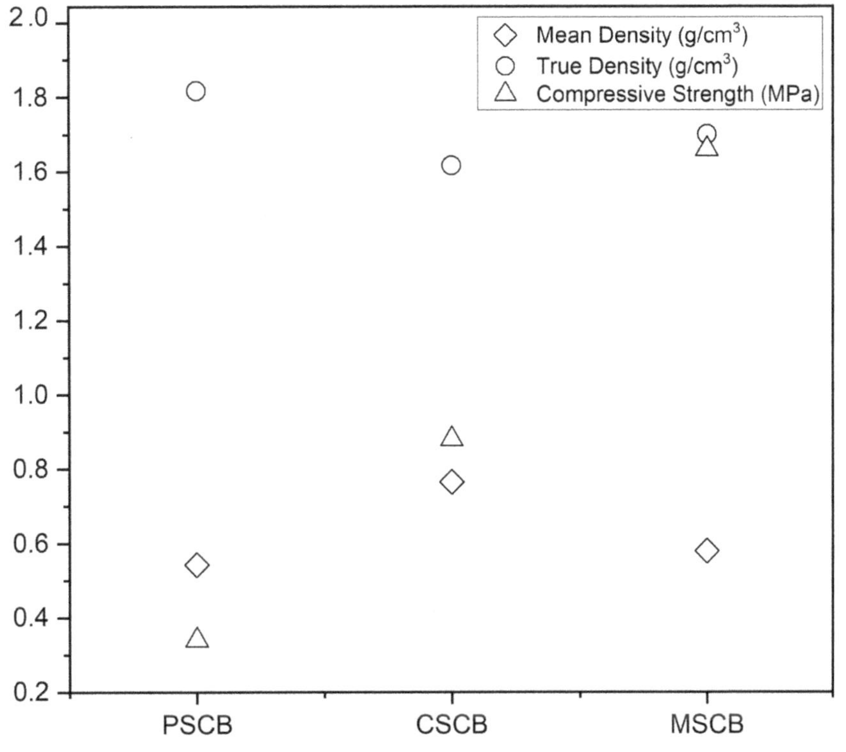

3. With respect to densification, these processed wastes can be briquetted directly with or without additives, and requires only very minimal binder dosage in case of de-oiled CNS wastes.
4. Looking into their fuel properties, increase in binder dosage and reduction in particle size enhanced the overall properties of these CNS briquettes.
5. Explaining this, increase in binder dosage and reduction in particle size increased their MC, AC, FC content, density and CS. Worth mentioning, calorific content of these briquettes remained same inspite of increase in their particle size.
6. Adequate amount of VM and FC content ensured the stable combustion of these CNS briquettes; and also produced burning characteristics similar to that of wood biomass.

Summing up, it is fairly evident that these briquettes serves well as a feasible supplement or alternative for fossil coal; and is seen as the most effective in handling the increasing volume of the CNS wastes. Moreover, this chapter sheds light on enriching the knowledge related to the life cycle of these CNS briquettes, guide the consumers in choosing their right application and handling techniques, and hinting individuals in identifying the possible opportunities for generating revenues, employment opportunities and trading. Ultimately, it can be deduced that usage of these renewable briquettes satisfies the energy demand renewably and help develop sustainably.

Table 5. Comparison of results from water boiling test on CNS briquettes and normal firewood

Parameters	Ordinary Firewood	CNS Briquettes
Net weight of the test sample (in grams)	500	500
Calorific content of the test samples used (in MJ/kg)	17.92	17.92
Initial temperature of water, before the start of heating (in ^{0}C)	32.5	32.5
Time taken for the water to start boiling (in minutes)	14.5	12
Total Time taken for complete combustion of test samples (in minutes)	50	42
Weight of water evaporated during the boiling test (in grams)	824.77	804.11
Weight of water remaining after the boiling test (in grams)	701.13	695.89
Burning Rate of the test samples (in g/min)	10	11.90
Heat Utilization Efficiency (in %)	11.65	18.01
Appearance of cracks in the test samples, if any	Yes	No
Remarks on overall combustion and performance of test samples	Good	Good
Traces of Fume, if any	Low	Low
Formation of soot particles, if any	Low	Low

Source: Sawekwiharee et al. (2015)

Table 6. Water boiling test (WBT) for CNS briquettes with varying concentration of binders

Briquette Sample	Weight (g)	Time Duration* (Minutes)	Burning Rate (g/min)
CNS wastes+ 30% binder	168.7	4	9.42
CNS wastes+ 20% binder	168.2	4.88	8.82
CNS wastes+ 10% binder	168.3	6.03	7.44
*For boiling 100 cm^3 of water			

Source: (Kimutai and Kimutai, 2019)

REFERENCES

Abdullahi, N., Sulaiman, F., & Safana, A. A. (2017). Bio-oil and biochar derived from the pyrolysis of palm kernel shell for briquette. *Sains Malaysiana*, *46*(12), 2441–2445. doi:10.17576/jsm-2017-4612-20

Agyemang, M., Zhu, Q., & Tian, Y. (2016). Analysis of opportunities for greenhouse emission reduction in the global supply chains of cashew industry in West Africa. *Journal of Cleaner Production*, *115*, 149–161. doi:10.1016/j.jclepro.2015.12.059

Akowuah, J. O., Kemausuor, F., & Mitchual, S. J. (2012). Physico-chemical characteristics and market potential of sawdust charcoal briquette. *International Journal of Energy and Environmental Engineering*, *3*(1), 1–6. doi:10.1186/2251-6832-3-20

Angulo-Mosquera, L. S., Alvarado-Alvarado, A. A., Rivas-Arrieta, M. J., Cattaneo, C. R., Rene, E. R., & García-Depraect, O. (2021). Production of solid biofuels from organic waste in developing countries: A review from sustainability and economic feasibility perspectives. *The Science of the Total Environment*, *795*, 148816. doi:10.1016/j.scitotenv.2021.148816 PMID:34237536

Basu, P. (2018). *Biomass gasification, pyrolysis and torrefaction: practical design and theory*. Academic press.

Bellamy, J. J., Ducos, C., Joly, N., & Joly, P. B. (1984). *Guide technique de la densification*. Association BOIS DE FEU.

Chen, W. H., Peng, J., & Bi, X. T. (2015). A state-of-the-art review of biomass torrefaction, densification and applications. *Renewable & Sustainable Energy Reviews*, *44*, 847–866. doi:10.1016/j.rser.2014.12.039

Chungcharoen, T., & Srisang, N. (2020). Preparation and characterization of fuel briquettes made from dual agricultural waste: Cashew nut shells and areca nuts. *Journal of Cleaner Production*, *256*, 120434. doi:10.1016/j.jclepro.2020.120434

Dendena, B., & Corsi, S. (2014). Cashew, from seed to market: A review. *Agronomy for Sustainable Development*, *34*(4), 753–772. doi:10.100713593-014-0240-7

FAO. (1985). Industrial Charcoal Making. FAO Forestry Paper. Rome, Food and Agriculture Organization of the United Nations.

Gandhi, T., Patel, M., & Dholakiya, B. K. (2012). Studies on effect of various solvents on extraction of cashew nut shell liquid (CNSL) and isolation of major phenolic constituents from extracted CNSL. *J. Nat. Prod. Plant Resour*, *2*(1), 135–142.

Gandhi, T. S., Dholakiya, B. Z., & Patel, M. R. (2013). Extraction protocol for isolation of CNSL by using protic and aprotic solvents from cashew nut and study of their physico-chemical parameter. *Polish Journal of Chemical Technology*, *15*(4), 24–27. doi:10.2478/pjct-2013-0062

Gimba, C. E., & Muyiwa, T. (2008). Optimum conditions for carbonisation of coconut shell. *Scientia Africana, 7*(2).

Glushankova, I., Ketov, A., Krasnovskikh, M., Rudakova, L., & Vaisman, I. (2018). Rice hulls as a renewable complex material resource. *Resources*, *7*(2), 31. doi:10.3390/resources7020031

Godjo, T., Tagutchou, J. P., Naquin, P., & Gourdon, R. (2015). Valorisation des coques d'anacarde par pyrolyse au Bénin. *Déchets Sciences et Techniques*, *70*, 11–18. doi:10.4267/dechets-sciences-techniques.3282

Himabindu, T., Raguram, V., & Kum, S. T. (2015). Review on various methods of extraction of cashew nut shell liquid and isolation of anacardic acid. *Int J Inst Pharm Life Sci*, *5*, 293–308.

Himbane, P. B., Grand Ndiaye, L., Napoli, A., & Kobor, D. (2018). *Physicochemical and mechanical properties of biomass coal briquettes produced by artisanal method*. Academic Press.

Huko, D., Kamau, D. N., & Ogola, W. O. (2015). Effects of varying particle size on mechanical and combustion characteristics of mango seed shell cashew nut shell composite briquettes. *Int. J. Eng. Sci. Invent*, *4*, 32–39.

Idah, P. A., Simeon, M. I., & Mohammed, M. A. (2014). *Extraction and characterization of cashew nut (Anacardium occidentale) oil and cashew shell liquid oil*. Academic Press.

Ifa, L. (2019). Production of bio-briquette from biochar derived from pyrolysis of cashew nut waste. *Ecology, Environment, and Conservation*, *25*, 125–131.

Ifa, L., Yani, S., Nurjannah, N., Darnengsih, D., Rusnaenah, A., Mel, M., Mahfud, M., & Kusuma, H. S. (2020). Techno-economic analysis of bio-briquette from cashew nut shell waste. *Heliyon*, *6*(9), e05009. doi:10.1016/j.heliyon.2020.e05009 PMID:33005808

Ismaila, A., Zakari, I. Y., Nasiru, R., Tijjani, B. I., Abdullahi, I., & Garba, N. N. (2013). Investigation on biomass briquettes as energy source in relation to their calorific values and measurement of their total carbon and elemental contents for efficient biofuel utilization. *Advances in Applied Science Research*, *4*(4), 303–309.

Kaliyan, N., & Morey, R. V. (2009). Factors affecting strength and durability of densified biomass products. *Biomass and Bioenergy*, *33*(3), 337–359. doi:10.1016/j.biombioe.2008.08.005

Karunanithy, C., Wang, Y., Muthukumarappan, K., & Pugalendhi, S. (2012). Physiochemical characterization of briquettes made from different feedstocks. *Biotechnology Research International*, *2012*, 2012. doi:10.1155/2012/165202 PMID:22792471

Kathuria, R. S., & Grover, S. (2012). *Using agricultural residues as a biomass briquetting: An alternative source of energy*. Academic Press.

Kimutai, S. K. (2021). Effects of Parameters on Cashew Nutshell and Cassava Binder Briquettes Densification: Structural Equation Modelling. *European Journal of Energy Research*, *1*(2), 12–16. doi:10.24018/ejenergy.2021.1.2.10

Kimutai, S. K., & Kimutai, I. K. (2019). Investigation of physical and combustion properties of briquettes from cashew nut shell and cassava binder. *International Journal of Educational Research*, *7*, 15–26.

Lubwama, M., & Yiga, V. A. (2017). Development of groundnut shells and bagasse briquettes as sustainable fuel sources for domestic cooking applications in Uganda. *Renewable Energy*, *111*, 532–542. doi:10.1016/j.renene.2017.04.041

Maia, J. G. S., Andrade, E. H. A., & Zoghbi, M. G. B. (2000). Volatile constituents of the leaves, fruits and flowers of cashew (Anacardium occidentale L.). *Journal of Food Composition and Analysis*, *13*(3), 227–232. doi:10.1006/jfca.2000.0894

Mgaya, J., Shombe, G. B., Masikane, S. C., Mlowe, S., Mubofu, E. B., & Revaprasadu, N. (2019). Cashew nut shell: A potential bio-resource for the production of bio-sourced chemicals, materials and fuels. *Green Chemistry*, *21*(6), 1186–1201. doi:10.1039/C8GC02972E

Mitchell, P. J., Kiel, J., Livingstone, B., & Dupont, R. (2007). *Torrefied Biomass: A Business Case for Pellets from Torrefied Biomass as New Solid Fuel*. Academic Press.

Mousa, E., Kazemi, M., Larsson, M., Karlsson, G., & Persson, E. (2019). Potential for developing biocarbon briquettes for foundry industry. *Applied Sciences (Basel, Switzerland)*, *9*(24), 5288. doi:10.3390/app9245288

Mubofu, E. B., & Mgaya, J. E. (2018). Chemical valorization of cashew nut shell waste. *Topics in Current Chemistry, 376*(2), 1–15. doi:10.100741061-017-0177-9 PMID:29442189

Ngusale, G. K., Luo, Y., & Kiplagat, J. K. (2014). Briquette making in Kenya: Nairobi and peri-urban areas. *Renewable & Sustainable Energy Reviews, 40*, 749–759. doi:10.1016/j.rser.2014.07.206

Njenga, M., Karanja, N., Karlsson, H., Jamnadass, R., Iiyama, M., Kithinji, J., & Sundberg, C. (2014). Additional cooking fuel supply and reduced global warming potential from recycling charcoal dust into charcoal briquette in Kenya. *Journal of Cleaner Production, 81*, 81–88. doi:10.1016/j.jclepro.2014.06.002

Oladeji, J. T. (2015). Theoretical aspects of biomass briquetting: A review study. *Journal of Energy Technologies and Policy, 5*(3), 72–81.

Olanders, B., & Steenari, B. M. (1995). Characterization of ashes from wood and straw. *Biomass and Bioenergy, 8*(2), 105–115. doi:10.1016/0961-9534(95)00004-Q

Onchieku, J. M., Chikamai, B. N., & Rao, M. S. (2012). Optimum parameters for the formulation of charcoal briquettes using bagasse and clay as binder. *European Journal of Sustainable Development, 1*(3), 477–477. doi:10.14207/ejsd.2012.v1n3p477

Onukak, I. E., Mohammed-Dabo, I. A., Ameh, A. O., Okoduwa, S. I., & Fasanya, O. O. (2017). Production and characterization of biomass briquettes from tannery solid waste. *Recycling, 2*(4), 17. doi:10.3390/recycling2040017

Pandiyan, C. V., Shylaja, G., Srinivasan, G. R., & Saravanan, S. (2020). Studies on use of Cashew Nut Shell Liquid (CNSL) in Biopesticide and Biofertilizer. *Nature Environment & Pollution Technology, 19*(1).

Patel, R. N., Bandyopadhyay, S., & Ganesh, A. (2006). Extraction of cashew (Anacardium occidentale) nut shell liquid using supercritical carbon dioxide. *Bioresource Technology, 97*(6), 847–853. doi:10.1016/j.biortech.2005.04.009 PMID:15939589

Quirino, R. L., Garrison, T. F., & Kessler, M. R. (2014). Matrices from vegetable oils, cashew nut shell liquid, and other relevant systems for biocomposite applications. *Green Chemistry, 16*(4), 1700–1715. doi:10.1039/C3GC41811A

Rabany, C., Rullier, N., & Ricau, P. (2015). *The African cashew sector in 2015. General trends and country profiles*. https://www.rongead.org

Raghavendra Prasada, S. A. (2014). A Review on CNSL Biodiesel as an Alternative fuel for Diesel Engine. *International Journal of Scientific Research, 3*(7), 2028–2038.

Rahman, M. M., Khan, K., Moral, M. N. A., & Rahman, A. N. M. (2003). *Biomass briquetting technology and environmental pollution. In The role of Renewable and Alternative Energy Sources for National Development. Khulna University of Engineering and Technology (KUET) and Asian Institute of Technology*. AIT.

Rodrigues, F. H., França, F. C., Souza, J. R., Ricardo, N. M., & Feitosa, J. (2011). Comparison between physico-chemical properties of the technical Cashew Nut Shell Liquid (CNSL) and those natural extracted from solvent and pressing. *Polímeros, 21*(2), 156–160. doi:10.1590/S0104-14282011005000028

Roy, M. M., & Corscadden, K. W. (2012). An experimental study of combustion and emissions of biomass briquettes in a domestic wood stove. *Applied Energy*, *99*, 206–212. doi:10.1016/j.apenergy.2012.05.003

Sanger, S. H., Mohod, A. G., Khandetode, Y. P., Shrirame, H. Y., & Deshmukh, A. S. (2011). Study of carbonization for cashew nut shell. *Research Journal of Chemical Sciences*, *1*(2), 43–55.

Sawadogo, M., Tanoh, S. T., Sidibé, S., Kpai, N., & Tankoano, I. (2018). Cleaner production in Burkina Faso: Case study of fuel briquettes made from cashew industry waste. *Journal of Cleaner Production*, *195*, 1047–1056. doi:10.1016/j.jclepro.2018.05.261

Sawekwiharee, S., Boonchoo, T., Kuttiyawong, A., & Chathirat, N. (2015). Heating energy briquettes from Cashew nut shell. *Applied Mechanics and Materials*, *804*, 283–286. doi:10.4028/www.scientific.net/AMM.804.283

Sen, R., Wiwatpanyaporn, S., & Annachhatre, A. P. (2016). Influence of binders on physical properties of fuel briquettes produced from cassava rhizome waste. *International Journal of Environment and Waste Management*, *17*(2), 158–175. doi:10.1504/IJEWM.2016.076750

Senthil Kumar, P., Arun Kumar, N., Sivakumar, R., & Kaushik, C. (2009). Experimentation on solvent extraction of polyphenols from natural waste. *Journal of Materials Science*, *44*(21), 5894–5899. doi:10.100710853-009-3834-8

Shahbandeh. (2022). https://www.statista.com/statistics/967702/global-cashew-nut-production

Sharma, M. K., Priyank, G., & Sharma, N. (2015). Biomass briquette production: A propagation of non-convention technology and future of pollution free thermal energy sources. *American Journal of Engineering Research*, *4*(02), 44–50.

Singh, R. N., Jena, U., Patel, J. B., & Sharma, A. M. (2006). Feasibility study of cashew nut shells as an open core gasifier feedstock. *Renewable Energy*, *31*(4), 481–487. doi:10.1016/j.renene.2005.04.010

Smith, R. L. Jr, Malaluan, R. M., Setianto, W. B., Inomata, H., & Arai, K. (2003). Separation of cashew (Anacardium occidentale L.) nut shell liquid with supercritical carbon dioxide. *Bioresource Technology*, *88*(1), 1–7. doi:10.1016/S0960-8524(02)00271-7 PMID:12573557

Sotannde, O. A., Oluyege, A. B., & Abah, G. A. (2010). Physical and combustion properties of charcoal briquettes from neem wood residues. *International Agrophysics*, *24*(2), 189–194.

Srinivasan, G. R., & Jambulingam, R. (2018). Comprehensive study on biodiesel produced from waste animal fats-a review. *Journal of Environmental Science and Technology*, *11*(3), 157–166. doi:10.3923/jest.2018.157.166

Srinivasan, G. R., Mahajan, A., Seth, R., & Mahajan, R. (2022). Impact of organic hydrocarbons on fuel properties and engine characteristics of thermally cracked cashew nut shell liquid. *Journal of Renewable Energy and Environment.*

Srinivasan, G. R., Shankar, V., Chandra Sekharan, S., Munir, M., Balakrishnan, D., Mohanam, A., & Jambulingam, R. (2020). Influence of fatty acid composition on process optimization and characteristics assessment of biodiesel produced from waste animal fat. *Energy Sources. Part A, Recovery, Utilization, and Environmental Effects*, 1–19. doi:10.1080/15567036.2020.1771477

Stolarski, M. J., Szczukowski, S., Tworkowski, J., Krzyżaniak, M., Gulczyński, P., & Mleczek, M. (2013). Comparison of quality and production cost of briquettes made from agricultural and forest origin biomass. *Renewable Energy*, *57*, 20–26. doi:10.1016/j.renene.2013.01.005

Subbarao, C. N., Krishna Prasad, K. M. M., & Prasad, V. S. R. K. (2011). Review on applications, extraction, isolation and analysis of cashew nut shell liquid (CNSL). *The Pharma Research Journal*, *6*(1), 21–41.

Suhartini, S., Hidayat, N., & Wijaya, S. (2011). Physical properties characterization of fuel briquette made from spent bleaching earth. *Biomass and Bioenergy, 35*(10), 4209-4214.

Sutrisno, A. W., Suprianto, F. D., Kasrun, A. W., & Siahaan, I. H. (2017). The effects of particle size and pressure on the combustion characteristics of Cerbera manghas leaf briquettes. *Arpn J. Eng. Appl. Sci*, *12*, 931–936.

Tamilvanan, A. (2013). Preparation of biomass briquettes using various agro-residues and waste papers. *Journal of Biofuels*, *4*(2), 47–55. doi:10.5958/j.0976-4763.4.2.006

Tinlot, M. (2010). *Intégration des filières dans la mitigation au changement climatique: évaluation carbone sur la filière anacarde au Burkina Faso. In Application de l'outil EXACT*. Étude pour la GTZ et la FAO.

Tuates, A. M. Jr, Suligan, J. M., Ligisan, A. R., Capariño, O. A., & Elauria, J. C. (2020). Development of De-Oiled Cashew Nut Shell as Fuel Briquettes. *Journal of the Japan Institute of Energy*, *99*(8), 123–128. doi:10.3775/jie.99.123

Tyman, J. H. P., Johnson, R. A., Muir, M., & Rokhgar, R. (1989). The extraction of natural cashew nut-shell liquid from the cashew nut (Anacardium occidentale). *Journal of the American Oil Chemists' Society*, *66*(4), 553–557. doi:10.1007/BF02885447

Ujjinappa, S., & Sreepathi, L. K. (2018). Evaluation of Physico-Mechanical-Combustion Characteristics of Fuel Briquettes Made from Blends of Areca Nut Husk, Simarouba Seed Shell and Black Liquor. *International Journal of Renewable Energy Development*, *7*(2), 131–137. doi:10.14710/ijred.7.2.131-137

Wilaipon, P. (2007). Physical characteristics of maize cob briquette under moderate die pressure. *American Journal of Applied Sciences*, *4*(12), 995–998. doi:10.3844/ajassp.2007.995.998

Yuliana, M., Tran-Thi, N. Y., & Ju, Y. H. (2012). Effect of extraction methods on characteristic and composition of Indonesian cashew nut shell liquid. *Industrial Crops and Products*, *35*(1), 230–236. doi:10.1016/j.indcrop.2011.07.007

Zafeer, M. K., & Bhat, K. S. (2023). Valorisation of Agro-waste Cashew Nut Husk (Testa) for Different Value-Added Products. *Sustainable Chemistry for Climate Action*, 100014.

Zapušek, A., Wirtgen, C., Weigandt, J., & Lenart, F. (2003). Characterisation of carbonizate produced from Velenje lignite in lab-scale reactor. *Acta Chimica Slovenica*, *50*, 789–798.

Zhang, Y., Ghaly, A. E., & Li, B. (2012). Physical properties of corn residues. *American Journal of Biochemistry and Biotechnology*, *8*(2), 44–53. doi:10.3844/ajbbsp.2012.44.53

Zhu, D., Chen, D., Pan, J., & Zheng, G. (2011). Chlorination behaviors of zinc phases by calcium chloride in high temperature oxidizing-chloridizing roasting. *ISIJ International*, *51*(11), 1773–1777. doi:10.2355/isijinternational.51.1773

KEY TERMS AND DEFINITIONS

Aromatic Hydrocarbons: A class of organic compounds having one or more rings of carbon atoms with alternating double bonds.

Binder: An organic material used for holding the biomass particles together under the influence of external force.

Bio-Char: Waste biomass carbonised or combusted partially at high temperatures in the absence of oxygen with their volatile content removed significantly.

Briquette: A small block or cylindrical piece of compressed biomass designed to burn efficiently with minimal smoke and ash.

Compaction: A mechanical process of increasing the density or reducing the volume of biomass under the application of external pressure or force.

Milling: A mechanical process of reducing the size of organic biomass under the influence of friction.

Press Cakes: Solid residues left behind after the extraction of oil from seed biomass, with high protein and fiber content.

Pyrolysis: Thermal Degradation of macro molecules in biomass at high temperatures in the absence of oxygen.

Testa: An outer layer of seed biomass that protects their kernel.

Compilation of References

Abbas, G., Asad, M. U., Gu, J., Alelyani, S., Balas, V. E., Hussain, M. R., Farooq, U., Awan, A. B., Raza, A., & Chang, C. (2021). Multivariable unconstrained pattern search method for optimizing digital PID controllers applied to isolated forward converter. *Energies, 14*(1), 77.

Abdullahi, H. S., Mahieddine, F., & Sheriff, R. E. (2015). *Technology impact on agricultural productivity: A review of precision agriculture using unmanned Aerial vehicles. Wireless and Satellite Systems*. Springer International Publishing.

Abdullahi, N., Sulaiman, F., & Safana, A. A. (2017). Bio-oil and biochar derived from the pyrolysis of palm kernel shell for briquette. *Sains Malaysiana*, *46*(12), 2441–2445. doi:10.17576/jsm-2017-4612-20

Abidoye, L. K., & Das, D. B. (2021). Carbon storage in portland cement mortar: Influences of hydration stage, carbonation time and aggregate characteristics. *Cleanroom Technology*, *3*(3), 563–580. doi:10.3390/cleantechnol3030034

Acevedo, F., Rubilar, M., Scheuermann, E., Cancino, B., Uquiche, E., Garcés, M., . . . Shene, C. (2013, March). Bioactive compounds of spent coffee grounds, a coffee industrial residue. In *Proceedings of the III Symposium on* (*Vol. 212*). Academic Press.

Acharya, N., Nanda, P., Panda, S., & Acharya, S. (2019). A comparative study of stability characteristics of mahua and jatropha biodiesel and their blends. *Journal of King Saud University-Engineering Sciences*, *31*(2), 184–190. doi:10.1016/j.jksues.2017.09.003

Adeleye, A. S., Conway, J. R., Garner, K., Huang, Y., Su, Y., & Keller, A. A. (2016). Engineered nanomaterials for water treatment and remediation: Costs, benefits, and applicability. *Chemical Engineering Journal*, *286*, 640–662. doi:10.1016/j.cej.2015.10.105

Afghoul, H., Krim, F., Babes, B., Beddar, A., & Kihel, A. (2018). Design and real time implementation of sliding mode supervised fractional controller for wind energy conversion system under sever working conditions. *Energy Conversion and Management, 167*, 91–101.

Agoro, M. A., Okoh, O. O., Adefisoye, M. A., & Okoh, A. I. (2018). Physicochemical Properties of Wastewater in Three Typical South African Sewage Works. *Journal of Environmental Studies (Northborough, Mass.)*, *27*, 1–9.

Agyemang, E. O., Awuah, E., Darkwah, L., Arthur, R., & Osei, G. (2013). Water quality assessment of a wastewater treatment plant in a Ghanaian Beverage Industry. *International Journal of Water Resources & Environmental Engineering*, *5*(5), 272–279.

Agyemang, M., Zhu, Q., & Tian, Y. (2016). Analysis of opportunities for greenhouse emission reduction in the global supply chains of cashew industry in West Africa. *Journal of Cleaner Production*, *115*, 149–161. doi:10.1016/j.jclepro.2015.12.059

Ahmad, H., & Oludele, J. A. (2022). Assessment of the level of conformity of Wupa sewage treatment plant to environmental standards. *Sokoto Journal of the Social Sciences*, *12*, 47–57.

Ahmed, S. F., Mofijur, M., Ahmed, B., Mehnaz, T., Mehejabin, F., Maliat, D., Hoang, A. T., & Shafiullah, G. M. (2022). Nanomaterials as a sustainable choice for treating wastewater. *Environmental Research, 214*, 113807. doi:10.1016/j.envres.2022.113807 PMID:35798266

Ai, L., & Jiang, J. (2012). Removal of methylene blue from aqueous solution with self-assembled cylindrical graphene–carbon nanotube hybrid. *Chemical Engineering Journal, 192*, 156–163. doi:10.1016/j.cej.2012.03.056

Akhihiero, E. T., Omorewa, Y. G., & Ebhodaghe, S. O. (2019). Effect of blending ratio on the properties of sunflower biodiesel. *J. Mater. Environ. Sci, 10*(10), 987–994.

Akinnagbe, O. M., & Irohibe, I. J. (2014). Agricultural adaptation strategies to climate change impacts in Africa: A review. *Bangladesh Journal of Agricultural Research, 39*(3), 407–418. doi:10.3329/bjar.v39i3.21984

Akowuah, J. O., Kemausuor, F., & Mitchual, S. J. (2012). Physico-chemical characteristics and market potential of sawdust charcoal briquette. *International Journal of Energy and Environmental Engineering, 3*(1), 1–6. doi:10.1186/2251-6832-3-20

Akpo, I. F., Zakari, F. T., Agani, F. O., Issaka, K., & Yabi, J. A. (2023). Effects of sustainable land management practices on food security on beneficiary households in context of adaptation to climate change in the communes of Malanville and Kouandé in Northern Benin. *International Journal of Agricultural Policy and Research, 11*(1), 16–26.

Al-Barwani, H. H., & Purnama, A. (2007). Re-assessing the Impact of Desalination Plants Brine Discharges on Eroding Beaches. *Desalination, 204*(1-3), 94–101. doi:10.1016/j.desal.2006.03.536

Ali, A., Zafar, H., Zia, M., ul Haq, I., Phull, A. R., Ali, J. S., & Hussain, A. (2016). Synthesis, characterization, applications, and challenges of iron oxide nanoparticles. *Nanotechnology, Science and Applications, 9*, 49–67. doi:10.2147/NSA.S99986 PMID:27578966

Ali, I., Alyona, S., Tatiana, K., Anastasiya, G., Albishri, H. M., & Alshitari, W. H. (2023). Facile adsorption-electroflotation method for the removal of heavy metal ions from water using carbon nanomaterials. *Environmental Science and Pollution Research International, 30*(13), 1–12. doi:10.100711356-022-24509-0 PMID:36593318

Allesina, G., Pedrazzi, S., Allegretti, F., & Tartarini, P. (2017). Spent coffee grounds as heat source for coffee roasting plants: Experimental validation and case study. *Applied Thermal Engineering, 126*, 730–736. doi:10.1016/j.applthermaleng.2017.07.202

Al-Soeidat, M. (2019). *A reconfigurable three-port DC–DC converter for integrated PV-battery system*. https://ieeexplore.ieee.org/abstract/document/8839067/

Alzahrani, A., Ferdowsi, M., & Shamsi, P. (2019). A family of scalable non-isolated interleaved DC-DC boost converters with voltage multiplier cells. *IEEE Access, 7*, 11707–11721.

Ambashta, R. D., & Sillanpää, M. (2010). Water purification using magnetic assistance: A review. *Journal of Hazardous Materials, 180*(1–3), 38–49. doi:10.1016/j.jhazmat.2010.04.105 PMID:20488616

Amin, M. T., Alazba, A. A., & Manzoor, U. (2014). A review of removal of pollutants from water/wastewater using different types of nanomaterials. *Advances in Materials Science and Engineering, 2014*, 1–24. Advance online publication. doi:10.1155/2014/825910

Ammarapala, V., Chinda, T., Pongsayaporn, P., Ratanachot, W., Punthutaecha, K., & Janmonta, K. (2018). Cross-border shipment route selection utilizing analytic hierarchy process (AHP) method. *Songklanakarin Journal of Science and Technology, 40*. Advance online publication. doi:10.14456jst-psu.2018.3

Amon, T., Amon, B., Kryvoruchko, V., Zollitsch, W., Mayer, K., & Gruber, L. (2007). Biogas production from maize and dairy cattle manure Influence of biomass composition on the methane yield. *Agriculture, Ecosystems & Environment, 118*(1-4), 173–182. doi:10.1016/j.agee.2006.05.007

Angulo-Mosquera, L. S., Alvarado-Alvarado, A. A., Rivas-Arrieta, M. J., Cattaneo, C. R., Rene, E. R., & García-Depraect, O. (2021). Production of solid biofuels from organic waste in developing countries: A review from sustainability and economic feasibility perspectives. *The Science of the Total Environment, 795*, 148816. doi:10.1016/j.scitotenv.2021.148816 PMID:34237536

Anjum, M., Miandad, R., Waqas, M., Gehany, F., & Barakat, M. A. (2019). Remediation of wastewater using various nano-materials. *Arabian Journal of Chemistry, 12*(8), 4897–4919. doi:10.1016/j.arabjc.2016.10.004

AQUARIUS. (2022). *Properties of fresh Water and Seawater*. AQUARIUS Educational and Public Outreach. Retrieved November 24, 2022 from https://aquarius.umaine.edu/activities/prop_fresh_sea.pdf

Argin, M., Yerci, V., Erdogan, N., Kucuksari, S., & Cali, U. (2019). Exploring the offshore wind energy potential of Turkey based on multi-criteria site selection. *Energy Strategy Rev., 23*, 33–46. doi:10.1016/j.esr.2018.12.005

Ariono, D., Purwasasmita, M., & Wenten, I. G. (2016). Brine Effluents: Characteristics, Environmental Impacts and their Hanling. *Journal of Engineering Technology and Science, 48*(4), 367–387. doi:10.5614/j.eng.technol.sci.2016.48.4.1

Ashbolt, N. J. (2015). Microbial contamination of drinking water and human health from community water systems. *Current Environmental Health Reports, 2*(1), 95–106. doi:10.100740572-014-0037-5 PMID:25821716

Ashok, B., Nanthagopal, K., Saravanan, B., Azad, K., Patel, D., Sudarshan, B., & Ramasamy, R. A. (2019b). Study on isobutanol and Calophyllum inophyllum biodiesel as a partial replacement in CI engine applications. *Fuel, 235*, 984–994. doi:10.1016/j.fuel.2018.08.087

Ashok, B., Nanthagopal, K., Saravanan, B., Somasundaram, P., Jegadheesan, C., Chaturvedi, B., & Patni, G. (2018). A novel study on the effect lemon peel oil as a fuel in CRDI engine at various injection strategies. *Energy Conversion and Management, 172*, 517–528. doi:10.1016/j.enconman.2018.07.037

Ashok, B., Raj, R. T. K., Nanthagopal, K., Krishnan, R., & Subbarao, R. (2017). Lemon peel oil–A novel renewable alternative energy source for diesel engine. *Energy Conversion and Management, 139*, 110–121. doi:10.1016/j.enconman.2017.02.049

Ashraf, S. N., & Singh, A. K. (2021). Impact of technological change on growth and agricultural sector in Gujarat state of India: A time series data study. *Asian Development Policy Review, 9*(3), 144–160. doi:10.18488/journal.107.2021.93.144.160

Asrat, P., & Simane, B. (2018). Farmers' perception of climate change and adaptation strategies in the Dabus watershed, North-West Ethiopia. *Ecological Processes, 7*(7), 1–13. doi:10.118613717-018-0118-8

Atabani, A. E., Ala'a, H., Kumar, G., Saratale, G. D., Aslam, M., Khan, H. A., ... Mahmoud, E. (2019). Valorization of spent coffee grounds into biofuels and value-added products: Pathway towards integrated bio-refinery. *Fuel, 254*, 115640. doi:10.1016/j.fuel.2019.115640

Aung, N. Z. (2008). Modification of diesel engine to producer gas engine. *Jurnal Teknologi Energi, 1*(6).

Awogbemi, O., & Adeyemo, S. B. (2013). Investigation of the influence of some parameters on biogas yield. *Elixir International Journal, 54A*, 12825–12828.

Aydin, N. Y., Kentel, E., & Duzgun, S. (2010). GIS-based environmental assessment of wind energy systems for spatial planning: A case study from Western Turkey. *Renewable & Sustainable Energy Reviews, 14*(1), 364–373. doi:10.1016/j.rser.2009.07.023

Ayedun, H., Daudu, O. O., & Eje, R. E. (2022). Laboratory treatment of industrial wastewater. Research Project of Olusegun Agagu University of Science and Technology, Ondo State, Nigeria.

Azer, P., & Emadi, A. (2020). Generalized state space average model for multi-phase interleaved buck, boost and buck-boost DC-DC converters: Transient, steady-state and switching dynamics. *IEEE Access, 8*, 77735–77745.

Aziz, N., Faraz, M., Pandey, R., Shakir, M., Fatma, T., Varma, A., Barman, I., & Prasad, R. (2015). Facile Algae-Derived Route to Biogenic Silver Nanoparticles: Synthesis, Antibacterial, and Photocatalytic Properties. *Langmuir*, *31*(42), 11605–11612. doi:10.1021/acs.langmuir.5b03081 PMID:26447769

Aziz, N., Faraz, M., Sherwani, M. A., Fatma, T., & Prasad, R. (2019). Illuminating the anticancerous efficacy of a new fungal chassis for silver nanoparticle synthesis. *Frontiers in Chemistry*, *7*(FEB), 65. doi:10.3389/fchem.2019.00065 PMID:30800654

Aziz, N., Fatma, T., Varma, A., & Prasad, R. (2014). Biogenic Synthesis of Silver Nanoparticles Using Scenedesmus abundans and Evaluation of Their Antibacterial Activity. *Journal of Nanoparticles*, *2014*, 1–6. doi:10.1155/2014/689419

Aziz, N., Pandey, R., Barman, I., & Prasad, R. (2016). Leveraging the attributes of mucor hiemalis-derived silver nanoparticles for a synergistic broad-spectrum antimicrobial platform. *Frontiers in Microbiology*, *7*(DEC), 1984. doi:10.3389/fmicb.2016.01984 PMID:28018316

Balasha, A. M., Munyahali, W., Kulumbu, J. T., Okwe, A. N., Fyama, J. N. M., Lenge, E. K., & Tabbwe, A. N. (2023). Understanding farmers' perception of climate change and adaptation practices in the marshland of South Kivu, Democratic Republic of Congo. *Climate Risk Management*, *39*(100469), 1–15. doi:10.1016/j.crm.2022.100469

Ban-Weiss, G. A., Chen, J. Y., Buchholz, B. A., & Dibble, R. W. (2007). A Numerical Investigation into the Anomalous Slight NOx Increase when Burning Biodiesel: A New (Old) Theory. *Fuel Processing Technology*, *88*(7), 659–667. doi:10.1016/j.fuproc.2007.01.007

Bari, S., Yu, C. W., & Lim, T. H. (2004). The effect of fuel injection timing with waste cooking oil as a fuel in a direct injection diesel engine. *Proc.IMechE, Part D. Journal of Automobile Engineering*, *218*(1), 93–104. doi:10.1243/095440704322829209

Barros, T. A. S., dos Santos Neto, P. J., Nascimento Filho, P. S., Moreira, A. B., & Ruppert Filho, E. (2017). An Approach for Switched Reluctance Generator in a Wind Generation System with a Wide Range of Operation Speed. *IEEE Transactions on Power Electronics, 32*(11), 8277–8292.

Bartolo, J.B., Degano, M., & Espina, J. (2017). Design and initial testing of high speed 45-kw switched reluctance drive for aerospace application. *IEEE Trans. Ind. Electron, 64*(2), 988-997.

Basha, Al Balushi, Al Shibli, Al Fazari, Al Rumhi, Al Mansouri, & Al Balushi. (2020). An emission control strategy in a low capacity single cylinder compression ignition engine powered with DEE blended fuels. *Materials Science for Energy Technologies*, *3,* 770-779.

Basha, J. S., Al Balushi, M., Al Shibli, K., Al Fazari, M., Al Rumhi, S., Al Mansouri, S., & Al Balushi, R. (2020). An emission control strategy in a low capacity single cylinder compression ignition engine powered with dee blended fuels. *Materials Science for Energy Technologies*, *3*, 770–779. doi:10.1016/j.mset.2020.09.004

Basha, J. S., Jafary, T., Vasudevan, R., Bahadur, J. K., Ajmi, M. A., Neyadi, A. A., Soudagar, M. E., Mujtaba, M. A., Hussain, A., Ahmed, W., Shahapurkar, K., Rahman, S. M., & Fattah, I. M. (2021). Potential of utilization of renewable energy technologies in Gulf Countries. *Sustainability*, *13*(18), 10261. doi:10.3390u131810261

Basu, P. (2018). *Biomass gasification, pyrolysis and torrefaction: practical design and theory*. Academic press.

Batool, M., & Shahzad, L. (2021). An analytical study on municipal wastewater to energy generation, current trends, and future prospects in South Asian developing countries (an update on Pakistan scenario). *Environmental Science and Pollution Research International*, *28*(25), 32075–32094. doi:10.100711356-021-14029-8 PMID:33914246

Bejenari, V., Marcu, A., Ipate, A. M., Rusu, D., Tudorachi, N., Anghel, I., Şofran, I.-E., & Lisa, G. (2021). Physicochemical characterization and energy recovery of spent coffee grounds. *Journal of Materials Research and Technology*, *15*, 4437–4451. doi:10.1016/j.jmrt.2021.10.064

Belitz, H. D., Grosch, W., & Schieberle, P. (2009). Coffee, tea, cocoa. *Food Chemistry*, •••, 938–970.

Bellamy, J. J., Ducos, C., Joly, N., & Joly, P. B. (1984). *Guide technique de la densification*. Association BOIS DE FEU.

Beyth, N., Houri-Haddad, Y., Domb, A., Khan, W., & Hazan, R. (2015). Alternative antimicrobial approach: Nano-antimicrobial materials. *Evidence-Based Complementary and Alternative Medicine*, *2015*, 1–16. Advance online publication. doi:10.1155/2015/246012 PMID:25861355

Bharti, M. K., Gupta, S., Chalia, S., Garg, I., Thakur, P., & Thakur, A. (2020). Potential of Magnetic Nano-ferrites in Removal of Heavy Metals from Contaminated Water: Mini Review. *Journal of Superconductivity and Novel Magnetism*, *33*(12), 3651–3665. doi:10.100710948-020-05657-1

Bhat, S. A., Pandit, B. A., Khan, J. N., Kumar, R., & Jan, R. (2017). Water Requirements and Irrigation Scheduling of Maize Crop using CROPWAT Model. *International Journal of Current Microbiology and Applied Sciences*, *6*(11), 1662–1670. doi:10.20546/ijcmas.2017.611.199

Bhuyan, T., Mishra, K., Khanuja, M., Prasad, R., & Varma, A. (2015). Biosynthesis of zinc oxide nanoparticles from Azadirachta indica for antibacterial and photocatalytic applications. *Materials Science in Semiconductor Processing*, *32*, 55–61. doi:10.1016/j.mssp.2014.12.053

Bielaczyc, P., Keskinen, J., Dzida, J., Sala, R., Ronkko, T., Kinnunen, T., Matilainen, P., Karjalainen, P., & Happonen, M. J. (2012). Performance of particle oxidation catalyst and particle formation studies with sulphur containing fuels. *SAE International Journal of Fuels and Lubricants*, *5*(2), 611–619. doi:10.4271/2012-01-0366

Bili, A., & Vagiona, D. G. (2018). Use of multi-criteria analysis and GIS for selecting sites for onshore wind farms: The case of Andros Island (Greece). *Eur. J. Environ. Sci.*, *2018*(8), 5–13.

Biswas, S., Chatterjee, S., & Roy, D. C. (2020). Understanding of farmers' perception of climate change and adaptation strategies: A case study in Jhargram district of West Bengal, India. *Journal of Applied and Natural Science*, *12*(2), 207–212. doi:10.31018/jans.vi.2241

Blinová, L., & Sirotiak, M. (2019). Utilization of spent coffee grounds for removal of hazardous substances from water: A Review. *Vedecké Práce Materiálovotechnologickej Fakulty Slovenskej Technickej Univerzity v Bratislave so Sídlom v Trnave, 27*(44), 145-152.

Bogue, R. (2011). Nanocomposites: A review of technology and applications. *Assembly Automation*, *31*(2), 106–112. doi:10.1108/01445151111117683

Bokare, V., Jung, J., Chang, Y.-Y., & Chang, Y.-S. (2013). Reductive dechlorination of octachlorodibenzo-p-dioxin by nanosized zero-valent zinc: Modeling of rate kinetics and congener profile. *Journal of Hazardous Materials*, *250–251*, 397–402. doi:10.1016/j.jhazmat.2013.02.020 PMID:23500419

Bok, J. P., Choi, H. S., Choi, Y. S., Park, H. C., & Kim, S. J. (2012). Fast pyrolysis of coffee grounds: Characteristics of product yields and biocrude oil quality. *Energy*, *47*(1), 17–24. doi:10.1016/j.energy.2012.06.003

Borkhade, R., Bhat, K. S., & Mahesha, G. T. (2022). Implementation of sustainable reforms in the Indian automotive industry: From Vehicle Emissions Perspective. *Cogent Engineering*, *9*(1), 2014024. Advance online publication. doi:10.1080/23311916.2021.2014024

Botes, M., & Cloete, T. E. (2010). *The potential of nanofibers and nanobiocides in water purification*. doi:10.3109/10408410903397332

Braeutigam, P., Franke, M., Schneider, R. J., Lehmann, A., Stolle, A., & Ondruschka, B. (2012, May). Degradation of carbamazepine in environmentally relevant concentrations in water by Hydrodynamic-Acoustic-Cavitation (HAC). *Water Research*, *46*(7), 2469–2477. doi:10.1016/j.watres.2012.02.013 PMID:22365175

Brazdis, R.-I., Fierascu, R. C., Avramescu, S.-M., & Fierascu, I. (2023). Nanomaterials for the removal of organic pollutants from agrofood wastewaters. *Advanced Technologies in Wastewater Treatment*, 355–381. doi:10.1016/B978-0-323-88510-2.00008-7

British Petroleum Company. (2020). *BP statistical review of world energy*. British Petroleum Co.

Brown, W. (2016). *Physical Properties of Seawater*. Springer Handbook of Ocean Engineering. doi:10.1007/978-3-319-16649-0_5

Brunerová, A., Roubík, H., Brožek, M., Haryanto, A., Hasanudin, U., Iryani, D. A., & Herák, D. (2019). Valorization of bio-briquette fuel by using spent coffee ground as an external additive. *Energies*, *13*(1), 54. doi:10.3390/en13010054

Bui, H. N., Do, H. Q., Duong, H. T. G., Perng, Y. S., Dam, V. N., Nguyen, V. T., & Bui, H. M. (2022). Taguchi optimization and life cycle assessment of biodiesel production from spent ground coffee. *Environment, Development and Sustainability*, *24*(11), 12900–12916. doi:10.100710668-021-01974-0

Burakov, A. E., Burakova, I. v., Galunin, E. v., & Kucherova, A. E. (2018). New Carbon Nanomaterials for Water Purification from Heavy Metals. In *Handbook of Ecomaterials* (pp. 1–20). Springer International Publishing. doi:10.1007/978-3-319-48281-1_166-1

Buros, O. K. (2000). *The ABCs of Desalting Topsfield*. International Desalination Association.

Cali, U., Erdogan, N., Kucuksari, E., & Argin, M. (2018). Techno-economic analysis of high potential offshore wind farm locations in Turkey. *Energy Strategy Reviews*, *22*, 325–336. doi:10.1016/j.esr.2018.10.007

Castrillon-Munoz, F. J., Gibson, J. J., & Birks, S. J. (2022). Carbon dissolution effects on pH changes of RAMP lakes in northeastern Alberta, Canada. *Journal of Hydrology. Regional Studies*, *40*, 101045. doi:10.1016/j.ejrh.2022.101045

Cavalcante, F. T. T., Neto, F. S., de Aguiar Falcão, I. R., da Silva Souza, J. E., de Moura, L. S. Junior, da Silva Sousa, P., Rocha, T. G., de Sousa, I. G., de Lima Gomes, P. H., & de Souza, M. C. M. (2021). Opportunities for improving biodiesel production via lipase catalysis. *Fuel*, *288*, 119577. doi:10.1016/j.fuel.2020.119577

Chandio, A. A., Jiang, Y., Amin, A., Ozturk, I., Sinha, A., & Ahmad, F. (2021). Modeling the impact of climatic and non-climatic factors on cereal production: Evidence from Indian agricultural sector. *Environmental Science and Pollution Research International*. Advance online publication. doi:10.100711356-021-16751-9 PMID:34617217

Chandra, R., Vijay, V. K., Subbarao, P. M. K., & Khura, T. K. (2011). Performance evaluation of a constant speed IC on CNG, methane enriched biogas and biogas. *Applied Energy*, *88*(11), 3969–3977. doi:10.1016/j.apenergy.2011.04.032

Chandrasekar, B. (2019). *Non-isolated unidirectional three-port Cuk-Cuk converter for fuel cell/solar PV systems*. https://www.koreascience.or.kr/article/JAKO201926838276891.page

Chandrasekar, B. (2020). *Non-isolated high-gain triple port DC–DC buck-boost converter with positive output voltage for photovoltaic applications*. https://ieeexplore.ieee.org/abstract/document/9120014/

Chang, D. Y., & Van Gerpan, J. H. (1997). *Fuel properties and engine performance for biodiesel prepared from modified feedstocks*. SAE Paper No. 971684.

Chen, X., Zhang, T., & Ye, W. (2020). *Blockchain-based electric vehicle incentive system for renewable energy consumption*. https://ieeexplore.ieee.org/abstract/document/9097307/

Chen, C. T. (2000). Extension of the TOPSIS for group decision-making under fuzzy environment. *Fuzzy Sets and Systems*, *114*(1), 1–9. doi:10.1016/S0165-0114(97)00377-1

Chen, C., Li, D., & Shao, X. (2014). High-performance nanocomposite films: Reinforced with chitosan nanofiber extracted from prawn shells. *Journal of Materials Science*, *49*(3), 1215–1221. doi:10.100710853-013-7803-x

Chen, W. H., Peng, J., & Bi, X. T. (2015). A state-of-the-art review of biomass torrefaction, densification and applications. *Renewable & Sustainable Energy Reviews*, *44*, 847–866. doi:10.1016/j.rser.2014.12.039

Chia, S. R., Ong, H. C., Chew, K. W., Show, P. L., Phang, S. M., Ling, T. C., Nagarajan, D., Lee, D.-J., & Chang, J. S. (2018). Sustainable approaches for algae utilisation in bioenergy production. *Renewable Energy*, *129*, 838–852. doi:10.1016/j.renene.2017.04.001

Cho, D. W., Cho, S. H., Song, H., & Kwon, E. E. (2015). Carbon dioxide assisted sustainability enhancement of pyrolysis of waste biomass: A case study with spent coffee ground. *Bioresource Technology*, *189*, 1–6. doi:10.1016/j.biortech.2015.04.002 PMID:25864025

Christidis, T., & Law, J. (2012). Review: The use of geographic information systems in wind turbine and wind energy research. *Journal of Renewable and Sustainable Energy*, *4*(1), 012701. Advance online publication. doi:10.1063/1.3673565

Chungcharoen, T., & Srisang, N. (2020). Preparation and characterization of fuel briquettes made from dual agricultural waste: Cashew nut shells and areca nuts. *Journal of Cleaner Production*, *256*, 120434. doi:10.1016/j.jclepro.2020.120434

Ciesielczuk, T., Karwaczyńska, U., & Sporek, M. (2015). The possibility of disposing of spent coffee ground with energy recycling. *Journal of Ecological Engineering*, *16*(4), 133–138. doi:10.12911/22998993/59361

Colak, T. A. I., Senel, G., & Goksel, C. (2021). GIS-based maritime spatial planning for site selection of offshore wind farms with exergy efficiency analysis: A case study. *International Journal of Exergy*, *34*(2), 255–268. doi:10.1504/IJEX.2021.113008

Colantoni, A., Paris, E., Bianchini, L., Ferri, S., Marcantonio, V., Carnevale, M., Palma, A., Civitarese, V., & Gallucci, F. (2021). Spent coffee ground characterization, pelletization test and emissions assessment in the combustion process. *Scientific Reports*, *11*(1), 1–14. doi:10.103841598-021-84772-y PMID:33664428

Corsi, I., Winther-Nielsen, M., Sethi, R., Punta, C., della Torre, C., Libralato, G., Lofrano, G., Sabatini, L., Aiello, M., Fiordi, L., Cinuzzi, F., Caneschi, A., Pellegrini, D., & Buttino, I. (2018). Ecofriendly nanotechnologies and nanomaterials for environmental applications: Key issue and consensus recommendations for sustainable and ecosafe nanoremediation. *Ecotoxicology and Environmental Safety*, *154*, 237–244. doi:10.1016/j.ecoenv.2018.02.037 PMID:29476973

Crini, G., & Lichtfouse, E. (2019). Advantages and disadvantages of techniques used for wastewater treatment. *Environmental Chemistry Letters*, *17*(1), 145–155. doi:10.100710311-018-0785-9

Curto, D. Franzitta, V., & Guercio, A. (2021) A Review of Water Desalination Techonologies. *Applied Science*, *11*, 670. doi:10.3390/app11020670

Damodharan, D., Sathiyagnanam, A. P., Kumar, B. R., & Ganesh, K. C. (2018). Cleaner emissions from a DI diesel engine fueled with waste plastic oil derived from municipal solid waste under the influence of n-pentanol addition, cold EGR, and injection timing. *Environmental Science and Pollution Research International*, *25*(14), 13611–13625. doi:10.100711356-018-1558-5 PMID:29497945

Daneshvar, B., Yildizbasi, A., & Arikan, Ü. (2018). *Using ıntuitionistic fuzzy TOPSIS in site selection of wind power plants in Turkey* (Vol. 2018). Advances in Fuzzy Systems.

Das, R., Vecitis, C. D., Schulze, A., Cao, B., Ismail, A. F., Lu, X., Chen, J., & Ramakrishna, S. (2017). Recent advances in nanomaterials for water protection and monitoring. *Chemical Society Reviews*, *46*(22), 6946–7020. doi:10.1039/C6CS00921B PMID:28959815

Datta, A., & Mandal, B. K. (2014). Use of Jatropha biodiesel as a future sustainable fuel. *Energy Technology & Policy*, *1*(1), 8–14. doi:10.1080/23317000.2014.930723

Dave, P. N., & Chopda, L. (2014). Application of iron oxide nanomaterials for the removal of heavy metals. *Journal of Nanotechnology*, *2014*, 1–14. Advance online publication. doi:10.1155/2014/398569

De aere, L. (2000). Anaerobic digestion of solid waste: state-of-theart. *Wat. Sci. Technol.*, *41*(3), 283-290.

De Luca, R., & Desideri, P. (2012). Wind energy: An application of Bernoulli's theorem generalized to isentropic flow of ideal gases. *European Journal of Physics*, *34*(1), 189–197. doi:10.1088/0143-0807/34/1/189

Den, W., Liu, H.-C., Chan, S.-F., Kin, K. T., & Huang, C. (2006). Adsorption of phthalate esters with multiwalled carbon nanotubes and its application. *Journal of Environmental Engineering and Management*, *16*(4). https://www.researchgate.net/publication/255583426

Dendena, B., & Corsi, S. (2014). Cashew, from seed to market: A review. *Agronomy for Sustainable Development*, *34*(4), 753–772. doi:10.100713593-014-0240-7

Depoures, M. V., Dillikannan, D., & Kaliyaperumal, G. (2019). Alcohol Contribution over Conventional Fuel. In *Introduction to Diesel Emissions*. IntechOpen.

Dey, S., Reang, N. M., Das, P. K., & Deb, M. (2021). A comprehensive study on prospects of economy, environment, and efficiency of palm oil biodiesel as a renewable fuel. *Journal of Cleaner Production*, *286*, 124981. doi:10.1016/j.jclepro.2020.124981

Dilmore, R. M., Howard, B. H., Soong, Y., Griffith, C., Hedges, S. W., DeGalbo, A. D., Morreale, B., Baltrus, J. P., Allen, D. E., & Fu, J. K. (2009). Sequestration of CO2 in Mixtures of Caustic Byproduct and Saline Waste Water. *Environmental Engineering Science*, *26*(8), 1325–1333. doi:10.1089/ees.2008.0395

Dinali, R., Ebrahiminezhad, A., Manley-Harris, M., Ghasemi, Y., & Berenjian, A. (2017). *Iron oxide nanoparticles in modern microbiology and biotechnology*. doi:10.1080/1040841X.2016.1267708

Dincer, I., & Rosen, M. (2007). *Exergy Analysis of Renewable Energy Systems.*, doi:10.1016/B978-008044529-8.50013-6

Dinesh Babu, G., & Shaafi, T. (2020). Use of animal fat methyl ester blend as a fuel in a CI engine. *IOP Conference Series. Materials Science and Engineering*, *992*(1), 012008. doi:10.1088/1757-899X/992/1/012008

Druckenmiller, M. L., & Maroto-Valer, M. M. (2005). Carbon sequestration using brine of adjusted pH to form mineral carbonates. *Fuel Processing Technology*, *86*(14–15), 1599–1614. doi:10.1016/j.fuproc.2005.01.007

Durgalakshmi, D., Rajendran, S., & Naushad, M. (2019). Current Role of Nanomaterials in Environmental Remediation. In *Advanced Nanostructured Materials for Environmental Remediation. Environmental Chemistry for a Sustainable World* (pp. 25–30). doi:10.1007/978-3-030-04477-0_1

Dutta, S., & Debnath, D. (2017). *A grid-connected single-phase transformerless inverter controlling two solar PV arrays operating under different atmospheric conditions*. https://ieeexplore.ieee.org/abstract/document/7938358/

DWAF-WRC. (2010). *Department of Water Affairs and Forestry and Water Research Commission. South African water quality management series*. Procedures to Assess Effluent Discharge Impacts.

Dwivedi, G., Jain, S., & Sharma, M. P. (2011). Pongamia as a source of biodiesel in India. *Smart Grid and Renewable Energy*, *2*(3), 184–189. doi:10.4236gre.2011.23022

Dwivedi, G., & Sharma, M. P. (2015). Investigation and improvement in cold flow properties of Pongamia biodiesel. *Waste and Biomass Valorization*, *6*(1), 73–79. doi:10.100712649-014-9330-7

Dwivedi, G., & Sharma, M. P. (2016). Experimental investigation on thermal stability of Pongamia Biodiesel by thermogravimetric analysis. *Egyptian Journal of Petroleum*, *25*(1), 33–38. doi:10.1016/j.ejpe.2015.06.008

Dwivedi, G., Sharma, M. P., & Kumar, M. (2014). Status and Policy of Biodiesel Development in India. *International Journal of Renewable Energy Research*, *4*(2), 246–254.

Echeverria, M. C., & Nuti, M. (2017). Valorisation of the residues of coffee agro-industry: Perspectives and limitations. *The Open Waste Management Journal*, *10*(1), 13–22. doi:10.2174/1876400201710010013

Ehler, C. (2018). In Offshore Energy and Marine Spatial Planning. In Marine Spatial Planning: an idea whose time has come. Routledge. doi:10.4324/9781315666877

Ehler, C., & Douvere, F. (2007). V*isions for a sea change: Report of the first international workshop on marine spatial planning*. UNESCO.

Ehler, C., & Douvere, F. (2010). An International Perspective on Marine Spatial Planning Initiatives, Environments. *Journal Volume Environments*, *37*(3), 9–20.

Elbeltagi, A., Azad, N., Arshad, A., Mohammed, S., Mokhtar, A., Pande, C., Etedali, H. R., Bhat, S. A., Islam, A. R. M. T., & Deng, J. (2021). Applications of Gaussian process regression for predicting blue water footprint: Case study in Ad Daqahliyah, Egypt. *Agricultural Water Management*, *255*, 107052. doi:10.1016/j.agwat.2021.107052

Emam, H. E., Badawi, A. K., Adeola, A. O., & Nomngongo, P. N. (2022). Advanced Polymeric Nanocomposites for Water Treatment Applications: A Holistic Perspective. *Polymers, 14*(12), 2462. doi:10.3390/polym14122462

Espuelas, S., Marcelino, S., Echeverría, A. M., Del Castillo, J. M., & Seco, A. (2020). Low energy spent coffee grounds briquetting with organic binders for biomass fuel manufacturing. *Fuel*, *278*, 118310. doi:10.1016/j.fuel.2020.118310

EU Commission Maritime Affairs and Fisheries. (2018). *Annual economic report on EU blue economy*. Directorate-General for Maritime Affairs and Fisheries European Commission.

European Commission. (2018). *Sewage Sludge*. Available at: https://ec.europa.eu/environment/waste/sludge/index.htm

Faisal, M., Mulana, F., Gani, A., & Daimon, H. (2015). Physical and chemical properties of wastewater discharged from Tofu industries in Banda Aceh city, Indonesia. *Research Journal of Pharmaceutical, Biological and Chemical Sciences*, *6*, 1053–1058.

Falco, S. D., & Veronesi, M. (2013). How can African agriculture adapt to climate change? A counterfactual analysis from Ethiopia. *Land Economics*, *89*(4), 743–766. doi:10.3368/le.89.4.743

Fan, X., Sun, H., Yuan, Z., Li, Z., Shi, R., & Ghadimi, N. (2020). High voltage gain DC/DC converter using coupled inductor and VM techniques. *IEEE Access, Vol, 8*, 131975–131987.

FAO. (1985). Industrial Charcoal Making. FAO Forestry Paper. Rome, Food and Agriculture Organization of the United Nations.

FAO. Food and Agriculture Organization of the United Nations. (2019). Available online: http://www.fao.org/ faostat/en/?#data

Faris, M. H., Mohammed, M. I., Alwash, R. S., & Kot, P. (2021). Phosphate removal from water using low-cost dolomite filters. IOP Conf. Series. *Materials Science and Engineering, 1058*, 012017.

Faust, S. D., & Aly, O. M. (2018). *Chemistry of water treatment*. CRC.

Fayad, M. A. (2019). Effect of fuel injection strategy on combustion performance and NO_x/smoke trade-off under a range of operating conditions for a heavy-duty DI diesel engine. *SN Appl. Sci., 1*(9), 1088. doi:10.100742452-019-1083-2

Fehse, F., Kummich, J., & Schröder, H. W. (2021). Influence of pre-treatment and variation of briquetting parameters on the mechanical refinement of spent coffee grounds. *Biomass and Bioenergy, 152*, 106201. doi:10.1016/j.biombioe.2021.106201

Fenati, R. A., Sherrell, P. C., Khodakov, D. A., Shapter, J. G., & Ellis, A. (2021). Spatially isolated redox processes enabled by ambipolar charge transport in multi-walled carbon nanotube mats. *Materials Horizons, 8*(4), 1304–1313. doi:10.1039/D0MH01967D PMID:34821923

Feng, R., Hu, X., Li, G., Sun, Z., & Deng, B. (2022). A comparative investigation between particle oxidation catalyst (POC) and diesel particulate filter (DPF) coupling aftertreatment system on emission reduction of a non-road diesel engine. *Ecotoxicology and Environmental Safety, 238*, 238. doi:10.1016/j.ecoenv.2022.113576 PMID:35512474

Feng, R., Sun, Z., Hu, X., Li, G., & Deng, B. (2022). Role of particle oxidation catalyst on emission reduction of a non-road diesel engine: A multi case study. *Chemical Engineering Science, 260*, 260. doi:10.1016/j.ces.2022.117914

Feng, X., Ge, Y., Ma, C., Tan, J., Yu, L., Li, J., & Wang, X. (2014). Experimental study on the nitrogen dioxide and particulate matter emissions from diesel engine retrofitted with particulate oxidation catalyst. *The Science of the Total Environment, 472*, 56–62. doi:10.1016/j.scitotenv.2013.11.041 PMID:24291555

Fermoso, J., & Mašek, O. (2018). Thermochemical decomposition of coffee ground residues by TG-MS: A kinetic study. *Journal of Analytical and Applied Pyrolysis, 130*, 358–367. doi:10.1016/j.jaap.2017.12.007

Fierros-González, I., & López-Feldman, A. (2021). Farmers' perception of climate change: A review of the literature for Latin America. *Frontiers in Environmental Science, 9*(672399), 1–7. doi:10.3389/fenvs.2021.672399

Firrisa, M. T., van Duren, I., & Voinov, A. (2014). Energy efficiency for rapeseed biodiesel production in different farming systems. *Energy Efficiency, 7*(1), 79–95. doi:10.100712053-013-9201-2

Fithratullah, R. (2022). The Study of Making Biomass Briquettes from Spent Coffee Ground. *Journal of Environmental Engineering and Waste Management, 7*(1), 34–53.

Fu, B. A., & Chen, M. Q. (2020). Microwave drying performance of spent coffee grounds briquette coupled with mineral additives. *Drying Technology, 38*(15), 2094–2101. doi:10.1080/07373937.2019.1692862

Fu, F., Dionysiou, D. D., & Liu, H. (2014). The use of zero-valent iron for groundwater remediation and wastewater treatment: A review. *Journal of Hazardous Materials, 267*, 194–205. doi:10.1016/j.jhazmat.2013.12.062 PMID:24457611

Fu, K., & Chen, W. (2019). Evaluation method of flyback converter behaviors on common-mode noise. *IEEE Access, Vol, 7*, 28019–28030.

Gadgil, S. (1995). Climate change and agriculture-an Indian perspective. *Current Science*, *69*(8), 649–659.

Gallareta-Olivares, G., Rivas-Sanchez, A., Cruz-Cruz, A., Hussain, S. M., González-González, R. B., Cárdenas-Alcaide, M. F., Iqbal, H. M. N., & Parra-Saldívar, R. (2023). Metal-doped carbon dots as robust nanomaterials for the monitoring and degradation of water pollutants. *Chemosphere*, *312*, 137190. doi:10.1016/j.chemosphere.2022.137190 PMID:36368530

Gan, C., Jin, N., & Sun, Q. (2018). Multiport bidirectional SRM drives for solar-assisted hybrid electric bus powertrain with flexible driving and self-charging functions. *IEEE Trans. Power Electron, 33*(10), 8231–8245.

Gandhi, T. S., Dholakiya, B. Z., & Patel, M. R. (2013). Extraction protocol for isolation of CNSL by using protic and aprotic solvents from cashew nut and study of their physico-chemical parameter. *Polish Journal of Chemical Technology*, *15*(4), 24–27. doi:10.2478/pjct-2013-0062

Gandhi, T., Patel, M., & Dholakiya, B. K. (2012). Studies on effect of various solvents on extraction of cashew nut shell liquid (CNSL) and isolation of major phenolic constituents from extracted CNSL. *J. Nat. Prod. Plant Resour*, *2*(1), 135–142.

Gangavarapu, S., & Akshay, K. R. G. (2019). A three-phase single-stage isolated flyback-based PFC converter with leakage energy recovery clamping circuit. *IEEE Transactions on Transportation Electrification, 5*(4), 1155–1168.

Gangavarapu, S., Rathore, A. K., & Khadkikar, V. (2019). High-Efficiency Three-Phase Single-Stage Isolated Flyback-Based PFC Converter With a Novel Clamping Circuit. *IEEE Transactions on Industry Applications, 56*(1), 718–729.

Ganthia & Barik. (2020). Steady-state and dynamic comparative analysis of PI and fuzzy logic controller in stator voltage oriented controlled DFIG fed wind energy conversion system. *Journal of The Institution of Engineers (India): Series B, 101*(3), 273-286.

Gaur, M., Misra, C., Yadav, A. B., Swaroop, S., Maolmhuaidh, F., Bechelany, M., & Barhoum, A. (2021). Biomedical Applications of Carbon Nanomaterials: Fullerenes, Quantum Dots, Nanotubes, Nanofibers, and Graphene. *Materials, 14*(20), 5978. doi:10.3390/ma14205978

Gaurav, D., & Sharma, M. P. (2014). Prospects of biodiesel from pongamia in India. *Renewable & Sustainable Energy Reviews*, *32*, 114–122. doi:10.1016/j.rser.2014.01.009

Gautam, R., & Kumar, S. (2020). Performance and Combustion Analysis Of Diesel and Tallow Biodiesel in CI Engine. *Energy Reports*, *6*(6), 2785–2793. doi:10.1016/j.egyr.2020.09.039

Ghadikolaei, M. A., Cheung, C. S., & Yung, K. F. (2018). Study of combustion, performance and emissions of diesel engine fueled with diesel/biodiesel/alcohol blends having the same oxygen concentration. *Energy*, *157*, 258–269. doi:10.1016/j.energy.2018.05.164

Ghafoor, A., Rauf, A., Arif, M., & Muzaffar, W. (1994). Chemical composition of effluents from different industries of the Faisalabad city. *Pakistan Journal of Agricultural Sciences*, *31*, 367–369.

Ghosh, A., Nayak, A. K., & Pal, A. (2016). Nano-Particle-Mediated Wastewater Treatment: a Review. *Current Pollution Reports, 3*(1), 17–30. doi:10.1007/s40726-016-0045-1

Gimba, C. E., & Muyiwa, T. (2008). Optimum conditions for carbonisation of coconut shell. *Scientia Africana, 7*(2).

Glazer, A. N. (1997). *Microbial biotechnology*. W.H. Freeman and Company.

Glushankova, I., Ketov, A., Krasnovskikh, M., Rudakova, L., & Vaisman, I. (2018). Rice hulls as a renewable complex material resource. *Resources*, *7*(2), 31. doi:10.3390/resources7020031

Godjo, T., Tagutchou, J. P., Naquin, P., & Gourdon, R. (2015). Valorisation des coques d'anacarde par pyrolyse au Bénin. *Déchets Sciences et Techniques*, *70*, 11–18. doi:10.4267/dechets-sciences-techniques.3282

Gogate, P. R., & Pandit, A. B. (2004). A review of imperative technologies for wastewater treatment I: Oxidation technologies at ambient conditions. *Advances in Environmental Research*, *8*(3-4), 501–551. doi:10.1016/S1093-0191(03)00032-7

Goh, B. H. H., Ong, H. C., Chong, C. T., Chen, W. H., Leong, K. Y., Tan, S. X., & Lee, X. J. (2020). Ultrasonic assisted oil extraction and biodiesel synthesis of Spent Coffee Ground. *Fuel*, *261*, 116121. doi:10.1016/j.fuel.2019.116121

Gómez-de la Cruz, F. J., Cruz-Peragón, F., Casanova-Pelaez, P. J., & Palomar-Carnicero, J. M. (2015). A vital stage in the large-scale production of biofuels from spent coffee grounds: The drying kinetics. *Fuel Processing Technology*, *130*, 188–196. doi:10.1016/j.fuproc.2014.10.012

Goncalves, A. A., & Gagnon, G. A. (2012). Recent Technologies for Ballast Water Treatment. *Ozone Science and Engineering*, *34*(3), 174–195. doi:10.1080/01919512.2012.663708

Gopakumar, D. A., Abdul, A. K., Pai, A. R., Thomas, S., Pasquini, D., & Ben, L. (2019). Nanomaterials—State of Art, New Challenges, and Opportunities. *Nanoscale Materials in Water Purification*, 1–24. doi:10.1016/B978-0-12-813926-4.00001-X

Gopal, G., Rao, P. S., Gopalakrishnan, K. V., & Murthy, B. S. (1982). Use of hydrogen in dual-fuel engines. *International Journal of Hydrogen Energy*, *7*(3), 267–272. doi:10.1016/0360-3199(82)90090-8

Grace, M. N., Wilson, G. M., & Leslie, P. F. (2012). Statistical testing of input factors in the carbonation of brine impacted fly ash. *Journal of Environmental Science and Health. Part A, Toxic/Hazardous Substances & Environmental Engineering*, *47*(2), 245–259. doi:10.1080/10934529.2012.640898 PMID:22242877

Granados-Luna, T.-R., Araujo-Vargas, I., Forsyth, A. J., Cano-Pulido, K., Velázquez-Elizondo, P.-E., Cervantes, I., Gómez-Olguín, F., & Villarruel-Parra, A. (2019). Two-Phase, Dual Interleaved Buck–Boost DC–DC Converter for Automotive Applications. *IEEE Transactions on Industry Applications, Vol*, *56*(1), 390–402.

Grigg, D. (2002). The worlds of tea and coffee: Patterns of consumption. *GeoJournal*, *57*(4), 283–294. doi:10.1023/B:GEJO.0000007249.91153.c3

Guerra, F. D., Attia, M. F., Whitehead, D. C., & Alexis, F. (2018). Nanotechnology for Environmental Remediation: Materials and Applications. *Molecules, 23*(7), 1760. doi:10.3390/molecules23071760

Guimaraes, J. S., Ricardo de Almeida, B., Tofoli, F. L., & Demercil de Souza, O. (2018). Three-phase grid-connected wecs with mechanical power control. *IEEE Transactions on Sustainable Energy, Vol*, *9*(4), 1508–1517.

Gumel, S. M., Usman, M. T., & Ado, A. (2015). Colouration Industry Wastewater Treatments in Nigeria–Hazard and Treatment: A Review. *International Journal of Chemical and Biomolecular Science*, *1*, 27–33. http://www.aiscience.org/journal/ijcbs

Guo, Q., Zhou, X., Li, X., Chen, S., Seema, A., Greiner, A., & Hou, H. (2009). Supercapacitors based on hybrid carbon nanofibers containing multiwalled carbon nanotubes. *Journal of Materials Chemistry*, *19*(18), 2810–2816. doi:10.1039/b820170f

Gürü, M., Koca, A., Can, Ö., Çınar, C., & Şahin, F. (2010). Biodiesel production from waste chicken fat based sources and evaluation with Mg based additive in a diesel engine. *Renewable Energy*, *35*(3), 637–643. doi:10.1016/j.renene.2009.08.011

Gutierrez, A. M., Dziubla, T. D., & Hilt, J. Z. (2017). Recent advances on iron oxide magnetic nanoparticles as sorbents of organic pollutants in water and wastewater treatment. *Reviews on Environmental Health*, *32*(1–2), 111–117. doi:10.1515/reveh-2016-0063 PMID:28231068

Gutu, L., Basitere, M., Harding, T., Ikumi, D., Njoya, M., & Gaszynski, C. (2021). Multi-Integrated Systems for Treatment of Abattoir Wastewater: A Review. *Water (Basel)*, *13*(18), 2462. doi:10.3390/w13182462

Hajlari, S. A., Najafi, B., & Ardabili, S. F. (2019). Castor oil, a source for biodiesel production and its impact on the diesel engine performance. *Renewable Energy Focus*, *28*, 1–10. doi:10.1016/j.ref.2018.09.006

HannahR.MaxR.RosadoP. (2022). *Energy*. Retrieved from: https://ourworldindata.org/energy

Hannan, M., Ghani, Z., & Hoque, M. (2019). *Fuzzy logic inverter controller in photovoltaic applications: Issues and recommendations*. https://ieeexplore.ieee.org/abstract/document/8642881/

Hao, M., Qiu, M., Yang, H., Hu, B., & Wang, X. (2021). Recent advances on preparation and environmental applications of MOF-derived carbons in catalysis. *The Science of the Total Environment*, *760*, 143333. doi:10.1016/j.scitotenv.2020.143333 PMID:33190884

Harini, S., & Chellammal, N. (2022). *A Novel High Gain Dual Input Single Output Z-Quasi Resonant (ZQR) DC/DC Converter for Off-Board EV Charging*. https://ieeexplore.ieee.org/abstract/document/9847236/

Harremöes, P. (1997). Integrated water and waste management. *Water Science and Technology*, *35*(9), 11–12. doi:10.2166/wst.1997.0327

Hashim, A., & Hajjaj, M. (2005). Impact of Desalination Plant of Alicante (SE Spain). *Desalination*, *182*, 373–393. doi:10.1016/j.desal.2005.04.020

Hashimoto, J., Ustun, T. S., Suzuki, M., Sugahara, S., Hasegawa, M., & Otani, K. (2021). Advanced Grid Integration Test Platform for Increased Distributed Renewable Energy Penetration in Smart Grids. *IEEE Access, Vol*, *9*, 34040–34053.

Hassan, Abbassi, Jerbi, Mehmood, Tahir, Cheema, Elavarasan, Ali, & Khan. (2020). A novel algorithm for MPPT of an isolated PV system using push pull converter with fuzzy logic controller. *Energies, 13*(15), 4007.

Heiss, A., & Kuster, T. (2015). A regulatory perspective on prioritization of emerging pollutants in the context of the Water Framework Directive. *Environmental Toxicology and Chemistry*, *34*(10), 2181–2187. doi:10.1002/etc.3047 PMID:26414540

Heywood, I., Cornelius, S., & Carver, S. (2006). *An Introduction to Geographical Information Systems* (3rd ed.). Pearson Education Limited.

Heywood, J. B. (1988). *Internal combustion engine fundamentals. McGraw-Hill series in mechanical engineering*. McGraw-Hill Inc.

Himabindu, T., Raguram, V., & Kum, S. T. (2015). Review on various methods of extraction of cashew nut shell liquid and isolation of anacardic acid. *Int J Inst Pharm Life Sci*, *5*, 293–308.

Himbane, P. B., Grand Ndiaye, L., Napoli, A., & Kobor, D. (2018). *Physicochemical and mechanical properties of biomass coal briquettes produced by artisanal method*. Academic Press.

Hoang, A. T. (2021). Prediction of the density and viscosity of biodiesel and the influence of biodiesel properties on a diesel engine fuel supply system. *Journal of Marine Engineering and Technology*, *20*(5), 299–311. doi:10.1080/20464177.2018.1532734

Hoepner, T., & Lattemann, S. (2003). Chemical Impacts of Seawater Desalination Plants– A Case Study of Northern red Sea. *Desalination*, *152*(1-3), 133–140. doi:10.1016/S0011-9164(02)01056-1

Holechek, J. L., Geli, H. M. E., Sawalhah, M. N., & Valdez, R. (2022). A Global Assessment: Can Renewable Energy Replace Fossil Fuels by 2050? *Sustainability*, *14*(8), 4792. doi:10.3390u14084792

Huang, F., Singer, G., Zhou, Y., Sha, Z., Chen, J., Han, Z., Brown, S. A., Zhang, J., & Wang, C. H. (2021). Creating ionic pathways in solid-state polymer electrolyte by using PVA-coated carbon nanofibers. *Composites Science and Technology, 207*, 108710. doi:10.1016/j.compscitech.2021.108710

Huang, J., & Crookes, R. (1998). Assessment of simulated biogas as a fuel for the spark ignition engine. *Fuel, 77*(15), 1793–1801. doi:10.1016/S0016-2361(98)00114-8

Huang, Y., Finell, M., Larsson, S., Wang, X., Zhang, J., Wei, R., & Liu, L. (2017). Biofuel pellets made at low moisture content–Influence of water in the binding mechanism of densified biomass. *Biomass and Bioenergy, 98*, 8–14. doi:10.1016/j.biombioe.2017.01.002

Huang, Z., Lu, H., Jiang, D., Zeng, K., Liu, B., Zhang, J., & Wang, X. (2004). Combustion behaviors of a compression-ignition engine fueled with diesel/methanol blends under various fuel delivery advance angles. *Bioresource Technology, 95*(3), 331–341. doi:10.1016/j.biortech.2004.02.018 PMID:15288277

Hu, B., Wang, H., Liu, R., & Qiu, M. (2021). Highly efficient U(VI) capture by amidoxime/carbon nitride composites: Evidence of EXAFS and modeling. *Chemosphere, 274*, 129743. doi:10.1016/j.chemosphere.2021.129743 PMID:33540307

Huko, D., Kamau, D. N., & Ogola, W. O. (2015). Effects of varying particle size on mechanical and combustion characteristics of mango seed shell cashew nut shell composite briquettes. *Int. J. Eng. Sci. Invent, 4*, 32–39.

Huong, N. T. L., Bo, Y. S., & Fahad, S. (2019). Economic impact of climate change on agriculture using Ricardian approach: A case of northwest Vietnam. *Journal of the Saudi Society of Agricultural Sciences, 18*(4), 449–457. doi:10.1016/j.jssas.2018.02.006

Hwang, C. L., & Yoon, K. P. (1981). *Multiple attribute decision making: methods and applications, a state-of-the-art survey*. SpringerVerlang. doi:10.1007/978-3-642-48318-9

Ibhadon, A. O., & Fitzpatrick, P. (2013). Heterogeneous Photocatalysis: Recent Advances and Applications. *Catalysts, 3*(1), 189–218. doi:10.3390/catal3010189

ICO. (2010). Available online: http://www.ico.org/historical/1990%20onwards/PDF/1a-total-production.pdf

ICO. (2018). *Coffee market recovers slightly from December slump*. http:// www.ico.org

ICO. (2019). *Coffee Imports by Selected Importing Countries, 2019*. http://www.ico.org/historical/1990 onwards/PDF/2b-imports.pdf

ICO. (2020). *Crop Year Production by Country, 2020*. http://www.ico.org/prices/po-production.pdf

ICO. (2021). *Coffee Market Report January 2021*. https://www.ico.org/documents/cy2020-21/cmr-0121-e.pdf

Idah, P. A., Simeon, M. I., & Mohammed, M. A. (2014). *Extraction and characterization of cashew nut (Anacardium occidentale) oil and cashew shell liquid oil*. Academic Press.

Idris-Nda, A., Aliyu, H. K., & Dalil, M. (2013). The challenges of domestic wastewater management in Nigeria: A case study of Minna, central Nigeria. *International Journal of Development and Sustainability, 2*, 1169–1182. www.isdsnet.com/ijds

IEA. (n.d.). *Energy in India Today – india energy outlook 2021 – analysis*. IEA. Retrieved December 1, 2022, from https://www.iea.org/reports/india-energy-outlook-2021/energy-in-india-today

Ifa, L. (2019). Production of bio-briquette from biochar derived from pyrolysis of cashew nut waste. *Ecology, Environment, and Conservation, 25*, 125–131.

Ifa, L., Yani, S., Nurjannah, N., Darnengsih, D., Rusnaenah, A., Mel, M., Mahfud, M., & Kusuma, H. S. (2020). Techno-economic analysis of bio-briquette from cashew nut shell waste. *Heliyon, 6*(9), e05009. doi:10.1016/j.heliyon.2020.e05009 PMID:33005808

Iheukwumere, S. O., Phil-Eze, P. O., & Nkwocha, K. F. (2021). Water resource management in Nigeria: Acceptability of treated municipal wastewater reuse in Federal Capital City, Abuja. *British Journal of Environmental Science, 9*, 26–39.

İkbal, F. (2010). *Rüzgâr Enerjisi Sistemlerinde Optimizasyon.* İstanbul Üniversitesi Fen Bilimleri Enstitüsü Elektrik-Elektronik Mühendisliği Anabilim Dalı.

Ilankoon, N. (2014). *Use of iron oxide magnetic nanosorbe...* Retrieved February 9, 2023, from https://scholar.google.com/scholar?hl=en&as_sdt=0%2C5&q=Ilankoon%2C+N.+%282014%29.+Use+of+iron+oxide+magnetic+nanosorbents+for+Cr%28VI%29+removal+from+aqueous+solutions%3A+A+review.+J.+Eng.+Res+Appl.%2C+4%2C+55%E2%80%9363.&btnG=

Ilkiliç, C., & Aydin, H. (2015). Wind power potential and usage in the coastal regions of Turkey. *Renewable & Sustainable Energy Reviews, 44*, 78–86. doi:10.1016/j.rser.2014.12.010

India Brand Equity Foundation. (n.d.). *Indian oil and Gas Industry Analysis Presentation: IBEF.* Retrieved November 30, 2022, from https://www.ibef.org/industry/indian-oil-and-gas-industry-analysis-presentation

IOP Publishing doi:10.1088/1757-899X/1058/1/01201

IPCC. (2018). Global Warming of 1.5°C. An IPCC Special Report on the impacts of global warming of 1.5°C above pre-industrial levels and related global greenhouse gas emission pathways, in the context of strengthening the global response to the threat of climate change, sustainable development, and efforts to eradicate poverty. IPCC.

Irshad, M. A., Nawaz, R., Wojciechowska, E., Mohsin, M., Nawrot, N., Nasim, I., & Hussain, F. (2023). Application of Nanomaterials for Cadmium Adsorption for Sustainable Treatment of Wastewater: A Review. *Water, Air, and Soil Pollution, 234*(1), 1–17. doi:10.100711270-023-06064-7

Islam, R., Banerjee, G., & Ali, F. (2014). Design and Benefit Analysis of Biogas Plant for Rural Development in Bangladesh. *International Journal of Engineering and Advanced Technology, 3*(3).

Ismaila, A., Zakari, I. Y., Nasiru, R., Tijjani, B. I., Abdullahi, I., & Garba, N. N. (2013). Investigation on biomass briquettes as energy source in relation to their calorific values and measurement of their total carbon and elemental contents for efficient biofuel utilization. *Advances in Applied Science Research, 4*(4), 303–309.

Jacobs, A. D., & Hitch, M. (2011). Experimental mineral carbonation: Approaches to accelerate CO2 sequestration in mine waste materials. *International Journal of Mining, Reclamation and Environment, 25*(4), 321–331. doi:10.1080/17480930.2011.608883

Jain, S., & Sharma, M. P. (2011). Oxidation stability of blends of Jatropha biodiesel with diesel. *Fuel, 90*(10), 3014–3020. doi:10.1016/j.fuel.2011.05.003

Jamal, A. M., Antwi-Agyei, P., Baffour-Ata, F., Nkiaka, E., Antwi, K., & Gbordzor, A. (2021). Gendered perceptions and adaptation practices of smallholder cocoa farmers to climate variability in the Central Region of Ghana. *Environmental Challenges, 5*(100293), 1–7. doi:10.1016/j.envc.2021.100293

Jat, M. L., Dagar, J. C., Singh, Y., Govaerts, B., Ridaura, S. L., Saharawat, Y. S., Sharma, R. K., Tetarwal, J. P., Jat, R. K., Hobbs, H., & Stirling, C. (2016). Climate change and agriculture: Adaptation strategies and mitigation opportunities for food security in South Asia and Latin America. *Advances in Agronomy, 137*(1), 127–235. doi:10.1016/bs.agron.2015.12.005

Jhansi, S. C., & Mishra, S. K. (2013). Wastewater treatment and reuse: Sustainability options. *Journal of Sustainable Development, 10*, 1–15.

Jindal, S., Nandwana, B. P., Rathore, N. S., & Vashistha, V. (2010). Experimental investigation of the effect of compression ratio and injection pressure in a direct injection diesel engine running on Jatropha methyl ester. *Applied Thermal Engineering, 30*(5), 442–448. doi:10.1016/j.applthermaleng.2009.10.004

Jo, H. Y., Kim, J. H., Lee, Y. J., Lee, M., & Choh, S.-J. (2012). Evaluation of factors affecting mineral carbonation of CO2 using coal fly ash in aqueous solutions under ambient conditions. *Chemical Engineering Journal, 183*, 77–87. doi:10.1016/j.cej.2011.12.023

John, B. (1988). *Heywood Internal Combustion Engine Fundamentals*. Mc-Graw-Hill.

Joseph, E. I. (2007). Wastewater treatment in the textile industry. *Dyeing Printing Finishing, 10*, 60-66.

Ju-Nam, Y., & Lead, J. (2016). Properties, Sources, Pathways, and Fate of Nanoparticles in the Environment. *Engineered Nanoparticles and the Environment: Biophysicochemical Processes and Toxicity*, 95–117. doi:10.1002/9781119275855.ch6

Jyoti, B., & Singh, A. K. (2020). Projected sugarcane yield in different climate change scenarios in Indian states: A state-wise panel data exploration. *International Journal of Food and Agricultural Economics, 8*(4), 343–365.

Kadam, A., Saratale, R. G., Shinde, S., Yang, J., Hwang, K., Mistry, B., Saratale, G. D., Lone, S., Kim, D. Y., Sung, J. S., & Ghodake, G. (2019). Adsorptive remediation of cobalt oxide nanoparticles by magnetized α-cellulose fibers from waste paper biomass. *Bioresource Technology, 273*, 386–393. doi:10.1016/j.biortech.2018.11.041 PMID:30458408

Kalaitzidou, K., Zouboulis, A., & Mitrakas, M. (2020). Cost evaluation for Se(IV) removal, by applying common drinking water treatment processes: Coagulation/precipitation or adsorption. *Journal of Environmental Chemical Engineering, 8*(5), 104209. doi:10.1016/j.jece.2020.104209

Kalam, M. A., & Masjuki, H. H. (2002). Biodiesel from palmoil—An analysis of its properties and potential. *Biomass and Bioenergy, 23*(6), 471–479. doi:10.1016/S0961-9534(02)00085-5

Kaliyan, N., & Morey, R. V. (2009). Factors affecting strength and durability of densified biomass products. *Biomass and Bioenergy, 33*(3), 337–359. doi:10.1016/j.biombioe.2008.08.005

Kangogo, D., Dentoni, D., & Bijman, J. (2021). Adoption of climate-smart agricultural among smallholder farmers: Does farmers entrepreneurship matter? *Land Use Policy, 109*(1), 1–13.

Kang, S. B., Oh, H. Y., Kim, J. J., & Choi, K. S. (2017). Characteristics of spent coffee ground as a fuel and combustion test in a small boiler (6.5 kW). *Renewable Energy, 113*, 1208–1214. doi:10.1016/j.renene.2017.06.092

Kan, T., Strezov, V., & Evans, T. (2014). Catalytic pyrolysis of coffee grounds using NiCu-impregnated catalysts. *Energy & Fuels, 28*(1), 228–235. doi:10.1021/ef401511u

Kareem, F. (2018). Applied second law of thermodynamics on the wind energy over Iraq. *Journal of Engineering and Sustainable Development., 22*(02), 75–92. Advance online publication. doi:10.31272/jeasd.2018.2.20

Karmee, S. K. (2018). A spent coffee grounds based biorefinery for the production of biofuels, biopolymers, antioxidants and biocomposites. *Waste Management (New York, N.Y.), 72*, 240–254. doi:10.1016/j.wasman.2017.10.042 PMID:29150258

Karthikeyan, V., Sundaramoorthy, K., Kumar, G. G., & Babaei, E. (2019). Regenerative switched-inductor/capacitor type DC–DC converter with large voltage gain for PV applications. *Wiley Online Library, 13*(1), 68–77. doi:10.1049/iet-pel.2019.0408

Karunanithy, C., Wang, Y., Muthukumarappan, K., & Pugalendhi, S. (2012). Physiochemical characterization of briquettes made from different feedstocks. *Biotechnology Research International, 2012*, 2012. doi:10.1155/2012/165202 PMID:22792471

Kathuria, R. S., & Grover, S. (2012). *Using agricultural residues as a biomass briquetting: An alternative source of energy*. Academic Press.

Kaur, R., Wani, S. P., Singh, A. K., & Lal, K. (n.d.). *Wastewater production, treatment and use in India*. Academic Press.

Keera, S. T., El Sabagh, S. M., & Taman, A. R. (2018). Castor oil biodiesel production and optimization. *Egyptian Journal of Petroleum, 27*(4), 979-984.

Kelkar, S., Saffron, C. M., Chai, L., Bovee, J., Stuecken, T. R., Garedew, M., Li, Z., & Kriegel, R. M. (2015). Pyrolysis of spent coffee grounds using a screw-conveyor reactor. *Fuel Processing Technology, 137*, 170–178. doi:10.1016/j.fuproc.2015.04.006

Keskin, A., Gürü, M., & Altiparmak, D. (2007). Biodiesel production from tall oil synthesized Mn and Ni-based additives: Effects of the additives on fuel consumption and emissions. *Fuel, 86*(7-8), 139–143. doi:10.1016/j.fuel.2006.10.021

Khan, A., Chatterjee, S., & Weng, Y. (2021). *3-UHI drivers and mapping the urban thermal environment. Urban Heat Island Modeling for Tropical Climates*. Elsevier.

Khan, S. T., & Malik, A. (2019). Engineered nanomaterials for water decontamination and purification: From lab to products. *Journal of Hazardous Materials, 363*, 295–308. doi:10.1016/j.jhazmat.2018.09.091 PMID:30312926

Khatun, K., Vakacharla, V. R., Kizhakkan, A. R., & Rathore, A. K. (2020). Small-signal analysis and control of snubberless naturally clamped soft-switching current-fed push–pull DC/DC converter. *IEEE Transactions on Industry Applications, Vol, 56*(4), 4299–4308.

Kidd, J., Westerhoff, P., & Maynard, A. (2021). Survey of industrial perceptions for the use of nanomaterials for in-home drinking water purification devices. *NanoImpact, 22*, 100320. doi:10.1016/j.impact.2021.100320 PMID:35559977

Kilingo, F., Bernard, Z., & Hong–bin, C. (2021). The Analysis of Wastewater Treatment System Efficiencies in Kenya: A Review. *International Journal of Scientific and Research Publications, 11*(5), 204–214. doi:10.29322/IJSRP.11.05.2021.p11322

Kimutai, S. K. (2021). Effects of Parameters on Cashew Nutshell and Cassava Binder Briquettes Densification: Structural Equation Modelling. *European Journal of Energy Research, 1*(2), 12–16. doi:10.24018/ejenergy.2021.1.2.10

Kimutai, S. K., & Kimutai, I. K. (2019). Investigation of physical and combustion properties of briquettes from cashew nut shell and cassava binder. *International Journal of Educational Research, 7*, 15–26.

Kim, Y., Park, T., & Hong, D. (2022). Heating and emission characteristics of briquettes developed from spent coffee grounds. *Environmental Engineering Research, 27*(4), 210063. doi:10.4491/eer.2021.063

Kolincio, M., Chrzan, P. J., & Musznicki, P. (2019). Multitransformer primary-side regulated flyback converter for supplying isolated IGBT and MOSFET drivers. *IEEE Transactions on Industrial Electronics, Vol, 67*(2), 1005–1012.

Kondamudi, N., Mohapatra, S. K., & Misra, M. (2008). Spent coffee grounds as a versatile source of green energy. *Journal of Agricultural and Food Chemistry, 56*(24), 11757–11760. doi:10.1021/jf802487s PMID:19053356

Konur, O. (2021). Rapeseed oil-based biodiesel fuels: A review of the research. *Biodiesel Fuels Based on Edible and Nonedible Feedstocks, Wastes, and Algae*, 497-514.

Kourmentza, C., Economou, C. N., Tsafrakidou, P., & Kornaros, M. (2018). Spent coffee grounds make much more than waste: Exploring recent advances and future exploitation strategies for the valorization of an emerging food waste stream. *Journal of Cleaner Production*, *172*, 980–992. doi:10.1016/j.jclepro.2017.10.088

Kumar, V., Singh, S., Singh, J., & Upadhyay, N. (2015). Erratum: Potential of plant growth promoting traits by bacteria isolated from heavy metal contaminated soils. Bulletin of Environmental Contamination and Toxicology, 91(6). doi:10.100700128-015-1551-3

Kumar, A., Ahmad, M. M., & Sharma, P. (2017). Influence of climatic and non-climatic factors on sustainable food security in India: A statistical investigation. *International Journal of Sustainable Agricultural Management and Informatics*, *3*(1), 1–30. doi:10.1504/IJSAMI.2017.082917

Kumar, A., & Sharma, P. (2014). Climate change and sugarcane productivity in India: An econometric analysis. *Journal of Social and Development Sciences*, *5*(2), 111–122. doi:10.22610/jsds.v5i2.811

Kumar, A., Sharma, P., & Ambrammal, S. K. (2014). Climatic effects on food grain productivity in India: A crop wise analysis. *Journal of Studies in Dynamics and Change*, *1*(1), 38–48.

Kumar, A., Sharma, P., & Ambrammal, S. K. (2015). Climatic effects on sugarcane productivity in India: A stochastic production function application. *International Journal of Economics and Business Research*, *10*(2), 179–203. doi:10.1504/IJEBR.2015.070984

Kumar, A., Sharma, P., & Joshi, S. (2016). Assessing the impacts of climate change on land productivity in Indian crop agriculture: An evidence from panel data analysis. *Journal of Agricultural Science and Technology*, *18*(1), 1–13.

Kumar, B. R., Muthukkumar, T., Krishnamoorthy, V., & Saravanan, S. (2016). A comparative evaluation and optimization of performance and emission characteristics of a DI diesel engine fueled with n-propanol/diesel', n-butanol/diesel and n-pentanol/diesel blends using response surface methodology. *RSC Advances*, *6*(66), 61869–61890. doi:10.1039/C6RA11643D

Kurtaran, A. (2019). İsdemir Kuvvet Santrali Enerji ve Ekserji Analizi. Batman Üniversitesi Fen Bilimleri Enstitüsü Otomotiv Mühendisliği Anabilim Dalı, Batman.

Labeckas, G., Slavinskas, S., & Kanapkiene, I. (2019). Study of the Effects of Biofuel-Oxygen of Various Origins on a CRDI Diesel Engine Combustion and Emissions. *Energies*, 12, 1241. *Int*, *24*(4), 389–403.

Lacour, S., Chinese, T., Alkandee, D., Perilhon, C., & Descombes, G. (2012). Energy and environmental balance of biogas for dual-fuel mobile applications. *Renewable & Sustainable Energy Reviews*, *16*(3), 1745–1753. doi:10.1016/j.rser.2011.11.029

Lamastra, L., Balderacchi, M., & Trevisan, M. (2016). Inclusion of emerging organic contaminants in groundwater monitoring plans. *MethodsX*, *3*, 459–476. doi:10.1016/j.mex.2016.05.008 PMID:27366676

Lam, E., & Luong, J. H. T. (2014). Carbon materials as catalyst supports and catalysts in the transformation of biomass to fuels and chemicals. *ACS Catalysis*, *4*(10), 3393–3410. doi:10.1021/cs5008393

Lapuerta, M., Rodríguez-Fernández, J., & Font de Mora, E. (2009). Correlation for the estimation of the cetane number of biodiesel fuels and implications on the iodine number. *Energy Policy*, *37*(11), 4337–4344. doi:10.1016/j.enpol.2009.05.049

Larramendy, M. L., Soloneski, S., Larramendy, M. L., & Soloneski, S. (2015). *Emerging Pollutants in the Environment - Current and Further Implications*. Emerging Pollutants in the Environment - Current and Further Implications. doi:10.5772/59332

Latinopoulos, D., & Kechagia, K. (2015). GIS-based multi-criteria evaluation for wind farm site selection. A regional scale application in Greece. *Renewable Energy*, *78*, 550–560. doi:10.1016/j.renene.2015.01.041

Lawrence, D., Kenchington, R., & Woodley, S. (2002). *The Great Barrier Reef: Finding the Right Balance*. Melbourne University Press.

Lee, X. J., Lee, L. Y., Gan, S., Thangalazhy-Gopakumar, S., & Ng, H. K. (2017). Biochar potential evaluation of palm oil wastes through slow pyrolysis: Thermochemical characterization and pyrolytic kinetic studies. *Bioresource Technology*, *236*, 155–163. doi:10.1016/j.biortech.2017.03.105 PMID:28399419

Lee, X. J., Ong, H. C., Gao, W., Ok, Y. S., Chen, W. H., Goh, B. H. H., & Chong, C. T. (2021). Solid biofuel production from spent coffee ground wastes: Process optimisation, characterisation and kinetic studies. *Fuel*, *292*, 120309. doi:10.1016/j.fuel.2021.120309

Leyland, D., Chivavava, J., & Lewis, A. E. (2019). Investigations into Ice Scaling during Eutectic Freeze Crystallization of Brine Streams at Low Scraper Speeds and High Supersaturation. *Separation and Purification Technology*, *220*, 33–41. doi:10.1016/j.seppur.2019.03.025

Liang, X., Zheng, Z., Zhang, H., Wang, Y., & Yu, H. (2019). A Review of Early Injection Strategy in Premixed Combustion Engines. *Applied Sciences (Basel, Switzerland)*, *9*(18), 3737. doi:10.3390/app9183737

Liao, R., Li, Q., & Mao, X. (2019). A Prototype for Detection of Particles in Seawater by using Polarized-Light Scattering. *Oceans*, 1–4. Advance online publication. doi:10.1109/OCEANSE.2019.8867414

Li, J., Wang, X., Zhao, G., Chen, C., Chai, Z., Alsaedi, A., Hayat, T., & Wang, X. (2018). Metal–organic framework-based materials: Superior adsorbents for the capture of toxic and radioactive metal ions. *Chemical Society Reviews*, *47*(7), 2322–2356. doi:10.1039/C7CS00543A PMID:29498381

Li, M., Yin, J. J., Wamer, W. G., & Lo, Y. M. (2014). Mechanistic characterization of titanium dioxide nanoparticle-induced toxicity using electron spin resonance. *Yao Wu Shi Pin Fen Xi*, *22*(1), 76–85. doi:10.1016/j.jfda.2014.01.006 PMID:24673905

Limousy, L., Jeguirim, M., Dutournié, P., Kraiem, N., Lajili, M., & Said, R. (2013). Gaseous products and particulate matter emissions of biomass residential boiler fired with spent coffee grounds pellets. *Fuel*, *107*, 323–329. doi:10.1016/j.fuel.2012.10.019

Li, Q., Mahendra, S., Lyon, D. Y., Brunet, L., Liga, M., Li, D., & Alvarez, P. J. J. (2008). Antimicrobial nanomaterials for water disinfection and microbial control: Potential applications and implications. *Water Research*, *42*(18), 4591–4602. doi:10.1016/j.watres.2008.08.015 PMID:18804836

Lisowski, A., Olendzki, D., Świętochowski, A., Dąbrowska, M., Mieszkalski, L., Ostrowska-Ligęza, E., Stasiak, M., Klonowski, J., & Piątek, M. (2019). Spent coffee grounds compaction process: Its effects on the strength properties of biofuel pellets. *Renewable Energy*, *142*, 173–183. doi:10.1016/j.renene.2019.04.114

Liu, J., Wang, K., Zheng, Z., & Li, C. (2019). A dual-active-clamp quasi-resonant isolated boost converter for PV integration to medium-voltage DC grids. doi:10.1109/JESTPE.2019.2924145

Liu, F., Zhu, S., Li, D., Chen, G., & Ho, S. H. (2020). Detecting Ferric Iron by Microalgal Residue-Derived Fluorescent Nanosensor with an Advanced Kinetic Model. *iScience*, *23*(6), 101174. doi:10.1016/j.isci.2020.101174 PMID:32498017

Liu, Q., & Maroto-Valer, M. M. (2011). Parameters affecting mineral trapping of CO 2 sequestration in brines: Parameters affecting mineral trapping of CO 2 sequestration in brines. Greenhouse Gases. *Science and Technology*, *1*(3), 211–222. doi:10.1002/ghg.29

Liu, X., Pang, H., Liu, X., Li, Q., Zhang, N., Mao, L., Qiu, M., Hu, B., Yang, H., & Wang, X. (2021). Orderly Porous Covalent Organic Frameworks-based Materials: Superior Adsorbents for Pollutants Removal from Aqueous Solutions. *The Innovation*, *2*(1), 100076. doi:10.1016/j.xinn.2021.100076 PMID:34557733

Liu, Z., Ge, Y., Tan, J., He, C., Shah, A. N., Ding, Y., Yu, L., & Zhao, W. (2012). Impacts of continuously regenerating trap and particle oxidation catalyst on the NO2 and particulate matter emissions emitted from diesel engine. *Journal of Environmental Sciences (China)*, *24*(4), 624–631. doi:10.1016/S1001-0742(11)60810-3 PMID:22894096

Liyanaarachchi, S., Shu, L., Muthukumaran, S., Jegatheesan, V., & Baskaran, K. (2014). *Problems in seawater industrial desalination processes and potential sustainable solutions: A review*. doi:10.1007/s11157-013-9326-y

Loganathan, P., Kandasamy, V. J., & Bolan, N. S. (2014). Removal and Recovery of Phosphate from Water Using Sorption. *Critical Reviews in Environmental Science and Technology*, *44*(8), 847–907. doi:10.1080/10643389.2012.741311

Lubwama, M., & Yiga, V. A. (2017). Development of groundnut shells and bagasse briquettes as sustainable fuel sources for domestic cooking applications in Uganda. *Renewable Energy*, *111*, 532–542. doi:10.1016/j.renene.2017.04.041

Luna-Lama, F., Rodríguez-Padrón, D., Puente-Santiago, A. R., Muñoz-Batista, M. J., Caballero, A., Balu, A. M., Romero, A. A., & Luque, R. (2019). Non-porous carbonaceous materials derived from coffee waste grounds as highly sustainable anodes for lithium-ion batteries. *Journal of Cleaner Production*, *207*, 411–417. doi:10.1016/j.jclepro.2018.10.024

Lungkhimba, H. M., Karki, A. B., & Shrestha, J. N. (2010). Biogas Production from Anaerobic Digestion of Biodegradable Household Wastes. *Nepal Journal of Science and Technology*, *11*, 167–172. doi:10.3126/njst.v11i0.4140

Lu, S. Y., Jin, M., Zhang, Y., Niu, Y. B., Gao, J. C., & Li, C. M. (2018). Chemically Exfoliating Biomass into a Graphene-like Porous Active Carbon with Rational Pore Structure, Good Conductivity, and Large Surface Area for High-Performance Supercapacitors. *Advanced Energy Materials*, *8*(11), 1702545. doi:10.1002/aenm.201702545

Luu, A. T., Hguyen, A. T., Trinh, Q. A., Pham, V. T., Le, B. B., Nguyen, D. T., Hoang, Q. N., Pham, H. T. T., Nguyen, T. K., Luu, V. N., & Hens, L. (2019). Farmers' intention to climate change adaptation in agriculture in the red river Delta Biosphere Reserve (Vietnam): A combination of structural equation modelling (SEM) and protection motivation theory (PMT). *Sustainability*, *11*(2993), 1–17. doi:10.3390u11102993

Luz, F. C., Volpe, M., Fiori, L., Manni, A., Cordiner, S., Mulone, V., & Rocco, V. (2018). Spent coffee enhanced biomethane potential via an integrated hydrothermal carbonization-anaerobic digestion process. *Bioresource Technology*, *256*, 102–109. doi:10.1016/j.biortech.2018.02.021 PMID:29433044

Ma, J. (2012). *CO2 Mineralization Using Reactive Species* [MSc Thesis].

Machado Corrêa, S., & Arbilla, G. (2008). Carbonyl emissions in diesel and biodiesel exhaust. *Atmospheric Environment*, *42*(4), 769–775. doi:10.1016/j.atmosenv.2007.09.073

Madannejad, R., Shoaie, N., Jahanpeyma, F., Darvishi, M. H., Azimzadeh, M., & Javadi, H. (2019). Toxicity of carbon-based nanomaterials: Reviewing recent reports in medical and biological systems. *Chemico-Biological Interactions*, *307*, 206–222. doi:10.1016/j.cbi.2019.04.036 PMID:31054282

Mahdy, M., & Bahaj, A. S. (2018). Multi-criteria decision analysis for offshore wind energy potential in Egypt. *Renewable Energy*, *118*, 278–289. doi:10.1016/j.renene.2017.11.021

Maia, J. G. S., Andrade, E. H. A., & Zoghbi, M. G. B. (2000). Volatile constituents of the leaves, fruits and flowers of cashew (Anacardium occidentale L.). *Journal of Food Composition and Analysis*, *13*(3), 227–232. doi:10.1006/jfca.2000.0894

Makopondo, R. O., Rotich, L. K., & Kamau, C. G. (2020). Potential Use and Challenges of Constructed Wetlands for Wastewater Treatment and Conservation in Game Lodges and Resorts in Kenya. *TheScientificWorldJournal*, *6*, 4–15. doi:10.1155/2020/9184192 PMID:33122972

Malczewski, J. (1999). *GIS and multi-criteria decision analysis*. John Wiley & Sons.

Malczewski, J. (2006). GIS-based multi-criteria decision analysis: A survey of the literature. *International Journal of Geographical Information Science*, *20*(7), 703–726. doi:10.1080/13658810600661508

Malczewski, J., & Rinner, C. (2015). *Multicriteria decision analysis in geographic ınformation science* (1st ed.). Springer-Verlag Berlin Heidelberg. doi:10.1007/978-3-540-74757-4

Malhi, G. S., Kaur, M., & Kaushik, P. (2021). Impact of climate change on agriculture and its mitigation strategies. *Sustainability*, *13*(1318), 1–21. doi:10.3390u13031318

Manikandan, R., &Saha, N. (2015). A control algorithm for nonlinear offshore structural dynamical systems. *Proceedings of the Royal Society of London A: Mathematical, Physical and Engineering Sciences, The Royal Society, 471*(2184).

Man, P., He, B., Zhang, Q., Zhou, Z., Li, C., Li, Q., Wei, L., & Yao, Y. (2019). A one-dimensional channel self-standing MOF cathode for ultrahigh-energy-density flexible Ni–Zn batteries. *Journal of Materials Chemistry. A, Materials for Energy and Sustainability*, *7*(48), 27217–27224. doi:10.1039/C9TA11759H

Manwell, J., Mcgowan, J., & Rogers, A. (2006). *Wind Energy Explained: Theory* (2nd ed.). Design and Application., doi:10.1260/030952406778055054

Maroti, P., & Padmanaban, S. (2019). *A new structure of high voltage gain SEPIC converter for renewable energy applications*. https://ieeexplore.ieee.org/abstract/document/8750785/

Maroto-Valer, M. M., Kuchta, M. E., Zhang, Y., Andrésen, J. M., & Fauth, D. J. (n.d.). *Comparison of Physical and Chemical Activation of Serpentine for Enhanced CO2 Sequestration*. Academic Press.

Mata, T. M., Martins, A. A., & Caetano, N. S. (2018). Bio-refinery approach for spent coffee grounds valorization. *Bioresource Technology*, *247*, 1077–1084. doi:10.1016/j.biortech.2017.09.106 PMID:28969966

Mathur, M. L., & Sharma, R. P. (1997). *Internal combustion engines*. Dhanpat Rai & Sons.

Mavukkandy, M.O., Chabib, C.M., Mustafa, I., Al Ghaferi, M., & AlMarzooqi, F. (2019). Brine Management in Desalination Industry: From Waste to Resources Generation. *Desalination, 472*. doi:10.1016/j.desal.2019.114187

McCormick, R. L., Hayes, R. R., Black, S., Ireland, J., McDaniel, T., & Tennant, C. J. (2005) *Regulated emissions from biodiesel tested in heavy duty engines meeting 2004 emission standards.* SAE Paper No. 2005-01-2200.

Mendez, S., Martinez, A., & Millan, W. (2014). Design, characterization, validation of a 1-kw AC self-excited switched reluctance generator. *IEEE Trans. Ind. Electron, 61*(2), 846-855.

Mgaya, J., Shombe, G. B., Masikane, S. C., Mlowe, S., Mubofu, E. B., & Revaprasadu, N. (2019). Cashew nut shell: A potential bio-resource for the production of bio-sourced chemicals, materials and fuels. *Green Chemistry*, *21*(6), 1186–1201. doi:10.1039/C8GC02972E

Milazzo, M. F., Spina, F., Cavallaro, S., & Bart, J. C. J. (2013). Sustainable soy biodiesel. *Renewable & Sustainable Energy Reviews*, *27*, 806–852. doi:10.1016/j.rser.2013.07.031

Milazzo, M. F., Spina, F., Primerano, P., & Bart, J. C. J. (2013). Soy biodiesel pathways: Global prospects. *Renewable & Sustainable Energy Reviews*, *26*, 579–624. doi:10.1016/j.rser.2013.05.056

Misra, A., Tajudeen, S., Venkataramani, G., Ayyasamy, E., & Ramalingam, V. (2017). Role of biodiesel with nanoadditives in port owned trucks and other vehicles for emission reduction. *Thermal Science*, *21*(1 Part B), 605–614. doi:10.2298/TSCI160613295M

Mitchell, P. J., Kiel, J., Livingstone, B., & Dupont, R. (2007). *Torrefied Biomass: A Business Case for Pellets from Torrefied Biomass as New Solid Fuel*. Academic Press.

Mitchell, E. J. S., Lea-Langton, A. R., Jones, J. M., Williams, A., Layden, P., & Johnson, R. (2016). The impact of fuel properties on the emissions from the combustion of biomass and other solid fuels in a fixed bed domestic stove. *Fuel Processing Technology*, *142*, 115–123. doi:10.1016/j.fuproc.2015.09.031

Mittelbach & Remschmidt. (2004). *Biodiesel—the Comprehensive Handbook* (2nd ed.). Boersedruck,Ges.m.b.H.

Mittelbach, M., & Gangl, S. (2001). Long storage stability of biodiesel made from rapeseed and used frying oil. *Journal of the American Oil Chemists' Society*, *78*(6), 573–577. doi:10.100711746-001-0306-z

Mligo, I., Misana, S., & Pauline, N. M. (2020). The effectiveness of adaptation strategies to climate change and variability in enhancing rural smallholder farmers' food security in Mvomero district, Tanzania. *Journal of the Geographical Association of Tanzania*, *42*(1), 45–63. doi:10.56279/jgat.v42i1.216

Moghaddam, I. N., Chowdhury, B., & Doostan, M. (2019). Optimal sizing and operation of battery energy storage systems connected to wind farms participating in electricity markets. *IEEE Trans. Sustain. Energy, 10*(3), 1184–1193.

Mohamed, M. B. (n.d.). *Low Cost Nanomaterials for Water Desalination and Purification*. Academic Press.

Mohammed, A. M. O. (2005). Impact of Land Disposal of Rejects Brine from Desalination Plants on Soil and Groundwater. *Desalination*, *182*(1-3), 411–433. doi:10.1016/j.desal.2005.02.035

Mojiri, A., & Bashir, M. J. K. (2022). Wastewater Treatment: Current and Future Techniques. *Water (Basel)*, *14*(3), 448. Advance online publication. doi:10.3390/w14030448

Mojiri, A., Zhou, J., Vakili, M., & Van Le, H. (2020). Removal performance and optimisation of pharmaceutical micropollutants from synthetic domestic wastewater by hybrid treatment. *Journal of Contaminant Hydrology*, *235*, 103736. doi:10.1016/j.jconhyd.2020.103736 PMID:33142212

Mondal, P., Jain, M., DeFries, R. S., Galford, G. L., & Small, C. (2015). Sensitivity of crop cover to climate variability: Insights from two Indian agro-ecoregions. *Journal of Environmental Management*, *148*(1), 21–30. doi:10.1016/j.jenvman.2014.02.026 PMID:24680541

Mondejar, M. E., Avtar, R., Diaz, H. L. B., Dubey, R. K., Esteban, J., Gómez-Morales, A., Hallam, B., Mbungu, N. T., Okolo, C. C., Prasad, K. A., She, Q., & Garcia-Segura, S. (2021). Digitalization to achieve sustainable development goals: Steps towards a smart green planet. *The Science of the Total Environment*, *794*(148539), 1–20. doi:10.1016/j.scitotenv.2021.148539 PMID:34323742

Moon, G. H., Kim, D. H., il Kim, H., Bokare, A. D., & Choi, W. (2014). Platinum-like Behavior of Reduced Graphene Oxide as a Cocatalyst on TiO 2 for the Efficient Photocatalytic Oxidation of Arsenite. *Environmental Science & Technology Letters*, *1*(2), 185–190. doi:10.1021/ez5000012

Moon, Y., & Hwang, J. (2018). Crowdfunding as an alternative means for funding sustainable appropriate technology: Acceptance determinants of backers. *Sustainability*, *10*(1456), 1–18. doi:10.3390u10051456

Moreno-Ariasa, C. A., Chamarraví-Guerraa, Ó., Navarro-Árqueza, E. F., & Jiménez-Jiméneza, M. F. (2021). International overview of the use of spent coffee ground as solid fuel. A review of worldwide studies. *Gestion y Ambiente*, *24*(2), 95261.

Morrison, E. S., Shields, M. R., Bianchi, T. S., Liu, Y., Newman, S., Tolic, N., & Chu, R. K. (2020). Multiple biomarkers highlight the importance of water column processes in treatment wetland organic matter cycling. *Water Research, 168*, 115153. doi:10.1016/j.watres.2019.115153 PMID:31655441

Mosaad, M. I. (2020). Direct power control of SRG-based WECSs using optimised fractional-order PI controller. *IET Electric Power Applications, 14*(3), 409–417.

Mosaad, M. I., Abu-Siada, A., & El-Naggar, M. F. (2019). Application of superconductors to improve the performance of DFIG-based WECS. *IEEE Access, 7*, 103760–103769.

Mousa, E., Kazemi, M., Larsson, M., Karlsson, G., & Persson, E. (2019). Potential for developing biocarbon briquettes for foundry industry. *Applied Sciences (Basel, Switzerland), 9*(24), 5288. doi:10.3390/app9245288

Msomba, B. H., Ndaki, P. M., & Nassary, E. K. (2021). Assessing adaptation measures to climate change applied by out-growers for the sustainable production of sugarcane in Tanzania. *Journal of Global Agriculture and Ecology, 12*(1), 18–29.

Mubofu, E. B., & Mgaya, J. E. (2018). Chemical valorization of cashew nut shell waste. *Topics in Current Chemistry, 376*(2), 1–15. doi:10.100741061-017-0177-9 PMID:29442189

Mucha, Z., & Kułakowski, P. (2016). Turbidity measurements as a tool of monitoring and control of the SBR effluent at the small wastewater treatment plant – preliminary study. *Archives of Environmental Protection, 42*(3), 33–36. doi:10.1515/aep-2016-0030

Mudasir, P., & Naqash, J. A. (2021). The effect of water cement ratio on the characteristics of multi-walled carbon nanotube reinforced concrete. *Materials Today: Proceedings, 43*, 3852–3855. doi:10.1016/j.matpr.2020.12.1176

Murthy, P. S., & Naidu, M. M. (2012). Sustainable management of coffee industry by-products and value addition—A review. *Resources, Conservation and Recycling, 66*, 45–58. doi:10.1016/j.resconrec.2012.06.005

Musafiri, C. M., Kiboi, M., Macharia, J., Ng'etich, O. K., Kosgei, D. K., Mulianga, B., Okoti, M., & Ngetich, F. K. (2022). Smallholders' adaptation to climate change in Western Kenya: Considering socioeconomic, institutional and biophysical determinants. *Environmental Challenges, 7*(100489), 1–11. doi:10.1016/j.envc.2022.100489

Musa, M. A., & Idrus, S. (2021). Physical and Biological Treatment Technologies of Slaughterhouse Wastewater: A Review. *Sustainability, 13*(9), 46–56. doi:10.3390u13094656

Mussatto, S. I., Dragone, G., & Roberto, I. C. (2006). Brewers' spent grain: Generation, characteristics and potential applications. *Journal of Cereal Science, 43*(1), 1–14. doi:10.1016/j.jcs.2005.06.001

Musunuri, D. (2014). Appropriate technology and economic development of emerging economies— A myth or a reality. *Journal of Technology Management for Growing Economies, 5*(1), 51–62. doi:10.15415/jtmge.2014.51003

Nabais, J. M. V., Nunes, P., Carrott, P. J., Carrott, M. M. L. R., García, A. M., & Díaz-Díez, M. A. (2008). Production of activated carbons from coffee endocarp by CO2 and steam activation. *Fuel Processing Technology, 89*(3), 262–268. doi:10.1016/j.fuproc.2007.11.030

Naidoo, S., & Olaniran, A. O. (2013). Treated Wastewater Effluent as a Source of Microbial Pollution of Surface Water Resources. *International Journal of Environmental Research and Public Health, 11*(1), 249–270. doi:10.3390/ijerph110100249

Nanthagopal, K., Ashok, B., Saravanan, B., Patel, D., Sudarshan, B., & Ramasamy, R. A. (2018). An assessment on the effects of 1-pentanol and 1-butanol as additives with Calophyllum Inophyllum biodiesel. *Energy Conversion and Management, 158*, 70–80. doi:10.1016/j.enconman.2017.12.048

Narayana Reddy, J., & Ramesh, A. (2006). Parametric studies for improving the performance of a Jatropha oil-fuelled compression ignition engine. *Renewable Energy, 31*(12), 1994–2016. doi:10.1016/j.renene.2005.10.006

Narayan, R. (2010). Use of nanomaterials in water purification. *Materials Today, 13*(6), 44–46. doi:10.1016/S1369-7021(10)70108-5

Nasrollahzadeh, M., Sajjadi, M., Iravani, S., & Varma, R. S. (2021a). Carbon-based sustainable nanomaterials for water treatment: State-of-art and future perspectives. *Chemosphere, 263*, 128005. doi:10.1016/j.chemosphere.2020.128005 PMID:33297038

Nasrollahzadeh, M., Sajjadi, M., Iravani, S., & Varma, R. S. (2021b). Green-synthesized nanocatalysts and nanomaterials for water treatment: Current challenges and future perspectives. *Journal of Hazardous Materials, 401*, 123401. doi:10.1016/j.jhazmat.2020.123401 PMID:32763697

Nathan, S. S., Mallikarjuna, J. M., & Ramesh, A. (2010). An experimental study of the biogas–diesel HCCI mode of engine operation. *Energy Conversion and Management, 51*(7), 1347–1353. doi:10.1016/j.enconman.2009.09.008

Ndamani, F., & Watanabe, T. (2016). Determinants of farmers' adaptation to climate change: A micro level analysis in Ghana. *Scientia Agrícola, 73*(3), 201–208. doi:10.1590/0103-9016-2015-0163

Neto, Barros, & Paula. (2018). Design of computational experiment for performance optimization of a switched reluctance generator in wind systems. *IEEE Trans. Energy Convers., 33*(1), 406-419.

Neto, D. S., Jose, P., Andre dos Santos Barros, T., Hancco Catata, E., & Ruppert Filho, E. (2021). Grid-connected SRG interfaced with bidirectional DC-DC converter in WECS. *IEEE Transactions on Energy Conversion.*

Ngusale, G. K., Luo, Y., & Kiplagat, J. K. (2014). Briquette making in Kenya: Nairobi and peri-urban areas. *Renewable & Sustainable Energy Reviews, 40*, 749–759. doi:10.1016/j.rser.2014.07.206

Nirendra, N. M., Robert, R. R., & Sebastian, V. (2013). Combustion and emissions characteristics of a dual fuel engine operated on alternative gaseous fuels. *Fuel, 109*, 669–678. doi:10.1016/j.fuel.2013.03.007

Njenga, M., Karanja, N., Karlsson, H., Jamnadass, R., Iiyama, M., Kithinji, J., & Sundberg, C. (2014). Additional cooking fuel supply and reduced global warming potential from recycling charcoal dust into charcoal briquette in Kenya. *Journal of Cleaner Production, 81*, 81–88. doi:10.1016/j.jclepro.2014.06.002

Noble, M. M., Harasti, D., Pittock, J., & Doran, B. (2019). Linking the social to the ecological using GIS methods in marine spatial planning and management to support resilience: A review. *Marine Policy, 108*, 103657. Advance online publication. doi:10.1016/j.marpol.2019.103657

Nowack, B., & Bucheli, T. D. (2007). Occurrence, behavior and effects of nanoparticles in the environment. *Environmental Pollution, 150*(1), 5–22. doi:10.1016/j.envpol.2007.06.006 PMID:17658673

NRC. (2016). *Backgrounder on tritium, radiation protection limits, and drinking water standards*. https://www.nrc.gov/reading-rm/doc-collections/factsheet

Oji, I. S., Chukwuma, N. P., Friday, N. K., & Philip, P.-E. O. (2018). Domestic Wastewater Treatment and Reuse in Awka Urban, Anambra State, Nigeria. *IIARD International Journal of Geography and Environmental Management, 4*(2). www.iiardpub.org

Oji, I. S., Chukwuma, N. P., & Friday, N. K. (2018). Domestic Wastewater Treatment and Reuse in Awka Urban, Anambra State, Nigeria. *International Journal of Geography and Environmental Management, 4*, 16–24. www.iiardpub.org

Oki, T., & Kanae, S. (2006). Global hydrological cycles and world water resources. Science, 313(5790), 1068–1072. doi:10.1126cience.1128845

Okullo, A. A., Temu, A. K., Ogwok, P., & Ntalikwa, J. W. (2012). Physico-chemical properties of biodiesel from jatropha and castor oils. *International Journal of Renewable Energy Research, 2*(1), 47-52.

Oladeji, J. T. (2015). Theoretical aspects of biomass briquetting: A review study. *Journal of Energy Technologies and Policy, 5*(3), 72–81.

Olanders, B., & Steenari, B. M. (1995). Characterization of ashes from wood and straw. *Biomass and Bioenergy, 8*(2), 105–115. doi:10.1016/0961-9534(95)00004-Q

Oliveira, M., Machado, A. V., & Nogueira, R. (2012). Phosphorus Removal from Eutrophic Waters with an Aluminium Hybrid Nanocomposite. *Phosphorus Removal from Eutrophic Waters with an Aluminium Hybrid Nanocomposite Water Air Soil Pollut, 2012*(223), 4831–4840. doi:10.100711270-012-1239-9

Omid, R., Nima, M., & Mohsen, A. (2011). Literature review and road map for using biogas in internal combustion engines. *Third International Conference on Applied Energy.*

Omosa, I. B., Wang, H., Cheng, S., & Li, F. (2012). *Sustainable tertiary wastewater treatment is required for water resources pollution control in Africa.* Oregon Water Science Center. http://or.water.usgs.gov/grapher/fnu. html

Onchieku, J. M., Chikamai, B. N., & Rao, M. S. (2012). Optimum parameters for the formulation of charcoal briquettes using bagasse and clay as binder. *European Journal of Sustainable Development, 1*(3), 477–477. doi:10.14207/ejsd.2012.v1n3p477

Onukak, I. E., Mohammed-Dabo, I. A., Ameh, A. O., Okoduwa, S. I., & Fasanya, O. O. (2017). Production and characterization of biomass briquettes from tannery solid waste. *Recycling, 2*(4), 17. doi:10.3390/recycling2040017

Pakmehr, S., Yazdanpanah, M., & Baradaran, M. (2020). Explaining farmers' response to climate change-induced water stress through cognitive theory of stress: An Iranian perspective. *Environment, Development and Sustainability, 23*(1), 5776–5793.

Pallavi, H. V., Srikantaswamy, S., Kiran, B. M., Vyshnavi, D. R., & Ashwin, C. A. (2013). Briquetting agricultural waste as an energy source. *Journal of Environmental Science. Computer Science and Engineering & Technology, 2*(1), 160–172.

Panahi, Y., Mellatyar, H., Farshbaf, M., Sabet, Z., Fattahi, T., & Akbarzadehe, A. (2018). Biotechnological applications of nanomaterials for air pollution and water/wastewater treatment. *Materials Today: Proceedings, 5*(7), 15550–15558. doi:10.1016/j.matpr.2018.04.162

Pandiyan, C. V., Shylaja, G., Srinivasan, G. R., & Saravanan, S. (2020). Studies on use of Cashew Nut Shell Liquid (CNSL) in Biopesticide and Biofertilizer. *Nature Environment & Pollution Technology, 19*(1).

Paradkar, M. M., & Irudayaraj, J. (2002). A rapid FTIR spectroscopic method for estimation of caffeine in soft drinks and total methylxanthines in tea and coffee. *Journal of Food Science, 67*(7), 2507–2511. doi:10.1111/j.1365-2621.2002.tb08767.x

Parkhurst, D.L., & Appelo, C.A.J. (2013). *Description of input and examples for PHREEQC version 3—A computer program for speciation, batch-reaction, one-dimensional transport, and inverse.* Academic Press.

Patel, R. N., Bandyopadhyay, S., & Ganesh, A. (2006). Extraction of cashew (Anacardium occidentale) nut shell liquid using supercritical carbon dioxide. *Bioresource Technology, 97*(6), 847–853. doi:10.1016/j.biortech.2005.04.009 PMID:15939589

Peechattukudy, L. A., & Dhoble, R. M. (2017). Removal of Nitrate from Water by Adsorption – A Review. *International Journal of Science Technology & Engineering*, *3*, 322–325.

Peshev, D., Mitev, D., Peeva, L., & Peev, G. (2018). Valorization of spent coffee grounds–A new approach. *Separation and Purification Technology*, *192*, 271–277. doi:10.1016/j.seppur.2017.10.021

Poirier, K., Al Mhanna, N., & Patchiolla, K. (2022). Techno-Economic Analysis of Brine Treatment by Multi- Crystallization Separation Process for Zero Liquid Discharge. *Separations*, *9*(10), 295. doi:10.3390eparations9100295

Pokharel, M., & Hildebrandt, N. (2019). *A fast-dynamic unipolar switching control scheme for single-phase inverters in DC microgrids*. https://ieeexplore.ieee.org/abstract/document/8322262/

Porpatham, E., Ramesh, A., & Nagalingam, B. (2008). Investigation on the effect of concentration of methane in biogas when used as a fuel for a spark ignition engine. *Fuel*, *87*(8-9), 1651–1659. doi:10.1016/j.fuel.2007.08.014

Potip, S., & Wongwuttanasatian, T. (2018). Combustion characteristics of spent coffee ground mixed with crude glycerol briquette fuel. *Combustion Science and Technology*, *190*(11), 2030–2043. doi:10.1080/00102202.2018.1482888

Pourjafar, S., Sedaghati, F., Shayeghi, H., & Maalandish, M. (2019). High step-up DC–DC converter with coupled inductor suitable for renewable applications. *IET Power Electronics*, *12*(1), 92–101.

Prabhu, A., Venkata Ramanan, M., & Jayaprabakar, J. (2021). Production, properties and engine characteristics of Jatropha biodiesel–a review. *International Journal of Ambient Energy*, *42*(15), 1810–1814. doi:10.1080/01430750.2018.1557548

Pradeep, T., & Anshup. (2009). Noble metal nanoparticles for water purification: A critical review. *Thin Solid Films*, *517*(24), 6441–6478. doi:10.1016/j.tsf.2009.03.195

Prakash, T., Geo, V. E., Martin, L. J., & Nagalingam, B. (2019). Evaluation of pine oil blending to improve the combustion of high viscous (castor oil) biofuel compared to castor oil biodiesel in a CI engine. *Heat and Mass Transfer*, *55*(5), 1491–1501. doi:10.100700231-018-2519-8

Prasad, R., Bhattacharyya, A., & Nguyen, Q. D. (2017). Nanotechnology in sustainable agriculture: Recent developments, challenges, and perspectives. *Frontiers in Microbiology*, *8*(JUN), 1014. doi:10.3389/fmicb.2017.01014 PMID:28676790

Prasad, R., Kumar, V., & Prasad, K. S. (2014). Nanotechnology in sustainable agriculture: Present concerns and future aspects. *African Journal of Biotechnology*, *13*(6), 705–713. doi:10.5897/AJBX2013.13554

Pronost, J., Beaudoin, G., Tremblay, J., Larachi, F., Duchesne, J., Hébert, R., & Constantin, M. (2011). Carbon Sequestration Kinetic and Storage Capacity of Ultramafic Mining Waste. *Environmental Science & Technology*, *45*(21), 9413–9420. doi:10.1021/es203063a PMID:21919443

PTI. (n.d.). *India's oil demand likely to jump 8% in 2022 - Times of India*. The Times of India. Retrieved December 1, 2022, from https://timesofindia.indiatimes.com/business/india-business/indias-oil-demand-likely-to-jump-8-in-2022/articleshow/90293976.cms

Puay, N. Q., Qiu, G., & Ting, Y. P. (2015). Effect of Zinc oxide nanoparticles on biological wastewater treatment in a sequencing batch reactor. *Journal of Cleaner Production*, *88*, 139–145. doi:10.1016/j.jclepro.2014.03.081

Pujol, D., Liu, C., Gominho, J., Olivella, M. À., Fiol, N., Villaescusa, I., & Pereira, H. (2013). The chemical composition of exhausted coffee waste. *Industrial Crops and Products*, *50*, 423–429. doi:10.1016/j.indcrop.2013.07.056

Qamar, M. A., Javed, M., Shahid, S., Shariq, M., Fadhali, M. M., Ali, S. K., & Khan, M. S. (2023). Synthesis and applications of graphitic carbon nitride (g-C3N4) based membranes for wastewater treatment: A critical review. *Heliyon*, *9*(1), e12685. doi:10.1016/j.heliyon.2022.e12685 PMID:36660457

Qian, T., & Wu, Q. (2017). A scheme of a resonant forward-flyback converter with suppressed frequency variation. *IEEE Transactions on Power Electronics, 33*(5), 3711–3716.

Qi, D. H., Chen, H., Geng, L. M., & Bian, Y. Z. (2011). Effect of diethyl ether and ethanol additives on the combustion and emission characteristics of biodiesel-diesel blended fuel engine. *Renewable Energy*, *36*(4), 1252–1258. doi:10.1016/j.renene.2010.09.021

Qi, D. H., Chen, H., Matthews, R. D., & Bian, Y. Z. H. (2010). Combustion and emission characteristics of ethanol–biodiesel–water micro–emulsions used in a direct injection compression ignition engine. *Fuel*, *89*(5), 958–964. doi:10.1016/j.fuel.2009.06.029

Quirino, R. L., Garrison, T. F., & Kessler, M. R. (2014). Matrices from vegetable oils, cashew nut shell liquid, and other relevant systems for biocomposite applications. *Green Chemistry*, *16*(4), 1700–1715. doi:10.1039/C3GC41811A

Qu, X., Alvarez, P. J. J., & Li, Q. (2013). Applications of nanotechnology in water and wastewater treatment. *Water Research*, *47*(12), 3931–3946. doi:10.1016/j.watres.2012.09.058 PMID:23571110

Rabany, C., Rullier, N., & Ricau, P. (2015). *The African cashew sector in 2015. General trends and country profiles.* https://www.rongead.org

Rabiu, K. O., Han, L., & Bhusan Das, D. (2017). CO2 Trapping in the Context of Geological Carbon Sequestration. In *Encyclopedia of Sustainable Technologies* (pp. 461–475). Elsevier. doi:10.1016/B978-0-12-409548-9.10124-1

Raghavendra Prasada, S. A. (2014). A Review on CNSL Biodiesel as an Alternative fuel for Diesel Engine. *International Journal of Scientific Research*, *3*(7), 2028–2038.

Rahimzadeh, M. R., Rahimzadeh, M. R., Kazemi, S., & Moghadamnia, A. A. (2017). Cadmium toxicity and treatment: An update. *Journal of Internal Medicine*, *8*, 135–145. PMID:28932363

Rahmanian, E., Akbari, H., & Sheisi, G. H. (2017). Maximum Power Point Tracking in Grid Connected Wind Plant by Using Intelligent Controller and Switched Reluctance Generator. *IEEE Transactions on Sustainable Energy*, *8*(3), 1313–1320.

Rahman, M. M., Khan, K., Moral, M. N. A., & Rahman, A. N. M. (2003). *Biomass briquetting technology and environmental pollution. In The role of Renewable and Alternative Energy Sources for National Development. Khulna University of Engineering and Technology (KUET) and Asian Institute of Technology*. AIT.

Raiker, G. (2021). *Current control of boost converter for PV interface with momentum-based perturb and observe MPPT.* https://ieeexplore.ieee.org/abstract/document/9435036/

Rajaseenivasan, T., Srinivasan, V., Qadir, G. S. M., & Srithar, K. (2016). An investigation on the performance of sawdust briquette blending with neem powder. *Alexandria Engineering Journal*, *55*(3), 2833–2838. doi:10.1016/j.aej.2016.07.009

Rajasekaran, S., Damodharan, D., Gopal, K., Kumar, B. R., & De Poures, M. V. (2020). Collective influence of 1-decanol addition, injection pressure and EGR on diesel engine characteristics fueled with diesel/LDPE oil blends. *Fuel*, *277*, 118166. doi:10.1016/j.fuel.2020.118166

Rajesh, A., Gopal, K., Kumar, B. R., Sathiyagnanam, A. P., & Damodharan, D. (2020). Effect of anisole addition to waste cooking oil methyl ester on combustion', emission and performance characteristics of a DI diesel engine without any modifications. *Fuel*, *278*, 118315. doi:10.1016/j.fuel.2020.118315

Rakesh Singh Asiwal, R. S., Sar, S. K., Singh, S., & Sahu, M. (2016). Wastewater Treatment by Effluent Treatment Plants. *SSRG International Journal of Civil Engineering, 3*, 29-35. www.internationaljournalssrg.org

Rakopoulos, D. C., Rakopoulos, C. D., Mavropoulos, G. C., & Kosmadakis, G. M. (2018). Investigating the EGR rate and temperature impact on diesel engine combustion and emissions under various injection timings and loads by comprehensive two-zone modeling. *Energy*, *157*, 990–1014. doi:10.1016/j.energy.2018.05.178

Raman, P., & Ram, N. K. (2013). Performance analysis of an internal combustion engine operated on producer gas, in comparison with the performance of the natural gas and diesel engines. *Energy*, *63*, 317–333. doi:10.1016/j.energy.2013.10.033

Rambabu, K., Bharath, G., Banat, F., & Show, P. L. (2021). Green synthesis of zinc oxide nanoparticles using Phoenix dactylifera waste as bioreductant for effective dye degradation and antibacterial performance in wastewater treatment. *Journal of Hazardous Materials*, *402*, 123560. doi:10.1016/j.jhazmat.2020.123560 PMID:32759001

Rana, N., Ghosh, K. S., Chand, S., & Gathania, A. K. (2018). Investigation of ZnO nanoparticles for their applications in wastewater treatment and antimicrobial activity. *IJPAP*, *56*, 19–25. https://nopr.niscpr.res.in/handle/123456789/43515

Rasalingam, S., Peng, R., & Koodali, R. T. (2014). Removal of hazardous pollutants from wastewaters. *Journal of Nanomaterials*, *2014*, 1–42. Advance online publication. doi:10.1155/2014/617405

Rasi, S., Veijanen, A., & Rintala, J. (2007). Trace compounds of biogas from different biogas production plants. *Energy*, *32*(8), 1375–1380. doi:10.1016/j.energy.2006.10.018

Ravina, M., Galletta, S., Dagbetin, A., Kamaleldin, O. A. H., Mng'ombe, M., Mnyenyembe, L., Shanko, A., & Zanetti, M. (2021). Urban Wastewater Treatment in African Countries: Evidence from the Hydroaid Initiative. *Sustainability*, *13*(22), 8–28. doi:10.3390u132212828

Reşitoğlu, İ. A., Altinişik, K., & Keskin, A. (2014). The pollutant emissions from diesel-engine vehicles and exhaust aftertreatment systems. *Clean Technologies and Environmental Policy*, *17*(1), 15–27. doi:10.100710098-014-0793-9

Rezk, H., Aly, M., & Al-Dhaifallah, M. (2019). *Design and hardware implementation of new adaptive fuzzy logic-based MPPT control method for photovoltaic applications*. https://ieeexplore.ieee.org/abstract/document/8786102/

Rezkallah, M., & Hamadi, A. (2018). *Design and implementation of active power control with improved P&O method for wind-PV-battery-based standalone generation system*. https://ieeexplore.ieee.org/abstract/document/8119845/

Rivera, X. C. S., Gallego-Schmid, A., Najdanovic-Visak, V., & Azapagic, A. (2020). Life cycle environmental sustainability of valorisation routes for spent coffee grounds: From waste to resources. *Resources, Conservation and Recycling*, *157*, 104751. doi:10.1016/j.resconrec.2020.104751

Road Transport Year Book. (n.d.). Retrieved November 30, 2022, from https://morth.gov.in/sites/default/files/RTYB-2017-18-2018-19.pdf

Roberts, D. A., Johnson, E. L., & Knott, N. A. (2010). Impacts of Desalination Plants Discharges on Marine Environments: A Critical Review of Published Studies. *Water Research*, *44*(18), 5117–5128. doi:10.1016/j.watres.2010.04.036 PMID:20633919

Robinson, K. G., Robinson, C. H., & Hawkins, S. A. (2005). Assessment of public perception regarding wastewater reuse. *Water Science and Technology: Water Supply*, *5*(1), 59–65. doi:10.2166/ws.2005.0008

Rodrigues, F. H., França, F. C., Souza, J. R., Ricardo, N. M., & Feitosa, J. (2011). Comparison between physico-chemical properties of the technical Cashew Nut Shell Liquid (CNSL) and those natural extracted from solvent and pressing. *Polímeros*, *21*(2), 156–160. doi:10.1590/S0104-14282011005000028

Rodrigues, S. M., Demokritou, P., Dokoozlian, N., Hendren, C. O., Karn, B., Mauter, M. S., Sadik, O. A., Safarpour, M., Unrine, J. M., Viers, J., Welle, P., White, J. C., Wiesner, M. R., & Lowry, G. (2017). Nanotechnology for sustainable food production: Promising opportunities and scientific challenges. *Environmental Science. Nano*, *4*(4), 767–781. doi:10.1039/C6EN00573J

Romeiro, G. A., Salgado, E. C., Silva, R. V. S., Figueiredo, M. K., Pinto, P. A., & Damasceno, R. N. (2012). A study of pyrolysis oil from soluble coffee ground using low temperature conversion (LTC) process. *Journal of Analytical and Applied Pyrolysis*, *93*, 47–51. doi:10.1016/j.jaap.2011.09.006

Rossati, A. (2017, January). Global Warming and Its Health Impact. *The International Journal of Occupational and Environmental Medicine*, *8*(1), 7–20. doi:10.15171/ijoem.2017.963 PMID:28051192

Rotmans, J., Den Elzen, M.G.J., Krol, M.S., Swart, R.J., & Van Der Woerd, H.J. (1992). Stabilizing atmospheric concentrations – Towards international methane control. *Ambio, 21*(6), 404–413.

Roubaud, A., Rothlisberge, R., & Favrat, D. (2005). Improving performances of a lean burn cogeneration biogas engine equipped with precombustion chambers. *Applied Thermodynamics*, *5*, 169–175.

Roy, M. M., & Corscadden, K. W. (2012). An experimental study of combustion and emissions of biomass briquettes in a domestic wood stove. *Applied Energy*, *99*, 206–212. doi:10.1016/j.apenergy.2012.05.003

Rudakiya, D. M., & Pawar, K. (2017). Bactericidal potential of silver nanoparticles synthesized using cell-free extract of Comamonas acidovorans: in vitro and in silico approaches. *Biotech, 7*(2). doi:10.1007/s13205-017-0728-3

Saadat, A. H. M., Islam, M. S., Islam, M. S., Parvin, F., & Sultana, A. (2018). Desalination Techniques for Developing Countries: A Review. *Journal of Scientific Research*, *10*(1), 77–97. doi:10.3329/jsr.v10i1.33179

Saaty, T. L. (1977). Scaling Method for Priorities in Hierarchical Structures. *Journal of Mathematical Psychology*, *15*(3), 234–281. doi:10.1016/0022-2496(77)90033-5

Saaty, T. L. (1980). *The analytic hierarchy process*. McGraw-Hill.

Sadhik Basha, J. (2018). Impact of Carbon Nanotubes and Di-Ethyl Ether as additives with biodiesel emulsion fuels in a diesel engine – An experimental investigation. *Journal of the Energy Institute*, *91*(2), 289–303. doi:10.1016/j.joei.2016.11.006

Sadhik Basha, J. (2022). Applications of Nano-Additives in Internal Combustion Engines: A Critical Review. *Journal of Thermal Analysis and Calorimetry*, *147*(17), 9383–9403. doi:10.100710973-022-11199-6

Sadhik Basha, J., & Anand, R. B. (2014). Performance, emission and combustion characteristics of a diesel engine using carbon nanotubes blended jatropha methyl ester emulsions. *Alexandria Engineering Journal*, *53*(2), 259–273. doi:10.1016/j.aej.2014.04.001

Saharan, P., Chaudhary, G. R., Mehta, S. K., & Umar, A. (2014). Removal of water contaminants by iron oxide nanomaterials. *Journal of Nanoscience and Nanotechnology*, *14*(1), 627–643. doi:10.1166/jnn.2014.9053 PMID:24730287

Şahin, A., Dincer, I., & Rosen, M. (2006). New Spatio-Temporal Wind Exergy Maps. Journal of Energy Resources Technology-transactions of The Asme - J Energ. *Resources and Technology*, *128*(3), 194–202. Advance online publication. doi:10.1115/1.2213271

Sahin, M., Bilgili, H., & Akilli, H. (2005). The wind power potential of the eastern Mediterranean region of Turkey. *Journal of Wind Engineering and Industrial Aerodynamics*, *93*(2), 171–183. doi:10.1016/j.jweia.2004.11.005

Sajjad, M., & Rasul, M. G. (2015). Simulation and optimization of solar desalination plant using Aspen Plus simulation software. *Procedia Engineering*, *105*, 739–750. doi:10.1016/j.proeng.2015.05.065

Salvador, M. A., Lazzarin, T. B., & Coelho, R. F. (2017). High step-up DC–DC converter with active switched-inductor and passive switched-capacitor networks. *IEEE Transactions on Industrial Electronics, 65*(7), 5644–5654.

Sanger, S. H., Mohod, A. G., Khandetode, Y. P., Shrirame, H. Y., & Deshmukh, A. S. (2011). Study of carbonization for cashew nut shell. *Research Journal of Chemical Sciences*, *1*(2), 43–55.

Sani, Y. M., Daud, W. M. A. W., & Aziz, A. R. A. (2012, December 3). *Biodiesel feedstock and production technologies: Successes, challenges and prospects.* IntechOpen.

Santos, C., Fonseca, J., Aires, A., Coutinho, J., & Trindade, H. (2017). Effect of different rates of spent coffee grounds (SCG) on composting process, gaseous emissions and quality of end-product. *Waste Management (New York, N.Y.)*, *59*, 37–47. doi:10.1016/j.wasman.2016.10.020 PMID:28340969

Sargani, G. R., Jiang, Y., Chandio, A. A., Shen, Y., Ding, Z., & Ali, A. (2022). Impacts of livelihood assets on adaptation strategies in response to climate change: Evidence from Pakistan. *Environment, Development and Sustainability*. Advance online publication. doi:10.100710668-022-02296-5

Sarkinnoma, A., Yarkasuwa, C. I., & Modu, K. A. (2013). Analysis of Physicochemical Parameters of Sewage Water used for Irrigation in Bauchi Metropolis – Nigeria. *Journal of Environment and Earth Science*, *10*, 37–41.

Sathasivam, M., Shanmugapriya, S., Yogeshwaran, V., & Priya, A. K. (2019). Industrial Wastewater Treatment Using Advanced Oxidation Process – A Review. *International Journal of Engineering and Advanced Technology*, *8*, 485–488.

Sattar, Al-Durra, Caruana, Debouza, & Muyeen. (2020). Testing the Performance of Battery Energy Storage in a Wind Energy Conversion System. *IEEE Transactions on Industry Applications, 56*(3), 3196 – 3206.

Savage, N., & Diallo, M. S. (2005). Nanomaterials and water purification: Opportunities and challenges. *Journal of Nanoparticle Research*, *7*(4-5), 331–342. doi:10.100711051-005-7523-5

Sawadogo, M., Tanoh, S. T., Sidibé, S., Kpai, N., & Tankoano, I. (2018). Cleaner production in Burkina Faso: Case study of fuel briquettes made from cashew industry waste. *Journal of Cleaner Production*, *195*, 1047–1056. doi:10.1016/j.jclepro.2018.05.261

Sawekwiharee, S., Boonchoo, T., Kuttiyawong, A., & Chathirat, N. (2015). Heating energy briquettes from Cashew nut shell. *Applied Mechanics and Materials*, *804*, 283–286. doi:10.4028/www.scientific.net/AMM.804.283

Saxena, N., & Hussain, I. (2017). *Implementation of a grid-integrated PV-battery system for residential and electrical vehicle applications.* https://ieeexplore.ieee.org/abstract/document/8023763/

Saydut, A., Erdogan, S., Kafadar, A. B., Kaya, C., Aydin, F., & Hamamci, C. (2016). Process optimization for production of biodiesel from hazelnut oil, sunflower oil and their hybrid feedstock. *Fuel*, *183*, 512–517. doi:10.1016/j.fuel.2016.06.114

Sayin, C., Ilhan, M., Canakci, M., & Gumus, M. (2009). Effect of injection timing on the exhaust emissions of a diesel engine using diesel-methanol blends. *Renewable Energy*, *34*(5), 1261–1269. doi:10.1016/j.renene.2008.10.010

Schattman, R. E., Hurley, S. E., Greenleaf, H. L., Niles, M. T., & Caswell, M. (2020). Visualizing climate change adaptation: An effective tool for agricultural outreach? *Weather, Climate, and Society*, *12*(1), 47–61. doi:10.1175/WCAS-D-19-0049.1

Schmidt, K., & Van Garpen, J. H. (1996). *The Effect of Biodiesel Fuel Composition on Diesel Combustion and Emission.* SAE Paper No. 961086.

Seco, A., Espuelas, S., Marcelino, S., Echeverría, A. M., & Prieto, E. (2020). Characterization of biomass briquettes from spent coffee grounds and xanthan gum using low pressure and temperature. *BioEnergy Research, 13*(1), 369–377. doi:10.100712155-019-10069-8

Selvakumar Ramalingam & Mahalakshmi. (2020). Influence Of *Moringa Oleifera* Biodiesel–Diesel–Hexanol And Biodiesel–Diesel–Ethanol Blends On Compression Ignition Engine Performance, Combustion And Emission Characteristics. *RSC Advances*, (10), 4274–4285. PMID:35495259

Şengül, Ü., Sibel, T., Şermin, A., & Şengül, A. (2014). Türkiye Gökçeada'da yenilenebilir enerji kaynaklarinin potansiyeli. *Akademik Araştırmalar ve Çalışmalar Dergisi, 6*(11), 41–55.

Sen, R., Wiwatpanyaporn, S., & Annachhatre, A. P. (2016). Influence of binders on physical properties of fuel briquettes produced from cassava rhizome waste. *International Journal of Environment and Waste Management, 17*(2), 158–175. doi:10.1504/IJEWM.2016.076750

Senthil Kumar, P., Arun Kumar, N., Sivakumar, R., & Kaushik, C. (2009). Experimentation on solvent extraction of polyphenols from natural waste. *Journal of Materials Science, 44*(21), 5894–5899. doi:10.100710853-009-3834-8

Sermyagina, E., Martinez, C. L. M., Nikku, M., & Vakkilainen, E. (2021). Spent coffee grounds and tea leaf residues: Characterization, evaluation of thermal reactivity and recovery of high-value compounds. *Biomass and Bioenergy, 150*, 106141. doi:10.1016/j.biombioe.2021.106141

Setiawan, B., Himawanto, D. A., Budiana, E. P., & Widodo, P. J. (2016). Analisa thermogravimetry pada pirolisis limbah pertanian. *ETHOS: Jurnal Penelitian dan Pengabdian kepada Masyarakat*, 49-56.

Setter, C., Silva, F. T. M., Assis, M. R., Ataíde, C. H., Trugilho, P. F., & Oliveira, T. J. P. (2020). Slow pyrolysis of coffee husk briquettes: Characterization of the solid and liquid fractions. *Fuel, 261*, 116420. doi:10.1016/j.fuel.2019.116420

Sferlazza, A., Albea-Sanchez, C., Martínez-Salamero, L., Garcia, G., & Alonso, C. (2019). Min-type control strategy of a DC–DC synchronous boost converter. *IEEE Transactions on Industrial Electronics, 67*(4), 3167–3179.

Shaafi, T., Sairam, K., Gopinath, A., Kumaresan, G., & Velraj, R. (2015). Effect of dispersion of various nanoadditives on the performance and emission characteristics of a CI engine fuelled with diesel, biodiesel and blends—A review. *Renewable & Sustainable Energy Reviews, 49*, 563–573. doi:10.1016/j.rser.2015.04.086

Shaafi, T., & Velraj, R. (2015). Influence of alumina nanoparticles, ethanol and isopropanol blend as additive with diesel–soybean biodiesel blend fuel: Combustion, engine performance and emissions. *Renewable Energy, 80*, 655–663. doi:10.1016/j.renene.2015.02.042

Shabbir, M. M., & Yaqoob, N. (2019). The impact of technological advancement on total factor productivity of cotton: A comparative analysis between Pakistan and India. *Journal of Economic Structures, 4*(1), 1–24. doi:10.118640008-019-0160-4

Shah, A. I., Din Dar, M. U., Bhat, R. A., Singh, J. P., Singh, K., & Bhat, S. A. (2020). Prospectives and challenges of wastewater treatment technologies to combat contaminants of emerging concerns. *Ecological Engineering, 152*, 105882. doi:10.1016/j.ecoleng.2020.105882

Shahbandeh. (2022). https://www.statista.com/statistics/967702/global-cashew-nut-production

Shan, X., & Que, L. Jr. (2006). High-valent nonheme iron-oxo species in biomimetic oxidations. *Journal of Inorganic Biochemistry, 100*(4), 421–433. doi:10.1016/j.jinorgbio.2006.01.014 PMID:16530841

Shao, M., Han, Z., Sun, J., Xiao, C., Zhang, S., & Zhao, Y. (2020). A review of multi-criteria decision-making applications for renewable energy site selection. *Renewable Energy, 157*, 377–403. doi:10.1016/j.renene.2020.04.137

Shariff, S., Alam, M., & Ahmad, F. (2019). *System design and realization of a solar-powered electric vehicle charging station.* https://ieeexplore.ieee.org/abstract/document/8822448/

Sharma, M. K., Priyank, G., & Sharma, N. (2015). Biomass briquette production: A propagation of non-convention technology and future of pollution free thermal energy sources. *American Journal of Engineering Research, 4*(02), 44–50.

Sharma, P., & Singh, A. K. (2017). Association of state-wise food security index with climatic factors in India: Evidence from state-wise panel data. *Journal of Global Agriculture and Ecology, 6*(3), 196–205.

Sharma, V. K., Filip, J., Zboril, R., & Varma, R. S. (2015). Natural inorganic nanoparticles – formation, fate, and toxicity in the environment. *Chemical Society Reviews, 44*(23), 8410–8423. doi:10.1039/C5CS00236B PMID:26435358

Sheanut Oil blends in Compression Ignition Engine. (2010). *Global Journal of Emerging Research in Engineering, 10*(1), 84–92.

Sheng, G. P., Yu, H.-Q., & Li, X.-Y. (2010). Extracellular polymeric substances (EPS) of microbial aggregates in biological wastewater treatment systems: A review. *Biotechnology Advances, 28*(6), 882–8946. doi:10.1016/j.biotechadv.2010.08.001 PMID:20705128

Shiraki, R., Rock, P. A., & Casey, W. H. (2000). Dissolution Kinetics of Calcite in 0.1 M NaCl Solution at Room Temperature: An Atomic Force Microscopic (AFM) Study. *Aquatic Geochemistry, 6*(1), 87–108. doi:10.1023/A:1009656318574

Shivanna, K. R. (2022). Climate change and its impact on biodiversity and human welfare. *Proc.Indian Natl. Sci. Acad., 88*(2), 160–171. doi:10.100743538-022-00073-6

Shrirame, H. Y., Panwar, N. L., & Bamniya, B. R. (2011). Bio diesel from castor oil-a green energy option. *Low Carbon Economy, 2*(1), 1–6. doi:10.4236/lce.2011.21001

Siddick, S. A. (2019). Appropriate technologies for improving yield and income of small holders growing rice paddy in rainfed low lands of agro-biodiversity hotspots in India. *Agricultural Sciences, 10*(1), 1497–1505. doi:10.4236/as.2019.1011110

Sikder, C., Husain, I., & Sozer, Y. (2014). Switched reluctance generator control for optimal power generation with current regulation. *IEEE Trans. Ind. Appl., 50*(1), 307-316.

Simpson, T., & Depcik, C. (2022). Multiple Fuel Injection Strategies for Compression Ignition Engines. *Energies, 15*(14), 5214. doi:10.3390/en15145214

Singh, A. K., & Jyoti, B. (2021). Projected food-grain production and yield in India: An evidence from state-wise panel data investigation during 1977-2014. *The Journal of Agricultural Sciences – Sri Lanka, 16*(1), 108-125.

Singh, A. K., Kumar, S., & Jyoti, B. (2022b). Association of sustainable food security with climatic and non-climatic factors in Gujarat state of India: A district-wise panel data investigation. In Sustainable Agriculture and Food Security. World Sustainability Series. Springer Nature Switzerland. doi:10.1007/978-3-030-98617-9_16

Singh, A., & Mishra, A. (2020). *An integrated converter with reduced components for electric vehicles utilizing solar and grid power sources.* https://ieeexplore.ieee.org/abstract/document/9104749/

Singh, A. K. (2017). *Climate Change and Productivity of Food-grain and Cash Crops in India.* LAP LAMBERT Academic Publishing.

Singh, A. K., & Issac, J. (2018). Impact of climatic and non-climatic factors on sustainable livelihood security in Gujarat state of India: A statistical exploration. *Agriculture and Food Sciences Research, 5*(1), 30–46. doi:10.20448/journal.512.2018.51.30.46

Singh, A. K., & Jyoti, B. (2019). Measuring the climate variability impact on cash crops farming in India: An Empirical Investigation. *Agriculture and Food Sciences Research*, *6*(2), 155–164. doi:10.20448/journal.512.2019.62.155.165

Singh, A. K., Jyoti, B., Kumar, S., & Lenka, S. K. (2021). Assessment of global sustainable development, environmental sustainability, economic development and social development index in selected economies. *International Journal of Sustainable Development and Planning*, *16*(1), 123–138. doi:10.18280/ijsdp.160113

Singh, A. K., & Kumar, S. (2022). Exploring the impact of sustainable development on social-economic, and science and technological development in selected countries: A panel data analysis. *Society & Sustainability*, *4*(1), 55–83. doi:10.38157s.v4i1.405

Singh, A. K., Kumar, S., Ashraf, S. N., & Jyoti, B. (2022a). Implications of farmer's adaptation strategies to climate change in agricultural sector of Gujarat: Experience from farm level data. *Research on World Agricultural Economy*, *3*(1), 42–57. doi:10.36956/rwae.v3i1.498

Singh, A. K., Kumar, S., & Jyoti, B. (2022c). Does sustainable livelihood security have a relationship with climatic and geographical factors? evidence from a state-wise panel data investigation in India. *The IUP Journal of Applied Economics*, *21*(1), 7–35.

Singh, A. K., Kumar, S., & Jyoti, B. (2022d). Influence of climate change on agricultural sustainability in India: A state-wise panel data analysis. *Asian Journal of Agriculture*, *6*(1), 15–27. doi:10.13057/asianjagric/g060103

Singh, A. K., Kumar, S., & Jyoti, B. (2022e). Impact of the COVID-19 on food security and sustainable development goals in India: Evidence from existing literature. *GNOSI: An Interdisciplinary Journal of Human Theory and Praxis*, *5*(2), 94–109.

Singh, A. K., Narayanan, K. G. S., & Sharma, P. (2017). Effect of climatic factors on cash crop farming in India: An application of Cobb-Douglas production function model. *International Journal of Agricultural Resources, Governance and Ecology*, *13*(2), 175–210. doi:10.1504/IJARGE.2017.086452

Singh, A. K., Narayanan, K. G. S., & Sharma, P. (2019). Measurement of technical efficiency of climatic and non-climatic factors in sugarcane farming in Indian states: Use of stochastic frontier production function approach. *Climatic Change*, *5*(19), 150–166.

Singh, A. K., & Sharma, P. (2018). Implications of climatic and non-climatic variables on food security in developing economies: A conceptual review. *MOJ Food Processing & Technology*, *6*(1), 1–12. doi:10.15406/mojfpt.2018.06.00138

Singh, A. K., Sharma, P., & Singh, D. K. (2016). Measuring the influence of weather variables on productivity of food-grain crops in India: An application of Just & Pope's production technique. *ABBS Management Business and Entrepreneurship Review*, *7*(2), 29–46. doi:10.23874/amber/2016/v7/i2/121780

Singh, A. K., & Singh, B. J. (2020). Assessing the infectious diseases of students in different weather seasons in DIT University Dehradun, Uttarakhand (India). *Asian Journal of Multidimensional Research*, *9*(3), 34–48. doi:10.5958/2278-4853.2020.00055.5

Singh, A. K., & Singh, B. J. (2021). Projected productivity of cash crops in different climate change scenarios in India: Use of marginal impact analysis technique. *Finance & Economics Review*, *3*(1), 63–87.

Singh, A. K., Singh, B. J., & Negi, V. (2020b). Does sustainable development have a causal relationship with environmental development? Evidence from a country-wise panel data analysis. *International Journal of Technology Management & Sustainable Development*, *19*(2), 147–171. doi:10.1386/tmsd_00020_1

Singh, A. K., Singh, J. B., & Sharma, P. (2020a). Assessing the climate change impact on rice and wheat production in Uttar Pradesh and Haryana states of India. *Climatic Change*, *6*(21), 74–93.

Singh, B., & Kushwaha, R. (2019). A PFC based EV battery charger using a bridgeless isolated SEPIC converter. doi:10.1109/TIA.2019.2951510

Singh, K. K., Singh, A., & Rai, S. (2022). A study on nanomaterials for water purification. *Materials Today: Proceedings*, *51*, 1157–1163. doi:10.1016/j.matpr.2021.07.116

Singh, R. N., Jena, U., Patel, J. B., & Sharma, A. M. (2006). Feasibility study of cashew nut shells as an open core gasifier feedstock. *Renewable Energy*, *31*(4), 481–487. doi:10.1016/j.renene.2005.04.010

Singh, S. (2020). Farmers' perception of climate change and adaptation decisions: A micro-level evidence from Bundelkhand region, India. *Ecological Indicators*, *116*(106475), 1–13. doi:10.1016/j.ecolind.2020.106475

Singh, S., & Singh, A. (2019). Farmers' perception of climate change and livelihood vulnerability in rainfall regions of India: A gender-environment perspective. *International Journal of Environment and Climate Change*, *9*(12), 878–889. doi:10.9734/ijecc/2019/v9i1230168

Sita Rama Raju, A. V. (2001). *Experimental investigations on the performance of a lean burn spark ignited gas engine* [PhD thesis]. IIT Madras, Chennai, India.

Sitko, R., Zawisza, B., & Malicka, E. (2013). Graphene as a new sorbent in analytical chemistry. *Trends in Analytical Chemistry*, *51*, 33–43. doi:10.1016/j.trac.2013.05.011

Slah, Mansour, Hajer, & Faouzi. (2017). Analysis, modeling and implementation of an interleaved boost DC-DC converter for fuel cell used in electric vehicle. *International Journal of Hydrogen Energy, 42*(48), 28852-28864.

Smit, B., & Skinner, M. W. (2002). Adaptation options in agriculture to climate change: A typology. *Mitigation and Adaptation Strategies for Global Change*, *7*(1), 85–114. doi:10.1023/A:1015862228270

Smit, H. C., & Meincken, M. (2012). Time/temperature combustion profiles of various wood-based biofuels. *Biomass and Bioenergy*, *39*, 317–323. doi:10.1016/j.biombioe.2012.01.021

Smith, R. L. Jr, Malaluan, R. M., Setianto, W. B., Inomata, H., & Arai, K. (2003). Separation of cashew (Anacardium occidentale L.) nut shell liquid with supercritical carbon dioxide. *Bioresource Technology*, *88*(1), 1–7. doi:10.1016/S0960-8524(02)00271-7 PMID:12573557

Smith, S. C., & Rodrigues, D. F. (2015). Carbon-based nanomaterials for removal of chemical and biological contaminants from water: A review of mechanisms and applications. *Carbon*, *91*, 122–143. doi:10.1016/j.carbon.2015.04.043

Snickars, M., & Pitkänen, T. (2007). *GIS tools for marine planning and management: BALANCE Interim (Report No. 28)*. Interreg Baltic Sea Region. Retrieved from https://balance-eu.org/xpdf/balance-interim-report-no-28.pdf

Soares, A. (2020). Wastewater treatment in 2050: Challenges ahead and future vision in a European context. *Environ. Sci. Ecotechnol.*, *2*, 100030. doi:10.1016/j.ese.2020.100030 PMID:36160927

Society of Indian Automobile Manufacturers. (n.d.). Retrieved December 1, 2022, from https://www.siam.in/uploads/filemanager/159SIAMWhitePaperonAlternativeFuelsforvehicles.pdf

Solis, K. L. B., Kwon, Y. H., Kim, M. H., An, H. R., Jeon, C., & Hong, Y. (2020). Metal organic framework UiO-66 and activated carbon composite sorbent for the concurrent adsorption of cationic and anionic metals. *Chemosphere*, *238*, 124656. doi:10.1016/j.chemosphere.2019.124656 PMID:31472345

Somnuk, K., Eawlex, P., & Prateepchaikul, G. (2017). Optimization of coffee oil extraction from spent coffee grounds using four solvents and prototype-scale extraction using circulation process. *Agriculture and Natural Resources (Bangkok), 51*(3), 181–189. doi:10.1016/j.anres.2017.01.003

Sotannde, O. A., Oluyege, A. B., & Abah, G. A. (2010). Physical and combustion properties of charcoal briquettes from neem wood residues. *International Agrophysics, 24*(2), 189–194.

Spyridonidou, S., & Vagiona, D. G. (2020). Systematic Review of Site-Selection Processes in Onshore and Offshore Wind Energy Research. *Energies, 13*(22), 5906. Advance online publication. doi:10.3390/en13225906

Spyridonidou, S., Vagiona, D. G., & Loukogeorgaki, E. (2020). Strategic Planning of OffshoreWind Farms in Greece. *Sustainability, 12*(3), 905. doi:10.3390u12030905

Srinivasan, G. R., Mahajan, A., Seth, R., & Mahajan, R. (2022). Impact of organic hydrocarbons on fuel properties and engine characteristics of thermally cracked cashew nut shell liquid. *Journal of Renewable Energy and Environment.*

Srinivasan, G. R., Mahajan, A., Seth, R., & Mahajan, R. (2022).A Critical Review on Briquettes Developed from Spent Coffee Ground Wastes. Preprints, 2022120138

Srinivasan, G. R., & Jambulingam, R. (2018). Comprehensive study on biodiesel produced from waste animal fats-a review. *Journal of Environmental Science and Technology, 11*(3), 157–166. doi:10.3923/jest.2018.157.166

Srinivasan, G. R., Shankar, V., Chandra Sekharan, S., Munir, M., Balakrishnan, D., Mohanam, A., & Jambulingam, R. (2020). Influence of fatty acid composition on process optimization and characteristics assessment of biodiesel produced from waste animal fat. *Energy Sources. Part A, Recovery, Utilization, and Environmental Effects*, 1–19. doi:10.1080/15567036.2020.1771477

Stelzenmüller, V., Lee, J., South, A., Foden, J., & Rogers, S. I. (2013). Practical tools to support marine spatial planning: A review and some prototype tools. *Marine Policy, 38*, 214–227. doi:10.1016/j.marpol.2012.05.038

Stolarski, M. J., Szczukowski, S., Tworkowski, J., Krzyżaniak, M., Gulczyński, P., & Mleczek, M. (2013). Comparison of quality and production cost of briquettes made from agricultural and forest origin biomass. *Renewable Energy, 57*, 20–26. doi:10.1016/j.renene.2013.01.005

Subbarao, C. N., Krishna Prasad, K. M. M., & Prasad, V. S. R. K. (2011). Review on applications, extraction, isolation and analysis of cashew nut shell liquid (CNSL). *The Pharma Research Journal, 6*(1), 21–41.

Suhartini, S., Hidayat, N., & Wijaya, S. (2011). Physical properties characterization of fuel briquette made from spent bleaching earth. *Biomass and Bioenergy, 35*(10), 4209-4214.

Sukumar Puhan, R. (2009). Effect of injection pressure on performance, emission and combustion characteristics of high linolenic linseed oil methyl ester in a DI engine. *Renewable Energy, 34*(5), 1227–1233. doi:10.1016/j.renene.2008.10.001

Sutrisno, A. W., Suprianto, F. D., Kasrun, A. W., & Siahaan, I. H. (2017). The effects of particle size and pressure on the combustion characteristics of Cerbera manghas leaf briquettes. *Arpn J. Eng. Appl. Sci, 12*, 931–936.

Swain, B. B. (2016). Does technological linkage in contract farming increase farm productivity and efficiency? The case of hybrid paddy seed cultivation in undivided Andhra Pradesh. *Agricultural Economics Research Review, 29*(2), 211–224. doi:10.5958/0974-0279.2016.00049.5

Swami, D., & Parthasarathy, D. (2022). Assessing effectiveness of agricultural adaptation strategies in context of crop loss: A case study of the Indian subcontinent. *Regional Environmental Change, 22*(71), 23–35. doi:10.100710113-022-01921-3

Tajudeen, S., & Velraj, R. (2014). Nanoadditives: Enhancement of combustion and performance characteristics of a CI diesel engine. *International Journal of Applied Environmental Sciences*, *9*(4), 1727–1741.

Tamilvanan, A. (2013). Preparation of biomass briquettes using various agro-residues and waste papers. *Journal of Biofuels*, *4*(2), 47–55. doi:10.5958/j.0976-4763.4.2.006

Tang, J., & Wang, J. (2020). Iron-copper bimetallic metal-organic frameworks for efficient Fenton-like degradation of sulfamethoxazole under mild conditions. *Chemosphere*, *241*, 125002. doi:10.1016/j.chemosphere.2019.125002 PMID:31590027

Tang, Y., Hu, S., He, Y., Wang, Y., Wan, X., Cui, S., & Long, K. (2021). Experiment on CO2-brine-rock interaction during CO2 injection and storage in gas reservoirs with aquifer. *Chemical Engineering Journal*, *413*, 127567. doi:10.1016/j.cej.2020.127567

Tan, K. T., Lee, K. T., Mohamed, A. R., & Bhatia, S. (2009). Palm oil: Addressing issues and towards sustainable development. *Renewable & Sustainable Energy Reviews*, *13*(2), 420–427. doi:10.1016/j.rser.2007.10.001

Tan, S. X., Lim, S., Ong, H. C., & Pang, Y. L. (2019). State of the art review on development of ultrasound-assisted catalytic transesterification process for biodiesel production. *Fuel*, *235*, 886–907. doi:10.1016/j.fuel.2018.08.021

Tarekegn, M. M., Hiruy, A. M., & Dekebo, A. H. (2021). Nano zero valent iron (nZVI) particles for the removal of heavy metals (Cd2+, Cu2+ and Pb2+) from aqueous solutions. *RSC Advances*, *11*(30), 18539–18551. doi:10.1039/D1RA01427G PMID:35480950

Tarzamni, H., Babaei, E., Esmaeelnia, F. P., Dehghanian, P., Tohidi, S., & Mohammad, B. B. S. (2019). Analysis and reliability evaluation of a high step-up soft switching push–pull dc–dc converter. *IEEE Transactions on Reliability*, *69*(4), 1376–1386.

Tatoulis, T. I., Michailides, M. K., Tekerlekopoulou, A. G., Akratos, C. S., Pavlou, S., & Vayenas, D. V. (2018). Simultaneous Treatment of Agro-Industrial and Industrial Wastewaters: Case Studies of Cr(VI)/Second Cheese Whey and Cr(VI)/Winery Effluents. *Water (Basel)*, *10*(4), 382–393. doi:10.3390/w10040382

Teja Muvva, R., Shaafi, T., & Arunkumar, M. (2020). Experimental investigation by utilizing nano alumina with waste cooking oil biodiesel fuel in CI engine. *Materials Today: Proceedings*. Advance online publication. doi:10.1016/j.matpr.2020.10.185

Tercan, E., Tapkın, S., Latinopoulos, D., Dereli, M., Tsiropoulos, A., & Ak, A. (2020). A GIS-based multi-criteria model for offshore wind energy power plants site selection in both sides of the Aegean Sea. *Environmental Monitoring and Assessment*, *192*(10), 652. doi:10.100710661-020-08603-9 PMID:32964332

Thines, R. K., Mubarak, N. M., Nizamuddin, S., Sahu, J. N., Abdullah, E. C., & Ganesan, P. (2017). Application potential of carbon nanomaterials in water and wastewater treatment: A review. *Journal of the Taiwan Institute of Chemical Engineers*, *72*, 116–133. doi:10.1016/j.jtice.2017.01.018

Thirumarimurugan, M., Sivakumar, V. M., Xavier, A. M., Prabhakaran, D., & Kannadasan, T. (2012). Preparation of biodiesel from sunflower oil by transesterification. *International Journal of Bioscience, Biochemistry, Bioinformatics*, *2*(6), 441–444. doi:10.7763/IJBBB.2012.V2.151

Thoniyot, P., Tan, M. J., Karim, A. A., Young, D. J., & Loh, X. J. (2015). Nanoparticle–Hydrogel Composites: Concept, Design, and Applications of These Promising, Multi-Functional Materials. *Advancement of Science*, *2*(1–2), 1400010. Advance online publication. doi:10.1002/advs.201400010 PMID:27980900

Thostenson, E. T., Ren, Z., & Chou, T. W. (2001). Advances in the science and technology of carbon nanotubes and their composites: A review. *Composites Science and Technology*, *61*(13), 1899–1912. doi:10.1016/S0266-3538(01)00094-X

Tinlot, M. (2010). *Intégration des filières dans la mitigation au changement climatique: évaluation carbone sur la filière anacarde au Burkina Faso. In Application de l'outil EXACT.* Étude pour la GTZ et la FAO.

Tiwari, J. N., Mahesh, K., Le, N. H., Kemp, K. C., Timilsina, R., Tiwari, R. N., & Kim, K. S. (2013). Reduced graphene oxide-based hydrogels for the efficient capture of dye pollutants from aqueous solutions. *Carbon*, *56*, 173–182. doi:10.1016/j.carbon.2013.01.001

Tongcumpou, C., Usapein, P., & Tuntiwiwattanapun, N. (2019). Complete utilization of wet spent coffee grounds waste as a novel feedstock for antioxidant, biodiesel, and bio-char production. *Industrial Crops and Products*, *138*, 111484. doi:10.1016/j.indcrop.2019.111484

Tsai, W. T., Liu, S. C., & Hsieh, C. H. (2012). Preparation and fuel properties of biochars from the pyrolysis of exhausted coffee residue. *Journal of Analytical and Applied Pyrolysis*, *93*, 63–67. doi:10.1016/j.jaap.2011.09.010

Tsarev, S., Collins, R. N., Fahy, A., & Waite, T. D. (2016). Reduced Uranium Phases Produced from Anaerobic Reaction with Nanoscale Zerovalent Iron. *Environmental Science & Technology*, *50*(5), 2595–2601. doi:10.1021/acs.est.5b06160 PMID:26840619

Tuates, A. M. Jr, Suligan, J. M., Ligisan, A. R., Capariño, O. A., & Elauria, J. C. (2020). Development of De-Oiled Cashew Nut Shell as Fuel Briquettes. *Journal of the Japan Institute of Energy*, *99*(8), 123–128. doi:10.3775/jie.99.123

TUREB. (2021). *Türkiye Rüzgar Enerjisi İstatistik Raporu.* Türkiye Rüzgar Enerjisi Birliği.

Tyman, J. H. P., Johnson, R. A., Muir, M., & Rokhgar, R. (1989). The extraction of natural cashew nut-shell liquid from the cashew nut (Anacardium occidentale). *Journal of the American Oil Chemists' Society*, *66*(4), 553–557. doi:10.1007/BF02885447

U.S. Energy Information Administration. (2013). *International Energy Outlook 2013 with Projections to 2040.* U.S. Energy Information Administration Office of Energy Analysis, U.S. Department of Energy.

Ujjinappa, S., & Sreepathi, L. K. (2018). Evaluation of Physico-Mechanical-Combustion Characteristics of Fuel Briquettes Made from Blends of Areca Nut Husk, Simarouba Seed Shell and Black Liquor. *International Journal of Renewable Energy Development*, *7*(2), 131–137. doi:10.14710/ijred.7.2.131-137

Uliasz-Bocheńczyk, A., Mokrzycki, E., Piotrowski, Z., & Pomykała, R. (2009). Estimation of CO2 sequestration potential via mineral carbonation in fly ash from lignite combustion in Poland. *Energy Procedia*, *1*(1), 4873–4879. doi:10.1016/j.egypro.2009.02.316

UNEP. (2019). *Towards Sustainable Desalination.* Retrieved August 7, 2022 from https://www.unep.org/news-and-stories/story/towards0sustainable-desalination

Vagiona, D. G., & Kamilakis, M. (2018). Sustainable site selection for offshore wind farms in the South Aegean— Greece. *Sustainability*, *10*(3), 749. doi:10.3390u10030749

Valdez-Resendiz, J. E., Rosas-Caro, J. C., Mayo-Maldonado, J. C., & Llamas-Terres, A. (2018). Quadratic boost converter based on stackable switching stages. *IET Power Electronics, 11*(8), 1373–1381.

Van Duren, I., Voinov, A., Arodudu, O., & Firrisa, M. T. (2015). Where to produce rapeseed biodiesel and why? Mapping European rapeseed energy efficiency. *Renewable Energy*, *74*, 49–59. doi:10.1016/j.renene.2014.07.016

Vasileiou, M., Loukogeorgaki, E., & Vagiona, D. G. (2017). GIS-based multi-criteria decision analysis for site selection of hybrid offshore wind and wave energy systems in Greece. *Renewable & Sustainable Energy Reviews, 73*, 745–757. doi:10.1016/j.rser.2017.01.161

Vellaiyan, S., & Partheeban, C. A. (2018). Emission analysis of diesel engine fueled with soybean biodiesel and its water blends. *Energy Sources. Part A, Recovery, Utilization, and Environmental Effects, 40*(16), 1956–1965. doi:10.1080/15567036.2018.1489911

Velmurugan, K., & Shaafi, T. (2019). Performance and emission characteristics of a twin cylinder di diesel engine running on diesel-waste cooking methyl ester blend. *International Journal of Mechanical and Production Engineering Research and Development, 9*(6), 183–190.

Venkatramanan, D. (2019). *Dynamic modeling and analysis of buck converter based solar PV charge controller for improved MPPT performance.* https://ieeexplore.ieee.org/abstract/document/8815948/

Vîlcu, A. D., & Vîlcu, G. E. (2013). On homogeneous production functions with proportional marginal rate of substitution. *Mathematical Problems in Engineering, 2013*(732643), 1–5. doi:10.1155/2013/732643

Vindis, P., Mursec, B., Rozman, C., Janzekovic, M., & Cus, F. (2008). Biogas production with the use of mini digester. *Journal Of Achievements in Materials and Manufacturing Engineering, 28*(1), 99–102.

Voisin, H., Bergström, L., Liu, P., & Mathew, A. P. (2017). Nanocellulose-Based Materials for Water Purification. *Nanomaterials, 7*(3), 57. doi:10.3390/nano7030057

Wang, H., Wang, T., & Zhang, B. (2017). Constructed wetlands for wastewater treatment in cold climate. A review. *Journal of Environmental Sciences (China), 57*, 112–127. doi:10.1016/j.jes.2016.12.019 PMID:28647250

Wang, H., Wang, T., Zhang, B., Li, F., Toure, B., Omosa, I. B., Chiramba, T., Abdel-Monem, M., & Pradhan, M. (2014). Water and Wastewater Treatment in Africa - Current Practices and Challenges. *Clean - Soil, Air. Water (Basel), 42*(8), 1029–1035. doi:10.1002/clen.201300208

Wang, H., Wang, T., Zhang, B., Li, F., Toure, B., Omosa, I. S., Chiramba, T., Abdel-Monem, M., & Pradhan, M. (2014). Water and Wastewater Treatment in Africa – Current Practices and Challenges Clean – *Soil, Air. Water (Basel), 42*, 1029–1035. www.clean-journal.com

Wang, J., Tang, X., Xu, Y., Cheng, X., Li, G., & Liang, H. (2020). Hybrid UF/NF process treating secondary effluent of wastewater treatment plants for potable water reuse: Adsorption vs. coagulation for removal improvements and membrane fouling alleviation. *Environmental Research, 188*, 109833. doi:10.1016/j.envres.2020.109833 PMID:32798951

Wang, J., Zhang, S., Pan, B., Zhang, W., & Lv, L. (2011). Hydrous ferric oxide–resin nanocomposites of tunable structure for arsenite removal: Effect of the host pore structure. *Journal of Hazardous Materials, 198*, 241–246. doi:10.1016/j.jhazmat.2011.10.036 PMID:22047722

Wang, M. M., Wang, J., Cao, R., Wang, S. Y., & Du, H. (2017). Natural Transformation of Zinc Oxide Nanoparticles and Their Cytotoxicity and Mutagenicity. *Journal of Nanomaterials, 2017*, 1–12. Advance online publication. doi:10.1155/2017/8457960

Wang, S., Ng, C. W., Wang, W., Li, Q., & Li, L. (2012). A comparative study on the adsorption of acid and reactive dyes on multiwall carbon nanotubes in single and binary dye systems. *Journal of Chemical & Engineering Data, 57*(5), 1563–1569. doi:10.1021/je3001552

Wawa, A. I. (2021). Challenges facing Wastewater Management in Fast Growing Cities in Tanzania: A Case of Dodoma City Council. *Huria: Journal of the Open University of Tanzania, 27*(1). https://www.ajol.info/index.php/huria/article/view/204348

Wawa, A. I. (2020). Challenges facing Wastewater Management in Fast Growing Cities in Tanzania: A Case of Dodoma City Council. *Huria Journal*, *27*, 168–185.

Wei, Y., Chen, M., Li, Q., Niu, S., & Li, Y. (2019). Isothermal combustion characteristics of anthracite and spent coffee grounds briquettes. *Journal of Thermal Analysis and Calorimetry*, *136*(3), 1447–1456. doi:10.100710973-018-7790-x

Weldesilassie, A. B., Assefa, B., & Hagos, A. (2015). Productivity and welfare impact of climate change in sugarcane and cotton producing regions of Ethiopia. EDRI Research Report 23, Addis Ababa: Ethiopian Development Research Institute.

WFO. (2020). *Global offshore wind report 2019*. World Forum Offshore Wind.

WFO. (2021). *Global offshore wind report 2020*. World Forum Offshore Wind.

Wilaipon, P. (2007). Physical characteristics of maize cob briquette under moderate die pressure. *American Journal of Applied Sciences*, *4*(12), 995–998. doi:10.3844/ajassp.2007.995.998

World Bank. (2017). *Water Resources Management*. Retrieved August 6, 2022 from https://www.worldbank.org/en/topic/wwaterresourcesmanagement

Xia, Y., Wei, W., & Yu, M. (2018). *Stability analysis of PV generators with consideration of P&O-based power control*. doi:10.1109/TIE.2018.2864695

Xiao, H., Zou, H., Liu, S., & Li, C. (2019). An investigation of the friction and wear behavior of soybean biodiesel. *Tribology International*, *131*, 377–385. doi:10.1016/j.triboint.2018.10.037

Xu, S., & Chang, L. (2019). *Single-phase voltage source inverter with voltage boosting and power decoupling capabilities*. https://ieeexplore.ieee.org/abstract/document/8805155/

Xue, X., Cheng, R., Shi, L., Ma, Z., & Zheng, X. (2017). Nanomaterials for water pollution monitoring and remediation. *Environmental Chemistry Letters*, *15*(1), 23–27. doi:10.100710311-016-0595-x

Xu, Y., Dai, Y., Zhou, J., Xu, Z. P., Qian, G., & Lu, G. Q. M. (2010). Removal efficiency of arsenate and phosphate aqueous solution using layered double hydroxide materials: Intercalation vs. precipitation. *Journal of Materials Chemistry*, *20*(22), 4684–4691. doi:10.1039/b926239c

Yadav, A., Chowdhary, P., Kaithwas, G., & Bharagava, R. N. (2017). Toxic Metals in the Environment : Threats on Ecosystem and Bioremediation Approaches. Handbook of Metal-Microbe Interactions and Bioremediation, 128–141. doi:10.1201/9781315153353-11

Yadav, M. K., Singh, R. S., Singh, K. K., Mall, R. K., Patel, C. B., Yadav, S. K., & Singh, M. K. (2015). Assessment of climate change impact on productivity of different cereal crops in Varanasi, India. *Journal of Agrometeorology*, *17*(2), 179–184. doi:10.54386/jam.v17i2.1000

Yamaguchi, D. (2018). *A new PV converter for a high-leg delta transformer using cooperative control of boost converters and inverters*. https://ieeexplore.ieee.org/abstract/document/8249823/

Yang, H. H., Lee, W. J., Mi, H. H., Wong, C. H., & Chen, C. B. (1998). *PAH emissions influenced by Mn based additive and turbo charging from a heavy-duty diesel engine*. Environ.

Yang, X., Guo, N., Yu, Y., Li, H., Xia, H., & Yu, H. (2020). Synthesis of magnetic graphene oxide-titanate composites for efficient removal of Pb(II) from wastewater: Performance and mechanism. *Journal of Environmental Management, 256*, 109943. doi:10.1016/j.jenvman.2019.109943 PMID:31989978

Yang, Y., Xie, Y., Pang, L., Li, M., Song, X., Wen, J., & Zhao, H. (2013). Preparation of reduced graphene oxide/poly(acrylamide) nanocomposite and its adsorption of Pb(II) and methylene blue. *Langmuir, 29*(34), 10727–10736. doi:10.1021/la401940z PMID:23895359

Yan, W., Lien, H. L., Koel, B. E., & Zhang, W. X. (2012). Iron nanoparticles for environmental clean-up: Recent developments and future outlook. *Environmental Science. Processes & Impacts, 15*(1), 63–77. doi:10.1039/C2EM30691C PMID:24592428

Yaqoob, A. A., Parveen, T., Umar, K., & Ibrahim, M. N. M. (2020). Role of Nanomaterials in the Treatment of Wastewater: A Review. *Water, 12*(2), 495. doi:10.3390/w12020495

YEGM. (2018). *Türkiye Rüzgâr Enerjisi Potansiyel Atlası.* http://www.yegm.gov.tr/YEKrepa/REPA-duyuru_01.html

Yin, X., Ma, H., You, Q., Wang, Z., & Chang, J. (2012). Comparison of four different enhancing methods for preparing biodiesel through transesterification of sunflower oil. *Applied Energy, 91*(1), 320–325. doi:10.1016/j.apenergy.2011.09.016

Yin, Z., Cui, C., Chen, H., Duoni, Yu., Yu, X., & Qian, W. (2020). The Application of Carbon Nanotube/Graphene-Based Nanomaterials in Wastewater Treatment. *Small, 16*(15), 1902301. doi:10.1002mll.201902301 PMID:31788946

Yuliana, M., Tran-Thi, N. Y., & Ju, Y. H. (2012). Effect of extraction methods on characteristic and composition of Indonesian cashew nut shell liquid. *Industrial Crops and Products, 35*(1), 230–236. doi:10.1016/j.indcrop.2011.07.007

Zabaniotou, A., & Kamaterou, P. (2019). Food waste valorization advocating Circular Bioeconomy-A critical review of potentialities and perspectives of spent coffee grounds biorefinery. *Journal of Cleaner Production, 211*, 1553–1566. doi:10.1016/j.jclepro.2018.11.230

Zafeer, M. K., & Bhat, K. S. (2023). Valorisation of Agro-waste Cashew Nut Husk (Testa) for Different Value-Added Products. *Sustainable Chemistry for Climate Action*, 100014.

Zahan, K. A., & Kano, M. (2018). Biodiesel production from palm oil, its by-products, and mill effluent: A review. *Energies, 11*(8), 2132. doi:10.3390/en11082132

Zaman, M. F., & Akter, S. M. (2015). Pharmaceutical Wastewater Treatment and the Efficiency of ETP in Context of Bangladesh. *Proceedings of the Waste Safe*, 1-4.

Zapušek, A., Wirtgen, C., Weigandt, J., & Lenart, F. (2003). Characterisation of carbonizate produced from Velenje lignite in lab-scale reactor. *Acta Chimica Slovenica, 50*, 789–798.

Zarrinbakhsh, N., Wang, T., Rodriguez-Uribe, A., Misra, M., & Mohanty, A. K. (2016). Characterization of wastes and coproducts from the coffee industry for composite material production. *BioResources, 11*(3), 7637–7765. doi:10.15376/biores.11.3.7637-7653

Zhang, Y., Shi, J., Zhou, L., & Li, J. (2016). *Wide input-voltage range boost three-level DC–DC converter with quasi-Z source for fuel cell vehicles.* https://ieeexplore.ieee.org/abstract/document/7736069/

Zhang, G., Iu, H. H.-C., Zhang, B., Li, Z., Fernando, T., Chen, S.-Z., & Zhang, Y. (2017). An impedance network boost converter with a high-voltage gain. *IEEE Transactions on Power Electronics, 32*(9), 6661–6665.

Zhang, Y., Chi, H., Zhang, W., Sun, Y., Liang, Q., Gu, Y., & Jing, R. (2014). Highly Efficient Adsorption of Copper Ions by a PVP-Reduced Graphene Oxide Based On a New Adsorptions Mechanism. *Nano-Micro Letters*, *6*(1), 80–87. doi:10.1007/BF03353772 PMID:30464959

Zhang, Y., Ghaly, A. E., & Li, B. (2012). Physical properties of corn residues. *American Journal of Biochemistry and Biotechnology*, *8*(2), 44–53. doi:10.3844/ajbbsp.2012.44.53

Zhang, Y., Wu, B., Xu, H., Liu, H., Wang, M., He, Y., & Pan, B. (2016). Nanomaterials-enabled water and wastewater treatment. *NanoImpact*, *3*, 22–39. doi:10.1016/j.impact.2016.09.004

Zhang, Z., Liao, M., Jiang, D., Yang, X., & Li, S. (2021). High step-up isolated forward-flyback DC/DC converter based on resonance with pulse frequency modulation. *Journal of Power Electronics, 21*(2), 483–493.

Zhao, H., Cao, Y., Orndorff, W., Cheng, Y. H., & Pan, W. P. (2012). Thermal behaviors of soy biodiesel. *Journal of Thermal Analysis and Calorimetry*, *109*(3), 1145–1150. doi:10.100710973-012-2551-8

Zheng, X., Wu, R., & Chen, Y. (2011). Effects of ZnO nanoparticles on wastewater biological nitrogen and phosphorus removal. *Environmental Science & Technology*, *45*(7), 2826–2832. doi:10.1021/es2000744 PMID:21381661

Zhou, S., Qiu, Y., Zou, F., He, D., Yu, P., & Du, J. (2019). *Dynamic EV charging pricing methodology for facilitating renewable energy with consideration of highway traffic flow*. https://ieeexplore.ieee.org/abstract/document/8928490/

Zhu, D., Chen, D., Pan, J., & Zheng, G. (2011). Chlorination behaviors of zinc phases by calcium chloride in high temperature oxidizing-chloridizing roasting. *ISIJ International*, *51*(11), 1773–1777. doi:10.2355/isijinternational.51.1773

Zhu, L. (2010). Emission characteristics of a diesel engine operating on biodiesel and biodiesel blended with ethanol and methanol. *The Science of the Total Environment*, *408*(4), 914–921. doi:10.1016/j.scitotenv.2009.10.078 PMID:19926114

Zhu, Y., Murali, S., Cai, W., Li, X., Suk, J. W., Potts, J. R., & Ruoff, R. S. (2010). Graphene and Graphene Oxide: Synthesis, Properties, and Applications. *Advanced Materials*, *22*(35), 3906–3924. doi:10.1002/adma.201001068 PMID:20706983

Zilberman, D., Lipper, L., McCarthy, N., & Gordon, B. (2018). Innovation in response to climate change. Natural Resource Management and Policy, 52(1), 49-74.

Zouboulis, A. I., & Katsoyiannis, I. A. (2019). Recent Advances in Water and Wastewater Treatment with Emphasis in Membrane Treatment Operations. *Water (Basel)*, *11*(1), 45–51. doi:10.3390/w11010045

Zupanc, M., Kosjek, T., Kompare, B., Blažeka, Z., Ješe, U., Dular, M., Širok, B., & Heath, E. (2022). *Hydrodynamic cavitation: A technique for augmentation of removal of persistent pharmaceuticals?* Assessed on 03/09/2022 https://www.researchgate.net/publication/267554856

About the Contributors

J. Sadhik Basha (Dr.) obtained his Doctorate Degree in Mechanical Engineering at the National Institute of Technology, Trichy, India in the year 2011. He is currently working as a Professor in the Process Engineering Department, National University of Science & Technology, Sohar, Oman. Dr. Basha is an experienced educator and researcher for more than two decades. His current research interests include water-diesel/biodiesel emulsions as alternative fuels, thermal desalination, combustion modeling, spray combustion, nano-additive emulsion fuels, thermal sciences, and renewable energy. He has received various research supported & consultancy projects totaling over 85,000 $ in the fields of thermal desalination, waste plastic oil synthesis, alternative fuels, nano-additive fuel catalysts, renewable energy, etc. He is an expert referee, editorial board member, and advisory board member for more than fifty reputed international journals in the fields of energy, combustion, and power engineering. He has more than five patents, and has published more than eighty research articles and participated in several national / international conferences. Further, the contents of his research work have been cited more than two thousand times in various technical articles, magazines, and international journals. He has more than fifty professional memberships (such as Combustion Institute, USA; IFPASA, etc.). He is a Recipient of "Best Supervisor Award" by MEDRC (Middle East Desalination Research Centre) and "Best Researcher Award" by IMCO in the year 2021. He has won "Best Teaching Cash Awards" several times. He is one of the top 2% of the academically cited researchers as per global Stanford University Ranking, 2022 in Oman. Dr. Sadhik is currently working in the area of zero emission techniques, and thermal desalination techniques to produce potable water from waste heat recovery systems.

Jimoh Adewole (Dr.) is presently an Associate Professor of Chemical Engineering and Head of Process Engineering Department, National University of Science & Technology, Sohar, Sultanate of Oman. He is also the Leader of the Process Design & Computational Engineering Research Group. He was previously a senior researcher in the Center for Integrative Petroleum Research, King Fahd University of Petroleum & Minerals (KFUPM), Saudi Arabia. He received his PhD and MS in Chemical Engineering from Universiti Sains, Malaysia, and King Fahd University of Petroleum & Minerals in Saudi Arabia, respectively. He also holds a BS in Chemical Engineering from Obafemi Awolowo University, Nigeria. Dr. Adewole's research focuses on membrane technology for varieties of applications (gas separation, water and wastewater treatment, biomedical, energy generation and controlled release of oilfield chemicals); process intensification, modeling and simulation. Over the years, he has been working on cutting-edge research that involved experimental, modelling and simulation studies. Some of his projects were funded by Saudi Aramco, King AbdulAziz City for Science and Technology (KACST), College of Petroleum and Geosciences (CPG), National Science, Technology and Innovation Plan (NSTIP), and The Research

Council of Oman (TRC). He has published more than 75 articles in about 25 prestigious international journals including the Journal of Membrane Science, the International Journal of Greenhouse Gas Control, Waste and Biomass Valorization, the Journal of Polymer Research, book chapters, patents, and in many conference proceedings including the conferences of American Institute of Chemical Engineers, Society of Petroleum Engineers, Nigerian Society of Chemical Engineers, etc. Dr Adewole is a member of the American Institute of Chemical Engineers, Nigerian Society of Chemical Engineers, Society of Petroleum Engineers, Nigerian Society of Engineers, and American Society of Engineering Education.

Anteneh Mesfin Yeneneh is a chemical and environmental engineer with extensive experience in academic and industry oriented research. He completed his PhD in Chemical Engineering at Curtin University, Perth, Australia. He has over 14 years of teaching and industrial research experience in the chemical and environmental engineering fields, particularly in the areas of wastewater treatment, renewable energy, climate response, environmental and bioprocess and waste to energy technologies, optimization and modelling of environmental technologies and environmental system analysis. He has executed several industry oriented research projects in Australia and Oman. He has extensively published in various areas including noble environmental and wastewater treatment technologies, wastewater and slurry rheological studies. Furthermore, he is also experienced in the areas of risk analysis and incident investigation. Researched on Technology Need Assessment (TNA) for the transfer of climate response technologies. Currently, he is an associate professor of chemical and environmental engineering in International Maritime College, National University of Science and Technology Oman and visiting research fellow at Curtin University, Western Australia.

* * *

L. K. Abidoye is an expert in Chemical Engineering and industrial processes with specialization in carbon emission mitigation and utilization.

Shah N. Ashraf is working as Research and Education Fellow (Entrepreneurship) in the department of Entrepreneur Academy, Mohammed VI Polytechnic University, Morocco. He received doctoral degree from Entrepreneurship Development Institute of India, Ahmedabad, Gujarat, India. He received Master degree in Appropriate Technology and Entrepreneurship from NIT Arunachal Pradesh & National Institute of Rural Development and Panchayati Raj, Hyderabad, India. He has published research papers in peer-reviewed journals including in ABDC Indexed Journal and presented various research papers on Entrepreneurship and its associated determinants in International Conferences. His research area primarily includes Social Entrepreneurship, Agri-entrepreneurship, Innovation, Incubation & Technology Entrepreneurship.

Peter Ayadi has a bachelor's degree in Industrial Chemistry with Second Class Upper division from Adekunle Ajasin University, Akungba-Akoko, Ondo State, Nigeria in 2012. He obtained Masters of Science degree in Chemistry from the University of Lagos, Lagos, Nigeria in 2016. He joined the services of Olusegun Agagu University of Science and Technology, Okitipupa as an Assistant Lecturer in 2019. He is currently a Ph. D student at the University of Ilorin, Ilorin, Kwara State.

Hassan Ayedun obtained B. Sc. in Chemical Sciences in the year 1993 from the University of Agriculture Abeokuta, Ogun State, Nigeria. He bagged M. Sc. in Chemistry in the year 2002 from the University of Ibadan, Oyo State, Nigeria, and, a Ph. D in Environmental Management and Toxicology (with specialization in water Chemistry, quality monitoring, and risk evaluation) from the Federal University of Agriculture Abeokuta, Ogun State, Nigeria in 2012. He worked as Chemistry Lecturer at Federal Polytechnic Ilaro, Ogun State, Nigeria between 2003 – 2014. He left the Polytechnic as a Senior Lecturer and joined the services of Olusegun Agagu University, Okitipupa in 2014 as Lecturer I. He was appointed Acting Head of the Department of Chemical Sciences between 2018 to 2021. He is presently a Senior Lecturer in the Department. His major area of research is on water quality, treatment, and risk evaluation. He has 28 journal articles, 2 book chapters, and 2 conference proceedings to his credit.

Arife Isiacik Colak graduated from Maritime Transportation Engineering at 2004- Near East University Maritime Faculty. Graduated from MSc at 2011- Istanbul Technical University Maritime Transportation and Management Engineering- Thesis: Monitoring Ship Based Oil Pollution from Satellite Systems Graduated from PhD- Information Institue, Advanced Technology, Remote Sensing and Satellite Communication, Istanbul Technical University-GIS Based Maritime Spatial Planning for Offshore Wind Farms Site Selection with Exergy and Energy Efficiency Calculation. Working at National University, International Maritime College Oman, Maritime Department as Assis. Prof. Capt.

Abdul Hameed Kalifullah is currently working as an Assistant Professor in the department of Marine Engineering at National University of Science and Technology (IMCO), Oman. He has 16 years of teaching and industrial experience in India and abroad. He received his B.E(EEE) from Bharathidasan university and M. Tech in control systems and instrumentation from SASTRA University and Ph.D. in Electrical engineering from Anna university Chennai. He has three patents and around 25 research publications in national and international journals. He received several awards which include Best faculty, Best research paper & Best presentation in India and abroad. His area of interest is soft computing, PID controller Design, IOT, and Renewable energy.

G. R. Kannan received his bachelor's degree in Mechanical engineering from the PSNA College of Engineering and Technology, Madurai Kamaraj University, INDIA in the year 2001 and his Master's Degree in Thermal Engineering from Karunya Institute of Technology, Bharathiar University, Coimbatore, India and in December 2002. He obtained his Ph.D. Degree in Mechanical engineering from the National Institute of Technology, Trichy, India in August 2012. His areas of interest include Alternate fuels, Internal combustion engines, and Emission control systems. Dr. G.R. Kannan has more than 55 papers in International journal/conference publications. He is currently working as Professor in Mechanical Engineering at PSNA College of Engineering and Technology, DIndigul, Tamil Nadu, India.

Sanjeev Kumar is currently working as an Assistant Professor in Department of Economics, Hansraj College, University of Delhi, India. He holds a Ph.D. degree in Economics from Indian Institute of Technology (IIT) Roorkee, India. He received MPhil in Economics from Central University of Gujarat, Gandhinagar. His MA Economic is form Central University of Himachal Pradesh, Dharamshala.

Hamid Motlagh received his BSc in marine engineering and PGDip in marine energy from the World Maritime University and MSc in Marine Engineering Professionals from the University of Plymouth. He received his PhD course certificate in Advance Energy Plan Model from Alborg University and Certificate of professional education in Offshore Wind Farm Design from TU Delft university. He began his career as a marine engineer on ocean-going vessels and continued to work as a chief engineer after obtaining this certificate. After gaining the necessary experience, he worked in a shipping company as a technical superintendent and then as a technical manager. At the same time, working part-time in a classification society as a surveyor and inspector for approving new projects. He received awards from the Lloyd's Maritime Academy as the best student. His areas of interest are offshore wind power plant, advanced energy modeling, machine learning and data analysis, decarburization, wave energy, renewable energy, energy management, and alternative fuels.

M. Mohamed Musthafa graduated in Mechanical Engineering, obtained a postgraduate degree in Thermal Plant Engineering and secured a PhD from the National Institute of Technology, Trichy. At present, he is in the faculty of Mechanical Engineering and a member of the centre of Bio-Energy at SASTRA Deemed University, Thanjavur. He has a vast teaching experience of more than two decades at undergraduate and postgraduate levels in different reputed institutions. He has a large number of research publications to his credit. His area of interest includes alternative fuels for IC engines, LHR engines, combustion, heat transfer, and solar.

Abdulrasaq Olalekan Oyedeji holds a Bachelor's degree in Chemical Sciences from the Federal University of Agriculture, Abeokuta, Ogun State, Nigeria, a Master's degree in Analytical Chemistry from the University of Lagos, Nigeria and a PhD degree in Analytical Chemistry from Covenant University, Ota, Ogun State, Nigeria. He is currently a Chief Lecturer in the Department of Science Laboratory Technology of the Federal Polytechnic, Ilaro, Ogun State, Nigeria. His area of research includes food residue and contaminant analyses, research into under - utilized plant materials as food and industrial resources, exploitation of plant oils for the production of bioplastic, and environmental analysis.

Gizem Senel received her B.Sc. degree in 2017 and M.Sc. degree in 2019 from the Department of Geomatics Engineering, Istanbul Technical University, Turkey, where she is working toward a Ph.D. degree. Her research focuses are mapping and monitoring of land use land cover and its changes using remote sensing images through different image processing approaches.

Rajiv Seth is an Aerospace Engineer by training. B Tech -IIT, Kanpur; M Tech - IIT, Madras. Ph D in Business Administration (Finance) -AMU. * 27 years in the Indian Air Force (retired Group Captain)- worked on various types of aero engines and was involved in technical evaluation, field testing and technology transfer. * 17 years with the TERI University -worked as Registrar, Dean(Admin) and as Pro Vice Chancellor - instrumental in the establishment of the University and the post-graduate programs it offers. He was responsible for the overall administration of the University * Taught courses in Financial Management to students of MBA programs. * Research interests: Education for Sustainable Education, Renewable Energy, Financial Derivatives, SD in Armed Forces * Presently Advisor with Steamax Envirocare Pvt Ltd. - a company working on Biofuels.

Sandeep K. Sharma is currently working as an Assistant Professor (English) in the School of Liberal Arts & Management, DIT University Dehradun. He received Ph.D. from IIT Patna and master degree from BHU Banaras (India). He has published several research articles in linguistics and its applications in different types of communications. His research area primarily in the domain of linguistics and its socio-economic implications in human livelihood.

Ajay K. Singh is working as an Assistant Professor (Economics) in the School of Liberal Arts & Management, DIT University Dehradun. He did Post-Doctorate Research with EDI of India, Ahmedabad, Gujarat (India). He received MPhil (Economics) from DAVV Indore (India), and PhD (Economics) from IIT Indore (India. He has published several research papers in the diversified area such as climate change, agricultural productivity, assessment of food security; estimation of GFSI, development of environmental sustainability index (ESI) and its association with socio-economic indicators; measurement and determinants of entrepreneurship ecosystem, and dimension of sustainable development and its interlinkages with economic development.

Ashok Kumar T. secured PhD from SASTRA Deemed University, Thanjavur. At present, he is a faculty of the school of Mechanical Engineering and a member of the centre for Bio-Energy, at SASTRA University. He has teaching experience of more than 25 years at undergraduate and postgraduate levels in different institutions. He has a large number of research publications to his credit. His research area includes the Exergy analysis and development computational model for thermal systems

Shaafi Tajudeen is presently working as a Professor, Institute of Agricultural Engineering, Saveetha School of Engineering, Saveetha Institute of Medical and Technical Sciences, Chennai. He has more than 22 years of experience in industry and teaching and has guided several project works for Undergraduate and Postgraduate Engineering students in the field of energy engineering. His areas of interest include alternative fuels and energy applications.

Thomas Thangam was born in Pudukkottai, India in 1973. Received a Bachelor of Engineering in Electrical and Electronics Engineering from Government College of Engineering, Manonmaniam Sundaranar University, India, in 1995, and a Master of Engineering in Power Electronics and Drives from Alagappa Chettiar College of Engineering and Technology, Anna University, India, in 2005. In 2002, he became an assistant professor in the Department of Electronics and Communication Engineering at Mount Zion College of Engineering and Technology in Pudukkottai. Currently a Lecturer at the International Maritime College of Oman in Sohar, Sultanate of Oman. Power electronics, photovoltaic systems, power systems, and renewable energy sources are among his areas of interest. He has 24 years of teaching experience and 3 years of industry experience. Seven of his articles have been published in international journals and conferences.

Index

F

G

I

K

L

M

N

O

P

R

S

T

W

www.ingramcontent.com/pod-product-compliance
Lightning Source LLC
LaVergne TN
LVHW081328021026
849161LV00010BA/2551